The Human
Impact

ON THE NATURAL
ENVIRONMENT

Andrew Goudie

The Human Impact

ON THE NATURAL ENVIRONMENT

Past, Present, and Future

Sixth Edition

Blackwell
Publishing

BLACKWELL PUBLISHING
350 Main Street, Malden, MA 02148-5020, USA
9600 Garsington Road, Oxford OX4 2DQ,
550 Swanston Street, Carlton, Victoria 3053, Australia

First published 1981 by Blackwell Publishing Ltd
Second edition 1986
Third edition 1990
Fourth edition 1993
Fifth edition 2000
Sixth edition 2006
2   2006

Library of Congress Cataloging-in-Publication Data

Goudie, Andrew.
  The human impact on the natural environment : past, present, and future / Andrew S. Goudie.—6th ed.
    p. cm.
  Includes bibliographical references and index.
  ISBN-13: 978-1-4051-2704-2 (pbk. : alk. paper)
  ISBN-10: 1-4051-2704-X (pbk. : alk. paper)
  1. Nature—Effect of human beings on.   I. Title.

  GF75.G68 2005
  304.2—dc22

                                                        2005004138

A catalogue record for this title is available from the British Library.

Set in 10 on 13pt Palatino
by Graphicraft Limited, Hong Kong
Printed and bound in the United Kingdom
by TJ International, Padstow

The publisher's policy is to use permanent paper from mills that operate a sustainable forestry policy, and which has been manufactured from pulp processed using acid-free and elementary chlorine-free practices. Furthermore, the publisher ensures that the text paper and cover board used have met acceptable environmental accreditation standards.

For further information on
Blackwell Publishing, visit our website:
http://www.blackwellpublishing.com

CONTENTS

PREFACE TO THE SIXTH EDITION

It is now a quarter of a century since the first edition of this book appeared. This period has seen a remarkable transformation in interest in the impact that humans are having on the environment, together with an explosion of knowledge. In this edition, I have made substantial changes to the text, figures, tables, and references, and have tried to provide updated statistical information. The biggest change, however, has been to add four new chapters that explore the ways in which global climate change may have an impact on Earth.

ASG

ACKNOWLEDGMENTS

I am most grateful to Mary Thornbush for her assistance in preparing this sixth edition. I have also benefited greatly from the comments of Stan Trimble, Rob Wilby and Tim Burt on an earlier draft. The publisher and author are grateful for permission to use figures and photographs from the following publications: Anisimov, O. A. 'Projection of changes in permafrost with global warming', from *Physical geography*, 10, 1989; Arnell, N. W. and Reynard, N. S. 'Monthly runoff by the 2050s under two scenarios for six British catchments' from M. C. Acreman (ed.), '*The Hydrology of the UK: a study of change*' (Routledge, London, 2000); Atkinson, B. W. 'Thunder in south-east England', 'Total thunder rain in south-east England' and 'Number of days with thunder overhead in south-east England', from *Transactions of the Institute of British Geographers*, 44, 1968; Birkeland, P. W. and Larson, E. E. 'Correlation between quantity of waste water pumped into a deep well and the number of earthquakes near Denver, Colorado', from *Putnam's geology* (Oxford University Press, New York, reprinted by permission of Oxford University Press, 1978); Böckh, A. 'Variations in the level of Lake Valencia, Venezuela to 1968', from M. T.

Farvar and J. P. Milton (eds), *The careless technology* (Tom Stacey, London, 1973); Brampton, A. H. 'A selection of "hard engineering" structures designed to afford coastal protection', from A. H. Brampton, 'Cliff conservation and protection: methods and practices to resolve conflicts', in J. Hooke (ed.), *Coastal defence and earth science conservation* (Geological Society Publishing House, 1998); Brimblecombe, P. 'Thunder in south-east England' and 'Thunder storms per year in London', from 'London air pollution 1500–1900', *Atmospheric environment*, 11, 1977; Brimblecombe, P. and Camuffo, D. 'Decadal means of the freeze–thaw cycles in Central England', from *The effects of air pollution on the built environment* (Imperial College Press, London, 2003); Brookes, A. 'Principal types of adjustment in straightened river channels', from 'The distribution and management of channelized streams in Denmark., *Regulated rivers*, 1, 1987; Brown, A. A. and Davis, K. P. 'The reduction in area burned per area protected for the USA between 1926 and 1969 as a result of fire-suppression policies', from *Forest fire control and its use*, 2nd edn (McGraw-Hill, New York, 1973, reproduced with permission of the McGraw-Hill Companies); Chapin,

F. S. and Danell, K. 'Changes in areas of boreal forest' from F. S. Chapin, O. E. Sala and E. Huber-Sannwald, *Global biodiversity in a changing environment* (Springer, New York, 2001); Cochrane, R. 'The changing state of the vegetation cover in New Zealand', from E. G. Anderson (ed.), *New Zealand in maps* (Hodder & Stoughton Educational, London, 1977); Cohen and Rushton, 'The effect of air quality on plant growth in Leeds, England, 1913', data from R. Barrass, *Biology, food and people* (Hodder & Stoughton Educational, London, 1974); Cole, S. 'A Neolithic chert axe-blade from Denmark, of the type which has been shown to be effective at cutting forest in experimental studies', from *The Neolithic revolution*, 5th edn (The British Museum [Natural History] , London, 1970); Cooke, R. U. and Reeves, R. W. 'A model for the formation of arroyos (gullies) in the south-western USA', from *Arroyos and environmental change in the American south-west* (Clarendon Press, Oxford, 1976, reprinted with permission of Oxford University Press); Council on Environmental Quality, 'Changes in water pollution in the Great Lakes of North America', from 17th (1986) and 22nd (1992) Annual Reports; Darby, H. C. 'The changing distribution of forest in Central Europe between (a) AD 900 and (b) AD 1900', from W. L. Thomas (ed.), *Man's role in changing the face of the Earth* (University of Chicago Press © The University of Chicago, 1956); Dolan, R., Godfrey, P. J. and Odum, W. E. 'Cross-sections of two barrier islands in North Carolina, USA', from *Man's impact on the barrier islands of North Carolina, American scientist*, 61, 1973; Doughty, R. W. 'The spread of the English house sparrow in the New World', from *The English sparrow in the American landscape: a paradox in nineteenth century wildlife conservation*, Research paper 19, School of Geography, University of Oxford, 1978; Edmonson, W. T. 'Changes in the state of Lake Washington, USA, associated with levels of untreated sewage from 1933 to 1973', from W. W. Murdoch (ed.), *Environment* (Sinauer Associates, Sunderland, 1975); Ehrlich, P. R., Ehrlich, A. H. and Holdren, J. P. 'The growth of human numbers for the past half million years', from *Ecoscience: population, resources, environment* (W. H. Freeman, San Francisco, 1977); Elton, C. S. 'The spread of the Japanese beetle, *Popilla japonica*, in the eastern USA', from *The ecology of invasion by plants and animals* (Methuen, London, 1958); Fenger, J. 'Schematic presentation of a typical development of urban air pollution levels' from *Atmospheric environ-

ment*, 33, 1999; Forman, S. L., Oglesby, R. and Webb, R. S. 'Summary of dune field activity' from *Global and planetary change*, 29, 2001; French, H. M. 'Projection of changes in permafrost with global warming' from *The periglacial environment* (2nd edn) (Longman, Harlow, 1996); Gameson, A. L. H. and Wheeler, A. 'The average dissolved oxygen content of the River Thames at half-tide in the July–September quarter since 1890', from J. Cairns, K. L. Dickson and E. E. Herricks (eds), *Recovery and restoration of damaged ecosystems* (University Press of Virginia, Charlottesville, 1977, reprinted with permission of the University Press of Virginia); Gilbert, O. L. 'Air pollution in north-east England and its impact upon growth area for lichens', from R. Barrass, *Biology, food and people* (Hodder & Stoughton Educational, London, 1974); Gorman, M. 'A set of general design rules for nature reserves based on theories of island biogeography', from *Island ecology* (Chapman and Hall Publishers Ltd, London, 1979, copyright with kind permission of Kluwer Academic Publishers); Gorman, M. 'Some relationships between the size of "islands" and numbers of species', from *Island ecology* (Chapman and Hall Publishers Ltd, London, 1979, copyright with kind permission of Kluwer Academic Publishers); Goudie, A. S. 'A schematic representation of some of the possible influences causing climatic change', from *Environmental change*, 3rd edn (Clarendon Press, Oxford, 1992, reprinted by permission of Oxford University Press); Goudie, A. S. 'The concentration of dust storms in the USA in 1939', from 'Dust storms in space and time', *Progress in physical geography*, 7, 1983; Goudie, A. S. and Wilkinson, J. C. 'The Ghyben–Herzberg relationship between fresh and saline ground water and the effect of excessive pumping from the well', from *The warm desert environment* (Cambridge University Press, Cambridge, 1977); Green, F. H. W. 'Annual total area of field drainage by the tile drains in England and Wales', from *Recent changes in land use and treatment, Geographical journal*, 142, 1996; Green, F. H. W. 'Percentages of drained agricultural land in Europe', from *Field drainage in Europe, Geographical journal*, 144, 1978; Gregory, K. J. 'Relation between drainage density and mean annual precipitation', from E. Derbyshire (ed.), *Geomorphology and climate* (Chichester: John Wiley and Sons. © 1976); Gupta, H. K. 'Worldwide distribution of reservoir-triggered changes in seismicity' from *Earth-science reviews*, 58, 2002; Haagen-Smit, A. J. 'Possible reactions involving primary and secondary pollutants', from R. A. Bryson and J. E. Kutzbach (eds), *Air pollution* (Commission on College Geography Resource Paper 2. Washington DC: Association of American Geographers); Haggett, P. 'Land rotation and population density', from *Geography: a modern synthesis,* 3rd edn (Prentice-Hall, London, 1979); Haigh, M. J., 'Some shapes produced by shale tipping', from *Evolution of slopes on artificial landforms – Blaenavon, UK,* Research Paper 183, Department of Geography, University of Chicago; Harris, J. M., Oltmans, S. J., Bodeker, G. E., Storlaski, R., Evans, R. D. and Quincy, D. M. 'Global ozone trends' from *Atmospheric environment,* 37, 2003; Hollis, G. E. 'Some hydrological consequences of urbanization', from *The effects of urbanization on floods of different recurrence intervals, Water resources research,* 11, 1975; Hughes, R. J., Sullivan, M. E. and York, D. 'Rates of erosion in Papua New Guinea in the Holocene derived from rates of sedimentation in Kuk Swamp', from *Human-induced erosion in a highlands catchment in Papua New Guinea: the prehistoric and contemporary records, Zeitschrift für geomorphologie, supplementband,* 83, 1991; Johansen, H. E. 'The spread of contour-strip soil conservation methods in Wisconsin, USA, between 1939 and 1967', from S. W. Trimble and S. W. Lund (eds), *Soil conservation and the reduction of erosion and sediment in the Coon Creek Basin, Wisconsin, US Geological Survey professional paper,* 1234, 1982; Judd, W. R. 'Relationships between reservoir levels and earthquake frequencies', from *Seismic effects of reservoir impounding, Engineering geology,* 8, 1974; Keller, E. A. 'Comparison of the natural channel morphology and hydrology with that of a channelized stream', from D. R. Coates (ed.), *Geomorphology and engineering* (Hutchinson and Ross, 1976); Kirby, C. 'Lead concentration (annual means) in UK sites' from *Geography,* 80, 1995; Komar, P. D. 'Examples of the effects of shoreline installations on beach and shoreline morphology', from *Beach processes and sedimentation* (Prentice-Hall, Englewood Cliffs, 1976); Laporte, L. F. 'Maximum temperatures for the spawning and growth of fish', from *Encounter with the Earth* (copyright © by permission of Harper and Row Publishers, Inc., 1975); MacGrimmon, 'The original area of distribution of the brown trout and areas where it has been artificially naturalized', from J. Illies (ed.), *Introduction to zoogeography* (Macmillan, London, 1974); Manshard, W. 'The irrigated areas in Sind (Pakistan) along the Indus Valley', from *Tropical agriculture* (Addison-Wesley-Longman, 1974, ©

Bibliographisches Institute A. G., Mannheim); McGlone, M. S. and Wilmshurst, J. M. 'Summary percentage pollen diagram', from *Quaternary international,* 59, 1999; Meade, R. H. and Trimble, S. W. 'The decline in suspended sediment discharge to the eastern seaboard of the USA between 1910 and 1970 as a result of soil conservation measures . . .', from *Changes in sediment loads in rivers of the Atlantic drainage of the United States since 1900, Publication of the International Association of Hydrological Science,* 113, 1974; Mellanby, K. 'The increase of lichen cover on trees outside the city of Belfast, Northern Ireland (after Fenton)', from *Pesticides and pollution* (Fontana, London, 1967); Meybeck, M. 'Recent trends of nitrate concentration in some rivers', from B. von Bodungen and R. K. Turner (eds), *Science and integrated coastal management* (Dahlem University Press, Dahlem, 2001); Midgley, G. F., Hannak, L., Millar, D., Thiuller, W. and Booth, A. 'Current mapped Fynbos biome', from *Biological conservation,* 112, 2003; Nature Conservancy Council, 'The changing range of the little ringed plover, related to habitat change, especially as a result of the increasing number' and Trimble, S. W. 'The decline in suspended sediment discharge to the eastern seaboard of the USA between 1910 and 1970 as a result of soil conservation measures of gravel pits', from *Nature conservation and agriculture* (Nature Conservancy Council, Her Majesty's Stationery Office, London, 1977); Nature Conservancy Council, 'Losses of lowland heath in southern England', from *Nature conservation in Great Britain* (Nature Conservancy Council, Shrewsbury, 1984); Nature Conservancy Council, 'Reduction in the range of the silver spotted skipper butterfly (*Hesperia comma*)', from *Nature conservation and agriculture* (Nature Conservancy Council, Her Majesty's Stationery Office, London, 1977); Nature Conservancy Council, 'Reduction in the range of species related to habitat loss', from *Nature conservation and agriculture* (Nature Conservancy Council: Her Majesty's Stationery Office, London, 1977); Oerlemans, J. 'Generalised curves of ablation and accumulation' from R. A. Warwick, E. M. Barrows and T. M. L. Wigley (eds), *Climate and sea level change: observations, projections and implications* (Cambridge University Press, Cambridge, 1993); Oppenheimer, M. 'Cross section of an ice stream' from *Nature,* 393 (© Macmillan Publishers Ltd, 1998); Parker, A. G., Goudie, A. S., Anderson, D. E., Robinson, M. A. and Bonsall, C. 'Relations between factors influencing the mid-Holocene elm decline' from *Progress in physical geography*, 26, 2002; Pereira, H. C. 'The increase of water yield after clear-felling a forest: a unique confirmation from the Coweeta catchment in North Carolina', from *Land use and water resources in temperate and tropical climates* (Cambridge University Press, Cambridge, 1973); Rapp, A., Le Houerou, H. N. and Lundholm, B. 'Desert encroachment in the northern Sudan 1958–75, as represented by the position of the boundary between sub-desert scrub and grassland in the desert', from *Ecological bulletin*, 24, 1976; Rapp, A. 'Relation between spacing of wells and over-grazing', from *A review of desertization in Africa – water, vegetation and man* (Secretariat for International Ecology, Stockholm, Report no 1, 1974); Roberts, N. 'The human colonization of Ice-Age earth', from *The Holocene: an environmental history* (Blackwell Publishers, Oxford, 1989); Schwartz, M. W., Porter, D. J., Randall, J. M. and Lyons, K. G. 'Number of non-indigenous plant species by date as reported in botanical treatments of the California flora', from Sierra Nevada Ecosystems project: final report for Congress Vol. II, Davis: University of California, 1996; Shiklomanov, A. I. 'Changes in annual runoff in the CIS due to human activity during 1936–2000', from J. C. Rodda (ed.), *Facets of hydrology II* (copyright © 1985 John Wiley & Sons, Inc., reprinted by permission of John Wiley & Sons Inc.); Shiklomanov, A. I. 'Some major schemes proposed for large-scale inter-basin water transfers', from J. C. Rodda (ed.), *Facets of hydrology II* (copyright © 1985 John Wiley & Sons, Inc., reprinted by permission of John Wiley & Sons Inc.); Spate, O. H. K. and Learmonth, A. T. A. 'The Madurai–Ramanthapuram tank country in south India', from *India and Pakistan* (Methuen, London, 1967); Spencer, J. E. and Thomas, W. L. 'The diffusion of mining and smelting in the Old World', from *Introducing cultural geography*, 2nd edn (copyright © 1978 John Wiley & Sons, Inc., reprinted by permission of John Wiley & Sons, Inc.); Strandberg, C. H. 'Biological concentration occurs when relatively indestructible substances (DDT for example) are ingested by lesser organisms at the base of the food pyramid', from G. H. Smith (ed.), *Conservation of natural resources*, 4th edn (copyright © 1971 John Wiley & Sons, Inc., reprinted by permission of John Wiley & Sons, Inc.); Titus, J. G. 'Overwash: natural response of undeveloped barrier islands to sea level rise' from *Coastal management*, 18, 1990; Trimble, S. W. 'Changes in the evolution of fluvial landscapes in the Piedmont of Georgia, USA in response to land-use change between 1700 and 1970', from *Man-induced soil erosion on the southern Piedmont* (Soil Conservation Society of America, © Soil Conservation Society of America, 1974); Viles, H. A. 'Conceptual model of the impact of effective precipitation' from *Journal of nature conservation*, 11, 2003; Vinnikov, K. Y. 'Observed decrease of Northern Hemisphere sea ice extent' from *Science*, 286, 1999; US Geological Survey, 'Historical sediment and water discharge trends for the Colorado river', H. E. Schwarz, J. Emel, W. J. Dickens, P. Rogers and J. Thompson (eds), 'Water quality and flows', from B. L. Turner, W. C. Clark, R. W. Kates, J. F. Richards, J. T. Matthews and W. B. Meyer (eds), *The Earth as transformed by human action* (Cambridge University Press, Cambridge, 1991); Wallwork, K. L. 'Subsidence in the salt area of mid-Cheshire, England, in 1954', from *Subsidence in the mid-Cheshire industrial area, Geographical journal*, 122, 1956; Watson, A. 'The distribution of pits and ponds in a portion of north-western England', from 'The origin and distribution of closed depressions in south-west Lancashire and north-west Cheshire', unpublished BA dissertation, University of Oxford; Wilkinson, W. B. and Brassington, F. C. 'Groundwater levels in the London area', from R. A. Downing and W. B. Wilkinson (eds), *Applied groundwater hydrology – a British perspective* (Clarendon Press, Oxford, 1991, by permission of Oxford University Press); Wolfe, S. A. 'Dune mobility for various stations' from *Journal of arid environments*, 36, 1997; Woo, M. K., Lewkowicz, A. G. and Rouse, W. R. 'Ground settlement in response to a thickening of the active layer' from *Physical geography*, 13, 1992; Wooster, W. S. 'Decrease in the salinity of the Great Bitter Lake, Egypt, resulting from the intrusion of fresher water by way of the Suez Canal', from *The ocean and man, Scientific American*, 221, 1969; Ziswiler, V. 'The former and present distribution of the bison in North America', from J. Illies (ed.), *Introduction of zoogeography* (Macmillan, London, 1974); Ziswiler, V. 'Time series of animal extinctions since the seventeenth century, in relation to the human population increase', figure (b) from *The earth and human affairs* (National Academy of Sciences, 1972).

The publisher and author are grateful to the following for permission to use photographs: Frank Lane Picture Agency/B. S. Turner (3.2); NASA (7.4); Telegraph Colour Library (7.3); University of Cambridge, Committee for Aerial Photography (9.1).

Part I

The Past and Present

1 INTRODUCTION

The development of ideas

To what extent have humans transformed their natural environment? This is a crucial question that intrigued the eighteenth century French natural historian, Count Buffon. He can be regarded as the first Western scientist to be concerned directly and intimately with the human impact on the natural environment (Glacken, 1963, 1967). He contrasted the appearance of inhabited and uninhabited lands: the anciently inhabited countries have few woods, lakes or marshes, but they have many heaths and scrub; their mountains are bare, and their soils are less fertile because they lack the organic matter which woods, felled in inhabited countries, supply, and the herbs are browsed. Buffon was also much interested in the **domestication** of plants and animals – one of the major transformations in nature brought about by human actions.

Studies of the torrents of the French and Austrian Alps, undertaken in the late eighteenth and early nineteenth centuries, deepened immeasurably the realization of human capacity to change the environment. Fabre and Surell studied the flooding, siltation, erosion and division of watercourses brought about by **deforestation** in the Alps. Similarly de Saussure showed that Alpine lakes had suffered a lowering of water levels in recent times because of deforestation. In Venezuela, von Humboldt concluded that the lake level of Lake Valencia in 1800 (the year of his visit) was lower than it had been in previous times, and that deforestation, the clearing of plains, and the cultivation of indigo were among the causes of the gradual drying up of the basin.

Comparable observations were made by the French rural economist, Boussingault (1845). He returned to Lake Valencia some 25 years after von Humboldt and noted that the lake was actually rising. He described this reversal to political and social upheavals following the granting of independence to the colonies of the erstwhile Spanish Empire. The freeing of slaves had led to a decline in agriculture, a reduction in the application of irrigation water, and the re-establishment of forest.

Boussingault also reported some pertinent hydrological observations that had been made on Ascension Island in the South Atlantic:

In the Island of Ascension there was an excellent spring situated at the foot of a mountain originally covered with wood; the spring became scanty and dried up after the trees which covered the mountain had been felled. The loss of the spring was rightly ascribed to the cutting down of the timber. The mountain was therefore planted anew. A few years afterwards the spring reappeared by degrees, and by and by followed with its former abundance. (p. 685)

Charles Lyell, in his *Principles of geology*, one of the most influential of all scientific works, referred to the human impact and recognized that tree felling and drainage of lakes and marshes tended 'greatly to vary the state of the habitable surface'. Overall, however, he believed that the forces exerted by people were insignificant in comparison with those exerted by nature:

If all the nations of the earth should attempt to quarry away the lava which flowed from one eruption of the Icelandic volcanoes in 1783, and the two following years, and should attempt to consign it to the deepest abysses of the ocean they might toil for thousands of years before their task was accomplished. Yet the matter borne down by the Ganges and Burrampooter, in a single year, probably very much exceeds, in weight and volume, the mass of Icelandic lava produced by that great eruption. (Lyell, 1835: 197)

Lyell somewhat modified his views in later editions of the *Principles* (see e.g., Lyell, 1875), largely as a result of his experiences in the USA, where recent deforestation in Georgia and Alabama had produced numerous ravines of impressive size.

One of the most important physical geographers to show concern with our theme was Mary Somerville (1858) (who clearly appreciated the unexpected results that occurred as man 'dextrously avails himself of the powers of nature to subdue nature'):

Man's necessities and enjoyments have been the cause of great changes in the animal creation, and his destructive propensity of still greater. Animals are intended for our use, and field-sports are advantageous by encouraging a daring and active spirit in young men; but the utter destruction of some races in order to protect those destined for his pleasure, is too selfish, and cruelty is unpardonable: but the ignorant are often cruel. A farmer sees the rook pecking a little of his grain, or digging at the roots of the springing corn, and poisons all his neighbourhood. A few years after he is surprised to find his crop destroyed by grubs. The works of the Creator are nicely balanced, and man cannot infringe his Laws with impunity. (Somerville, 1858: 493)

This is in effect a statement of one of the basic laws of **ecology**: that everything is connected to everything else and that one cannot change just one thing in nature.

Considerable interest in conservation, climatic change and extinctions arose amongst European colonialists who witnessed some of the consequences of western-style economic development in tropical lands (Grove, 1997). However, the extent of human influence on the environment was not explored in detail and on the basis of sound data until George Perkins Marsh published *Man and nature* (1864), in which he dealt with human influence on the woods, the waters and the sands. The following extract illustrates the breadth of his interests and the ramifying connections he identified between human actions and environmental changes:

Vast forests have disappeared from mountain spurs and ridges; the vegetable earth accumulated beneath the trees by the decay of leaves and fallen trunks, the soil of the alpine pastures which skirted and indented the woods, and the mould of the upland fields, are washed away; meadows, once fertilized by irrigation, are waste and unproductive, because the cisterns and reservoirs that supplied the ancient canals are broken, or the springs that fed them dried up; rivers famous in history and song have shrunk to humble brooklets; the willows that ornamented and protected the banks of lesser watercourses are gone, and the rivulets have ceased to exist as perennial currents, because the little water that finds its way into their old channels is evaporated by the droughts of summer, or absorbed by the parched earth, before it reaches the lowlands; the beds of the brooks have widened into broad expanses of pebbles and gravel, over which, though in the hot season passed dryshod, in winter sealike torrents thunder, the entrances of navigable streams are obstructed by sandbars, and harbours, once marts of an extensive commerce, are shoaled by the deposits of the rivers at whose mouths they lie; the elevation of the beds of estuaries, and the consequently diminished velocity of the streams which flow into them, have converted thousands of leagues of shallow sea and fertile lowland into unproductive and miasmatic morasses. (Marsh, 1965: 9)

More than a third of the book is concerned with 'the woods'; Marsh does not touch upon important themes such as the modifications of mid-latitude grasslands, and he is much concerned with Western civilization. Nevertheless, employing an eloquent style and copious footnotes, Marsh, the versatile Vermonter, stands as a landmark in the study of environment (Thomas, 1956; Lowenthal, 2000).

Marsh, however, was not totally pessimistic about the future role of humankind or entirely unimpressed by positive human achievements (1965: 43–4):

New forests have been planted; inundations of flowing streams restrained by heavy walls of masonry and other constructions; torrents compelled to aid, by depositing the slime with which they are charged, in filling up lowlands, and raising the level of morasses which their own overflows had created; ground submerged by the encroachment of the ocean, or exposed to be covered by its tides, has been rescued from its dominion by diking; swamps and even lakes have been drained, and their beds brought within the domain of agricultural industry; drifting coast dunes have been checked and made productive by plantation; sea and inland waters have been repeopled with fish, and even the sands of the Sahara have been fertilized by artesian fountains. These achievements are far more glorious than the proudest triumphs of war . . .

Reclus (1873), one of the most prominent French geographers of his generation, and an important influence in the USA, also recognized that the 'action of man may embellish the earth, but it may also disfigure it; according to the manner and social condition of any nation, it contributes either to the degradation or glorification of nature' (p. 522). He warned rather darkly (p. 523) that 'in a spot where the country is disfigured, and where all the grace of poetry has disappeared from the landscape, imagination dies out, and the mind is impoverished; a spirit of routine and servility takes possession of the soul, and leads it on to torpor and death'. Reclus (1871) also displayed a concern with the relationship between forests, torrents and sedimentation.

In 1904 Friedrich coined the term 'Raubwirtschaft', which can be translated as economic plunder, robber economy or, more simply, devastation. This concept has been extremely influential but is open to criticism. He believed that destructive exploitation of resources leads of necessity to foresight and to improvements, and that after an initial phase of ruthless exploitation and resulting deprivation human measures would, as in the old countries of Europe, result in conservation and improvement. This idea was opposed by Sauer (1938) and Whitaker (1940), the latter pointing out that some soil erosion could well be irreversible (p. 157):

It is surely impossible for anyone who is familiar with the eroded loessial lands of northwestern Mississippi, or the burned and scarred rock hills of north central Ontario, to accept so complacently the damage to resources involved in the process of colonization, or to be so certain that resource depletion is but the forerunner of conservation.

Nonetheless Friedrich's concept of robber economy was adopted and modified by the great French geographer, Jean Brunhes, in his *Human geography* (1920). He recognized the interrelationships involved in anthropogenic environmental change (p. 332): 'Devastation always brings about, not a catastrophe, but a series of catastrophes, for in nature things are dependent one upon the other.' Moreover, Brunhes acknowledged that the 'essential facts' of human geography included 'Facts of Plant and Animal Conquest' and 'Facts of Destructive Exploitation'. At much the same time other significant studies were made of the same theme. Shaler of Harvard (*Man and the earth*, 1912) was very much concerned with the destruction of mineral resources (a topic largely neglected by Marsh).

Sauer led an effective campaign against destructive exploitation (Speth, 1977), reintroduced Marsh to a wide public, recognized the ecological virtues of some so-called primitive peoples, concerned himself with the great theme of domestication, concentrated on the landscape changes that resulted from human action, and gave clear and far-sighted warnings about the need for conservation (Sauer, 1938: 494):

We have accustomed ourselves to think of ever expanding productive capacity, of ever fresh spaces of the world to be filled with people, of ever new discoveries of kinds and sources of raw materials, of continuous technical progress operating indefinitely to solve problems of supply. We have lived so long in what we have regarded as an expanding world, that we reject in our contemporary theories of economics and of population the realities which contradict such views. Yet our modern expansion has been affected in large measure at the cost of an actual and permanent impoverishment of the world.

The theme of the human impact on the environment has, however, been central to some historical geographers studying the evolution of the cultural landscape. The clearing of woodland (Darby, 1956; Williams, 1989, 2003), the domestication process (Sauer, 1952), the draining of marshlands (Williams, 1970), the introduction of alien plants and animals (McKnight, 1959), and the transformation of the landscape of North America (Whitney, 1994) are among some of the recurrent themes of a fine tradition of historical geography.

In 1956, some of these themes were explored in detail in a major symposium volume, *Man's role in changing the face of the earth* (Thomas, 1956). Kates et al. (1990: 4) write of it:

Man's role seems at least to have anticipated the ecological movement of the 1960s, although direct links between the two have not been demonstrated. Its dispassionate, academic approach was certainly foreign to the style of the movement . . . Rather, *Man's role* appears to have exerted a much more subtle, and perhaps more lasting, influence as a reflective, broad-ranging and multidimensional work.

In the past three decades many geographers have contributed to, and been affected by, the phenomenon which is often called the environmental revolution or the ecological movement. The subject of the human impact on the environment, dealing as it does with such matters as environmental degradation, pollution and **desertification**, has close links with these developments, and is once again a theme in many textbooks and research monographs in geography (see e.g., Manners and Mikesell, 1974; Wagner, 1974; Cooke and Reeves, 1976; Gregory and Walling, 1979; Simmons, 1979; Tivy and O'Hare, 1981; Turner et al., 1990; Bell and Walker, 1992; Middleton, 1995; Meyer, 1996; Mannion, 1997, 2002).

Concerns about the human impact have become central to many disciplines and to the public, particularly since the early 1970s, when a range of major developments in the literature and in legislation have taken place (Table 1.1). The concepts of **global change** or global environmental change have developed. These phrases are much used, but seldom rigorously defined. Wide use of the term global change seems to have emerged in the 1970s but in that period was used principally, although by no means invariably, to refer to changes in international social, economic, and political systems (Price, 1989). It included such issues as proliferation of nuclear weapons, population growth, inflation, and matters relating to international insecurity and decreases in the quality of life.

Since the early 1980s the concept of global change has taken on another meaning which is more geocentric in focus. The geocentric meaning of global change can be seen in the development of the International Geosphere–Biosphere Program: a Study of Global Change (IBGP). This was established in 1986 by the International Council of Scientific Unions, 'to describe

Table 1.1 Some environmental milestones

1864	George Perkins Marsh, *Man and nature*
1892	John Muir founds Sierra Club in USA
1935	Establishment of Soil Conservation Service in USA
1956	Man's role in changing the face of the earth
1961	Establishment of World Wildlife Fund
1962	Rachel Carson's *Silent spring*
1969	Friends of the Earth established
1971	Greenpeace established
1971	Ramsar Treaty on International Wetlands
1972	United Nations Environmental Program (UNEP) established
1972	*Limits to growth* published by Club of Rome
1973	Convention on International Trade in Endangered Species (CITES)
1974	F. S. Rowland and M. Molina warn about CFCs and ozone hole
1975	Worldwatch Institute established
1979	Convention on Long-range Transboundary Air Pollution
1980	IUCN's (International Union for the Conservation of Nature and Natural Resources) World Conservation Strategy
1985	British Antarctic Survey finds ozone hole over Antarctic
1986	International Geosphere Biosphere Program (IGBP)
1986	Chernobyl nuclear disaster
1987	World Commission on Environment and Development (Brundtland Commission). *Our common future*
1987	Montreal Protocol on substances that deplete the ozone layer
1988	Intergovernmental Panel on Climate Change (IPCC)
1989	Global Environmental Facility
1992	Earth Summit in Rio and Agenda 21
1993	United Nations Commission on Sustainable Development
1994	United Nations Convention to Combat Desertification
1996	International Human Dimensions Program on Global Environmental Change
1997	Kyoto Protocol on greenhouse gas emissions
2001	Amsterdam Declaration
2002	Johannesburg Earth Summit

and understand the interactive physical, chemical and biological processes that regulate the total Earth system, the unique environment that it provides for life, the changes that are occurring in this system, and the manner in which they are influenced by human activities'.

The term 'global environmental change' has in many senses come to be used synonymously with the more geocentric use of 'global change'. Its validity and wide currency were recognized when *Global environmental change* was established in 1990 as

an international journal that addresses the human ecological and public policy dimensions of the environmental

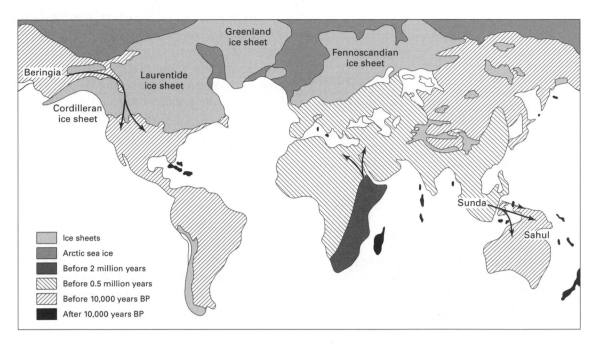

Figure 1.1 The human colonization of Ice Age Earth (after Roberts, 1989, figure 3.7).

processes which are threatening the sustainability of life of Earth. Topics include, but are not limited to, deforestation, desertification, soil degradation, species extinction, sea-level rise, acid precipitation, destruction of the ozone layer, atmospheric warming/cooling, nuclear winter, the emergence of new technological hazards, and the worsening effects of natural disasters.

In addition to the concept of global change, there is an increasing interest in the manner in which biogeochemical systems interact at a global scale, and an increasing appreciation of the fact that Earth is a single system. Earth system science has emerged in response to this realization (see Steffen et al., 2004).

The huge increase in interest in the study of the human impact on the environment and of global change has not been without its great debates and controversies, and some have argued that environmentalists have overplayed their hand (see e.g., Lomborg's *The skeptical environmentalist*, 2001) and have exaggerated the amount of environmental harm that is being caused by human activities. In this book, I take a long-term perspective and seek to show the changes that humankind has caused to a wide spectrum of environmental phenomena.

The development of human population and stages of cultural development

Some six or so million years ago, primitive human precursors or **hominids** appear in the fossil record (Wood, 2002). However, the first recognizable human, *Homo habilis*, evolved about 2.4 million years ago, more or less at the time that the ice ages were developing in mid-latitudes. The oldest remains have been found either in sediments from the Rift Valleys of East Africa, or in cave deposits in South Africa. Since that time the human population has spread over virtually the entire land surface of the planet (Oppenheimer, 2003) (Figure 1.1). *Homo* may have reached Asia by around two million years ago (Larick and Ciochan, 1996) and Europe not much later. In Britain the earliest fossil evidence, from Boxgrove, is from around half a million years ago. Modern humans, *Homo sapiens*, appeared in Africa around 160,000 years ago (Crow, 2002, Stringer, 2003; White et al., 2003).

Table 1.2 gives data on recent views of the dates for the arrival of humans in selected areas. Some of these dates are controversial, and this is especially true of Australia, where they range from *c.* 40,000 years to as much as 150,000 years (Kirkpatrick, 1994: 28–30). There

Table 1.2 Dates of human arrivals

Area	Source	Date (years BP)
Africa	Klein (1983)	2,700,000–2,900,000
China	Huang et al. (1995)	1,900,000
Georgian Republic	Gabunia and Vekua (1995)	1,600,000–1,800,000
Java	Swisher et al. (1994)	1,800,000
Europe	Champion et al. (1984)	c. 1,600,000 but most post-350,000
Britain	Roberts et al. (1995)	c. 500,000
Japan	Ikawa-Smith (1982)	c. 50,000
New Guinea	Bulmer (1982)	c. 50,000
Australia	Bowler et al. (2003)	c. 40,000–50,000
North America	Irving (1985)	15,000–40,000
South America	Guidon and Delibrias (1986)	32,000
Peru	Keefer et al. (1998)	12,500–12,700
Ireland	Edwards (1985)	9000
Caribbean	Morgan and Woods (1986)	4500
Polynesia	Kirch (1982)	2000
Madagascar	Battistini and Verin (1972)	c. AD 500
New Zealand	Green (1975)	AD 700–800

is also considerable uncertainty about the dates for humans arriving in the Americas. Many authorities have argued that the first colonizers of North America, equipped with so-called Clovis spears, arrived via the Bering landbridge from Asia around 12,000 years ago. However, some earlier dates exist for South America and these perhaps imply an earlier phase of colonization (Dillehay, 2003).

There are at least three interpretations of global population trends over the past two to three million years (Whitmore et al., 1990). The first, described as the 'arithmetic-exponential' view, sees the history of the global population as a two-stage phenomenon: the first stage is one of slow growth, while the second stage, related to the industrial revolution, displays a staggering acceleration in growth rates. The second view, described as 'logarithmic-logistic', sees the past million years or so in terms of three revolutions – the tool, agricultural and industrial revolutions. In this view, humans have increased the carrying capacity of Earth at least three times. There is also a third view, described as 'arithmetic-logistic', which sees the global population history over the past 12,000 years as a set of three cycles: the 'primary cycle', the 'medieval cycle' and the 'modernization cycle'; these three alternative models are presented graphically in Figure 1.2.

Estimates of population levels in the early stages of human development are difficult to make with any degree of certainty (Figure 1.3a). Before the agricultural 'revolution' some 10,000 years ago, human groups lived by hunting and gathering in parts of the world where this was possible. At that time the world population may have been of the order of five million people (Ehrlich et al., 1977: 182) and large areas would only recently have witnessed human migration. The Americas and Australia, for example, were probably virtually uninhabited until about 11,000 and 40,000 years ago respectively.

The agricultural revolution probably enabled an expansion of the total human population to about 200 million by the time of Christ, and to 500 million by AD 1650. It is since that time, helped by the medical and industrial revolutions and developments in agriculture and colonization of new lands, that human population has exploded, reaching about 1000 million by AD 1850, 2000 million by AD 1930 and 4000 million by AD 1975. The figure had reached over 6000 million by the end of the millennium. Victory over malaria, smallpox, cholera and other diseases has been responsible for marked decreases in death rates throughout the non-industrial world, but death-rate control has not in general been matched by birth control. Thus the annual population growth rate in the late 1980s in South Asia was 2.64%, Africa 2.66% and Latin America (where population increased sixfold between 1850 and 1950) 2.73%. The global annual growth in population over the past decade has been around 80 million people (Figure 1.3b).

The history of the human impact, however, has not been a simple process of increasing change in response to linear population growth over time, for in specific places at specific times there have been periods of reversal in population growth and ecological change as cultures collapsed, wars occurred, disease struck and **habitats** abandoned. Denevan (1992), for example, has pointed to the decline of native American populations in the new world following European entry into the Americas. This created what was 'probably the greatest demographic disaster ever'. The overall population of the western hemisphere in 1750 was perhaps less than a third of what it may have been in 1492, and the ecological consequences were legion.

Clearly, this growth of the human population of Earth is in itself likely to be a highly important cause of the transformation of nature. Of no lesser importance,

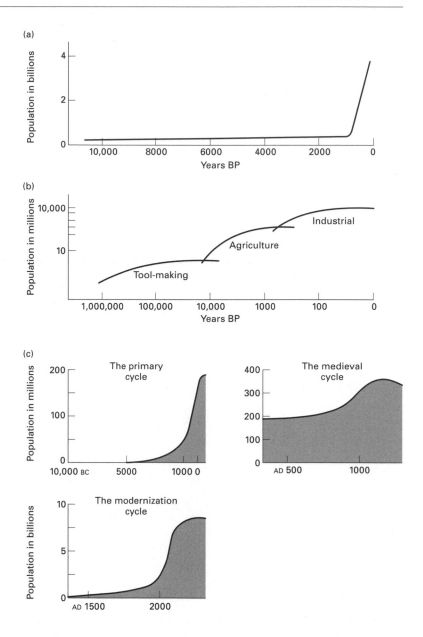

Figure 1.2 Three interpretations of global population trends over the millennia (billion = thousand million): (a) the arithmetic-exponential; (b) the logarithmic-logistic; (c) the arithmetic-logistic (after Whitmore et al., 1990, figure 2.1).

however, has been the growth and development of culture and technology. Sears (1957: 51) has put the power of humankind into the context of other species:

Man's unique power to manipulate things and accumulate experience presently enabled him to break through the barriers of temperature, aridity, space, seas and mountains that have always restricted other species to specific habitats within a limited range. With the cultural devices of fire, clothing, shelter, and tools he was able to do what no other organism could do without changing its original character. Cultural change was, for the first time, substituted for biological evolution as a means of adapting an organism to new habitats in

a widening range that eventually came to include the whole earth.

The evolving impact of humans on the environment has often been expressed in terms of a simple equation:

$$I = P\,A\,T$$

where I is the amount of pressure or impact that humans apply on the environment, P is the number of people, A is the affluence (or the demand on resources per person), and T is a technological factor (the power that humans can exert through technological change).

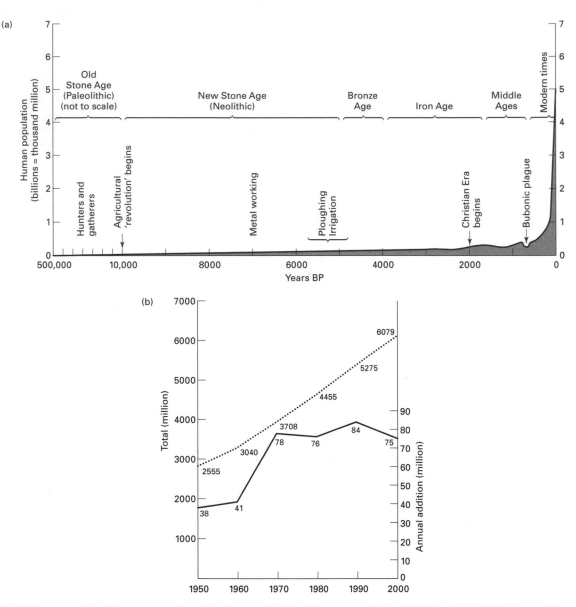

Figure 1.3 (a) The growth of human numbers for the past half million years (after Ehrlich et al., 1977, figure 5.2). (b) Annual growth of population since 1950.

The variables P, A and T have been seen by some as 'the three horsemen of the environmental apocalypse' (Meyer, 1996: 24). There may be considerable truth in the equation and in that sentiment; but as Meyer points out, the formula cannot be applied in too mechanistic a way. The 'cornucopia view', indeed, sees population not as the ultimate depleter of resources but as itself the ultimate resource capable of causing change for the better (see e.g., Simon, 1981, 1996). There are cases where strong population growth has appeared to lead to a reduction in environmental degradation (Tiffen et al., 1994). Likewise, there is debate about whether it is poverty or affluence that creates deterioration in the environment. On the one hand many poor countries have severe environmental problems and do not have the resources to clear them up, whereas affluent countries do. Conversely it can be argued that affluent countries have plundered and fouled less fortunate countries, and that it would be environmentally catastrophic if all countries used resources at the rate that the rich countries do. Similarly, it would be naïve to see all technologies as malign, or indeed benign.

Technology can be a factor either of mitigation and improvement or of damage. Sometimes it is the problem (as when ozone depletion has been caused by a new technology – the use of **chlorofluorocarbons**) and sometimes it can be the solution (as when renewable energy sources replace the burning of polluting lignite in power stations).

In addition to the three factors of population, affluence, and technology, environmental changes also depend on variations in the way in which different societies are organized and in their economic and social structures (see Meyer, 1996: 39–49 for an elaboration of this theme). For example, the way in which land is owned is a crucial issue.

The controls of environmental changes caused by the human impact are thus complex and in many cases contentious, but all the factors discussed play a role of some sort, at some places, and at some times.

We now turn to a consideration of the major cultural and technical developments that have taken place during the past two to three million years. Three main phases will form the basis of this analysis: the phase of hunting and gathering; the phase of plant cultivation, animal keeping and metal working; and the phase of modern urban and industrial society. These developments are treated in much greater depth by Ponting (1991) and Simmons (1996).

Hunting and gathering

The definition of 'human' is something of a problem, not least because, as is the case with all existing organisms, new forms tend to emerge by perceptible degrees from antecedent ones. Moreover, the fossil evidence is scarce, fragmentary and can rarely be dated with precision. Although it is probably justifiable to separate the hominids from the great apes on the basis of their assumption of an upright posture, it is much less justifiable or possible to distinguish on purely zoological grounds between those hominids that remained pre-human and those that attained human status. To qualify as a human, a hominid must demonstrate cultural development: the systematic manufacture of implements as an aid to manipulating the environment.

The oldest records of human activity and technology, pebble tools (crude stone tools which consist of a pebble with one end chipped into a rough cutting edge), have been found with human bone remains in various parts of Africa (Gosden, 2003). For example, at Lake Turkana in northern Kenya, and the Omo Valley in southern Ethiopia, a tool-bearing bed of volcanic material called tuff has been dated by isotopic means at about 2.6 million years old, another from Gona in the northeast of Ethiopia at about 2.5 million years old (Semaw et al., 1997), while another bed at the Olduvai Gorge in Tanzania has been dated by similar means at 1.75 million years. Indeed, these very early tools are generally termed 'Oldowan'.

As the Stone Age progressed the tools became more sophisticated, varied and effective. Greater exploitation of plant and animal resources became feasible. Stone may not, however, have been the only material used. Sticks and animal bones, the preservation of which is less likely than stone, are among the first objects that may have been used as implements, although the sophisticated utilization of antler and bone as materials for weapons and implements appears to have developed surprisingly late in pre-history. There is certainly a great deal of evidence for the use of wood throughout the Paleolithic Age, for ladders, fire, pigment (charcoal), the drying of wood and digging sticks. Tyldesley and Bahn (1983: 59) go so far as to suggest that 'The Palaeolithic might more accurately be termed the "Palaeoxylic" or "Old Wood Age".'

The building of shelters and the use of clothing became a permanent feature of human life as the Paleolithic period progressed, and permitted habitation in areas where the climate was otherwise not congenial. European sites from the Mousterian of the Middle Paleolithic have revealed the presence of purposefully made dwellings as well as caves, and by the Upper Paleolithic more complex shelters were in use, allowing people to live in the **tundra** lands of central Europe and Russia.

Another feature of early society that seems to have distinguished humans from the surviving non-human primates was their seemingly omnivorous diet. Biological materials recovered from settlements in many different parts of the world indicate that in the Paleolithic Age humans secured a wide range of animal meats, whereas the great apes, although not averse to an occasional taste of animal food, are predominately vegetarian. One consequence of enlarging the range of their diet was that, in the long run, humans

Figure 1.4 Fire was one of the first and most powerful tools of environmental transformation employed by humans. The high grasslands of southern Africa may owe much of their character to regular burning, as shown here in Swaziland.

were able to explore a much wider range of environment (G. Clark, 1977: 19). Another major difference that set humankind above the beasts was the development of communicative skills such as speech. Until hominids had developed words as symbols, the possibility of transmitting, and so accumulating, culture hardly existed. Animals can express and communicate emotions, but they never designate or describe objects.

At an early stage humans discovered the use of fire (Figure 1.4). This, as we shall see (Chapter 2), is a major agent by which humans have influenced their environment. The date at which fire was first deliberately employed is a matter of ongoing controversy (Bogucki, 1999: 51–54; Caldararo, 2002). It may have been employed very early in East Africa, where Gowlett et al. (1981) have claimed to find evidence for deliberate manipulation of fire from over 1.4 million years ago. However, it is not until after around 400,000 years ago that evidence for the association between human and fire becomes compelling. Nonetheless, as Pyne (1982: 3) has written:

It is among man's oldest tools, the first product of the natural world he learned to domesticate. Unlike floods, hurricanes or windstorms, fire can be initiated by man; it can be combated hand to hand, dissipated, buried, or 'herded' in ways unthinkable for floods or tornadoes.

He goes on to stress the implications that fire had for subsequent human cultural evolution (p. 4):

It was fire as much as social organisation and stone tools that enabled early big game hunters to encircle the globe and to begin the extermination of selected species. It was fire that assisted hunting and gathering societies to harvest insects, small game and edible plants; that encouraged the spread of agriculture outside the flood plains by allowing for rapid landclearing, ready fertilization, the selection of food grains, the primitive herding of grazing animals that led to domestication, and the expansion of pasture and grasslands against climate gradients; and that, housed in machinery, powered the prime movers of the industrial revolution.

Overall, compared with later stages of cultural development, early hunters and gatherers had neither the numbers nor the technological skills to have a very substantial effect on the environment. Besides the effects of fire, early cultures may have caused some diffusion of seeds and nuts, and through hunting activities (see Chapter 3) may have had some dramatic effects on animal populations, causing the extinction of many great mammals (the so-called 'Pleistocene overkill'). Locally some **eutrophication** may have occurred, and around some archaeological sites phosphate and nitrate levels may be sufficiently raised to make them an indicator of habitation to archaeologists today (Holliday, 2004). Equally, although we often assume that early humans were active and effective hunters, they may well have been dedicated scavengers of carcasses of animals that had either died natural deaths or been killed by carnivores such as lion.

It is salutary to remember, however, just how significant this stage of our human cultural evolution has been. As Lee and DeVore (1968: 3) wrote:

Of the estimated 80,000,000,000 men who have ever lived out a life span on earth, over 90% have lived as hunters and gatherers, about 6% have lived by agriculture and the remaining few per cent have lived in industrial societies. To date, the hunting way of life has been the most successful and persistent adaptation man has ever achieved.

Figure 1.5 indicates the very low population densities of hunter/gatherer/scavenger groups in comparison with those that were possible after the development of pastoralism and agriculture.

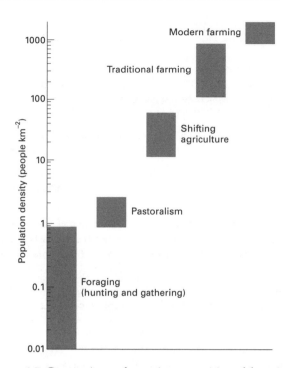

Figure 1.5 Comparison of carrying capacities of foraging, pastoralist, and agricultural societies.

Humans as cultivators, keepers, and metal workers

It is possible to identify some key stages of economic development that have taken place since the end of the Pleistocene (Table 1.3). First, around the beginning of the Holocene, about 10,000 years ago, humans started in various parts of the world to domesticate rather than to gather food plants and to keep, rather than just hunt, animals. This phase of human cultural development is well reviewed in Roberts (1998). By taking up farming and domesticating food plants, they reduced enormously the space required for sustaining each individual by a factor of the order of 500 at least (Sears, 1957: 54). As a consequence we see shortly thereafter, notably in the Middle East, the establishment of the first major settlements – towns. So long as man had

to subsist on the game animals, birds and fish he could catch and trap, the insects and eggs he could collect and the foliage, roots, fruits and seeds he could gather, he was limited in the kind of social life he could develop; as a rule he could only live in small groups, which gave small scope for specialization and the subdivision of labour, and in the course

of a year he would have to move over extensive tracts of country, shifting his habitation so that he could tap the natural resources of successive areas. It is hardly to be wondered at that among communities whose energies were almost entirely absorbed by the mere business of keeping alive, technology remained at a low ebb. (Clark, 1962: 76)

Although it is now recognized that some hunters and gathers had considerable leisure, there is no doubt that through the controlled breeding of animals and plants humans were able to develop a more reliable and readily expandable source of food and thereby create a solid and secure basis for cultural advance, an advance which included civilization and the 'urban revolution' of Childe (1936) and others. Indeed, Isaac (1970) has termed domestication 'the single most important intervention man had made in his environment'; and Harris (1996) has termed the transition from foraging to farming as 'the most fateful change in the human career'. Diamond (2002) termed it 'the most momentous change in Holocene human history'.

A distinction can be drawn between cultivation and domestication. Whereas cultivation involves deliberate sowing or other management, and entails plants that do not necessarily differ genetically from wild populations of the same species, domestication results in genetic change brought about through conscious or unconscious human selection. This creates plants that differ morphologically from their wild relatives and which may be dependent on humans for their survival. Domesticated plants are thus necessarily cultivated plants, but cultivated plants may or may not be domesticated. For example, the first plantations of *Hevea* rubber and quinine in the Far East were established from seed that had been collected from the wild in South America. Thus at this stage in their history these crops were cultivated but not yet domesticated.

The origin of agriculture remains controversial (Harris, 1996). Some early workers saw agriculture as a divine gift to humankind, while others thought that animals were domesticated for religious reasons. They argued that it would have been improbable that humans could have predicted the usefulness of domestic cattle before they were actually domesticated. Wild cattle are large, fierce beasts, and no one could have foreseen their utility for labor or milk until they were tamed – tamed perhaps for ritual sacrifice in connection with lunar goddess cults (the great curved horns being

Table 1.3 Five stages of economic development. Source: adapted from Simmons (1993: 2–3)

Economic stage	Dates and characteristics
Hunting–gathering and early agriculture	Domestication first fully established in southwestern Asia around 7500 BC; hunter–gatherers persisted in diminishing numbers until today. Hunter–gatherers generally manipulate the environment less than later cultures, and adapt closely to environmental conditions
Riverine civilizations	Great irrigation-based economies lasting from c. 4000 BC to 1st century AD in places such as the Nile Valley and Mesopotamia. Technology developed to attempt to free civilizations from some of the constraints of a dry season
Agricultural empires	From 500 BC to around AD 1800 a number of city-dominated empires existed, often affecting large areas of the globe. Technology (e.g., terracing and selective breeding) developed to help overcome environmental barriers to increased production
The Atlantic industrial era	From c. AD 1800 to today a belt of cities from Chicago to Beirut, and around the Asian shores to Tokyo, form an economic core area based primarily on fossil fuel use. Societies have increasingly divorced themselves from the natural environment, through air conditioning for example. These societies have also had major impacts on the environment
The Pacific global era	Since the 1960s there has been a shifting emphasis to the Pacific Basin as the primary focus of the global economy, accompanied by globalization of communications and the growth of multinational corporations

the reason for the association). Another major theory was that domestication was produced by crowding, possibly brought on by a combination of climatic deterioration (alleged post-glacial progressive desiccation) and population growth. Such pressure may have forced communities to intensify their methods of food production. Current paleoclimatological research tends not to support this interpretation, but that is not to say that other severe climatic changes could not have played a role (Sherratt, 1997).

Sauer (1952), a geographer, believed that plant domestication was initiated in Southeast Asia by fishing folk, who found that lacustrine and riverine resources would underwrite a stable economy and a sedentary or semi-sedentary life style. He surmises that the initial domesticates would be multi-purpose plants set around small fishing villages to provide such items as starch foods, substances for toughening nets and lines and making them water-resistant, and drugs and poisons. He suggested that 'food production was one and perhaps not the most important reason for bringing plants under cultivation.'

Yet another model was advanced by Jacobs (1969) who turned certain more traditional models upside down. Instead of following the classic pattern whereby farming leads to village, which leads to town, which leads to civilization, she proposed that one could be a hunter–gatherer and live in a town or city, and that agriculture originated in and around such cities rather

than in the countryside. Her argument suggests that even in primitive hunter–gatherer societies particularly valuable commodities such as fine stones, pigments and shells could create and sustain a trading center which would possibly become large and stable. Food would be exchanged for goods, but natural produce brought any distance would have to be durable, so meat would be transported on the hoof for example, but not all the animals would be consumed immediately; some would be herded together and might breed. This might be the start of domestication.

The process of domestication and cultivation was also once considered a revolutionary system of land procurement that had evolved in only one or two hearths and diffused over the face of Earth, replacing the older hunter–gathering systems by stimulus diffusion. It was felt that the deliberate rearing of plants and animals for food was a discovery or invention so radical and complex that it could have developed only once (or possibly twice) – the so-called 'Eureka model'. In reality, however, the domestication of plants occurred at approximately the same time in widely separated areas (Table 1.4). This might be construed to suggest that developments in one area triggered experiments with local plant materials in others. The balance of botanical and archaeological evidence seems to suggest that humans started experimenting with domestication and cultivation of different plants at different times in different parts of the world (Figure 1.6).

Table 1.4 Dates that indicate that there may have been some synchroneity of plant domestication in different centers

Center	Dates (000 years BP)	Plant
Mesoamerica	10.7–9.8	Squash-pumpkin
	9.0	Bottle gourd
Near East	10.0–9.3	Emmer wheat
		Two-rowed barley
	9.8–9.6	Einkorn wheat
		Pea
		Lentil
Far East	8.0–7.0	Broomcorn millet
		Rice
		Gourd
		Water chestnut
Andes	9.4–8.0	Chile pepper
		Common bean
		Ullucu
		White potato
		Squash and gourd

The locations and dates for domestication of some important domestic animals are shown in Figure 1.7.

The Near East, and in particular the Fertile Crescent, was especially important for both plant and animal domestication (Lev-Yadun et al., 2000; Zohary and Hopf, 2000), and wild progenitors were numerous in the area, including those of wheat, barley, lentils, peas, sheep, goats, cows, and pigs – a list that includes what are still the most valuable crops and livestock of the modern world (Diamond, 2002).

One highly important development in agriculture, because of its rapid and early effects on environment, was irrigation (Figure 1.8) and the adoption of riverine agriculture. This came rather later than domestication. Amongst the earliest evidence of artificial irrigation is the mace-head of the Egyptian Scorpion King, which shows one of the last pre-dynastic kings ceremonially cutting an irrigation ditch around 5050 years ago (Butzer, 1976), although it is possible that irrigation in Iraq started even earlier.

A major difference has existed in the development of agriculture in the Old and New Worlds; in the New World there were few counterparts to the range of domesticated animals which were an integral part of Old World systems (Sherratt, 1981). A further critical difference was that in the Old World the secondary applications of domesticated animals were explored. The plow was particularly important in this process (Figure 1.9) – the first application of animal power to the mechanization of agriculture. Closely connected to this was the use of the cart, which both permitted more intensive farming and enabled the transportation of its products. Furthermore, the development of textiles from animal fibers afforded, for the first time, a commodity that could be produced for exchange in areas where arable farming was not the optimal form of land use. Finally, the use of animal milk provided a means whereby large herds could use marginal or exhausted land, encouraging the development of the pastoral sector with transhumance or nomadism.

This secondary utilization of animals therefore had radical effects, and the change took place over quite a short period. The plow was invented some 5000 years ago, and was used in Mesopotamia, Assyria and Egypt. The remains of plow marks have also been found beneath a burial mound at South Street, Avebury in

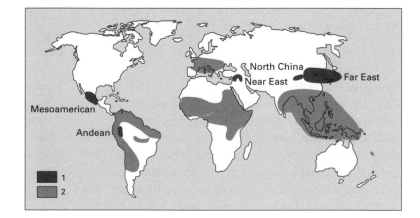

Figure 1.6 Major areas of domestication of plants identified by various workers. (1) The prime centers in which a number of plants were domesticated and which then diffused outwards into neighboring regions. (2) Broader regions in which plant domestication occurred widely and which may have received their first domesticated plants from the prime centers.

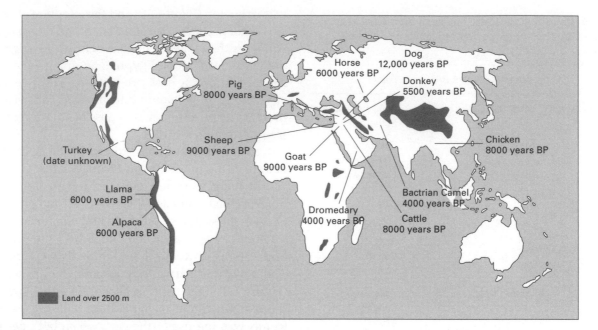

Figure 1.7 The places of origin, with approximate dates, for the most common domesticated animals.

Figure 1.8 Irrigation using animal power, as here in Rajasthan, India, is an example of the use of domesticated stock to change the environment.

England, dated at around 3000 BC, and ever since that time have been a dominant feature of the English landscape (Taylor, 1975). The wheeled cart was first produced in the Near East in the fourth millennium BC, and rapidly spread from there to both Europe and India during the course of the third millennium.

The development of other means of transport preceded the wheel. Sledge-runners found in Scandinavian bogs have been dated to the Mesolithic period (Cole, 1970: 42), while by the Neolithic era humans had developed boats, floats and rafts that were able to cross to Mediterranean islands and sail the Irish Sea. Dugout canoes could hardly have been common before polished stone axes and adzes came into general use during Neolithic times, although some paddle and canoe remains are recorded from Mesolithic sites in northern Europe. The middens of the hunter–fishers of the Danish Neolithic contain bones of deep-sea fish such as cod, showing that these people certainly had seaworthy craft with which to exploit ocean resources.

Both the domestication of animals and the cultivation of plants have been among the most significant causes of the human impact (see Mannion, 1995). Pastoralists have had many major effects – for example, on soil erosion – though Passmore (1974: 12) believed that nomadic pastoralists are probably more conscious than agriculturists that they share the earth with other living things. Agriculturists, on the other hand, deliberately transform nature in a sense which nomadic pastoralists do not. Their main role has been to simplify the world's ecosystems. Thus in the prairies of North America, by plowing and seeding the grasslands, farmers have eliminated a hundred species of native prairie herbs and grasses, which they replace with pure

Figure 1.9 The development of plows provided humans with the ability to transform soils. This simple type is in Pakistan.

Table 1.5 Estimated changes in the areas of the major land cover types between pre-agricultural times and the present*. Source: from J. T. Matthews (personal communication), in Meyer and Turner (1994). With permission of Cambridge University Press

Land cover type	Pre-agricultural area	Present area	Percent change
Total forest	46.8	39.3	−16.0
Tropical forest	12.8	12.3	−3.9
Other forest	34.0	27.0	−20.6
Woodland	9.7	7.9	−18.6
Shrubland	16.2	14.8	−8.6
Grassland	34.0	27.4	−19.4
Tundra	7.4	7.4	0.0
Desert	15.9	15.6	−1.9
Cultivation	0.0	17.6	+1760.0

*Figures are given in millions of square kilometers.

The spread of agriculture has transformed **land cover** at a global scale. As Table 1.5 shows, there have been great changes in the area covered by particular biomes since pre-agricultural times. Even in the past three hundred years the areas of cropland and pasture have increased by around five to sixfold (Goldewijk, 2001). It is possible (Ruddiman, 2003) that Holocene deforestation and land-cover change modified global climates by releasing carbon dioxide into the atmosphere.

One further development in human cultural and technological life that was to increase human power was the mining of ores and the smelting of metals. Neolithic cultures used native copper from the eighth millennium BC onwards, but evidence for its smelting occurs at Catal Hüyük in Turkey from the sixth millennium BC. The spread of metal working into other areas was rapid, particularly in the second half of the fifth millennium (Muhly, 1997) (Figure 1.10), and by 2500 BC bronze products were in use from Britain in the west to northern China in the east. The smelting of iron ores may date back to the late third millennium BC. Metalworking required enormous amounts of wood and so led to deforestation.

In recent decades fossil-fuelled machinery has allowed mining activity to expand to such a degree that in terms of the amount of material moved its effects are reputed to rival the natural processes of erosion. Taking overburden into account, the total

stands of wheat, corn or alfalfa. This simplification may reduce stability in the ecosystem (but see Chapter 13, section on 'The susceptibility to change'). Indeed, on a world basis (see Harlan, 1976) such simplification is evident. Whereas people once enjoyed a highly varied diet, and have used for food several thousand species of plants and several hundred species of animals, with domestication their sources are greatly reduced. For example, today four crops (wheat, rice, maize, and potatoes) at the head of the list of food supplies contribute more tonnage to the world total than the next twenty-six crops combined. Simmonds (1976) provides an excellent account of the history of most of the major crops produced by human society.

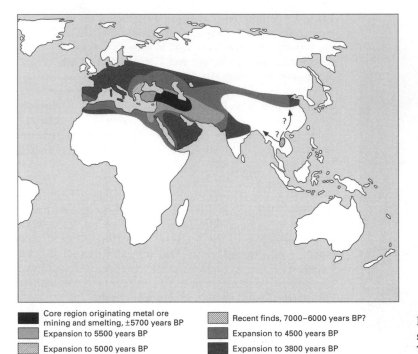

Core region originating metal ore mining and smelting, ±5700 years BP
Expansion to 5500 years BP
Expansion to 5000 years BP
Recent finds, 7000–6000 years BP?
Expansion to 4500 years BP
Expansion to 3800 years BP

Figure 1.10 The diffusion of mining and smelting in the Old World (after Spencer and Thomas, 1978, figure 4.4).

Table 1.6 Environmental impacts of mineral extraction. Source: Young (1992, table 5)

Activity	Potential impacts
Excavation and ore removal	Destruction of plant and animal habitat, human settlements, and other features (surface mining) Land subsidence (underground mining) Increased erosion: silting of lakes and streams Waste generation (overburden) Acid drainage (if ore or overburden contain sulfur compounds) and metal contamination of lakes, streams, and groundwater
Ore concentration	Waste generation (tailings) Organic chemical contamination (tailings often contain residues of chemicals used in concentrators) Acid drainage (if ore contains sulfur compounds) and metal contamination of lakes, streams, and groundwater
Smelting/refining	Air pollution (substances emitted can include sulfur dioxide, arsenic, lead, cadmium, and other toxic substances) Waste generation (slag) Impacts of producing energy (most of the energy used in extracting minerals goes into smelting and refining)

amount of material moved by the mining industry globally is probably at least 28 billion tonnes – about 1.7 times the estimated amount of sediment carried each year by the world's rivers (Young, 1992). The environmental impacts of mineral extraction are diverse but extensive, and relate not only to the process of excavation and removal, but also to the processes of mineral concentration, smelting, and refining (Table 1.6).

Modern industrial and urban civilizations

In ancient times, certain cities had evolved which had considerable human populations. It has been estimated that Nineveh may have had a population of 700,000, that Augustan Rome may have had a population of around one million, and that Carthage, at its fall in 146 BC, had 700,000 (Thirgood, 1981). Such cities would have already exercised a considerable influence on their environs, but this influence was never as extensive as that of the past few centuries; for the modern era, especially since the late seventeenth century, has witnessed the transformation of, or revolution in, culture and technology – the development of major industries.

Figure 1.11 Urbanization (and, in particular, the growth of large conurbations such as Toronto in Canada) is an increasingly important phenomenon. Urbanization causes and accelerates a whole suite of environmental problems.

Table 1.7 World's urban agglomerations of ten million or more inhabitants, estimated 1999

City and country	Includes	Population
Tokyo, Japan	Yokohama, Kawasaki	34,200,000
New York, USA	Newark, Patterson	20,150,000
São Paulo, Brazil		19,750,000
Seoul, South Korea	Inchon, Songnam	19,350,000
Mexico City, Mexico	Nezahualcoyotl, Ecatepec de Morelos	18,000,000
Osaka, Japan	Kobe, Kyoto	17,700,000
Bombay (Mumbai), India	Kalyan, Thane, Ulhasnagar	17,200,000
Los Angeles, USA	Riverside, Anaheim	15,950,000
Cairo, Egypt	El Giza	13,950,000
Moscow, Russia		13,200,000
Buenos Aires, Argentina	San Justo, La Plata	13,100,000
Manila, Philippines	Quezon City, Caloocan	13,000,000
Calcutta (Kolkata), India	Haora	12,650,000
Lagos, Nigeria		12,450,000
Jakarta, Indonesia		12,100,000
Karachi, Pakistan		11,800,000
London, England		11,750,000
Shanghai, China		11,600,000
Rio de Janeiro, Brazil		11,450,000
Delhi, India		11,100,000
Tehran, Iran	Karaj	10,150,000
Paris, France		10,050,000

This, like domestication, has reduced the space required to sustain each individual and has increased the intensity with which resources are utilized. Modern science and modern medicine have compounded these effects, leading to accelerating population increase even in non-industrial societies. Urbanization has gone on apace (Figure 1.11), and it is now recognized that large cities have their own environmental problems (Cooke et al., 1982), and environmental effects (Douglas, 1983). As Table 1.7 shows, the world now has some enormous urban agglomerations. These, in turn, have large **ecological footprints**.

The perfecting of sea-going ships in the sixteenth and seventeenth centuries was part of this industrial and economic transformation, and this was the time when mainly self-contained but developing regions of the world coalesced so that the ecumene became to all intents and purposes continuous. The invention of the steam engine in the late eighteenth century, and the internal combustion engine in the late nineteenth century, massively increased human access to energy and lessened dependence on animals, wind, and water.

Modern science, technology, and industry have also been applied to agriculture, and in recent decades some spectacular progress has been made through, for example, the use of fertilizers and the selective breeding of plants and animals.

The twentieth century was a time of extraordinary change (McNeill, 2003). Human population increased from 1.5 to 6 billion, the world's economy increased fifteenfold, the world's energy use increased thirteen to fourteenfold, freshwater use increased ninefold, and the irrigated area by fivefold. In the hundred centuries from the dawn of agriculture to 1900, McNeill calculates that humanity only used about two-thirds as much energy (most of it from biomass) as it used in the twentieth century. Indeed, he argued that humankind used more energy in the twentieth century than in all preceding human history put together. In addition he suggests that the seas surrendered more fish in the twentieth century than in all previous centuries, and that the forest and woodland area shrank by about 20%, accounting for perhaps half the net deforestation in world history.

To conclude, we can recognize certain trends in human manipulation of the environment which have taken place in the modern era. The first of these is that the ways in which humans are affecting the environment are proliferating, so that we now live on what some people have argued is a human dominated planet

Table 1.8 Some indicators of change in the global economy from 1950–2000

World indicator	1950	2000	Change (× n)
Grain production (million tons)	631	1863	2.95
Meat production (million tons)	44	232	5.27
Coal consumption (million tons of oil equivalent)	1074	2217	2.06
Oil consumption (million tons)	470	3519	7.49
Natural gas consumption (million tons of oil equivalent)	171	2158	12.62
Car production (million)	8.0	41.1	5.14
Bike production (million)	11	104	9.45
Human population (million)	2555	6079	2.38

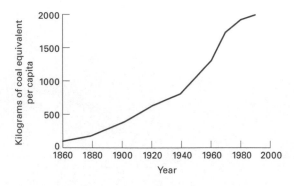

Figure 1.12 World per capita energy consumption since 1860, based on data from the United Nations.

(Vitousek et al., 1997). For example, nearly all the powerful pesticides post-date the Second World War, and the same applies to the construction of nuclear reactors. Second, environmental issues that were once locally confined have become regional or even global problems. An instance of this is the way in which substances such as DDT (dichlorodiphenyltrichloroethane), lead and sulfates are found at the poles, far removed from the industrial societies that produced them. This is one aspect of increasing globalization. Third, the complexity, magnitude, and frequency of impacts are probably increasing; for instance, massive modern dams such as at Aswan in Egypt and the Three Gorges Dam in China have very different impacts from a small Roman one. Finally, compounding the effects of rapidly expanding populations is a general increase in per capita consumption and environmental impact (Myers and Kent, 2003) (Table 1.8). Energy resources are being developed at an ever increasing rate, giving

humans enormous power to transform the environment. One index of this is world commercial energy consumption, which trebled in size between the 1950s and 1980. Figure 1.12 shows worldwide energy consumption since 1860 on a per capita basis. Nonetheless, it is important to recognize that there are huge differences in the likely environmental impacts of different economies in different parts of the world. As Table 1.9 indicates, the environmental impact, as measured by the so-called ecological footprint, is twelve times greater, for example, for the average American than for the average Indian (Wackernagel and Rees, 1995).

Modern technologies have immense power output. A pioneer steam engine in AD 1800 might rate at 8–16 kW. Modern railway diesels top 3.5 MW, and a large aero engine 60 MW. Figure 1.13 shows how the human impact on six 'component indicators of the **biosphere**' has increased over time. This graph is based on work by Kates et al. (1990). Each component indicator

Table 1.9 Comparing people's average consumption in Canada, USA, India, and the world. Source: Wackernagel and Rees (1995)

Consumption per person in 1991	Canada	USA	India	World
CO_2 emission (tonnes per year)	15.2	19.5	0.81	4.2
Purchasing power ($US)	19,320	22,130	1150	3800
Vehicles per 100 persons	46	57	0.2	10
Paper consumption (kg year^{-1})	247	317	2	44
Fossil energy use (GJ year^{-1})	250 (234)	287	5	56
Freshwater withdrawal (m^3 year^{-1})	1688	1868	612	644
Ecological footprint* (hectares per person)	4.3	5.1	0.4	1.8

*An ecological footprint is an accounting tool for ecological resources in which various categories of human consumption are translated into areas of productive land required to provide resources and assimilate waste products. It is thus a measure of how sustainable the lifestyles of different population groups are.

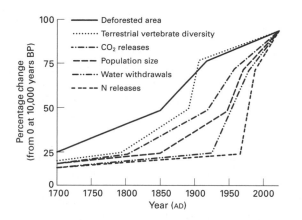

Figure 1.13 Percentage change (from assumed zero human impact at 10,000 years BP) of selected human impacts on the environment.

was taken to be 0% for 10,000 years ago (before the present = BP) and 100% for 1985. They then estimated the dates by which each component had reached successive quartiles (that is, 25, 50 and 75%) of its total change at 1985. They believe that about half of the components have changed more in the single generation since 1950 than in the whole of human history before that date. McNeill (2000) provides an exceptionally fine picture of all the changes in the environment that humans achieved in the twentieth century, and Carpenter (2001) examines the issue of whether civil engineering projects are environmentally sustainable.

Likewise, we can see stages in the pollution history of Earth. Mieck (1990), for instance, has identified a sequence of changes in the nature and causes of pollution: *pollution microbienne* or *pollution bacterielle*, caused by bacteria living and developing in decaying and putrefying materials and stagnant water associated with settlements of growing size; *pollution artisanale*, associated with small-scale craft industries such as tanneries, potteries, and other workshops carrying out various rather disagreeable tasks, including soap manufacture, bone burning, and glue-making; *pollution industrielle*, involving large-scale and pervasive pollution over major centers of industrial activity, particularly from the early nineteenth century in areas such as the Ruhr and the English 'Black Country'; *pollution fondamentale*, in which whole regions are affected by pollution, as with the desiccation and subsequent salination of the Aral Sea area; *pollution foncière*, in which vast quantities of chemicals are deliberately applied to the land as fertilizers and biocides; and finally, *pollution accidentale*, in which major accidents can cause pollution which is neither foreseen nor calculable (e.g., the Chernobyl nuclear disaster).

Above all, as a result of the escalating trajectory of environmental transformation it is now possible to talk about *global* environmental change. There are two components to this (Turner et al., 1990): systemic global change and cumulative global change. In the systemic meaning, 'global' refers to the spatial scale of operation and comprises such issues as global changes in climate brought about by atmospheric pollution. This is a topic discussed at length in Chapters 7–12. In the cumulative meaning, 'global' refers to the areal or substantive accumulation of localized change, and a change is seen to be 'global' if it occurs on a worldwide scale, or represents a significant fraction of the total environmental phenomenon or global resource. Both types of change are closely intertwined. For example, the burning of vegetation can lead to systemic change through such mechanisms as carbon dioxide release and albedo change, and to cumulative change through its impact on soil and biotic diversity (Table 1.10). It is for this reason that we now talk of

Table 1.10 Types of global environmental change. Source: from Turner et al. (1990, table 1)

Type	Characteristic	Examples
Systemic	Direct impact on globally functioning system	(a) Industrial and land use emissions of 'greenhouse' gases (b) Industrial and consumer emissions of ozone-depleting gases (c) Land cover changes in albedo
Cumulative	Impact through worldwide distribution of change	(a) Groundwater pollution and depletion (b) Species depletion/genetic alteration (biodiversity)
	Impact through magnitude of change (share of global resource)	(a) Deforestation (b) Industrial toxic pollutants (c) Soil depletion on prime agricultural lands

Earth system science and recognize the complex inter-actions that take place at a multitude of scales on our planet (Steffen et al., 2004).

We can conclude this introductory chapter by quoting from Kates et al. (1991: 1):

Most of the change of the past 300 years has been at the hands of humankind, intentionally or otherwise. Our ever-growing role in this continuing metamorphosis has itself essentially changed. Transformation has escalated through time, and in some instances the scales of change have shifted from the locale and region to the earth as a whole. Whereas humankind once acted primarily upon the visible 'faces' or 'states' of the earth, such as forest cover, we are now also altering the fundamental flows of chemicals and energy that sustain life on the only inhabited planet we know.

Points for review

What have been the main stages in the development of ideas about the human impact on the environment?

How have human population levels changed over the past few millions of years?

To what extent did early humans change their environment?

What have been the main changes in the environment wrought by humans over the past 300 years?

What do you think is meant by the term Earth system science?

Guide to reading

Baker, A. R. H., 2003, *Geography and history, bridging the divide*. Cambridge: Cambridge University Press. Chapter 3 contains a valuable and perceptive discussion of environmental geographies and histories.

Freedman, B., 1995, *Environmental ecology*, 2nd edn. San Diego: Academic Press. An enormously impressive and wide-ranging study with a strong ecological emphasis.

Goudie, A. S. (ed.), 1997, *The human impact reader. Readings and case studies*. Oxford: Blackwell. A collection of key papers on many of the themes discussed in this book.

Goudie, A. S. (ed.), 2002, *Encyclopedia of global change*. New York: Oxford University Press. A multi-author, two-volume compilation.

Goudie, A. S. and Viles, H., 1997, *The Earth transformed*. Oxford: Blackwell. An introductory treatment of the human impact, with many case studies.

Kemp, D. D., 2004, *Exploring environmental issues: an integrated approach*. London: Routledge. A balanced, accessible, and comprehensive analysis of many environmental issues.

Mannion, A. M., 2002, *Dynamic world: land-cover and land-use change*. London: Hodder Arnold. A new and comprehensive study of the important role that land use plays in land transformation.

Meyer, W. B., 1996, *Human impact on the Earth*. Cambridge: Cambridge University Press. A good point of entry to the literature that brims over with thought-provoking epigrams.

Middleton, N. J., 2003, *The global casino*. London: Edward Arnold. The third edition of an introductory text, by a geographer, which is well illustrated and clearly written.

Oppenheimer, S., 2002, *Out of Eden. Peopling of the world*. London: Constable. A very accessible account of human development in prehistory.

Pickering, K. T. and Owen, L. A., 1997, *Global environmental issues* (2nd edn.) London: Routledge. A well illustrated survey.

Ponting, C., 1991, *A green history of the world*. London: Penguin. An engaging and informative treatment of how humans have transformed Earth through time.

Simmons, I. G., 1996, *Changing the face of the Earth: culture, environment and history*, 2nd edn. Oxford: Blackwell. A characteristically amusing and perceptive review of many facets of the role of humans in transforming Earth, from an essentially historical perspective.

Simmons, I. G., 1997, *Humanity and environment: a cultural ecology*. A broad account of some major themes relating to humans and the environment.

State of the Environment Advisory Council, 1996, *Australia State of the Environment 1996*. Collingword, Australia: CSIRO Publishing. A large compendium which discusses major environmental issues in the context of Australia.

Steffen, W. and 10 others, 2004, *Global change and the Earth system*. Berlin: Springer-Verlag. A multi-author, high-level, earth system science based overview of environmental change at a global scale.

Turner, B. L. II (ed.), 1990, *The Earth as transformed by human action*. Cambridge: Cambridge University Press. A very good analysis of global and regional changes over the past 300 years.

2 THE HUMAN IMPACT ON VEGETATION

Introduction

In any consideration of the human impact on the environment it is probably appropriate to start with vegetation, for humankind has possibly had a greater influence on plant life than on any of the other components of the environment. Through the many changes humans have brought about in land use and land cover they have modified soils (Meyer and Turner 1994) (see Chapter 4), influenced climates (see Chapter 7), affected geomorphic processes (see Chapter 6), and changed the quality (see Chapter 5) and quantity of some natural waters. Indeed, the nature of whole landscapes has been transformed by human-induced vegetation change (Figure 2.1). Hannah et al. (1994) attempt to provide a map and inventory of human

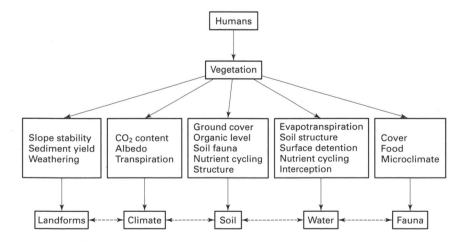

Figure 2.1 Some ramifications of human-induced vegetation change.

disturbance of world ecosystems, but many of their criteria for recognizing undisturbed areas are naïve. Large tracts of central Australia are classed as undisturbed, when plainly they are not (Fitzpatrick, 1994). Following Hamel and Dansereau (1949, cited by Frenkel, 1970) we can recognize five principal degrees of interference – each one increasingly remote from pristine conditions. These are:

1 *Natural habitats*: those that develop in the absence of human activities.
2 *Degraded habitats*: those produced by sporadic, yet incomplete, disturbances; for example, the cutting of a forest, burning and the non-intensive grazing of natural grassland.
3 *Ruderal habitats*: where disturbance is sustained but where there is no intentional substitution of vegetation. Roadsides are an example of a ruderal habitat.
4 *Cultivated habitats*: when constant disturbance is accompanied by the intentional introduction of plants.
5 *Artificial habitats*: which are developed when humans modify the ambient climate and soil, as in greenhouse cultivation.

An alternative model for classifying the extent of human influence on vegetation is provided by Westoff (1983), who adopts a four-part scheme:

1 *Natural*: a landscape or an ecosystem not influenced by human activity.
2 *Subnatural*: a landscape or ecosystem partly influenced by humans, but still belonging to the same (structural) formation type as the natural system from which it derives (e.g., a wood remaining a wood).
3 *Semi-natural*: a landscape or ecosystem in which flora and fauna are largely spontaneous, but the vegetation structure is altered so that it belongs to another formation type (e.g., a pasture, moorland or heath deriving from a wood).
4 *Cultural*: a landscape or ecosystem in which flora and fauna have been essentially affected by human agency in such a way that the dominant species may have been replaced by other species (e.g., arable land).

In this chapter we shall be concerned mainly with degraded and ruderal habitats, or subnatural and seminatural habitats, but first we need to consider some of the processes that human societies employ: notably fire, grazing and the physical removal of forest.

The use of fire

Humans are known to have used fire since Paleolithic times (see Chapter 1). As Sauer, one of the great proponents of the role of fire in environmental change, put it (1969: 10–11):

Through all ages the use of fire has perhaps been the most important skill to which man has applied his mind. Fire gave to man, a diurnal creature, security by night from other predators . . . The fireside was the beginning of social living, the place of communications and reflection.

People have utilized fire for a great variety of reasons (Bartlett, 1956; Stewart, 1956; Wertine, 1973): to clear forest for agriculture, to improve grazing land for domestic animals or to attract game; deprive game of cover; drive game from cover in hunting; kill or drive away predatory animals, ticks, mosquitoes, and other pests; repel the attacks of enemies, or burn them out of their refuges; cooking; expedite travel; burn the dead and raise ornamental scars on the living; provide light; transmit messages via smoke signaling; break up stone for tool-making: protect settlements or encampments from great fires by controlled burning; satisfy the sheer love of fires as spectacles; make pottery; smelt ores; harden spears; provide warmth; make charcoal; and assist in the collection of insects such as crickets for eating. Given this remarkable utility, it would be surprising if it had not been turned to account. Indeed it is still much used, especially by pastoralists such as the cattle-keepers of Africa, and by shifting agriculturists. For example, the Malaysian and Indonesian *ladang* and the *milpa* system of the Maya in Latin America involved the preparation of land for planting by felling or deadening forest, letting the debris dry in the hot season, and burning it before the commencement of the rainy season. With the first rains, holes were dibbled in the soft ash-covered earth with a planting stick. This system was suited to areas of low population density with sufficiently extensive forest to enable long intervals of 'forest fallow' between burnings. Land that was burned too frequently became overgrown with perennial grasses, which tended to make it difficult to farm with primitive tools. Land cultivated for too long rapidly suffered deterioration in fertility, while land recently burned was temporarily rich in nutrients (see Figure 13.1).

The use of fire, however, has not been restricted to primitive peoples in the tropics. Remains of charcoal are found in Holocene soil profiles in Britain (Moore, 2000); large parts of North America appear to have suffered fires at regular intervals prior to European settlement (Parshall and Foster, 2002); and in the case of South America the 'great number of fires' observed by Magellan during the historical passage of the Strait that bears his name resulted in the toponym, 'Tierra del Fuego'. Indeed, says Sternberg (1968: 718): 'for thousands of years, man has been putting the New World to the torch, and making it a "land of fire".' Given that the native American population may have been greater than once thought, their impact, even in Amazonia, may have been appreciable (Deneven, 1992). North American Indians were also not passive occupants of the land. They managed forests extensively creating 'a dynamic mosaic containing patches of young and old trees, interspersed with patches of grass and shrubs' (Bonnicksen, 1999: 442). As a result, some North American forests, prior to European settlement, were open and park-like, not dark and dense.

Fire (Figure 2.2) was also central to the way of life of the Australian aboriginals, including those of Tasmania (Hope, 1999), and the carrying of fire sticks was a common phenomenon. As Blainey (1975: 76) has put it, 'Perhaps never in the history of mankind was there a people who could answer with such unanimity the question: "have you got a light, mate?". There can have been few if any races who for so long were able to practise the delights of incendiarism.' That is not to say that aboriginal burning necessarily caused wholesale modification of Australian vegetation in pre-European times. That is a hotly debated topic (Kohen, 1995).

In neighboring New Zealand, Polynesians carried out extensive firing of vegetation in pre-European settlement times, and hunters used fire to facilitate travel and to frighten and trap a major food source – the flightless moa (Cochrane, 1977). The changes in vegetation that resulted were substantial (Figure 2.3). The forest cover was reduced from about 79% to 53%, and fires were especially effective in the drier forests of central and eastern South Island in the rain shadow of the Southern Alps. The fires continued over a period of about 1000 years up to the period of European settlement (Mark and McSweeney, 1990). Pollen analyses

Figure 2.2 The pindan bush of northwest Australia, composed of *Eucalyptus*, *Acacia*, and grasses, is frequently burned. The pattern of the burning shows up clearly on Landsat imagery.

(Figure 2.4) show dramatically the reduction in tall trees and the spread of open shrubland following the introduction of the widespread use of fire by the Maoris (McGlone and Wilmshurst, 1999).

Fires: natural and anthropogenic

Although people have used fire for all the reasons that have been mentioned, before one can assess the role of this facet of the human impact on the environment one must ascertain how important fires started by human action are in comparison with those caused

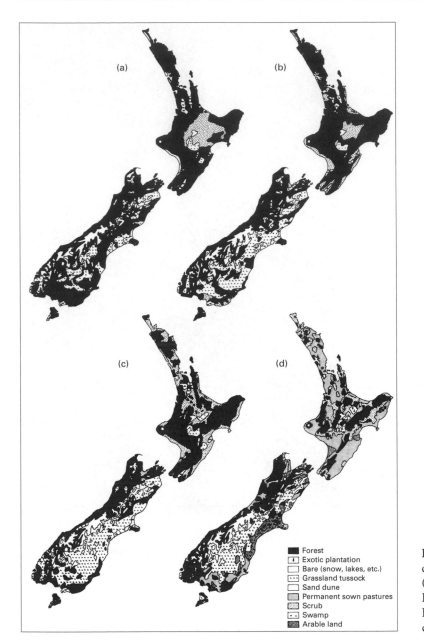

Forest
Exotic plantation
Bare (snow, lakes, etc.)
Grassland tussock
Sand dune
Permanent sown pastures
Scrub
Swamp
Arable land

Figure 2.3 The changing state of the vegetation cover in New Zealand (from Cochrane, 1977): (a) early Polynesian vegetation, *c.* AD 700; (b) Pre-Classical Maori vegetation, *c.* AD 1200; (c) Pre-European vegetation, *c.* AD 1800; (d) present-day vegetation.

naturally, especially by lightning, which on average strikes the land surface of the globe 100,000 times each day (Yi-Fu Tuan, 1971). Some natural fires may result from spontaneous combustion (Vogl, 1974), for in certain ecosystems heavy vegetal accumulations may become compacted, rotted and fermented, thus generating heat. Other natural fires can result from sparks produced by falling boulders and by landslides (Booysen and Tainton, 1984). In the forest lands of the western USA about half the fires are caused by lightning. Lightning starts over half the fires in the pine savanna of Belize, Central America, and about 8% of the fires in the bush of Australia, while in the south of France nearly all the fires are caused by people.

One method of gauging the long-term frequency of fires (Clarke et al., 1997) is to look at tree rings and lake sediments, for these are affected by them. Studies in the USA indicate that over wide areas fires occurred

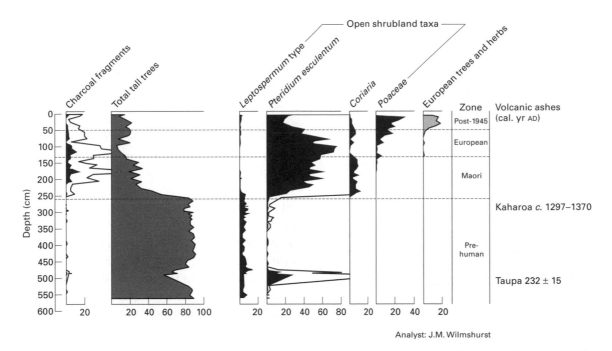

Figure 2.4 Summary percentage pollen diagram, Lake Rotonuiaha, eastern North Island (from McGlone and Wilmshurst, 1999, figure 1), showing the changes in vegetation brought about in Maori and European times.

with sufficient frequency to have effects on annual tree rings and lake cores every 7 to 80 years in pre-European times. The frequency of fires in different environments tends to show some variation. Rotation periods may be in excess of a century for tundra, 60 years for boreal pine forest, 100 years for spruce-dominated ecosystems, 5 to 15 years for savanna, 10 to 15 years for chaparral, and less than 5 years for semi-arid grasslands (Wein and Maclean, 1983).

The temperatures attained in fires

The effects which fires have on the environment depend very much on their size, duration, and intensity. Some fires are relatively quick and cool, and destroy only ground vegetation. Other fires, crown fires, affect whole forests up to crown level and generate very high temperatures In general forest fires are hotter than grassland fires. Perhaps more significantly in terms of forest management, fires that occur with great frequency do not attain too high a temperature because there is inadequate inflammable material to feed them. However, when humans deliberately suppress fire, as has frequently been normal policy in forest areas

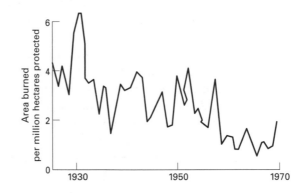

Figure 2.5 The reduction in area burned per million hectares protected for the USA between 1926 and 1969 as a result of fire-suppression policies (after Brown and Davis, 1973, figure 2.1).

(see Figure 2.5), large quantities of inflammable materials accumulate, so that when a fire does break out it is of the hot, crown type. Such fires can be disastrous ecologically and there is now much debate about the wisdom of fire suppression given that, in many forests, fires under so-called 'natural' conditions appear, as we have seen, to have been a relatively frequent and regular phenomenon.

Some consequences of fire suppression

Given that fire has long been a feature of many ecosystems, and irrespective of whether the fires were or were not caused by people, it is clear that any deliberate policy of fire suppression will have important consequences for vegetation.

Fire suppression, as has already been suggested, can magnify the adverse effects of fire. The position has been well stated by Sauer (1969: 14):

The great fires we have come to fear are effects of our civilization. These are the crown fires of great depths and heat, notorious aftermaths of the pyres of slash left by lumbering. We also increase fire hazard by the very giving of fire protection, which permits the indefinite accumulation of inflammable litter. Under the natural and primitive order, such holocausts, that leave a barren waste, even to the destruction of the organic soil, were not common.

Recent studies have indicated that rigid fire-protection policies have often had undesirable results (Bonnicksen et al., 1999) and as a consequence many foresters stress the need for 'environmental restoration burning' (Vankat, 1977). One of the best ways to prevent the largest forest fires is to allow the small and medium fires to burn (Malamud et al., 1998). So, for example, in the coniferous forests of the middle upper elevations in the Sierra Nevada mountains of California fire protection since 1890 has made the stands denser, shadier and less park-like, and sequoia seedlings have decreased in number. Likewise, at lower elevations the Mediterranean semi-arid shrubland, called chaparral, has had its character changed. The vegetation has increased in density, the amount of combustible fuels has risen, fire-intolerant species have encroached, and vegetation diversity has decreased, resulting in a monotony of old-age stands, instead of a mosaic of different successional stages. Unfavorable consequences of fire suppression have also been noted in Alaska (Oberle, 1969). It has been found that when fire is excluded from many lowland sites an insulating carpet of moss tends to accumulate and raise the permafrost level. **Permafrost** close to the surface encourages the growth of black spruce, a low-growing species with little timber or food value. In the Kruger National Park, in South Africa, fires have occurred less frequently after the establishment of the game reserve, when it became uninhabited by natives and hunters. As a result, bush encroachment has taken place in areas that were formerly grassland and the carrying capacity for grazing animals has declined. Controlled burning has been re-instituted as a necessary game-management operation.

However, following the severe fires that ravaged America's Yellowstone National Park in 1988, there has been considerable debate as to whether the inferno, the worst since the park was established in the 1870s, was the result of a policy of fire suppression. Without such a policy the forest would burn at intervals of 10–20 years because of lightning strikes. Could it be that the suppression of fires over long periods of, say, 100 years or more, allegedly to protect and preserve the forest, led to the build-up of abnormal amounts of combustible fuel in the form of trees and shrubs in the understory? Should a program of prescribed burning be carried out to reduce the amount of fuel available?

Fire suppression policies at Yellowstone did indeed lead to a critical build-up in flammable material. However, other factors must also be examined in explaining the severity of the fire. One of these was the fact that the last comparable fire had been in the 1700s, so that the Yellowstone forests had had nearly 300 years in which to become increasingly flammable. In other words, because of the way vegetation develops through time (a process called **succession**) very large fires may occur every 200–300 years as part of the natural order of things (Figure 2.6). Another crucial factor was that the weather conditions in the summer of 1988 were abnormally dry, bringing a great danger of fire. As Romme and Despain (1989: 28) remarked:

It seems that unusually dry, hot and windy weather conditions in July and August of 1988 coincided with multiple ignitions in a forest that was at its most flammable stage of succession. Yet it is unlikely that past suppression efforts greatly exacerbated the Yellowstone fire. If fires occur naturally at intervals ranging from 200 to 400 years, then 30 or 40 years of effective suppression is simply not enough for excessive quantities of fuel to build up. Major attempts at suppression in the Yellowstone forests may have merely delayed the inevitable.

Certainly, studies of the long-term history of fire in the western USA have shown the great importance of climatic changes at the decennial and centennial scales in controlling fire frequency during the Holocene (Meyer and Pierce, 2003; Whitlock et al., 2003).

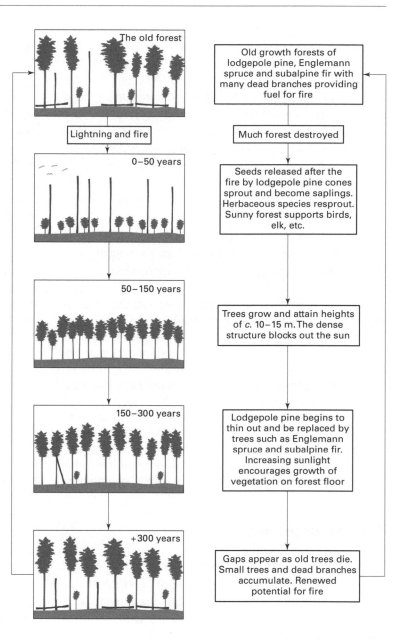

Figure 2.6 Ecological succession in response to fire in Yellowstone National Park (after Romme and Despain, 1989: 24–5, greatly modified).

Some effects of fire on vegetation

There is evidence that fire has played an important role in the formation of various major types of vegetation. This applies, for instance, to some tropical savannas, and mid-latitude grasslands and shrublands. Before examining these, however, it is worth looking at some of the general consequences of burning (Crutzen and Goldammer, 1993).

Fire may assist in seed germination. For example, the abundant germination of dormant seeds on recently burned **chaparral** sites has been reported by many investigators, and it seems that some seeds of chaparral species require scarification by fire. The better germination of those not requiring scarification may be related to the removal by fire of competition, litter and some substances in the soils which are toxic to plants (Hanes, 1971). Fire alters seedbeds. If litter and humus removals are substantial, large areas of rich ash, bare soil, or thin humus may be created. Some trees, such as Douglas fir and the giant sequoia, benefit from such seedbeds (Heinselman and Wright, 1973). Fire sometimes triggers the release of seeds (as with the Jack pine, *Pinus bankdiana*) and seems to stimulate the vegetative reproduction of many woody and herbaceous species. Fire can control forest insects, parasites and fungi – a process termed 'sanitization'. It also seems to stimulate the flowering and fruiting of many shrubs and herbs, and to modify the physicochemical environment of the plants. Mineral elements are released both as ash and through increased decomposition rates of organic layers. Above all, areas subject to fire often show greater species diversity, which is a factor that tends to favor stability.

One can conclude by quoting at length from Pyne (1982: 3), who provides a detailed and scholarly analysis of the history of cultural fires in America:

Hardly any plant community in the temperate zone has escaped fire's selective action, and, thanks to the radiation of *Homo sapiens* throughout the world, fire has been introduced to nearly every landscape on earth. Many biotas have consequently so adapted themselves to fire that, as with biotas frequented by floods and hurricanes, adaptation has become symbiosis. Such ecosystems do not merely tolerate fire, but often encourage it and even require it. In many environments fire is the most effective form of decomposition, the dominant selective force for determining the relative distribution of certain species, and the means for effective nutrient recycling and even the recycling of whole communities.

The role of grazing

Many of the world's grasslands have long been grazed by wild animals, such as the bison of North America or the large game of East Africa, but the introduction of pastoral economies also affects their nature and productivity (Figure 2.7) (Coupland, 1979).

Light grazing may increase the productivity of wild pastures (Warren and Maizels, 1976). Nibbling, for example, can encourage the vigor and growth of plants, and in some species, such as the valuable African grass, *Themeda triandra*, the removal of coarse, dead stems permits succulent sprouts to shoot. Likewise the seeds of some plant species are spread efficiently by being carried in cattle guts, and then placed in favorable seedbeds of dung or trampled into the soil surface. Moreover, the passage of herbage through the gut and out as feces modifies the nitrogen cycle, so that grazed pastures tend to be richer in nitrogen than ungrazed

Figure 2.7 Grazing by domestic animals has many environmental consequences and assists in the maintenance of grasslands. Excessive grazing can compact the soil and contribute to both wind and water erosion.

ones. Also, like fire, grazing can increase species diversity by opening out the community and creating more niches.

On the other hand, heavy grazing may be detrimental. Excessive trampling when conditions are dry will reduce the size of soil aggregates and break up plant litter to a point where they are subject to removal by eolian **deflation** processes. Trampling, by puddling the soil surface, can accelerate soil deterioration and erosion as infiltration capacity is reduced. Heavy grazing can kill plants or lead to a marked reduction in their level of photosynthesis. In addition, when relieved of competition from palatable plants or plants liable to trampling damage, resistant and usually unpalatable species expand their cover. Thus in the western USA poisonous burroweed (*Haploplappius* spp.) has become dangerously common, and many woody species have intruded. These include the mesquite (*Prosopis juliflora*), the big sagebrush (*Artemisia tridentata*) (Vale, 1974), the one-seed juniper (*Juniperus monosperma*) (Harris, 1966), and the Pinyon pine (Blackburn and Tueller, 1970).

Grover and Musick (1990) see shrubland encroachment by creosote bush (*Larrea tridentata*) and mesquite as part and parcel of desertification and indicate that in southern New Mexico the area dominated by them has increased several fold over the past century. This has been as a result of a corresponding decrease in the area's coverage of productive grasslands. They attribute both tendencies to excessive livestock overgrazing at the end of the nineteenth century, but point out that this was compounded by a phase of rainfall regimes that were unfavorable for perennial grass growth.

It is evident then that the semi-arid grasslands of southwestern North America have changed dramatically over the past 150 years as a result of the encroachment of native woody species such as mesquite. Van Auken (2000: 207) has concluded: 'The major cause of the encroachment of these woody species seems to be the reduction of grass **biomass** (fire fuel) by chronic high levels of domestic herbivory coupled to a reduction of grassland fires, which would have killed or suppressed the woody plants to the advantage of the grasses.'

There are many reasons why excessive grazing can lead to shrub dominance (Archer et al., 1999):

1 Livestock may effectively disperse woody plant seeds, particularly those of some leguminous shrubs and arborescents.
2 Utilization of grasses increases chances for germination and early establishment of woody seedlings.
3 Concomitant reductions in transpirational leaf area, root biomass, and root activity associated with grazing of grasses can:
  (a) increase superficial soil moisture to enhance woody seedling establishment and growth of shallow-rooted woody species;
  (b) increase the amount of water percolating to deeper depths and benefit established woody species with deep root systems;
  (c) increase nutrient availability to woody plants;
  (d) release suppressed populations of established tree or shrub 'seedling reserves'.
4 Grazing increases mortality rates and decreases plant basal area, seed production, and seedling establishment of palatable grasses.
5 Grazing may also increase susceptibility of grasses to other stresses such as drought. These factors would combine to increase the rate of gap formation and available area for woody plant seedling establishment, especially in post-drought periods.
6 Herbaceous species may be replaced by assemblages that compete less effectively with woody plants.
7 Reduction of the biomass and continuity of fine fuel may reduce fire frequency and intensity.
8 Invading woody species are often unpalatable relative to grasses and forbs and are thus not browsed

with sufficient regularity or severity to limit establishment or growth.
9 Lower soil fertility and alterations in physicochemical properties occur with loss of ground cover and subsequent erosion. This favors N_2-fixing woody plants (e.g., *Prosopis*, *Acacia*) and evergreen woody plant growth forms that are tolerant of low nutrient conditions.

In Australia the widespread adoption of sheep grazing led to significant changes in the nature of grasslands over extensive areas. In particular, the introduction of sheep led to the removal of kangaroo grass (*Themeda australis*) – a predominantly summer-growing species – and its replacement by essentially winter-growing species such as *Danthonia* and *Stipa*. Also in Australia, not least in the areas of tropical savanna in the north, large herds of introduced feral animals (e.g., *Bos taurus*, *Equus caballus*, *Camelus dromedarius*, *Bos banteng*, and *Cervus unicolor*) have resulted in overgrazing and alteration of native habitats. As they appear to lack significant control by predators and pathogens, their densities, and thus their effects, became very high (Freeland, 1990).

Similarly in Britain many plants are avoided by grazing animals because they are distasteful, hairy, prickly, or even poisonous (Tivy, 1971). The persistence and continued spread of bracken (*Pteridium aquilinium*) on heavily grazed rough pasture in Scotland is aided by the fact that it is slightly poisonous, especially to young stock. The success of bracken is furthered by its reaction to burning, for with its extensive system of underground stems (rhizomes) it tends to be little damaged by fire. The survival and prevalence of shrubs such as elder (*Sambucus nigra*), gorse (*Ulex* spp.), broom (*Sarothamus scoparius*), and the common weeds ragwort (*Senecio jacobaea*) and creeping thistle (*Cirsium arvense*) in the face of grazing can be attributed to their lack of palatability.

The role of grazing in causing marked deterioration of habitat has been the subject of further discussion in the context of upland Britain. In particular, Darling (1956) stressed that while trees bring up nutrients from rocks and keep minerals in circulation, pastoralism means the export of calcium phosphate and nitrogenous organic matter. The vegetation gradually deteriorates, the calcicoles disappear, and the herbage becomes deficient in both minerals and protein. Progressively more xerophytic plants come in: *Nardus stricta, Molinia caerulaea, Erica tetralix*, and then *Scripus caespitosa*. However, this is not a view that now receives universal support (Mather, 1983). Studies in several parts of upland Britain have shown that mineral inputs from precipitation are very much greater than the nutrient losses in wool and sheep carcasses. During the 1970s the idea of upland deterioration was gradually undermined.

In general terms it is clear that in many part of the world the grass family is well equipped to withstand grazing. Many plants have their growing points located on the apex of leaves and shoots, but grasses reproduce the bulk of fresh tissue at the base of their leaves. This part is least likely to be damaged by grazing and allows regrowth to continue at the same time that material is being removed.

Communities severely affected by the treading of animals (and indeed people) tend to have certain distinctive characteristics. These include diminutiveness (since the smaller the plant is, the more protection it will get from soil surface irregularities); strong ramification (the plant stems and leaves spread close to the ground); small leaves (which are less easily damaged by treading); tissue firmness (cell-wall strength and thickness to limit mechanical damage); a bending ability; strong vegetative increase and dispersal (e.g., by stolons); small hard seeds which can be easily dispersed, and the production of a large number of seeds per plant (which is particularly important because the mortality of seedlings is high under treading and trampling conditions).

Deforestation

Deforestation is one of the great processes of landscape transformation (Williams, 2003). However, controversy surrounds the meaning of the word 'deforestation' and this causes problems when it comes to assessing rates of change and causes of the phenomenon (Williams, 1994). It is best defined (Grainger, 1992) as 'the temporary or permanent clearance of forest for agriculture or other purposes'. According to this definition, if clearance does not take place then deforestation does not occur. Thus, much logging in the tropics, which is selective in that only a certain proportion of trees and only certain species are removed, does not involve clear felling and so cannot be said to constitute defor-

estation. However, how 'clear' is clear? Many shifting cultivators in the humid tropics leave a small proportion of trees standing (perhaps because they have special utility). At what point does the proportion of trees left standing permit one to say that deforestation has taken place? This is not a question to which there is a simple answer. There is not even a globally accepted definition of forest, and the UN's Food and Agriculture Organization (FAO) defines forest cover as greater than 10% canopy cover, whereas the International Geosphere Biosphere Program (IGBP) defines it as greater than 60%.

The deliberate removal of forest is one of the most longstanding and significant ways in which humans have modified the environment, whether achieved by fire or cutting. Pollen analysis shows that temperate forests were removed in Mesolithic and Neolithic times and at an accelerating rate thereafter. The example of Easter Island in the Pacific is salutary, for its deforestation at an unsustainable rate caused the demise of the culture that created its great statues. Since pre-agricultural times forests have declined by approximately one-fifth, from five to four billion hectares. The highest loss has occurred in temperate forests (32–5%), followed by subtropical woody savannas and deciduous forests (24–5%). The lowest losses have been in tropical evergreen forests (4–6%) because many of them have for much of their history been inaccessible and sparsely populated (World Resources Institute, 1992: 107).

In Britain, Birks (1988) analyzed dated pollen sequences to ascertain when and where tree pollen values in sediments from upland areas drop to 50% of their Holocene maximum percentages, and takes this as a working definition of 'deforestation'. From this work he identified four phases:

1 *3700–3900 years BP*: northwest Scotland and the eastern Isle of Skye;
2 *2100–2600 years BP (the pre-Roman iron age)*: Wales, England (except the Lake District), northern Skye, and northern Sutherland;
3 *1400–1700 years BP (post-Roman)*: Lake District, southern Skye, Galloway and Knapdale–Ardnamurchan;
4 *300–400 years BP*: the Grampians and the Cairngorms.

Sometimes forests are cleared to allow agriculture; at other times to provide fuel for domestic purposes, or

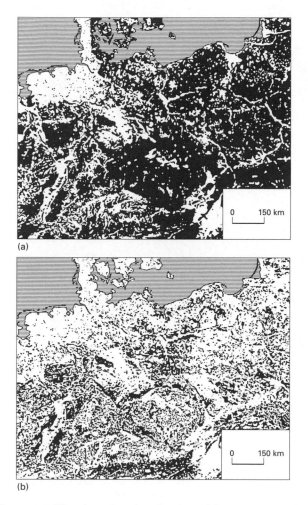

(a)

(b)

Figure 2.8 The changing distribution of forest in central Europe between (a) AD 900 and (b) AD 1900 (reprinted from Darby, 1956: 202–3, in *Man's role in changing the face of the Earth*, ed. W. L. Thomas, by permission of the University of Chicago Press © The University of Chicago 1956).

to provide charcoal or wood for construction; sometimes to fuel locomotives, or to smoke fish; and sometimes to smelt metals. The Phoenicians were exporting cedars as early as 4600 years ago (Mikesell, 1969), both to the Pharaohs and to Mesopotamia. Attica in Greece was laid bare by the fifth century BC and classical writers allude to the effects of fire, cutting and the destructive nibble of the goat. The great phase of deforestation in central and western Europe, described by Darby (1956: 194) as 'the great heroic period of reclamation', occurred properly from AD 1050 onwards for about 200 years (Figure 2.8). In particular, the Germans moved eastward: 'What the new west meant to

young America in the nineteenth century, the new east meant to Germany in the Middle Ages' (Darby, 1956: 196). The landscape of Europe was transformed, just as that of North America, Australia, New Zealand, and South Africa was to be as a result of the European expansions, especially in the nineteenth century.

Temperate North America underwent particularly brutal deforestation (Williams, 1989), and lost more woodland in 200 years than Europe did in 2000. The first colonialists arriving in the *Mayflower* found a continent that was wooded from the Atlantic seaboard as far as the Mississippi River (Figure 2.9). The forest originally occupied some 170 million hectares. Today only about 10 million hectares remain.

Fears have often been expressed that the mountains of High Asia (e.g., in Nepal) have been suffering from a wave of deforestation that has led to a whole suite of environmental consequences, which include accelerated landsliding, flooding in the Ganges Plain and sedimentation in the deltaic areas of Bengal. Ives and Messerli (1989) doubt that this alarmist viewpoint is soundly based and argue (p. 67) that 'the popular claims about catastrophic post-1950 deforestation of the Middle Mountain belt and area of the high mountains of the Himalayas are much exaggerated, if not inaccurate.'

Likewise, some workers in West Africa have interpreted 'islands' of dense forest in the savanna as the relics of a once more extensive forest cover that was being rapidly degraded by human pressure. More recent research (Fairhead and Leach, 1996) suggests that this is far from the truth and that villagers have in recent decades been extending rather than reducing forest islands (Table 2.1). There are also some areas of Kenya in East Africa where, far from recent population pressures promoting vegetation removal and land degradation, there has been an increase in woodland since the 1930s.

With regard to the equatorial rain forests, the researches of Flenley and others (Flenley, 1979) have indicated that forest clearance for agriculture has been going on since at least 3000 years BP in Africa, 7000 years BP in South and Central America, and possibly since 9000 years BP or earlier in India and New Guinea. Recent studies by archeologists and paleoecologists have tended to show that prehistoric human activities were rather more extensive in the tropical forests than originally thought (Willis et al., 2004).

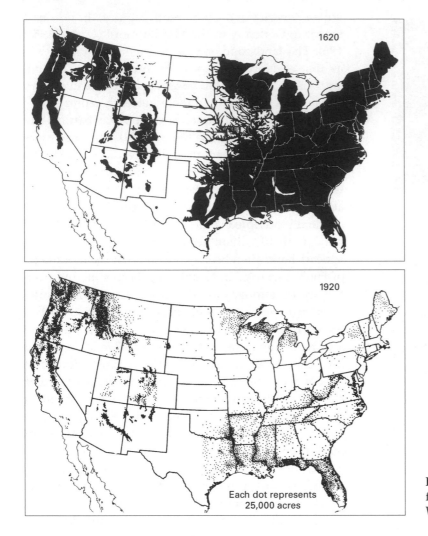

Each dot represents 25,000 acres

Figure 2.9 The distribution of American natural forest in 1620 and 1920 (modified from Williams, 1989).

Table 2.1 Estimates of deforestation since 1900 (millions of hectares)

Country	Orthodox estimate of forest area lost	Forest area lost according to World Conservation Monitoring Center	Forest area lost according to Fairhead and Leach (1996, table 9.1)
Côte d'Ivoire (Ivory Coast)	13	20.2	4.3–5.3
Liberia	4–4.5	5.5	1.3
Ghana	7	12.9	3.9
Benin	0.7	12.9	3.9
Togo	0	1.7	0
Sierra Leone	0.8–5	6.7	c. 0
Total	25.3–30.2	48.6	9.5–10.5

The causes of the present spasm of deforestation in the tropics (Figure 2.10) are complex and multifarious and are summarized in Grainger (1992) and Table 2.2. Grainger also provides an excellent review of the problems of measuring and defining loss of tropical forest area. Indeed, there are very considerable difficulties in estimating rates of deforestation, in part because different groups of workers use different definitions of what constitutes a forest (Allen and Barnes, 1985) and what distinguishes rain forest from other types of forest. There is thus some variability in views as to the present rate of rain forest removal and this is brought out in a recent debate (see Achard et al., 2002; Fearnside and Lawrence, 2003). The FAO estimates (Lanly et al., 1991) that the total annual deforestation in 1990 for 62

Figure 2.10 One of the most serious environmental problems in tropical areas is the removal of the rain forest. The felling of trees in Brazil on slopes as steep as these will cause accelerated erosion and loss of soil nutrients, and may promote lateritization.

Table 2.2 The causes of deforestation. Source: from Grainger (1992)

Immediate causes – land use	Underlying causes
Shifting agriculture:	Socio-economic mechanisms:
(a) traditional long-rotation shifting cultivation	(a) population growth
(b) short-rotation shifting cultivation	(b) economic development
(c) encroaching cultivation	Physical factors:
(d) pastoralism	(a) distribution of forests
Permanent agriculture:	(b) proximity of rivers
(a) permanent staple crop cultivation	(c) proximity of roads
(b) fish farming	(d) distance from urban centers
(c) Government sponsored resettlement schemes	(e) topography
(d) cattle ranching	(f) soil fertility
(e) tree crop and other cash crop plantations	Government policies:
Mining	(a) agriculture policies
Hydroelectric schemes	(b) forestry policies
Cultivation of illegal narcotics	(c) other policies

countries (representing some 78% of the tropical forest area of the world) was 16.8 million hectares, a figure significantly higher than the one obtained for these same countries for the period 1976–80 (9.2 million hectares per year). Myers (1992) suggests that there was an 89% increase in the tropical deforestation rate during the 1980s (compared with an FAO estimate of a 59% increase). He believed that the annual rate of loss in 1991 amounted to about 2% of the total forest expanse.

The change in forest area on a global basis in the 1990s is shown in Table 2.3a. These FAO data indicate the ongoing process of deforestation in Africa, Asia and the Pacific, Latin America and the Caribbean, and the stable or slightly reducing amount of deforestation in Europe, West Asia, and North America.

There is, however, a considerable variation in the rate of forest regression in different areas (Figure 2.11), with some areas under relatively modest threat (e.g., western Amazonia, the forests of Guyana, Surinam and French Guyana, and much of the Zaire Basin in central Africa (Myers, 1983, 1984)). Some other areas are being exploited so fast that minimal areas will soon be left, for example: the Philippines, peninsular Malaya, Thailand, Australia, Indonesia, Vietnam, Bangladesh, Sri Lanka, Central America, Madagascar, West Africa, and eastern Amazonia. Myers (1992) refers to particular 'hot spots' where the rates of deforestation are especially threatening, and presents data for certain locations where the percentage loss of forest is more than three times his global figure of 2%: southern Mexico (10%), Madagascar (10%), northern and eastern Thailand (9.6%), Vietnam (6.6%), and the Philippines (6.7%).

A more recent examination of rate of deforestation in selected hot spots is provided by Achard et al. (2002) (Table 2.3b). They suggest that the annual deforestation rate for rain forests is 0.38% for Latin America, 0.43% for Africa, and 0.91% for Southeast Asia, but that the hot spot areas can have rates that are much greater (e.g., up to 5.9% in central Sumatra and up to 4.7% in Madagascar). These rates are not as high as those of Myers, but are still extremely rapid.

The rapid loss of rain forest is potentially extremely serious, because as Poore (1976: 138) stated, these forests are a source of potential foods, drinks, medicines, contraceptives, abortifacients, gums, resins, scents, colorants, specific pesticides, and so on. Their removal may contribute to crucial global environmental concerns (e.g., climatic change and loss of biodiversity), besides causing regional and local problems, including lateritization, increased rates of erosion, and accelerated mass movements. The great range of potential impacts of tropical deforestation is summarized in Table 2.4.

Table 2.3

(a) Change in forested land 1990–2000 by region. Source: compiled from FAO (2001)

	Total land area (10⁶ hectares)	Total forest 1990 (10⁶ hectares)	Total forest 2000 (10⁶ hectares)	Percent of land forested in 2000	Change 1990–2000 (10⁶ hectares)	Percent change per year
Africa	2963.3	702.5	649.9	21.9	−52.6	−0.7
Asia and the Pacific	3463.2	734.0	726.3	21.0	−7.7	−0.1
Europe	2359.4	1042.0	1051.3	44.6	9.3	0.1
Latin America and the Caribbean	2017.8	1011.0	964.4	47.8	−46.7	−0.5
North America	1838.0	466.7	470.1	25.6	3.9	0.1
West Asia	372.4	3.6	3.7	1.0	0.0	0.0
World*	13,014.1	3960.0	3866.1	29.7	−93.9	−0.24

*Numbers may not add due to rounding.

(b) Annual deforestation rates as a percentage of the 1990 forest cover, for selected areas of rapid forest cover change (hot spots) within each continent. Source: Achard et al. (2002)

Hot-spot areas by continent	Annual deforestation rate of sample sites within hot-spot area (range), %
Latin America	0.38
Central America	0.8–1.5
Brazilian Amazonian belt	
Acre	4.4
Rondônia	3.2
Mato Grosso	1.4–2.7
Pará	0.9–2.4
Columbia–Ecuador border	c. 1.5
Peruvian Andes	0.5–1.0
Africa	0.43
Madagascar	1.4–4.7
Côte d'Ivoire	1.1–2.9
Southeast Asia	0.91
Southeastern Bangladesh	2.0
Central Myanmar	c. 3.0
Central Sumatra	3.2–5.9
Southern Vietnam	1.2–3.2
Southeastern Kalimantan	1.0–2.7

Some traditional societies have developed means of exploiting the rain forest environment which tend to minimize the problems posed by soil fertility deterioration, soil erosion, and vegetation degradation. Such a system is known as shifting agriculture (*swidden*). As Geertz (1963: 16) remarked:

In ecological terms, the most distinctive positive characteristic of swidden agriculture . . . is that it is integrated into and, when genuinely adaptive, maintains the general structure of the pre-existing natural ecosystem into which it is projected, rather than creating and sustaining one organized along novel lines and displaying novel dynamics.

The tropical rain forest and the swidden plots have certain common characteristics. Both are closed cover systems, in part because in swidden some trees are left standing, in part because some tree crops (such as banana, papaya, areca, etc.) are planted, but also because food plants are not planted in an open field,

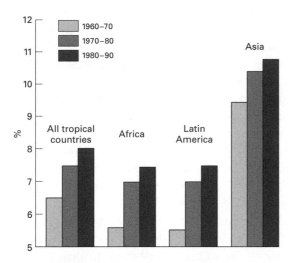

Figure 2.11 Estimated rate of tropical deforestation, 1960–1990, showing the particularly high rates in Asia (modified from *World Resources, 1996–97*, World Resources Institute, 1998).

crop-row manner, but helter-skelter in a tightly woven, dense botanical fabric. It is in Geertz's words (1963: 25) 'a miniaturized tropical forest'. Second, swidden agriculture normally involves a wide range of cultigens, thereby having a high diversity index like the rain forest itself. Third, both swidden plots and the rain forest have high quantities of nutrients locked up in the biotic community (Douglas, 1969) compared to that in the soil. The primary concern of 'slash-and-burn' activities is not merely the clearing of the land, but rather the transfer of the rich store of nutrients locked up in the prolific vegetation of the rain forest to a botanical complex whose yield to people is a great deal larger. If the period of cultivation is not too long and the period of fallow is long enough, an equilibrated, non-deteriorating and reasonable productive

farming regime can be sustained in spite of the rather impoverished soil base upon which it rests.

Unfortunately this system often breaks down, especially when population increase precludes the maintenance of an adequately long fallow period. When this happens, the rain forest cannot recuperate and is replaced by a more open vegetation assemblage ('derived savanna') that is often dominated by the notorious *Imperata* savanna grass, which has turned so much of Southeast Asia into a green desert. *Imperata cyclindrica* is a tall grass that springs up from rhizomes. Because of its rhizomes it is fire-resistant, but since it is a tall grass it helps to spread fire (Gourou, 1961).

One particular type of tropical forest ecosystem coming under increasing pressure from various human activities is the mangrove forest (Figure 2.12) characteristic of intertidal zones. These ecosystems constitute a reservoir, refuge, feeding ground, and nursery for many useful and unusual plants and animals (Mercer and Hamilton, 1984). In particular, because they export decomposable plant debris into adjacent coastal waters, they provide an important energy source and nutrient input to many tropical estuaries. In addition they can serve as buffers against the erosion caused by tropical storms – a crucial consideration in low-lying areas such as Bangladesh. In spite of these advantages, mangrove forests are being degraded and destroyed on a large scale in many parts of the world, either through exploitation of their wood resources or because of their conversion to single-use systems such as agriculture, aquaculture, salt-evaporation ponds, or housing developments. Data from FAO suggest that the world's mangrove forests, which covered 19.8 million hectares in 1980, have now been reduced to only 14.7 million hectares, with the annual loss running at about 1% per year (compared with 1.7 % a year from

Table 2.4 The consequences of tropical deforestation. Source: from Grainger (1992)

Reduced biological diversity	Changes in local and regional environments	Changes in global environments
Species extinction	More soil degradation	Reduction in carbon stored in terrestrial biota
Reduced capacity to breed improved crop varieties	Changes in water flows from catchments	Increase in carbon dioxide content of atmosphere
Inability to make some plants economic crops	Changes in buffering of water flows by wetland forests	Changes in global temperature and rainfall patterns by greenhouse effects
Threat to production of minor forest products	Increased sedimentation of rivers, reservoirs etc.	Other changes in global climate due to changes in land surface processes
	Possible changes in rainfall characteristics	

Figure 2.12 Mangrove swamps are highly productive and diverse ecosystems that are being increasingly abused by human activities.

1980 to 1990 (http://www.fao.org/DOCREP/005/Y7581E/y7581e04.htm).

On a global basis Richards (1991: 164) has calculated that since 1700 about 19% of the world's forests and woodlands have been removed. Over the same period the world's cropland area has increased by over four and a half times, and between 1950 and 1980 it amounted to well over 100,000 km² per year.

Deforestation is not an unstoppable or irreversible process. In the USA the forested area has increased substantially since the 1930s and 1940s. This 'rebirth of the forest' (Williams, 1988) has a variety of causes: new timber growth and planting was rendered possible because the old forest had been removed, forest fires have been suppressed and controlled, farmland has been abandoned and reverted to forest, and there has been a falling demand for lumber and lumber-derived products. It is, however, often very difficult to disentangle the relative importance of grazing impacts, fire, and fluctuating climates in causing changes in the structure of forested landscapes. In some cases, as for example the Ponderosa pine forests of the American southwest, all three factors may have contributed to the way in which the park-like forest of the nineteenth century has become significantly denser and younger today (Savage, 1991).

Given the perceived and actual severity of tropical deforestation it is evident that strategies need to be developed to reduce the rate at which it is disappearing. Possible strategies include the following:

- *research, training, and education* to give people a better understanding of how forests work and why they are important, and to change public opinion so that more people appreciate the uses and potential of forests;
- *land reform* to reduce the mounting pressures on landless peasants caused by inequalities in land ownership;
- *conservation of natural resources* by setting aside areas of rain forest as National Parks or nature reserves;
- *restoration and reforestation* of damaged forests;
- *sustainable development*, namely development which, while protecting the habitat, allows a type and level of economic activity that can be sustained into the future with minimum damage to people or forest (e.g., selective logging rather than clear felling; promotion of nontree products; small-scale farming in plots within the forest);
- *control of the timber trade* (e.g., by imposing heavy taxes on imported tropical forest products and outlawing the sale of tropical hardwoods from nonsustainable sources);
- *'debt-for-nature' swaps* whereby debt-ridden tropical countries set a monetary value on their ecological capital assets (in this case forests) and literally trade them for their international financial debt;
- *improvement of local peoples* in managing the remaining rain forests;
- *careful control of international aid* and development funds to make sure that they do not inadvertently lead to forest destruction;
- *reducing demand for wood products*.

Having considered the importance of the three basic processes of fire, grazing, and deforestation, we can now turn to a consideration of some of the major changes in vegetation types that have taken place over extensive areas as a result of such activities.

Secondary rain forest

When an area of rain forest that has been cleared for cultivation or timber exploitation is abandoned by humans, the forest begins to regenerate, but for an extended period of years the type of forest that occurs – **secondary forest** – is very different in character from the virgin forest it replaces. The features of such secondary forest, which is widespread in many tropical regions and accounts for as much as 40% of the total forest area in the tropics, have been summarized by Richards (1952), Ellenberg (1979), Brown and Lugo (1990), and Corlett (1995).

First, secondary forest is lower and consists of trees of smaller average dimensions than those of primary forest, but since it is comparatively rare that an area of primary forest is clear-felled or completely destroyed by fire, occasional trees much larger than average are usually found scattered through secondary forest. Second, very young secondary forest is often remarkably regular and uniform in structure, although the abundance of small climbers and young saplings gives it a dense and tangled appearance, which is unlike that of primary forest and makes it laborious to penetrate. Third, secondary forest tends to be much poorer in species than primary, and is sometimes, although by no means always, dominated by a single species, or a small number of species. Fourth, the dominant trees of secondary forest are light-demanding and intolerant of shade, most of the trees possess efficient dispersal mechanisms (having seeds or fruits well adapted for transport by wind or animals), and most of them can grow very quickly. The rate of net primary production of secondary forests exceeds that of the primary forests by a factor of two (Brown and Lugo, 1990). Some species are known to grow at rates of up to 12 m in 3 years, but they tend to be short-lived and to mature and reproduce early. One consequence of their rapid growth is that their wood often has a soft texture and low density.

Secondary forests should not be dismissed as useless scrub. All but the youngest secondary forests are probably effective at preventing soil erosion, regulating water supply and maintaining water quality. They provide refuge for some flora and fauna and can provide a source of timber, albeit generally of inferior quality than derived from primary forest. In addition, secondary forest regrowth can have a profound effect on water resources (see, e.g., Wilby and Gell, 1994).

The human role in the creation and maintenance of savanna

The **savannas** (Figure 2.13) can be defined, following Hills (1965: 218–19), as:

Figure 2.13 A tropical savanna in the Kimberley District of northwest Australia.

a plant formation of tropical regions, comprising a virtually continuous ecologically dominant stratum of more or less xeromorphic plants, of which herbaceous plants, especially grasses and sedges, are frequently the principal, and occasionally the only, components, although woody plants often of the dimension of trees or palms generally occur and are present in varying densities.

They are extremely widespread in low latitudes (Harris, 1980), covering about 18 million km^2 (an area about 2.6 million km^2 greater than that of the tropical rain forest). Their origin has been the subject of great contention in the literature of biogeography, although most savanna research workers agree that no matter what the savanna origins may be, the agent that seems to maintain them is intentional or inadvertent burning (Scott, 1977). As with most major vegetation types a large number of interrelated factors are involved in causing savanna (Mistry, 2000), and too many arguments about origins have neglected this fact (Figure 2.14). Confusion has also arisen because of the failure to distinguish clearly between predisposing, causal, resulting, and maintaining factors (Hills, 1965). It appears, for instance, that in the savanna regions around the periphery of the Amazon basin the climate *predisposes* the vegetation towards the development of savanna rather than forest. The geomorphic evolution of the landscape may be a *causal* factor; increased laterite development a *resulting* factor; and fire, *a maintaining* factor.

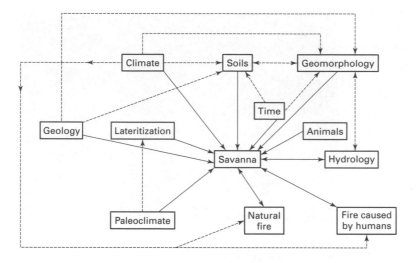

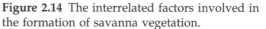

Figure 2.14 The interrelated factors involved in the formation of savanna vegetation.

Originally, however, savanna was envisaged as a predominantly natural vegetation type of climatic origin (see, e.g., Schimper, 1903), and the climatologist Köppen used the term 'savanna climate', implying that a specific type of climate is associated with all areas of savanna origin. According to supporters of this theory, savanna is better adapted than other plant formations to withstand the annual cycle of alternating soil moisture: rain forests could not resist the extended period of extreme drought, whereas dry forests could not compete successfully with perennial grasses during the equally lengthy period of large water surplus (Sarmiento and Monasterio, 1975).

Other workers have championed the importance of edaphic (soil) conditions, including poor drainage, soils which have a low water-retention capacity in the dry season, soils with a shallow profile due to the development of a lateritic crust, and soils with a low nutrient supply (either because they are developed on a poor parent rock such as quartzite, or because the soil has undergone an extended period of leaching on an old land surface). Associated with soil characteristics, ages of land surfaces, and degree of drainage is the geomorphology of an area (as stressed by, e.g., Cole, 1963). This may also be an important factor in savanna development.

Some other researchers – for example, Eden (1974) – found that savannas are the product of former drier conditions (such as late Pleistocene aridity) and that, in spite of a moistening climate, they have been maintained by fire. He pointed to the fact that the patches

of savanna in southern Venezuela occur in forest areas of similar humidity and soil infertility, suggesting that neither soil, nor drainage, nor climate was the cause. Moreover, the present 'islands' of savanna are characterized by species which are also present elsewhere in tropical American savannas and whose disjunct distribution conforms to the hypothesis of previous widespread continuity of that formation.

The importance of fire in maintaining and originating some savannas is suggested by the fact that many savanna trees are fire-resistant. Controlled experiments in Africa (Hopkins, 1965) demonstrated that some tree species, such as *Burkea africana* and *Lophira lanceolata*, withstand repeated burning better than others do. There are also many observations of the frequency with which, for example, African herdsmen and agriculturists burn over much of tropical Africa and thereby maintain grassland. Certainly the climate of savanna areas is conducive to fire for, as Gillian (1983: 617) put it, 'Large scale grass fires are more likely to take place in areas having a climate moist enough to permit the production of a large amount of grass, but seasonally dry enough to allow the dried material to catch fire and burn easily.' On the other hand Morgan and Moss (1965) express some doubts about the role of fire in western Nigeria, and point out that fire is not itself necessarily an independent variable:

The evidence suggests that some notions of the extent and destructiveness of savanna fires are rather exaggerated. In particular the idea of an annual burn that affects a large proportion of the area in each year could seem to be false. It is more likely that some patches, peculiarly susceptible to fire, as a result of edaphic or biotic influences upon the character of the community itself, are repeatedly burned, whereas others are hardly, if ever, affected . . . It is also important to note that there is no evidence anywhere along the forest fringe . . . to suggest that fire sweeps into the forest, effecting notable destruction of forest trees.

Some savannas are undoubtedly natural, for pollen analysis in South America shows that savanna vegetation was present before the arrival of human civilization. Nonetheless even natural savannas, when subjected to human pressures, change their characteristics. For example, the inability of grass cover to maintain itself over long periods in the presence of heavy stock grazing may be documented from many of the warm countries of the world (Johannessen, 1963). Heavy grazing tends to remove the fuel (grass) from much of the surface. The frequency of fires is therefore significantly reduced, and tree and bush invasion take place. As Johannessen (p. 111) wrote: 'Without intense, almost annual fires, seedlings of trees and shrubs are able to invade the savannas where the grass sod has been opened by heavy grazing . . . the age and size of the trees on the savannas usually confirm the relative recency of the invasion.' In the case of the savannas of interior Honduras, he reports that they only had a scattering of trees when the Spaniards first encountered them, whereas now they have been invaded by an assortment of trees and tall shrubs.

Elsewhere in Latin America changes in the nature and density of population have also led to a change in the nature and distribution of savanna grassland. C. F. Bennett (1968: 101) noted that in Panama the decimation of the Indian population saw the re-establishment of trees:

Large areas that today are covered by dense forest were in farms, grassland or low second growth in the early sixteenth century when the Spaniards arrived. At that time horses were ridden with ease through areas which today are most easily penetrated by river, so dense has the tree growth become since the Indians died away.

Such bush or shrub encroachment is a serious cause of rangeland deterioration in savanna regions (Rogues et al., 2001). Overgrazing reduces the vigor of favorable perennial grasses, which tend to be replaced, as we have seen, by less reliable annuals and by woody vegetation. Annual grasses do not adequately hold or cover the soil, especially early in the growing season. Thus runoff increases and topsoil erosion occurs. Less water is then available in the topsoil to feed the grasses, so that the woody species, which depend on deeper water, become more competitive relative to the grasses, which can use water only within their shallow root zones. The situation is exacerbated when there are few browsers in the herbivore population and, once established, woody vegetation competes effectively for light and nutrients. It is also extremely expensive to remove established dense scrub cover by mechanical means, although some success has been achieved by introducing animals that have a browsing or bulldozing effect, such as goats, giraffes, and elephants. Experiments in Zimbabwe have shown that, if bush clearance is carried out, a threefold increase in the sustainable carrying capacity can take place in semi-arid areas (Child, 1985).

Another example of human interventions modifying savanna character is through their effects on savanna-dwelling mammals such as the elephant. They are what is known as a '**keystone species**' because they exert a strong influence on many aspects of the environment in which they live (Waithaka, 1996). They diversify the ecosystems that they occupy and create a mosaic of habitats by browsing, trampling, and knocking over of bushes and trees. They also disperse seeds through their eating and defecating habits and maintain or create water holes by wallowing. All these roles are beneficial to other species. Conversely, where human interference prevents elephants from moving freely within their habitats and leads to their numbers exceeding the carrying capacity of the savanna, their effect can be environmentally catastrophic. Equally, if humans reduce elephant numbers in a particular piece of savanna, the savanna may become less diverse and less open, and its water holes may silt up. This will be to the detriment of other species.

In addition to its lowland savannas, Africa has a series of high-altitude grasslands called 'Afromontane grasslands'. They extend as a series of 'islands' from the mountains of Ethiopia to those of the Cape area of South Africa. Are they the result mainly of forest clearance by humans in the recent past? Or are they a long-standing and probably natural component of the pattern of vegetation (Meadows and Linder, 1993)?

Are they caused by frost, seasonal aridity, excessively poor soils, or an intensive fire regime? This is one of the great controversies of African vegetation studies.

Almost certainly a combination of factors has given rise to these grasslands. On the one hand current land management practices, including the use of fire, prevent forest from expanding. There has undoubtedly been extensive deforestation in recent centuries. On the other hand, pollen analysis from various sites in southern Africa suggests that grassland was present in the area as long ago as 12,000 years BP. This would mean that much grassland is not derived from forest through very recent human activities.

The spread of desert vegetation on desert margins

One of the most contentious and important environmental issues of recent years has been the debate on the question of the alleged expansion of deserts (Middleton and Thomas, 1997).

The term 'desertification' was first used but not formally defined by Aubréville (1949), and for some years the term 'desertization' was also employed, for example by Rapp (1974: 3) who defined it as: 'The spread of desert-like conditions in arid or semi-arid areas, due to man's influence or to climatic change'.

There has been some variability in how 'desertification' itself is defined. Some definitions stress the importance of human causes (e.g., Dregne, 1986: 6–7):

Desertification is the impoverishment of terrestrial ecosystems under the impact of man. It is the process of deterioration in these ecosystems that can be measured by reduced productivity of desirable plants, undesirable alterations in the biomass and the diversity of the micro and macro fauna and flora, accelerated soil deterioration, and increased hazards for human occupancy.

Others admit the possible importance of climatic controls but give them a relatively inferior role (e.g., Sabadell et al., 1982: 7):

The sustained decline and/or destruction of the biological productivity of arid and semi arid lands caused by man made stresses, sometimes in conjunction with natural extreme events. Such stresses, if continued or unchecked, over

the long term may lead to ecological degradation and ultimately to desert-like conditions.

Yet others, more sensibly, are even-handed or open-minded with respect to natural causes (e.g., Warren and Maizels, 1976: 1):

A simple and graphic meaning of the word 'desertification' is the development of desert like landscapes in areas that were once green. Its practical meaning . . . is a sustained decline in the yield of useful crops from a dry area accompanying certain kinds of environmental change, both natural and induced.

It is also by no means clear how extensive desertification is or how fast it is proceeding. Indeed, the lack of agreement on the former makes it impossible to determine the latter. As Grainger (1990: 145) has remarked in a well-balanced review, 'Desertification will remain an ephemeral concept to many people until better estimates of its extent and rate of increase can be made on the basis of actual measurements.' He continues (p. 157): 'The subjective judgements of a few experts are insufficient evidence for such a major component of global environmental change . . . Monitoring desertification in the drylands is much more difficult than monitoring deforestation in the humid tropics, but it should not be beyond the ingenuity of scientists to devise appropriate instruments and procedures.'

The United Nations Environment Program (UNEP) has played a central role in the promotion of desertification as a major environmental issue, as is made evident by the following remark by Tolba and El-Kholy (1992: 134):

Desertification is the main environmental problem of arid lands, which occupy more than 40% of the total global land area. At present, desertification threatens about 3.6 billion hectares – 70% of potentially drylands, or nearly one-quarter of the total land area of the world. These figures exclude natural hyper-arid deserts. About one sixth of the world's population is affected.

However, in their book *Desertification: exploding the myth*, Thomas and Middleton (1994) have discussed UNEP's views on the amount of land that is desertified. They state:

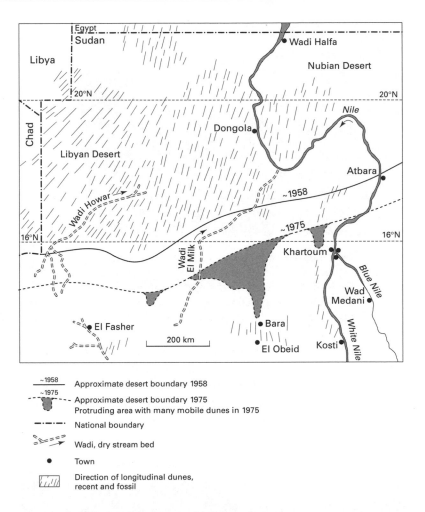

Figure 2.15 Desert encroachment in the northern Sudan 1958–1975, as represented by the position of the boundary between subdesert scrub and grassland in the desert (after Rapp et al., 1976, figure 8.5.3).

The bases for such data are at best inaccurate and at worst centred on nothing better than guesswork. The advancing desert concept may have been useful as a publicity tool but it is not one that represents the real nature of desertification processes. (Thomas and Middleton, 1994: 160)

Their views have been trenchantly questioned by Stiles (1995), but huge uncertainties do indeed exist.

There are relatively few reliable studies of the rate of supposed desert advance. Lamprey (1975) attempted to measure the shift of vegetation zones in the Sudan (see Figure 2.15) and concluded that the Sahara had advanced by 90 to 100 km between 1958 and 1975, an average rate of about 5.5 km per year. However, on the basis of analysis of remotely sensed data and ground observation, Helldén (1985) found sparse evidence that this had in fact happened. One problem is that there may be very substantial fluctuations in biomass production from year to year. This has been revealed by meteorological satellite observations of green biomass production levels on the south side of the Sahara (Dregne and Tucker, 1988).

The spatial character of desertification is also the subject of some controversy (Helldén, 1985). Contrary to popular rumor, the spread of desert-like conditions is not an advance over a broad front in the way that a wave overwhelms a beach. Rather, it is like a 'rash' which tends to be localized around settlements (Figure 2.16). It has been likened to Dhobi's itch – a ticklish problem in difficult places. Fundamentally, as Mabbutt (1985: 2) has explained, 'the extension of desert-like conditions tends to be achieved through a process of accretion from without, rather than through expansionary forces acting from within the deserts.' This distinction is important in that it influences perceptions of appropriate remedial or combative strategies, which are discussed in Goudie (1990).

Figure 2.16 On desert margins human activities such as wood collection, overgrazing and cultivation in drought-prone areas have led to the process of desertification. Here we see the effects of overgrazing around a water hole in the Molopo Valley region of the Kalahari, southern Africa.

Figure 2.17 shows two models of the spatial pattern of desertification. One represents the myth of a desert spreading outward over a wide front, whereas the other shows the more realistic pattern of degradation around routes and settlements.

Woodcutting is an extremely serious cause of vegetation decline around almost all towns and cities of the Sahelian and Sudanian zones of Africa. Many people depend on wood for domestic uses (cooking, heating, brick manufacture, etc.), and the collection of wood for charcoal and firewood is an especially serious problem in the vicinity of urban centers. This is illustrated for Khartoum in Sudan in Figure 2.18. Likewise, the installation of modern boreholes has enabled rapid multiplication of livestock numbers and large-scale destruction of the vegetation in a radius of 15–30 km around boreholes (Figure 2.19). Given this localization of degradation, amelioration schemes such as local tree-planting may be partially effective, but ideas of planting green belts as a 'cordon sanitaire' along the desert edge (whatever that is) would not halt deterioration of the situation beyond this Maginot line (Warren and Maizels, 1977: 222). The deserts are not invading from without; the land is deteriorating from within.

There has been considerable debate as to whether the vegetation change and environmental degradation associated with desertization is irreversible. In many cases, where ecological conditions are favorable because of the existence of such factors as deep sandy

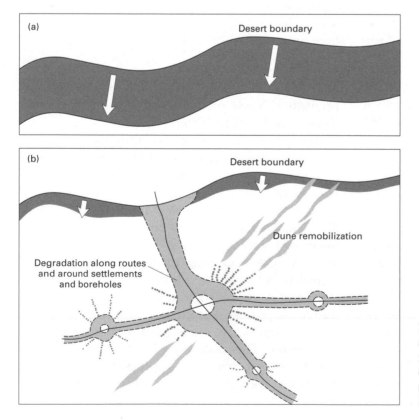

Figure 2.17 Two models of the spatial pattern of desertification: (a) the myth of desert migration over a wide front; (b) the more local development of degradation around routes, settlements, and boreholes (modified after Kadomura, 1994, figure 9).

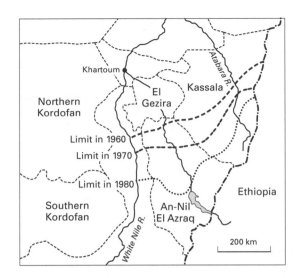

Figure 2.18 The expanding wood and charcoal exploitation zone south of Khartoum, Sudan (after Johnson and Lewis, 1995, figure 6.2).

soils or beneficial hydrological characteristics, vegetation recovers once excess pressures are eliminated. There is evidence of this in arid zones throughout the world where temporary or permanent enclosures have been set up (Le Houérou, 1977). The speed of recovery will depend on how advanced deterioration is, the size of the area that is degraded, the nature of the soils and moisture resources, and the character of local vegetation. It needs to be remembered in this context that much desert vegetation is adapted to drought and to harsh conditions, and that it often has inbuilt adaptations which enable a rapid response to improved circumstances.

Nonetheless, in certain specific circumstances recovery is slow and so limited that it may be appropriate to talk of 'irreversible desertization'. Le Houérou (1977: 419), for example, has pointed to such a case in

Figure 2.20 A lichen field, near Swakopmund in Namibia, shows the effects of vehicular traffic. Such scars can take a long time to recover.

North Africa, where, decades after the end of hostilities, wheel tracks and degraded vegetation produced in World War II were still present in the desert. Some desert surfaces, such as those covered by lichens, are easily destroyed by vehicular traffic but take a long time to recover (Figure 2.20).

The causes of desertification are also highly controversial and diverse (Table 2.5). The question has been asked whether this process is the result of temporary drought periods of high magnitude, whether it is due to long-term climatic change towards aridity (either as alleged post-glacial progressive desiccation or as part of a 200-year cycle), whether it is caused by anthropogenic climatic change, or whether it is the result of human action degrading the biological environments in arid zones. There is little doubt that severe droughts do take place (Nicholson, 1978), and that their effects become worse as human and domestic animal populations increase. The devastating drought in the African Sahel from 1968 to 1984 caused greater

Figure 2.19 Relation between spacing of wells and overgrazing: (a) original situation. End of dry season 1. Herd size limited by dry season pasture. Small population can live on pastoral economy. (b) After well-digging but no change in traditional herding. End of dry season 2. Larger total subsistence herd and more people can live there until a situation when grazing is finished in a drought year; then the system collapses (after Rapp, 1974).

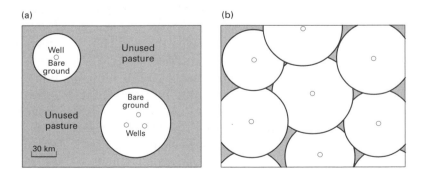

Table 2.5 Some examples of causes and consequences of desertification. Source: Williams (2000, table 2)

Trigger factor	Consequences
Direct land use:	Physical processes affected:
(a) overcultivation (decreased fallows, mechanized farming)	(a) decline in soil structure, water permeability, depletion of soil nutrients and organic matter, increased susceptibility to erosion, compaction of soil, sand dune mobilization
(b) overgrazing	(b) loss of biodiversity and biomass, increased soil erosion via wind and water, soil compaction from trampling, increased runoff, sand dune mobilization
(c) mismanagement of irrigated lands	(c) causes water logging and salinization of soil, hence lower crop yields, possible sedimentation of water reservoirs
(d) deforestation (burning, fuel and fodder collection)	(d) promotes artificial establishment of savanna vegetation, loss of soil-stabilizing vegetation, soil exposed and eroded, increases aridity, increased frequency of dust storms, sand dune mobilization (e.g., Ethiopia is losing c. 1000 million tonnes of topsoil per year)
(e) exclusion of fire	(e) promotes growth of unpalatable woody shrubs at the expense of herbage
Indirect government policies:	Drives overexploitation land-use practices:
(a) failed population planning policies	(a) increases need for food cultivation, hence overexploitation
(b) irrigation subsidies	(b) exacerbates flooding and salinization
(c) settlement policies/land tenure	(c) forces settlement of nomads, promotes concentrated use of land which often exceeds the carrying capacity
(d) improved infrastructure (e.g., roads, large-scale dams, canals, boreholes)	(d) although beneficial, can exacerbate the problem by attracting increased livestock and human populations or increasing risk from salinization, possible lowering of groundwater table below dams, problems of silting up reservoirs, water-logging; promotes large-scale commercial activity with little local benefit, flooding may displace people and perpetuate cycles of poverty
(e) promotion of cash crops and push towards national/international markets	(e) displaces subsistence cropping, pushes locals into marginal areas to survive, promotes less resilient monocultures, and promotes expansion and intensification of land use
(f) price increases on agricultural produce	(f) incentive to crop on marginal lands
(g) war	(g) valuable resources, both human and financial, are expended on war at the expense of environmental management and the needs of the people, large-scale migration with resultant increased pressure on receiving areas
(h) high interest rates	(h) forcing grazing or cultivation to levels beyond land capacity
Natural:	
(a) extreme drought	(a) decreased vegetation cover and increased land vulnerability for soil erosion. Creates an environment which exacerbates overexploitation
(b) ecological fragility	(b) impact of land-use practices – impact also depends on resilience of environment

ecological stress than the broadly comparable droughts of 1910–15 and 1944–8, largely because of the increasing anthropogenic pressures.

The venerable idea that climate is deteriorating through the mechanism of post-glacial progressive desiccation is now discredited (see Goudie, 1972a, for a critical analysis), although the idea that the Sahel zone is currently going through a 200-year cycle of drought has been proposed (Winstanley, 1973). However, numerous studies of available meteorological data (which in some cases date back as far as 130–150 years) do not allow any conclusions to be reached on the question of systematic long-term changes in rainfall, and the case for climatic deterioration – whether natural or aggravated by humans – is not proven. Indeed, in a judicious review, Rapp (1974: 29) wrote that after consideration of the evidence for the role of climatic change in desertization his conclusion was 'that the reported desertization northwards and southwards from the Sahara could not be explained by a general trend towards drier climate during this century'.

It is evident, therefore, that it is largely a combination of human activities (Figure 2.21) with occasional series of dry years that leads to presently observed desertification. The process also seems to be fiercest not in desert interiors, but on the less arid marginal areas around them. It is in semi-arid areas – where biological productivity is much greater than in extremely

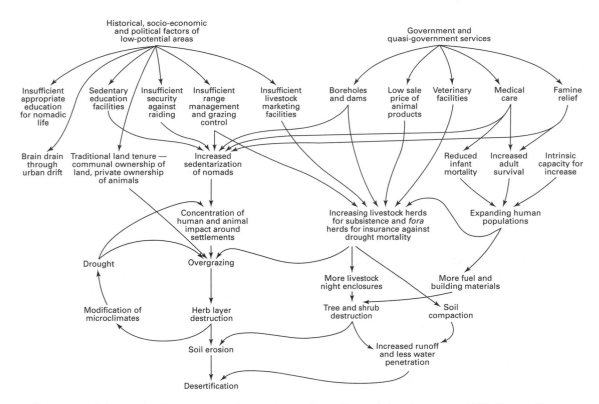

Figure 2.21 Some causal factors in desert encroachment in northern Kenya (after Lamprey, 1975, figure 2).

arid zones, where precipitation is frequent and intense enough to cause rapid erosion of unprotected soils, and where humans are prone to mistake short-term economic gains under temporarily favorable climatic conditions for long-term stability – that the combination of circumstances particularly conducive to desert expansion can be found. It is in these marginal areas that dry farming and cattle rearing can be a success in good years, so that susceptible areas are plowed and cattle numbers become greater than the vegetation can support in dry years. In this way, a depletion of vegetation occurs which sets in train such insidious processes as water erosion and deflation. The vegetation is removed by clearance for cultivation, by the cutting and uprooting of woody species for fuel, by overgrazing and by the burning of vegetation for pasture and charcoal.

These tendencies towards bad land-use practices result in part from the restrictions imposed on many nomadic societies through the imposition of national boundaries across their traditional migration routes, or through various schemes implemented for political and social reasons to encourage their establishment in settled communities. Some of their traditional grazing lands have been taken over by cash-crop farmers. In Niger, for example, there was a sixfold increase in the acreage of peanuts grown between 1934 and 1968. The traditional ability to migrate enabled pastoral nomads and their cattle to emulate the natural migration of such wild animals as wildebeest and kob, and thereby to make flexible use of available resources according to season and according to yearly variations in rainfall. They could also move away from regions that had become exhausted after a long period of use. As soon as migrations are stopped and settlements imposed, such options are closed, and severe degradation occurs (Sinclair and Fryxell, 1985).

The suggestion has sometimes been made that not only are deserts expanding because of human activity, but that deserts themselves are created by human activity. There are authors who have suggested, for example, that the Thar Desert of India is a post-glacial and possibly post-medieval creation (see Allchin et al., 1977, for a critique of such views), while Ehrlich and Ehrlich (1970) have written: 'The vast Sahara desert itself is largely man-made, the result of over-grazing, faulty irrigation, deforestation, perhaps combined with a shift in the course of a jet stream.' Nothing could be further from the truth. The Sahara, while it has fluctuated greatly in extent, is many millions of years old, pre-dates human life, and is the product of the nature of the general atmospheric circulation, occupying an area of dry descending air.

The maquis of the Mediterranean lands

Around much of the Mediterranean basin there is a plant formation called **maquis** (Figure 2.22). This consists of a stand of xerophilous nondeciduous bushes and shrubs which are evergreen and thick, and whose trucks are normally obscured by low-level branches. It includes such plants as holly oak (*Quercus ilex*), kermes oak (*Quercus coccifera*), tree heath (*Erica arborea*), broom heath (*Erica scoparia*), and strawberry trees (*Arbutus unedo*).

Some of the maquis may represent a stage in the evolution towards true forest in places where the **climax** has not yet been reached, but in large areas it represents the degeneration of forest. Considerable concern has been expressed about the speed with which degeneration to, and degeneration beyond, maquis is taking place as a result of human influences (Tomaselli, 1977), of which cutting, grazing, and fire

Figure 2.22 The maquis of the Mediterranean lands, illustrated here near Marseilles, south of France, is a vegetation type in which humans have played a major role. It represents the degeneration of the natural forest cover.

are probably the most important and long continued. Charcoal burners, goats, and frequent outbreaks of fires among the resinous plants in the dry Mediterranean summer have all taken their toll. Such aspects of degradation in Mediterranean environments are discussed in Conacher and Sala (1998).

On the other hand, in some areas, particularly marginal mountainous and semi-arid portions of the Mediterranean basin, agricultural uses of the land have declined in recent decades, as local people have sought easier and more remunerative employment. In such areas scrubland, sometimes termed 'post-cultural shrub formations', may start to invade areas of former cultivation (May, 1991), while maquis has developed to become true woodland. Mediterranean vegetation appears to be very resilient (Grove and Rackham, 2001).

There is considerable evidence that maquis vegetation is in part adapted to, and in part a response to, fire. One effect of fire is to reduce the frequency of standard trees and to favor species that after burning send up a series of suckers from ground level. Both *Quercus ilex* and *Quercus coccifera* seem to respond in this way. Similarly, a number of species (e.g., *Cistus albidus*, *Erica arborea*, and *Pinus halepensis*) seem to be distinctly advantaged by fire, perhaps because it suppresses competition or perhaps because (as with the comparable chaparral of the southwest USA) a short burst of heat encourages germination (Wright and Wanstall, 1977).

The prairie problem

The mid-latitude grasslands of North America – the prairies – are another major vegetation type that can be used to examine the human impact, although, as in the case of savanna grasslands in the tropics, the human role is the subject of controversy (Whitney, 1994).

It was once fairly widely believed that the prairies were essentially a climatically related phenomenon (see Changuon et al., 2002, for a historical review). Workers such as J. E. Weaver (1954) argued that under the prevailing conditions of soil and climate the invasion and establishment of trees was significantly hindered by the presence of a dense sod. High evapotranspiration levels combined with low precipitation were thought to give a competitive advantage to herbaceous plants with shallow, densely ramifying root systems, capable of completing their life cycles rapidly under conditions of pervasive drought.

An alternative view was, however, put forward by Stewart (1956: 128): 'The fact that throughout the tall-grass prairie planted groves of many species have flourished and have reproduced seedlings during moist years and, furthermore, have survived the most severe and prolonged period of drought in the 1930s suggests that there is no climatic barrier to forests in the area.'

Other arguments along the same lines have been advanced. Wells (1965), for example, has pointed out that in the Great Plains a number of woodland species, notably the junipers, are remarkably drought-resistant, and that their present range extends into the Chihuahua Desert where they often grow in association with one of the most xerophytic shrubs of the American deserts, the creosote bush (*Larrea divaricata*). He remarks (p. 247): 'There is no range of climate in the vast grassland climate of the central plains of North America which can be described as too arid for all species of trees native to the region.' Moreover, confirming Stewart, he points out that numerous plantations and shelterbelts have indicated that trees can survive for at least 50 years in a 'grassland' climate. One of his most persuasive arguments is that, in the distribution of vegetation types in the plains, a particularly striking vegetation feature is the widespread but local occurrence of woodlands along escarpments and other abrupt breaks in topography remote from fluvial irrigation. A probable explanation for this is the fact that fire effects are greatest on flat, level surfaces, where there are high wind speeds and no interruptions to the course of fire. It has also been noted that where burning has been restricted there has been extension of woodland into grassland.

The reasons why fire tends to promote the establishment of grassland have been summarized by Cooper (1961: 150–1).

In open country fire favours grass over shrubs. Grasses are better adapted to withstand fire than are woody plants. The growing point of dormant grasses from which issues the following years growth, lies near or beneath the ground, protected from all but the severest heat. A grass fire removes only one year's growth, and usually much of this is dried and dead. The living tissue of shrubs, on the other hand, stands well above the ground, fully exposed to fire. When it is burned, the growth of several years is destroyed.

Even though many shrubs sprout vigorously after burning, repeated loss of their top growth keeps them small. Perennial grasses, moreover, produce seeds in abundance one or two years after germination; most woody plants require several years to reach seed-bearing age. Fires that are frequent enough to inhibit seed production in woody plants usually restrict the shrubs to a relatively minor part of the grassland area.

Thus, as with savanna, anthropogenic fires may be a factor which maintains, and possibly forms, the prairies and there is now some paleoecological information to support this view (Boyd, 2002). Though again, following the analogy with savanna, it is possible that some of the American prairies may have developed in a post-glacial dry phase, and that with a later increase in rainfall re-establishment of forest cover was impeded by humans through their use of fire and by grazing animals. The grazing animals concerned were not necessarily domesticated, however, for Larson (1940) has suggested that some of the short-grass plains were maintained by wild bison. These, he believed, stocked the plains to capacity so that the introduction of domestic livestock, such as cattle, after the destruction of the wild game was merely a substitution so far as the effect of grazing on plants is concerned. There is indeed pollen analytical evidence that shows the presence of prairie in the western mid-west over 11,000 years ago, prior to the arrival of human settlers (Bernabo and Webb, 1977). Therefore some of it, at least, may be natural.

Comparable arguments have attended the origins of the great Pampa grassland of Argentina. The first Europeans who penetrated the landscape were much impressed by the treeless open country and it was always taken for granted, and indeed became dogma, that the grassland was a primary climax unit. However, this interpretation was successfully challenged, notably by Schmieder (1927a, b) who pointed out that planted trees thrived, that precipitation levels were quite adequate to maintain tree growth and that, in topographically favorable locations such as the steep gullies (*barrancas*) near Buenos Aires, there were numerous endemic representatives of the former forest cover (*monte*). Schmieder believed that Pampa grasslands were produced by a pre-Spanish aboriginal hunting and pastoral population, the density of which had been underestimated, but whose efficiency in the use of fire was proven.

In recent centuries mid-latitude grasslands have been especially rapidly modified by human activities. As Whitney (1994: 257) points out, they have been altered in ways that can be summarized in three phrases: 'plowed out', 'grazed out', and 'worn out'. Very few areas of natural North America prairies remain. The destruction of 'these masterpieces of nature' took less than a century. The preservation of remaining areas of North American grasslands is a major challenge for the conservation movement (Joern and Keeler, 1995). On a global basis it has been estimated that grasslands have been reduced by about 20% from their pre-agricultural extent (Graetz, 1994).

Post-glacial vegetation change in Britain and Europe

The classic interpretation of the vegetational changes of post-glacial (Holocene) times – that is, over the past 11,000 or so years – has been in terms of climate. That changes in the vegetation of Britain and other parts of western Europe took place was identified by pollen analysts and paleobotanists, and these changes were used to construct a model of climatic change – the Blytt–Sernander model. There was thus something of a chicken-and-egg situation: vegetational evidence was used to reconstruct past climates, and past climates were used to explain vegetational change. More recently, however, the importance of humanly induced vegetation changes in the Holocene has been described thus by Behre (1986: vii):

Human impact has been the most important factor affecting vegetation change, at least in Europe, during the last 7000 years. With the onset of agriculture, at the so-called Neolithic revolution, the human role changed from that of a passive component to an active element that impinged directly on nature. This change had dramatic consequences for the natural environment and landscape development. Arable and pastoral farming, the actual settlements themselves and the consequent changes in the economy significantly altered the natural vegetation and created the cultural landscape with its many different and varying aspects.

Recent paleoecological research has indicated that certain features of the post-glacial pollen record in Europe and Britain can perhaps be attributed to the action of Mesolithic and Neolithic peoples. It is, for example,

possible that the expansion of the alder (*Alnus glutinosa*) was not so much a consequence of supposed wetness in the Atlantic period as a result of Mesolithic colonizers. Their removal of natural forest cover helped the spread of alder by reducing competition, as possibly did the burning of reed swamp. It may also have been assisted in its spread by the increased runoff of surface waters occasioned by deforestation and burning of catchment areas. Furthermore, the felling of alder itself promotes vegetative sprouting and cloning, which could result in its rapid spread in swamp forest areas (Moore, 1986).

In the Yorkshire Wolds of northern England pollen analysis suggests that Mesolithic peoples may have caused forest disturbance as early as 8900 years BP. They may even have so suppressed forest growth that they permitted the relatively open landscapes of the early post-glacial to persist as grasslands even when climatic conditions favored forest growth (Bush, 1988).

One vegetational change that has occasioned particular interest is the fall in *Ulmus* pollen – the so-called elm decline – which appears in all pollen diagrams from northwest Europe, though not from North America. A very considerable number of radiocarbon dates have shown that this decline was approximately synchronous over wide areas and took place at about 5000 yr BP (Pennington, 1974). It is now recognized that various hypotheses can be advanced to explain this major event in vegetation history: the original climatic interpretation; progressive soil deterioration; the spread of disease; the role of people.

The climatic interpretation of the elm decline as being caused by cold, wet conditions has been criticized on various grounds (Rackham, 1980: 265):

A deterioration of climate is inadequate to explain so sudden, universal and specific a change. Had the climate become less favourable for elm, this would not have caused a general decline in elm and elm alone; it would have wiped out elm in areas where the climate had been marginal for it, but would not have affected elm at the middle of its climatic range unless the change was so great as to affect other species also. A climatic change universal in Europe ought to have some effect on North American elms.

Humans may have contributed to soil deterioration and affected elms thereby. Troels-Smith (1956), however, postulated that around 5000 years ago a new technique of keeping stalled domestic animals was introduced by Neolithic peoples, and that these animals were fed by repeated gathering of heavy branches from those trees known to be nutritious – elms. This, it was held, reduced enormously the pollen production of the elms. In Denmark it was found that the first appearance of the pollen of a weed, Ribwort plaintain (*Plantago lanceolata*), always coincided with the fall in elm pollen levels, confirming the association with human settlements.

Experiments have also shown that Neolithic peoples, equipped with polished stone axes, could cut down mature trees and clear by burning a fair-sized patch of established forest within about a week. Such clearings were used for cereal cultivation. In Denmark a genuine chert Neolithic axe was fitted into an ashwood shaft (Figure 2.23). Three men managed to clear about 600 m² of birch forest in four hours. Remarkably, more than 100 trees were felled with one axe head, which had not been sharpened for about 4000 years (Cole, 1970: 38). Rackham, however, doubts whether humans alone could have achieved the sheer extent of change in such a short period (see also Peglar and Birks, 1993), and postulates that epidemics of elm disease may have played a role, aided by the fact that the cause of the disease, a fungus called *Ceratocystis*, is particularly attracted to pollarded elms (Rackham, 1980: 266).

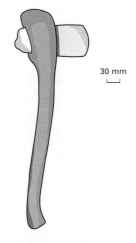

30 mm

Figure 2.23 A Neolithic chert axe-blade from Denmark, of the type which has been shown to be effective at cutting forest in experimental studies (after Cole, 1970, figure 25).

In reality the elm decline may have resulted from a complex cocktail of causes relating to climate change, soil deterioration, disease and human activity. A model of the interacting factors that could be involved (Parker et al., 2002) is shown in Figure 2.24.

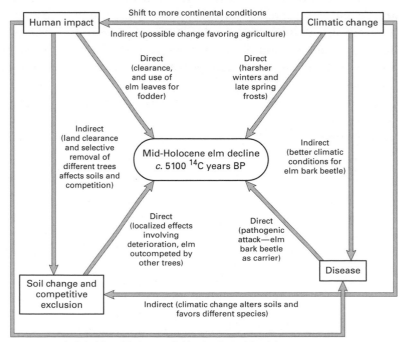

Figure 2.24 Relationships between factors influencing the mid-Holocene elm decline (after Parker et al., 2002, figure 9).

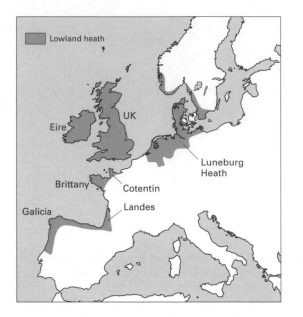

Figure 2.25 The lowland heath region of western Europe (modified after Gimingham and de Smidt, 1983, figure 2).

Lowland heaths

Heathland is characteristic of temperate, oceanic conditions on acidic substrates, and is composed of ericoid low shrubs, which form a closed canopy at heights that are usually less than 2 m. Trees and tall shrubs are absent or scattered. Some heathlands are natural: for example, communities at altitudes above the forest limit on mountains and those on exposed coasts. There are also well-documented examples of heath communities that appear naturally in the course of plant succession as, for example, where *Calluna vulgaris* (heather) colonizes *Ammophila arenaria* and *Carex arenaria* on coastal dunes.

However, at low and medium altitudes on the western fringes of Europe between Portugal and Scandinavia (Figure 2.25) extensive areas of heathland occur. The origin of these is strongly disputed (Gimingham and de Smidt, 1983). Some areas were once thought to have developed where there were appropriate edaphic conditions (e.g., well-drained loess or very sandy, poor soils), but pollen analysis showed that most heathlands occupy areas which were formerly tree-covered. This evidence alone, however, did not settle the question whether the change from forest to heath might have been caused by Holocene climatic change. However, the presence of human artifacts and buried charcoal, and the fact that the replacement of forest by heath has occurred at many different points in time between the Neolithic and the late nineteenth century, suggest that human actions established, and then maintained, most of the heathland areas. In particular, fire is an important management tool for heather in locations such as upland Britain, since the value of *Calluna* as a source of food for grazing animals increases if it is periodically burned.

The area covered by heathland in western Europe reached a peak around 1860, but since then there has been a very rapid decline. For example, by 1960 there had been a 60–70% reduction in Sweden and Denmark (Gimingham, 1981). Reductions in Britain averaged 40% between 1950 and 1984, and this was a continuation of a more long-term trend (Figure 2.26). The reasons for this fall are many, and include unsatisfactory burning practices, peat removal, drainage fertilization, replacement by improved grassland, conversion to forest, and sand and gravel abstraction. In England, the Dorset heathlands that were such a feature of Hardy's Wessex novels are now a fraction of their former extent.

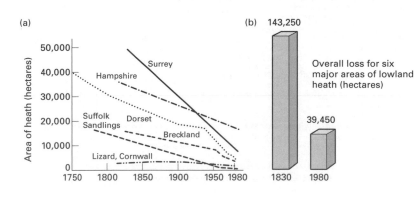

Figure 2.26 Losses of lowland heath in southern England (from Nature Conservancy Council, 1984).

Thus far we have considered the human impact on general assemblages of vegetation over broad zones. However, in turning to questions such as the range of individual plant species, the human role is no less significant.

Introduction, invasion, and explosion

People are important agents in the spread of plants and other organisms (Bates, 1956). Some plants are introduced deliberately by humans to new areas; these include crops, and ornamental and miscellaneous landscape modifiers (trees for reforestation, cover plants for erosion control, etc.). Indeed, some plants, such as bananas and breadfruit, have become completely dependent on people for reproduction and dispersal, and in some cases they have lost the capacity for producing viable seeds and depend on human-controlled vegetation propagation. Most cultigens are not able to survive without human attention, partly because of this low capacity for self-propagation, but also because they cannot usually compete with the better-adapted native vegetation.

However, some domesticated plants have, when left to their own devices, shown that they are capable of at least ephemeral colonization, and a small number have successfully naturalized themselves in areas other than their supposed region of origin (Gade, 1976). Examples of such plants include several umbelliferous annual garden crops (fennel, parsnip, and celery) which, though native to Mediterranean Europe, have colonized wastelands in California. The Irish potato, which is native to South America, grows unaided in the mountains of Lesotho. The peach (in New Zealand), the guava (in the Philippines), coffee (in Haiti), and the coconut palm (on Indian Ocean island strands) are perennials that have established themselves as wild-growing populations, although the last-named is probably within the hearth region of its probable domestication. In Paraguay, orange trees (originating in Southeast Asia and the East Indies) have demonstrated their ability to survive in direct competition with natural vegetation.

Plants that have been introduced deliberately because they have recognized virtues (Jarvis, 1979) can be usefully divided into an economic group (e.g., crops, timber trees, etc.) and an ornamental or amenity one.

In the British Isles, Jarvis believes that the great bulk of deliberate introductions before the sixteenth century had some sort of economic merit, but that only a handful of the species introduced thereafter were brought in because of their utility. Instead, plants were introduced increasingly out of curiosity or for decorative value.

A major role in such deliberate introductions was played by European botanic gardens (Figure 2.27) and those in the colonial territories from the sixteenth century onwards. Many of the 'tropical' gardens (such as those in Calcutta, Mauritius, and Singapore) were often more like staging posts or introduction centers than botanic gardens in the modern sense (Heywood, 1989).

Figure 2.27 A monument in the botanic gardens in Mauritius, drawing attention to the value of introduced plants.

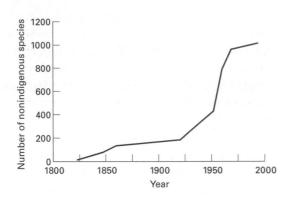

Figure 2.28 Number of nonindigenous plant species by date as reported in botanical treatments of the California flora (from Schwartz et al., 1996, figure 47.1).

Many plants, however, have been dispersed accidentally as a result of human activity: some by adhesion to moving objects, such as individuals themselves or their vehicles; some among crop seed; some among other plants (such as fodder or packing materials); some among minerals (such as ballast or road metal); and some by the carriage of seeds for purposes other than planting (as with drug plants). As illustrated in Figure 2.28, based on California, the establishment of alien species has proceeded rapidly over the past 150 years. In the Pampa of Argentina, Schmeider (1927b) estimates that the invasion of the country by European plants has taken place on such a large scale that at present only one-tenth of the plants growing wild in the Pampa are native.

The accidental dispersal of such plants and organisms can have serious ecological consequences (see Williamson, 1996, for a recent analysis). In Britain, for instance, many elm trees died in the 1970s because of the accidental introduction of the Dutch elm disease fungus which arrived on imported timber at certain ports, notably Avonmouth and the Thames Estuary ports (Sarre, 1978). There are also other examples of the dramatic impact of some introduced plant pathogens (von Broembsen, 1989). The American chestnut *Castanea dentata* was, following the introduction of the chestnut blight fungus *Cryphonectria parasitica* in ornamental nursery material from Asia late in the 1890s, almost eliminated throughout its natural range in less than 50 years. In western Australia the great jarrah forests have been invaded and decimated by a root fungus, *Phytophthora cinnamomi*. This was probably

introduced on diseased nursery material from eastern Australia, and the spread of the disease within the forests was facilitated by road building, logging, and mining activities that involved movement of soil or gravel containing the fungus. More than three million hectares of forest have been affected.

In the USA, Kudzu (*Pueraria montana*) has been a particularly difficult invader. It was introduced into the USA from Japan in 1876 and was promoted as an ornamental and forage crop plant. From 1935 to the mid-1950s it was recommended as a good means of reducing soil erosion. Unfortunately this large vine, with massive tap roots and the ability to spread at *c.* 30 cm per day, now smothers large expanses of the southeastern USA. It is fiendishly difficult to eradicate because of the nature of its root system.

Ocean islands have often been particularly vulnerable to plant invasions. The simplicity of their ecosystems inevitably leads to diminished stability, and introduced species often find that the relative lack of competition enables them to broaden their ecological range more easily than on the continents. Moreover, because the natural species inhabiting remote islands have been selected primarily for their dispersal capacity, they have not necessarily been dominant or even highly successful in their original continental setting. Therefore, introduced species may prove more vigorous and effective (Holdgate and Wace, 1961). There may also be a lack of indigenous species to adapt to conditions such as bare ground caused by humans. Many successful invasive plants have escaped the natural enemies that hold them in check, freeing them to utilize their full competitive potential (Callaway and Aschehoug, 2000). Thus introduced weeds may catch on.

Table 2.6 illustrates clearly the extent to which the flora of selected islands now contain alien species, with the percentage varying between about one-quarter and two-thirds of the total number of species present.

There are a number of major threats that invasive plants pose to natural ecosystems (Levine et al., 2003). These have been discussed by Cronk and Fuller (1995):

1 replacement of diverse systems with single species stands of aliens, leading to a reduction in biodiversity, as for example where Australian acacias have invaded the fynbos heathlands of South Africa (Le Maitre et al., 2000);

Table 2.6 Alien plant species on oceanic islands. Source: from data in Moore (1983)

Island	Number of native species	Number of alien species	Percent of alien species in flora
New Zealand	1200	1700	58.6
Campbell Island	128	81	39.0
South Georgia	26	54	67.5
Kerguelen	29	33	53.2
Tristan da Cunha	70	97	58.6
Falklands	160	89	35.7
Tierra del Fuego	430	128	23.0

2 direct threats to native faunas by change of habitat;

3 alteration of soil chemistry (e.g., the African *Mesembryanthemum crystallinum* accumulates large quantities of salt, and in this way it salinizes invaded areas in Australia and may prevent the native vegetation from establishment);

4 alteration of geomorphic processes, especially rates of sedimentation and movement of mobile landforms (e.g., dunes and salt marshes);

5 plant extinction by competition;

6 alteration of fire regime (e.g., in Florida, USA the introduction of the Australasian *Melaleuca quinquenervia* has increased the frequency of fires because of its flammability, and has damaged the native vegetation which is less well adapted to fire);

7 alteration of hydrological conditions (e.g., reduction in groundwater levels caused by some species having high rates of transpiration).

The introduction of new animals can have an adverse effect on plant species. A clear demonstration of this comes from the atoll of Laysan in the Hawaii group. Rabbits and hares were introduced in 1903 in the hope of establishing a meat cannery. The number of native species of plants at this time was 25; by 1923 it had fallen to four. In that year all the rabbits and hares were systematically exterminated to prevent the island turning into a desert, but recovery has been slower than destruction. By 1930 there were nine species, and by 1961 sixteen species, on the island (Stoddart, 1968).

Another example is the extensive introduction of pigs to the islands of the Pacific. Long-established feral populations are known on many islands. Like rabbits

they have caused considerable damage, not least because of their non-fastidious eating habits and their propensity for rooting into the soil. This is a theme that is reviewed by Nunn (1991).

It has often been proposed that the introduction of exotic terrestrial mammals has had a profound effect on the flora of New Zealand. Among the reasons that have been put forward for this belief are that the absence of native terrestrial mammalian herbivores permitted the evolution of a flora highly vulnerable to damage from browsing and grazing, and that the populations of wild animals (including deer and opossums) that were introduced in the nineteenth century grew explosively because of the lack of competitors and predators. It has, however, proved difficult to determine the magnitude of the effects that the introduced mammals had on the native forests (Veblen and Stewart, 1982).

Overall, invasive species may be one of the prime causes of loss of **biodiversity** because of their role in competition, predation, and **hybridization**, although this is a view that has been challenged by Theodoropoulos (2003). They also have enormous economic costs. On a global basis plant and animal invaders may cost as much as $1.4 trillion a year, representing 5% of the world economy (Pimentel, 2003).

There are many other examples of ecological explosions – bioinvasions – caused by humans creating new habitats. Some of the most striking are associated with the establishment of artificial lakes in place of rivers. Riverine species which cannot cope with the changed conditions tend to disappear, while others that can exploit the new sources of food, and reproduce themselves under the new conditions, multiply rapidly in the absence of competition (Lowe-McConnell, 1975). Vegetation on land flooded as the lake waters rise decomposes to provide a rich supply of nutrients, which allows explosive outgrowth of organisms as the new lake fills. In particular, floating plants that form dense mats of vegetation, which in turn support large populations of invertebrate animals, may cause fish deaths by deoxygenating the water, and can create a serious nuisance for turbines, navigators, and fishermen. On Lake Kariba in Central Africa there were dramatic growths in the communities of the South American water fern (*Salvinia molesta*), bladder-wort (*Utricularia*), and the African water lettuce (*Pistia stratiotes*); on the Nile behind the Jebel Aulia Dam there was a huge increase in the number of water hyacinths (*Eichhornia crassipes*); and in the Tennessee Valley lakes there was a massive outbreak of the Eurasian water-millfoil (*Myriophyllum*).

Roads have been of major importance in the spread of plants. As Frenkel (1970) has pointed out in a valuable survey of this aspect of anthropogenic biogeography:

By providing a route for the bearers of plant propagules – man, animal, and vehicle – and by furnishing, along their margins, a highly specialized habitat for plant establishment, roads may facilitate the entry of plants into a new area. In this manner roads supply a cohesive directional component, cutting across physical barriers, linking suitable habitat to suitable habitat.

Roadsides tend to possess a distinctive flora in comparison with the natural vegetation of an area. As Frenkel (1970: 1) has again written:

Roadsides are characterized by numerous ecological modification including: treading, soil compaction, confined drainage, increased runoff, removal of organic matter and sometimes additions of litter or waste material of frequently high nitrogen content (including urine and feces), mowing or crushing of tall vegetation but occasionally the addition of wood chips or straw, substrate maintained in an ecologically open condition by blading, intensified frost action, rill and sheetwash erosion, snow deposition (together with accumulated dirt, gravel, salt and cinder associated with winter maintenance), soil and rock additions related to slumping and rock-falls, and altered microclimatic conditions associated with pavement and right-of-way structures. Furthermore, road rights-of-way may be used for driving stock in which case unselective, hurried but often close grazing may constitute an additional modification. Where highway landscaping or stability of cuts and fills is a concern, exotic or native plants may be planted and nurtured.

The speed with which plants can invade roadsides is impressive. A study by Helliwell (1974) demonstrated that the M1 motorway in England, less than 12 years after its construction, had on its cuttings and embankments not only the 30 species that had been deliberately sown or planted, but also more than 350 species that had not been introduced there.

Railways have also played their role in plant dispersal. The classic example of this is provided by the Oxford ragwort, *Senecio squalidus*, a species native to Sicily and southern Italy. It spread from the Oxford Botanical Garden (where it had been established since at least 1690) and colonized the walls of Oxford. Much of its dispersal to the rest of Britain (Usher, 1973) was achieved by the Great Western Railway, in the vortices of whose trains and in whose cargoes of ballast and iron ore the plumed fruits were carried. The distribution of the plant was very much associated with railway lines, railway towns, and waste ground.

Indeed, by clearing forest, cultivating, depositing rubbish, and many other activities, humans have opened up a whole series of environments that are favorable to colonization by a particular group of plants. Such plants are generally thought of as weeds. In fact it has often been said that the history of weeds is the history of human society (although the converse might equally be true), and that such plants follow people like flies follow a ripe banana or a gourd of unpasteurized beer (see Harlan, 1975b).

One weed which has been causing especially severe problems in upland Britain in the 1980s is bracken (*Pteridium aquilinum*), although the problems it poses through rapid encroachment are of wider geographical significance (Pakeman et al., 1996). Indeed, Taylor (1985: 53) maintains that it 'may justifiably be dubbed as the most successful international weed of the twentieth century', for 'it is found, and is mostly expanding, in all of the continents.' This tolerant, aggressive opportunist follows characteristically in the wake of evacuated settlement, deforestation or reduced grazing pressure, and estimated encroachment rates in the upland parts of the UK average 1% per annum. This encroachment results from reduced use of bracken as a resource (e.g., for roofing) and from changes in grazing practices in marginal areas. As bracken is hostile to many other plants and animals, and generates toxins, including some carcinogens, this is a serious issue.

Air pollution and its effects on plants

Air pollutants exist in gaseous or particulate forms. The gaseous pollutants may be separated into primary and secondary forms. The primary pollutants, such as sulfur dioxide, most oxides of nitrogen, and carbon monoxide, are those directly emitted into the air from, for example, industrial sources. Secondary air pollutants, such as ozone and products of **photochemical** reactions, are formed as a consequence of subsequent chemical processes in the atmosphere, involving the primary pollutants and other agents such as sunlight. Particulate pollutants consist of very small solid- or liquid-suspended droplets (e.g., dust, smoke, and aerosolic salts), and contain a wide range of insoluble components (e.g., quartz) and of soluble components (e.g., various common cations, together with chloride, sulfate, and nitrate).

Some of the air pollutants that humans have released into the atmosphere have had detrimental impacts on plants (Yunus and Iqbal, 1996): sulfur dioxide, for example, is toxic to them. This was shown by Cohen and Rushton (in Barrass, 1974) who grew plants in containers of similar soil in different parts of Leeds, England, in 1913. They found a close relationship between the amount of sulfate in the air and in the plants, and in the yield obtained (Figure 2.29). In the more polluted areas of the city the leaves were blackened by soot and there was a smaller leaf area. The whole theme of urban vegetation has been studied in the North American context by Schmid (1975).

Lichens are also sensitive to air pollution and have been found to be rare in central areas of cities such as Bonn, Helsinki, Stockholm, Paris, and London. There appears to be a zonation of lichen types around big cities as shown for northeast England (Figure 2.30a) and Belfast (Figure 2.30b). Moreover, when a healthy lichen is transplanted from the country to a polluted atmosphere, the algae component gradually deteriorates and then the whole plant dies (see Gilbert, 1970).

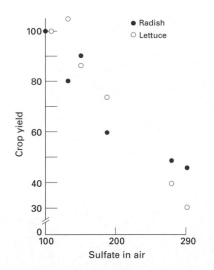

Figure 2.29 The effect of air quality on plant growth in Leeds, England, in 1913. Values for yield and for air sulfate in a low pollution area are taken as 100 and other values are scaled in proportion (data of Cohen and Rushton, in Barrass, 1974: 187).

Overall it has been calculated (Rose, 1970) that more than one-third of England and Wales, extending in a belt from the London area to Birmingham, broadening out to include the industrial Midlands and most of Lancashire and West Yorkshire, and reaching up to Tyneside, has lost nearly all its epiphytic lichen flora, largely because of sulfur dioxide pollution. Hawksworth (1990: 50), indeed, believes 'the evidence that sulphur dioxide is the major pollutant responsible

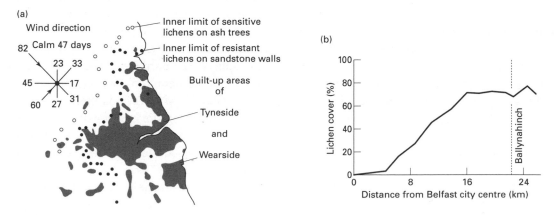

Figure 2.30 (a) Air pollution in northeast England and its impact upon growth area for lichens (after Gilbert, in Barrass, 1974, figure 73). (b) The increase of lichen cover on trees outside the city of Belfast, Northern Ireland (after Fenton, in Mellanby, 1967, figure 3).

Figure 2.31 The great smelter at Sudbury in Canada. Pollution derived from the fumes from such sources can have many deleterious impacts on vegetation in their neighborhood.

for impoverishment of lichen communities over wide areas of Europe is now overwhelming', but he also points out that sulfur dioxide is not necessarily always the cause of impoverishment. Other factors, such as fluoride or photochemical smog, acting independently or synergistically, can also be significant in particular locations.

Local concentrations of industrial fumes also kill vegetation (Freedman, 1995). In the case of the smelters of the Sudbury mining district of Canada (Figure 2.31), 2 million tonnes of noxious gases annually affect a 1900 km² area and white pine now exists only on about 7–8% of the productive land. Likewise in the lower Swansea Valley, Wales, the fumes from a century of

coal burning resulted in almost complete destruction of the vegetation, with concomitant soil erosion. The area became a virtual desert. In Norway a number of the larger Norwegian aluminum smelters built immediately after the Second World War were sited in deep, narrow, steep-sided valleys at the heads of fjords. The relief has not proved conducive to the rapid dispersal of fumes, particularly of fluoride. In one valley a smelter with a production of 1,110,000 tonnes per year causes the death of pines (*Pinus sylveststris*) for over 13 km in each direction up and down the valley. To about 6 km from the source all pines are dead. Birch, however, seems to be able to withstand these conditions, and to grow vigorously right up to the factory fence (Gilbert, 1975).

Photochemical smog is also known to have adverse effects on plants both within cities and on their outskirts. In California, ponderosa pines (*Pinus ponderosa*) in the San Bernadino mountains as much as 129 km to the east of Los Angeles have been damaged extensively by smog and ozone (Diem, 2003). In summer months in Britain ozone concentrations produced by photochemical reactions have reached around 17 pphm (parts per hundred million) compared with a maximum of 4 pphm associated with clean air, while in Los Angeles ozone concentrations may reach 70 pphm (Marx, 1975). Fumigation experiments in the USA show that plant injury can occur at levels only marginally above the natural maximum and well within the summertime levels now known to be present in Britain and the USA. Ozone appears to reduce photosynthesis and to inhibit flowering and germination (Smith, 1974). It also seems to predispose conifers to bark-beetle infestation and to microbial pathogens.

Vegetation will also be adversely affected by excessive quantities of suspended particulate matter in the atmosphere. The particles, by covering leaves and plugging plant stomata, reduce both the absorption of carbon dioxide from the atmosphere and the intensity of sunlight reaching the interior of the leaf. Both tendencies may suppress the growth of some plants. This and other consequences of air pollution are well reviewed by Elsom (1992).

The adverse effects of pollution on plants are not restricted to air pollution: water and soil pollution can also be serious. Excessive amounts of heavy metals may prove toxic to them (Hughes et al., 1980) and, as a consequence, distinctive patterns of plant species may occur in areas contaminated with the waste from copper, lead, zinc, and nickel mines (Cole and Smith, 1984). Heavy metals in soils may also be toxic to microbes, and especially to fungi, which may in turn change the environment by reducing rates of leaf-litter decomposition (Smith, 1974). In many areas it has proved extremely difficult to undertake effective re-establishment of plant communities on mining spoil tips, although toxicity is only one of the problems (Kent, 1982).

Other types of industrial effluents may smother and poison some species. Salt marshes, mangrove swamps, and other kinds of wetlands are particularly sensitive to oil spills, for they tend to be anaerobic environments in which the plants must ventilate their root systems through pores or openings that are prone to coating and clogging (Lugo et al., 1981). The situation is especially serious if the system is not subjected to flushing by, for example, frequent tidal inundation. There are may case studies of the consequences of oil spills. For example, at the Fawley Oil Refinery on Southampton Water in England, Dicks (1977) has shown how the salt marsh vegetation has been transformed by the pollution created by films of oil, and how, over extensive areas, *Spartina anglica* has been killed off.

Forest decline

Forest decline, now often called *Waldsterben* or *Waldschäden* (the German words for 'forest death' and 'forest decline'), is an environmental issue that attained considerable prominence in the 1980s. The common symptoms of this phenomenon (modified from World Resources Institute, 1986, table 12.9) are:

1 *Growth-decreasing symptoms*:
- discoloration and loss of needles and leaves;
- loss of feeder-root biomass (especially in conifers);
- decreased annual increment (width of growth rings);
- premature aging of older needles in conifers;
- increased susceptibility to secondary root and foliar pathogens;
- death of herbaceous vegetation beneath affected trees;
- prodigious production of lichens on affected trees;
- death of affected trees.

2 *Abnormal growth symptoms*:
- active shedding of needles and leaves while still green, with no indication of disease;
- shedding of whole green shoots, especially in spruce;
- altered branching habit;
- altered morphology of leaves.
3 *Water-stress symptoms*:
- altered water balance;
- increased incidence of wet wood disease.

The decline appeared to be widespread in much of Europe and to be particularly severe in Poland and the Czech Republic (see Table 2.7). The process was also thought to be undermining the health of North America's high-elevation eastern coniferous forests (World Resources Institute, 1986, chapter 12). In Germany it was the white fir, *Abies alba*, which was afflicted initially, but since then the symptoms spread to at least ten other species in Europe, including Norway spruce (*Picea abies*), Scotch pine (*Pinus sylvestris*), European larch (*Larix decidua*), and seven broad-leaved species.

Many hypotheses have been put forward to explain this dieback (Wellburn, 1988; Federal Research Center for Forestry and Forest Products, 2000): poor forest management practices, aging of stands, climatic change, severe climatic events (such as the severe

Table 2.7 Results from forest damage surveys in Europe: percentage of trees with > 25% defoliation (all species). Source: data in *Acid News* (1995: 7)

	Mean of 1993–94		Mean of 1993–94
Austria	8	Latvia	33
Belarus	33	Lithuania	26
Belgium	16	Luxembourg	29
Bulgaria	26	Netherlands	22
Croatia	24	Norway	26
Czech Republic	56	Poland	52
Denmark	35	Portugal	7
Estonia	18	Romania	21
Finland	14	Slovak Republic	40
France	8	Slovenia	18
Germany	24	Spain	16
Greece	22	Switzerland	20
Hungary	21	UK	15
Italy	19		

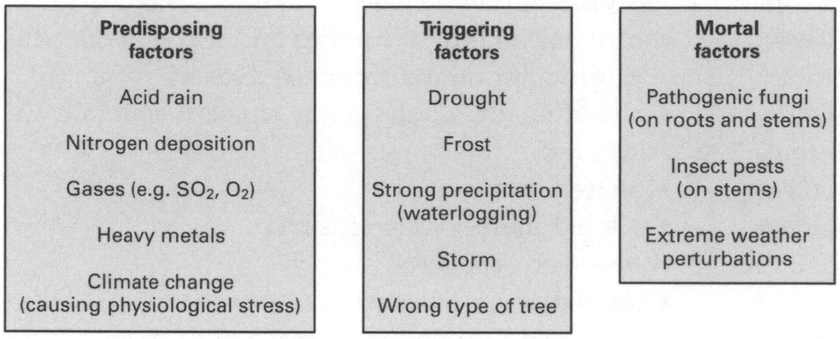
Predisposing factors
Acid rain
Nitrogen deposition
Gases (e.g. SO_2, O_2)
Heavy metals
Climate change (causing physiological stress)
Triggering factors
Drought
Frost
Strong precipitation (waterlogging)
Storm
Wrong type of tree
Mortal factors
Pathogenic fungi (on roots and stems)
Insect pests (on stems)
Extreme weather perturbations

Figure 2.32 Some of the different stress variables used to explain forest decline (modified from Nihlgård, 1997, figure 24–1).

droughts in Britain during 1976), nutrient deficiency, viruses, fungal pathogens, and pest infestation. However, particular attention is being paid to the role of pollution, either by gaseous pollutants (sulfur dioxide, nitrous oxide, or ozone), acid deposition on leaves and needles, soil acidification and associated aluminum toxicity problems and excess leaching of nutrients (e.g., magnesium), overfertilization by deposited nitrogen, and trace metal or synthetic organic compound (e.g., pesticide, herbicide) accumulation as a result of atmospheric deposition.

The arguments for and against each of these possible factors have been expertly reviewed by Innes (1987), who believes that in all probability most cases of forest decline are the result of the cumulative effects of a number of stresses. He draws a distinction between predisposing, inciting, and contributing stresses (p. 25):

> Predisposing stresses are those that operate over long time scales, such as climatic change and changes in soil properties. They place the tree under permanent stress and may weaken its ability to resist other forms of stress. Inciting stresses are those such as drought, frost and short-term pollution episodes, that operate over short time scales. A fully healthy tree would probably have been able to cope with these, but the presence of predisposing stresses interferes with the tree's mechanisms of natural recovery. Contributing stresses appear in weakened plants and are frequently classed as secondary factors. They include attack by some insect pests and root fungi. It is probable that all three types of stress are involved in the decline of trees.

An alternative categorization of the stresses leading to forest decline has been proposed by Nihlgård (1997). He also has three classes of stress that have a temporal dimension: predisposing factors (related primarily to pollution and long-term climatic change); triggering factors (including droughts, frosts, and inappropriate management or choice of trees); and mortal factors such as pests, pathogenic fungi, and extreme weather events, which can lead to plant death (Figure 2.32).

As with many environmental problems, interpretation of forest decline is bedeviled by a paucity of long-term data and detailed surveys. Given that forest condition oscillates from year to year in response to variability in climatic stress (e.g., drought, frost, wind throw) it is dangerous to infer long-term trends from short-term data. There may also be differences in causation in different areas. Thus while widespread forest death in eastern Europe may result from high concentrations of sulfur dioxide combined with extreme winter stress, this is a much less likely explanation in Britain, where sulfur dioxide concentrations have shown a marked decrease in recent years. Indeed, in Britain Innes and Boswell (1990: 46) suggest that the direct effects of gaseous pollutants appear to be very limited.

It is also important to recognize that some stresses may be particularly significant for a particular tree species. Thus, in 1987 a survey of ash trees in Great Britain showed extensive dieback over large areas of the country. Almost one-fifth of all ash trees sampled showed evidence of this phenomenon. Hull and Gibbs (1991) indicated that there was an association between dieback and the way the land is managed around the tree, with particularly high levels of damage being evident in trees adjacent to arable land. Uncontrolled stubble burning, the effects of drifting herbicides, and the consequences of excessive nitrate fertilizer applications to adjacent fields were seen as possible mechanisms. However, the prime cause of dieback was seen to be root disturbance and soil compaction by large agricultural machinery. Ash has shallow roots and if these are damaged repeatedly the tree's uptake

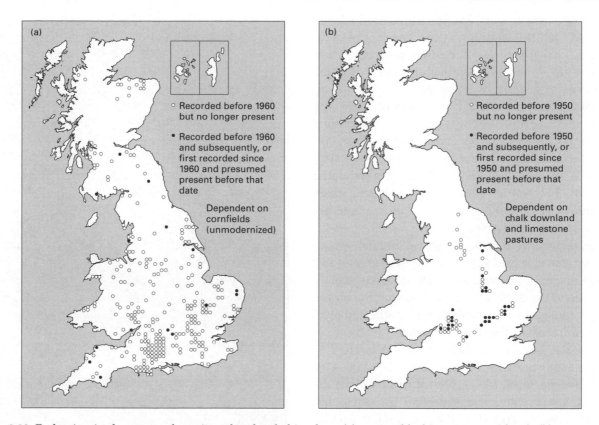

Figure 2.33 Reduction in the range of species related to habitat loss: (a) corncockle (*Agrostemma githago*); (b) pasque flower (*Pulsatilla vulgaris*) (after Nature Conservancy Council, 1977, figures 8 and 9).

of water and nutrients may be seriously reduced, while broken root surfaces would be prone to infection by pathogenic fungi.

Innes (1992: 51) also suggests that there has been some modification in views about the seriousness of the problem since the mid-1980s:

The extent and magnitude of the forest decline is much less than initially believed. The use of crown density as an index of tree health has resulted in very inflated figures for forest 'damage' which cannot now be justified . . . If early surveys are discounted on the basis of inconsistent methodology . . . then there is very little evidence for a large-scale decline of tree health in Europe.

. . . the term 'forest decline' is rather misleading in that there are relatively few cases where entire forest ecosystems are declining. Forest ecosystems are dynamic and may change through natural processes.

He also suggests that the decline of certain species has been associated with climatic stress for as long as records have been maintained.

Miscellaneous causes of plant decline

Some of the main causes of plant decline have already been referred to: deforestation, grazing, fire, and pollution. However, there are many records of species being affected by other forms of human interference. For example, casual flower-picking has resulted in local elimination of previously common species, and has been held responsible for decreases in species such as the primrose (*Primula vulgaris*) on a national scale in England. In addition, serious naturalists or plant collectors using their botanical knowledge to seek out rare, local, and unusual species can cause the eradication of rare plants in an area.

More significantly, agricultural 'improvements' mean that many types of habitat are disappearing or that the range of such habitats is diminishing. Plants associated with distinctive habitats suffer a comparable reduction in their range. This is certainly the case for two British plants (Figure 2.33a and b), the corncockle (*Agrostemma githago*) and the pasque flower (*Pulsatilla vulgaris*). The former was characteristic of unmodified cereal fields, while the latter was characteristic of traditional chalk downland and limestone pastures. These calcareous grasslands were semi-natural swards formed by centuries of grazing by sheep and rabbits, and covered large areas underlain by Cretaceous chalk and Jurassic limestone until the Napoleonic Wars. Since then they have been subjected to increasing amounts of cultivation, reclamation and reseeding, and the loss of such grasslands between 1934 and 1972 has been estimated at around 75% (Nature Conservancy Council, 1984). Other plants have suffered a reduction in range because of drainage activities.

The introduction of pests, either deliberately or accidentally, can also lead to a decrease in the range and numbers of a particular species. Reference has already been made to the decline in the fortunes of the elm in Britain because of the unintentional establishment of Dutch Elm disease. There are, however, many cases where 'pests' have been introduced deliberately to check the explosive invasion of a particular plant. One of the most spectacular examples of this involves the history of the prickly pear (*Opuntia*) imported into Australia from the Americas (Figure 2.34). It was introduced some time before 1839 (Dodd, 1959), and spread dramatically. By 1900, it covered 4 million hectares, and by 1925 more than 24 million. Of the latter figure approximately one-half was occupied by dense growth (1200–2000 tonnes per hectare) and other more useful plants were excluded. To combat this menace one of *Opuntia*'s natural enemies, a South American moth, *Cactoblastus*, was introduced to remarkable effect. By the year 1940 not less than 95% of the former 50 million acres (20 million hectares) of prickly pear in Queensland had been wiped out (Dodd, 1959: 575).

A further good example comes from Australia. By 1952 the aquatic fern, *Salvinia molesta*, which originated in southeastern Brazil, appeared in Queensland and spread explosively as a result of clonal growth, accompanied by fragmentation and dispersal. Significant pests and parasites appear to have been absent. Under optimal conditions *Salvinia* has a doubling time for biomass production of only 2.5 days. In June 1980 possible control agents from the *Salvinia*'s native range in Brazil – the black, long-snouted weevil (*Cyrobagous* sp.) – were released on to Lake Moon Darra (which carried an infestation of 50,000 tonnes fresh weight of *Salvinia*, covering an area of 400 hectares). By August 1981 there was estimated to be less than 1 tonne of the weed left on the lake.

Figure 2.34 The prickly pear (*Opuntia*) is a plant that has been introduced from the Americas to Africa and Australia. It has often spread explosively. Recently it has been controlled by the introduction of moths and beetles, an example of biological control.

Finally, the growth of leisure activities is placing greater pressure on increasingly fragile communities, notably in tundra and high-altitude areas. These areas tend to recover slowly from disturbance, and both the trampling of human feet and the actions of vehicles can be severe (see e.g., Bayfield, 1979).

The change in genetic and species diversity

The application of modern science, technology, and industry to agriculture has led to some spectacular progress in recent decades through such developments as the use of fertilizers and the selective breeding of plants and animals. The latter has caused some concern, for in the process of evolution domesticated plants have become strikingly different from their wild progenitors. Plant species that have been cultivated for a very long time and are widely distributed demonstrate this particularly clearly. Crop evolution through the millennia has been shaped by complex interactions reflecting the pressures of both artificial and natural selection. Alternate isolation of stocks followed by migration and seed exchanges brought distinctive stocks into new environments and permitted new hybridizations and the recombination of characteristics. Great genetic diversity resulted.

There are fears, however, that since the Second World War the situation has begun to change (Harlan, 1975a). Modern plant-breeding programs have been established in many parts of the developing world in the midst of genetically rich centers of diversity. Some of these programs, associated with the so-called **Green Revolution**, have been successful, and new, uniform high-yielding varieties have begun to replace the wide range of old, local strains that have evolved over the millennia. This may lead to a serious decline in the genetic resources that could potentially serve as reservoirs of variability. Ehrlich et al. (1977: 344) have warned:

Aside from nuclear war, there is probably no more serious environmental threat than the continued decay of the genetic variability of crops. Once the process has passed a certain point, humanity will have permanently lost the coevolutionary race with crop pests and diseases and will no longer be able to adapt crops to climatic change.

New, high-yielding crop varieties need continuous development if they are to avoid the effects of crop pests, and Ehrlich and Ehrlich (1982: 65) have summarized the situation thus:

The life of a new cultivated wheat variety in the American Northwest is about five years. The rusts (fungi) adapt to the strain, and a new resistant one must be developed. That development is done through artificial selection: the plant breeder carefully combines genetic types that show promise of giving resistance.

The impact of human activities on species diversity, while clearly negative on a global scale (as evidenced by extinction rates) is not so cut-and-dried on the local scale. Under certain conditions chronic stress caused

by humans can lead to extremely high numbers of coexisting species within small areas. By contrast, site enrichment or fertilization can result in a decline of species density (Peet et al., 1983).

Certain low productivity grasslands which have been grazed for long periods have high species densities in Japan, the UK and The Netherlands. The same applies to Mediterranean scrub vegetation in Israel, and to savanna in Sri Lanka and North Carolina (USA). Studies have confirmed that species densities may increase in areas subject to chronic mowing, burning, domestic grazing, rabbit grazing, or trampling. It is likely that in such ecosystems humans encourage a high diversity of plant growth by acting as a 'keystone predator', a species which prevents competitive exclusion by a few dominant species.

Grassland enrichment experiments employing fertilizers have suggested that in many cases high growth rates result in the competitive exclusion of many plants. Thus the tremendous increase in fertilizer use in agriculture has led to, with what in retrospect is the predictable result, a widespread decrease in the species diversity of grasslands.

One other area in which major developments will take place in the coming years, which will have implications for plant life, is the field of **genetic engineering**. This involves the manipulation of **DNA**, the basic chromosomal unit that exists in all cells and contains genetic information that is passed on to subsequent generations. Recombinant DNA technology (also known as *in vitro* genetic manipulation and gene cloning) enables the insertion of foreign DNA (containing the genetic information necessary to confer a particular target characteristic) into a vector. The advent of modern techniques of genetic modification (GM) enables the removal of individual genes from one species and their insertion into another, without the need for sexual compatibility. Once the new gene has been inserted, offspring that will contain copies of that new gene can be produced in the traditional manner. For example, a bacterial gene can be inserted into maize to give it resistance to certain insect pests. Indeed, genetic engineering is vastly different from conventional plant breeding methods because it allows scientists to insert a gene or genes from virtually any organism into any other organism. Fears have been expressed that this technology could produce pathogens that might interact detrimentally with naturally occurring species (see, e.g., Beringer, 2000; Letourneau and Burrows, 2001; Hails, 2002).

It has also been feared that GM could make a plant more invasive, leading to a new threat from super-weeds. Equally, some genetically modified crops might be highly effective at controlling or eliminating weeds, thereby disrupting essential food sources for a wide range of organisms. Conversely, genetically modified herbicide-tolerant crops would bring greater economy and feasibility in weed control, so that fewer environmentally damaging chemical herbicides would be used than in conventional farming (Squire et al., 2003). Other beneficial traits from GM technology include resistance to insects or pathogens, drought resistance and salt tolerance, the ability to fix nitrogen, and so forth. There is an urgent need to evaluate the costs and benefits of this new technology, for already GM crops are being widely grown. The number of hectares planted to them in the USA increased from 1.4 million in 1996 to around 40 million in 2003. An estimated 38 trillion GM plants have now been grown in American soil!

Points for review

How does the use of fire have an impact upon the environment?

In what ways does grazing by domestic stock modify vegetation?

Where is present day deforestation taking place and why?

What controversies surround desertification?

What role have humans played in the development of savannas and mid-latitude grasslands?

What are ecological explosions?

How does pollution affect plants?

Guide to reading

Crutzen, P. J. and Goldammer, J. G., 1993, *Fire in the environment*. Chichester: Wiley. A particularly useful study of fire's importance.

Grainger, A., 1990, *The threatening desert: controlling desertification*. London: Earthscan. A very readable and wide-ranging review of desertification.

Grainger, A., 1992, *Controlling tropical deforestation*. London: Earthscan. An up-to-date introduction with a global perspective.

Grove, A. T. and Rackham, O., 2002, *The nature of Mediterranean Europe: an ecological history*. New Haven and London: Yale University Press. A well illustrated account of natural and human changes in the Mediterranean lands.

Holzner, W., Werger, M. J. A., Werger, I., and Ikusima, I. (eds), 1983, *Man's impact on vegetation*. Hague: Junk. A wide-ranging edited work with examples from many parts of the world.

Manning, R., 1995, *Grassland: the history, biology, politics and promise of the American prairie*. New York: Viking Books. A 'popular' discussion of the great grasslands of North America.

Meyer, W. B. and Turner, B. L. (eds), 1994, *Changes in land use and land cover: a global perspective*. Cambridge: Cambridge University Press. An excellent edited survey of human transformation of the biosphere.

Pimentel, D. (ed.), 2003, *Biological invasions*. Washington, DC: CRC Press. An edited, international account of invasive species.

Simmons, I. G., 1979, *Biogeography: natural and cultural*. London: Arnold. A useful summary of natural and human impacts on ecosystems.

Stewart, C. N., 2004, *Genetically modified planet. Environmental impacts of genetically engineered plants*. Oxford: Oxford University Press. A readable review of the costs and benefits of GM plants.

Williams, M., 2003, *Deforesting the earth. from prehistory to global crisis*. Chicago: University of Chicago Press. A magisterial overview of the global history of deforestation.

3 HUMAN INFLUENCE ON ANIMALS

Introduction

The range of impacts that humans, who now have the greatest biomass of any species, have had on animals, although large, can be grouped conveniently into five main categories: domestication, dispersal, extinction, expansion, and contraction. Humans have domesticated many animals, to the extent that, as with many plants, those animals depend on humans for their survival and, in some cases, for their reproduction. As with plants, people have helped to disperse animals deliberately, although many have also been dispersed accidentally, for the number of animals that accompany people without their leave is enormous, especially if we include the clouds of microorganisms that infest their land, food, clothes, shelter, domestic animals, and their own bodies. The extinction of animals by human predators has been extensive over the past 20,000 years, and in spite of recent interest in conservation continues at a high rate. In addition, the presence of humans has led to the contraction in the distribution and welfare of many animals (because of factors such as pollution), although in other cases human alteration of

the environment and modification of competition has favored the expansion of some species, both numerically and spatially.

Domestication of animals

We have already referred briefly in Chapter 1 to one of the great themes in the study of human influence on nature: domestication. This has been one of the most profound ways in which humans have affected animals, for during the ten or eleven millennia that have passed since this process was initiated the animals that human societies have selected as useful to them have undergone major changes. Relatively few animal species have been domesticated in comparison with plant species, but for those that have been the consequences are so substantial that the differences between breeds of animals of the same species often exceed those between different species under natural conditions (Figure 3.1). A cursory and superficial comparison of the tremendous range of shapes and sizes of modern dog breeds (as, e.g., between a wolfhound

Figure 3.1 The consequences of domestication for the form of animals is dramatically illustrated by this large-horned but dwarf cow being displayed in a circus in Chamonix, France.

and a chihuahua) is sufficient to establish the extent of alteration brought about by domestication, and the speed at which domestication has accelerated the process of evolution. In particular humans have changed and enhanced the characteristics for which they originally chose to domesticate animals. For example, the wild ancestors of cattle gave no more than a few hundred milliliters of milk; today the best milk cow can yield up to 15,000 L of milk during its lactation period. Likewise, sheep have changed enormously (Ryder, 1966). Wild sheep have short tails, whereas modern domestic sheep have long tails, which may have arisen during human selection of a fat tail. Wild sheep also have an overall brown color, whereas domestic sheep tend to be mainly white. Moreover, the woolly undercoats of wild sheep have developed at the expense of bristly outer coats. With the ancestors of domestic sheep, wool (which served as protection for the skin and as insulation) consisted mainly of thick rough hairs and a small amount of down; the total weight of wool grown per year probably never reached 1 kg. The wool of present-day fine-fleeced sheep consists of uniform, thin down fibers and the total yearly weight may now reach 20 kg. Wild sheep also undergo a complete spring molt, whereas domestic sheep rarely shed wool.

Indeed, one of the most important consequences or manifestations of the domestication of animals consists of a sharp change in the seasonal biology. Whereas wild ancestors of domesticated beasts are often characterized by relatively strict seasonal reproduction and molting rhythms, most domesticated species can reproduce at almost any season of the year and tend not to molt to a seasonal pattern.

The most important center of animal domestication, shown in figure 1.6, was southwest Asia (cattle, sheep, goat, and pig) but other centers were important (e.g., the chicken in southeast Asia, the turkey in the Andes, and the horse in the Ukraine).

Dispersal and invasion of animals

Zoogeographers devote much time to dividing the world into regions with distinctive animal life – the great 'faunal realms' of Wallace and subsequent workers. This pattern of wildlife distribution evolved slowly over geologic time. The most striking dividing line between such realms is probably that between Australasia and Asia – **Wallace's Line**. Because of its isolation from the Asian landmass, Australasia developed a distinctive fauna characterized by its relative absence of placental mammals, its well-developed marsupials, and the egg-laying monotremes (echidna and the platypus).

Modern societies, by moving wildlife from place to place, consciously or otherwise, are breaking down these classic distinct faunal realms. Highly adaptable and dispersive forms are spreading, perhaps at the expense of more specialized organisms. Humans have introduced a new order of magnitude into distances over which dispersal takes place, and through the transport by design or accident of seeds or other propagules, through the disturbance of native plant and animal communities and of their habitat, and by the creation of new habitats and niches, the invasion and colonization by adventive species is facilitated.

Di Castri (1989) has identified three main stages in the process of biological invasions stimulated by human actions. In the first stage, covering several millennia up to about AD 1500, human historical events favored invasions and migrations primarily within the Old World. The second stage commenced about AD 1500, with the discovery, exploration, and colonization of new territories, and the initiation of 'the globalization of exchanges'. It shows the occurrence of flows of invaders from, to, and within the Old World. The

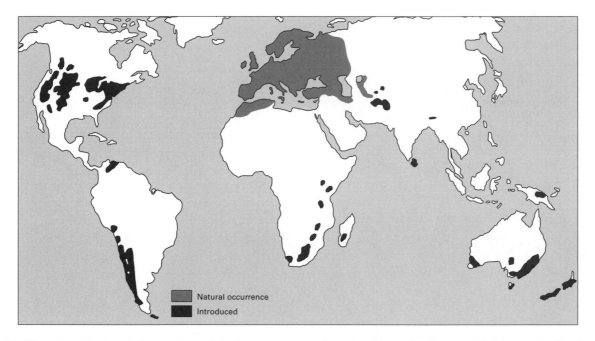

Figure 3.2 The original area of distribution of the brown trout and areas where it has been artificially naturalized (after MacCrimmon, in Illies, 1974, figure 3.2).

third stage, which covers only the last 100 to 150 years, sees an even more extensive 'multifocal globalization' and an increasing rate of exchanges. The Eurocentric focus has diminished.

Deliberate introductions of new animals to new areas have been carried out for many reasons (Roots, 1976): for food, for sport, for revenue, for sentiment, for control of other pests, and for aesthetic purposes. Such *deliberate* actions probably account, for instance, for the widespread distribution of trout (see Figure 3.2). There have, however, been many *accidental* introductions, especially since the development of ocean-going vessels. The rate is increasing, for whereas in the eighteenth century there were few ocean-going vessels of more than 300 tonnes, today there are thousands. Sea-borne trade now exceeds 5 billion tonnes per year and in 1998 much of it was carried in some 165 million containers. Many vessels dump ballast containing potentially invasive exotic species. Because of this, in the words of Elton (1958: 31), 'we are seeing one of the great historical convulsions in the world's fauna and flora.' Indeed, many animals are introduced with vegetable products, for 'just as trade followed the flag, so animals have followed the plants.' One of the most striking examples of the accidental introduction of animals given by Elton is the arrival of some chafer beetles, *Popillia japonica* (the Japanese beetle), in

New Jersey in a consignment of plants from Japan. From an initial population of about one dozen beetles in 1916, the center of population grew rapidly outwards to cover many thousands of square kilometers in only a few decades (Figure 3.3).

A more recent example of the spread of an introduced insect in the Americas is provided by the Africanized honeybee, which is popularly known as the 'killer bee' (Rinderer et al., 1993). A number of these were brought to Brazil from South Africa in 1957 as an experiment and some escaped. Since then (Figure 3.4) they have moved northwards to Central America and Texas, spreading at a rate of 300–500 km per year, and competing with established populations of European honey bees. In 1998 they reached as far north as Nevada. Goulson (2003) reviews some of the adverse environmental effects of introduced bees, including their pollination of exotic weeds.

Some animals arrive accidentally with other beasts that are imported deliberately. In northern Australia, for instance, water buffalo were introduced (McKnight, 1971) and brought their own bloodsucking fly, a species which bred in cattle dung and transmitted an organism sometimes fatal to cattle. Australia's native dung beetles, accustomed only to the small sheep-like pellets of the grazing marsupials, could not tackle the large dung pats of the buffalo. Thus untouched pats

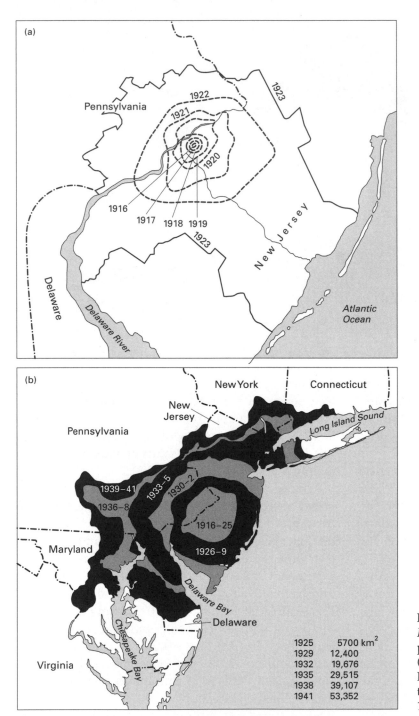

1925	5700 km²
1929	12,400
1932	19,676
1935	29,515
1938	39,107
1941	53,352

Figure 3.3 The spread of the Japanese beetle, *Popillia japonica*, in the eastern USA: (a) from its point of introduction in New Jersey, 1916–23 (after Elton, 1958, figure 14, and Smith and Hadley, 1926); (b) from its point of introduction to elsewhere, 1916–41 (after Elton, 1958, figure 15, and US Bureau of Entomology, 1941).

abounded and the flies were able to breed undisturbed. Eventually African dung beetles were introduced to compete with the flies (Roots, 1976). Indeed, animal invaders have had tragic consequences for the native Australian fauna and flora. Among the worst offenders are the rabbit, the fox, the feral pigs and goats, dogs, cats, mice, buffalo, camels, rats, cave toads, and various birds (Bomford and Hart, 2002). In neighboring New Zealand, Australian possums have become major pests.

While domesticated plants have, in most cases, been unable to survive without human help, the same is not so true of domesticated animals. There are a great

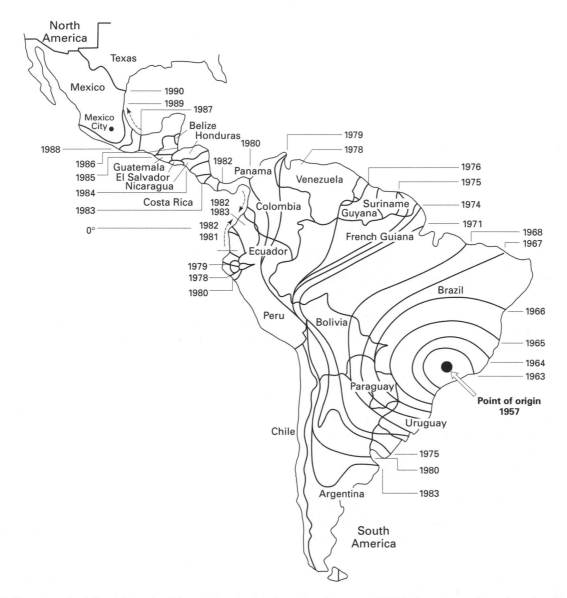

Figure 3.4 The spread of the Africanized honey bee in the Americas between 1957 (when it was introduced to Brazil) and 1990 (modified after Texas Agricultural Experiment Station, in *Christian science monitor*, September 1991).

many examples of cattle, horses (see, e.g., McKnight, 1959), donkeys, and goats which have effectively adapted to new environments and have virtually become wild (feral). Frequently feral animals have both ousted native animals and, particularly in the case of goats on ocean islands, caused desertification. Sometimes, however, introduced animals have spread so thoroughly and rapidly, and have led to such a change in the environment to which they were introduced, that they have sown the seeds of their own demise. Reindeer, for example, were brought from Lapland to

Alaska in 1891–1902 to provide a new resource for the Eskimos, and the herds increased and spread to over half a million animals. By the 1950s, however, there was less than a twentieth of that number left, since the reindeer had been allowed to eat the lichen supplies that are essential to winter survival; as lichen grows very slowly their food supply was drastically reduced (Elton, 1958: 129).

The accidental dispersal of animals can be facilitated by means other than transport on ships or introduction with plants. This applies particularly to aquatic

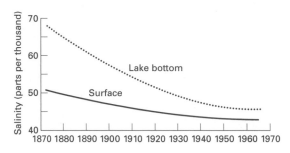

Figure 3.5 Decrease in the salinity of the Great Bitter Lake, Egypt, resulting from the intrusion of fresher water by way of the Suez Canal (after Wooster, 1969, with subsequent modification).

life, which can be spread through human alteration of waterways by methods such as canalization. Wooster (1969) gives the example of the way in which the construction of the Suez Canal has enabled the exchange of animals between the Red Sea and the eastern Mediterranean. Initially, the high salinity of the Great Bitter Lake acted to prevent movement, but the infusion of progressively fresher waters (Figure 3.5) through the Suez Canal has meant that this barrier has gradually become less effective. Some 39 Red Sea fish immigrants have now been identified in the Mediterranean, and they are especially important in the Levant basin, where they comprise about 12% of the fish population. Menacing jellyfish (*Rhopilema nomadica*) have invaded Levantine beaches (Spanier and Galil, 1991). This type of movement has recently been termed 'Lessepsian migration'. Similarly, the construction of the Welland Canal, linking the Atlantic with the Great Lakes of North America, has permitted similar movements with more disastrous consequences. Much of the native fish fauna has been displaced by alewife (*Alosa pseudoharengus*) through competition for food, and by the sea-lamprey (*Petromyzon marinum*) as a predator, so that once common Atlantic salmon (*Salmo salar*), lake trout (*Salvelinus namaycush*), and lake herring (*Leucichtys artedi*) have been nearly exterminated (Aron and Smith, 1971). The whole question of the presence and effects of exotic species in large lakes is reviewed by Hall and Mills (2000).

Can one make any generalizations about the circumstances that enable successful invasion by some exotic vertebrates and less successful invasion by others (Duncan et al., 2003)? Brown (1989) suggests that there may be 'five rules of biological invasions':

Rule 1
'Isolated environments with a low diversity of native species tend to be differentially susceptible to invasion.'
Rule 2
'Species that are successful invaders tend to be native to continents and to extensive, non-isolated habitats within continents.'
Rule 3
'Successful invasion is enhanced by similarity in the physical environment between the source and target areas.'
Rule 4
'Invading exotics tend to be more successful when native species do not occupy similar niches.'
Rule 5
'Species that inhabit disturbed environments and those with a history of close association with humans tend to be successful in invading man-modified habitats.'

Human influence on the expansion of animal populations

Although most attention tends to be directed towards the decline in animal numbers and distribution brought about by human agency, there are many circumstances where alterations of the environment and modification of competition has favored the expansion of some species. Such expansion is not always welcome or expected, as Marsh (1864: 34) appreciated:

Insects increase whenever the birds which feed upon them disappear. Hence in the wanton destruction of the robin and other insectivorous birds, the *Bipes implumis*, the featherless biped, man, is not only exchanging the vocal orchestra which greets the rising sun for the drowsy beetle's evening drone, and depriving his groves and his fields of their fairest ornament, but he is waging a treacherous warfare on his natural allies.

Human actions, however, are not invariably detrimental, and even great cities may have effects on animal life which can be considered desirable or tolerable. This has been shown in the studies of bird populations in several urban areas. For example, Nuorteva (in Jacobs, 1975) examined the bird fauna in the city of Helsinki (Finland), in agricultural areas near rural houses, and in uninhabited forests (see Table 3.1). The city supported by far the highest biomass and the highest number of birds, but exhibited the lowest number of species and the lowest diversity. In the artificially

Table 3.1 Dependence of bird biomass and diversity on urbanization. Source: data from Nuorteva (1971) modified after Jacobs (1975, table 2)

	City (Helsinki)	Near rural houses	Uninhabited forest
Biomass (kg km^{-2})	213	30	22
No. of birds (km^{-2})	1089	371	297
Number of species	21	80	54
Diversity	1.13	3.40	3.19

created rural areas the number of species, and hence diversity, were much higher than in the uninhabited forest, and so was biomass. Altogether, human civilization appeared to have brought about a very significant increase of diversity in the whole area: there were 37 species in city and rural areas that were not found in the forest. Similarly, after a detailed study of suburban neighborhoods in west central California, Vale and Vale (1976) found that in suburban areas the number of bird species and the number of individuals increased with time. Moreover, when compared with the pre-suburban habitats adjacent to the suburbs, the residential areas were found to support a larger number of both species and individuals. Horticultural activities appear to provide more luxuriant and more diverse habitats than do pre-suburban environments.

Some beasts other than the examples of birds given here are also favored by urban expansion (Schmid, 1974). Animals that can tolerate disturbance, are adaptable, utilize patches of open or woodland-edge habitat, creep about inside buildings, tap people's food supply surreptitiously, avoid recognizable competition with humans, or attract human appreciation and esteem may increase in the urban milieu. For these sorts of reasons the northeast megalopolis of the USA hosts thriving populations of squirrels, rabbits, raccoons, skunks, and opossums, while some African cities are now frequently blessed with the scavenging attention of hyenas.

Jacobs (1975) provides another apposite example of how humans can increase species diversity inadvertently. The saline Lake Nakuru in Kenya before 1961 was not a particularly diverse ecosystem. There were essentially one or two species of algae, one copepod, one rotifer, corixids, notonectids, and some 500,000

flamingos belonging virtually to one species. In 1962, however, a fish (*Tilapia grahami*) was introduced in order to check mosquitoes. In the event it established itself as a major consumer of algae, and its numbers increased greatly. As a consequence, some 30 species of fish-eating birds (pelicans, anhingas, cormorants, herons, egrets, grebes, terns and fish-eagles) have colonized the area to make Lake Nakuru a much more diverse system. Nonetheless, such fish introductions are not without their potential ecological costs. In Lake Atitlan in Central America the introduction of the game fish *Micropterus salmoides* (largemouth bass) and *Pomoxis nigromaculatus* (black crappie), both voracious eaters, led to the diminution of local fish and crab populations. Similarly, the introduction to the Gatun Lake in the Panama Canal Zone in 1967 of the cichlid fish, *Cichla ocellaris* (a native of the Amazon River), led to the elimination of six of the eight previously common fish species within five years, and the tertiary consumer populations such as the birds, formerly dependent on the small fishes for food, appeared less frequently (Zaret and Paine, 1973).

Human economic activities may lead to a rise in the number of examples of a particular habitat which can lead to an expansion in the distribution of certain species, though often humans have prompted a contraction in the range of a particular species because of the removal or modification of its preferred habitat. In Britain the range of the little ringed plover (*Charadrius dubius*), a species virtually dependent on anthropogenic habitats, principally wet gravel and sand pits, greatly expanded as mineral extraction accelerated (see Figure 3.6). Indeed, it needs to be stressed that the changes of habitat brought about by urbanization, industry (Ratcliffe, 1974; Davis, 1976), and mining need not be detrimental.

Very often human activities do not lead to species diversity, but to important increases in numbers of individuals, by creating new and favorable environments. Two especially serious examples of this from the human point of view are the explosions in the prevalence of both mosquitoes and bilharzia snails as a result of the extension of irrigation.

One of the most remarkable examples of the consequences of creating new environments is provided by the European rabbit (*Oryctolagus cuniculus*). Introduced into Britain in early medieval times, and originally an inhabitant of the western Mediterranean lands, it was

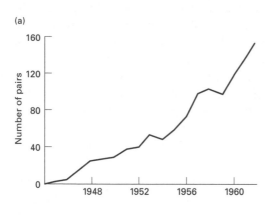

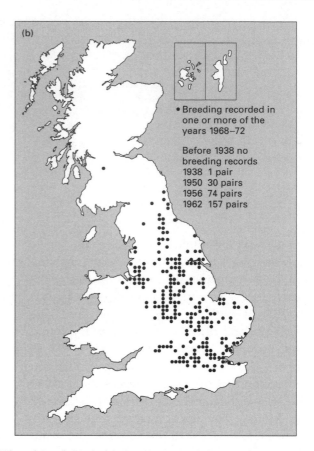

Figure 3.6 The changing range of the little ringed plover (*Charadrius dubius*): (a) the increase in the number summering in Britain (after Murton, 1971, figure 6); (b) the increase in the range, related to habitat change, especially as a result of the increasing number of gravel pits (after Nature Conservancy Council, 1977, figure 11).

kept for food and fur in carefully tended warrens. Agricultural improvements, especially to grassland, together with the increasing decline in the numbers of predators such as hawks and foxes, brought about by game-guarding landlords (Sheail, 1971), enabled the rabbit to become one of the most numerous of mammals in the British countryside. By the early 1950s there were 60–100 million rabbits in Britain. Frequently, as many contemporary reports demonstrated, it grazed the land so close that in areas of light soil, like the Breckland of East Anglia or in coastal dune areas, wind erosion became a serious problem. Similarly, the rabbit flourished in Australia, especially after the introduction of the merino sheep, which created favorable pasturelands. Erosion in susceptible lands such as the Mallee was severe. Both in England and Australia an effective strategy developed to control the rabbit was the introduction of a South American virus, *Myxoma*.

Some of the familiar British birds have benefited from agricultural expansion. Formerly, when Britain was an extensively wooded country, the starling (*Sturnus vulgaris*) was a rare bird. The lapwing (*Vanellus vanellus*) is yet another component of the grassland fauna of central Europe which has benefited from agriculture and the creation of open country with relatively sparse vegetation (Murton, 1971). There is also a large class of beasts which profit so much from the environmental conditions wrought by humans that they become very closely linked to them. These animals are often referred to as **synanthropes**. Pigeons and sparrows now form permanent and numerous populations in almost all the large cities of the world; human food supplies are the food supplies for many synanthropic rodents (rats and mice); a once shy forest-bird, the blackbird (*Turdus merula*), has in the course of a few generations become a regular and bold inhabitant of

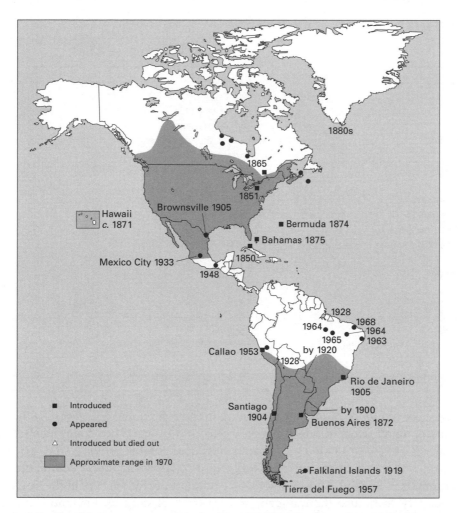

Figure 3.7 The spread of the English house sparrow in the New World (after Doughty, 1978: 14).

many gardens in the UK; and the squirrel (*Sciurus vulgaris*) in many places now occurs more frequently in parks than in forests (Illies, 1974: 101). The English house sparrow (*Passer domesticus*), a familiar bird in towns and cities, currently occupies approximately one-quarter of Earth's surface, and over the past 100 years it has doubled the area that it inhabits as settlers, immigrants, and others have carried it from one continent to another (Figure 3.7), although recently in Britain, for reasons that are not clear, it has suffered a decline in numbers. It is found around settlements both in Amazonia and the Arctic Circle (Doughty, 1978). The marked increase in many species of gulls in temperate regions over recent decades is largely attributable to their growing utilization of food scraps on refuse tips. This is, however, something of a mixed

blessing, for most of the gull species which feed on urban rubbish dumps breed in coastal areas, and there is now good evidence to show that their greatly increased numbers are threatening other less common species, especially the terns (*Sterna* spp.). It may seem incongruous that the niches for scavenging birds which exploit rubbish tips should have been filled by sea birds (Murton, 1971), but the gulls have proved ideal replacements for the kites which humans had removed, as they have the same ability to watch out for likely food sources from aloft and then to hover and plunge when they spy a suitable meal.

Most of the examples given so far to illustrate how human actions can lead to expansion in the numbers and distributions of certain animal species have been used to make the point that such expansion frequently

Table 3.2 Introduced mammals in Britain. Source: from Jarvis (1979: 188, table 1)

Species	Date of introduction of present stock	Reason for introduction
House mouse *Mus musculus*	Neolithic?	Accidental
Wild goat *Capra hircus*	Neolithic?	Food
Fallow deer *Dama dama*	Roman or earlier	Food, sport
Domestic cat *Felis catus*	Early Middle Ages	Ornament, pest control
Rabbit *Oryctolagus cuniculus*	Mid- to late twelfth century	Food, sport
Black rat *Rattus rattus*	Thirteenth century?	Accidental
Brown rat *Rattus norvegicus*	Early eighteenth century (1728–9)	Accidental
Sika deer *Cervus nippon*	1860	Ornament
Grey squirrel *Sciurus carolinensis*	1876	Ornament
Indian muntjak *Muntiacus muntjak*	1890	Ornament
Chinese muntjak *Muntiacus reevesi*	1900	Ornament
Chinese water deer *Hydropetes inermis*	1900	Ornament
Edible dormouse *Glis glis*	1902	Ornament
Musk rat *Ondatra zibethica*	1929–1937	Fur
Coypu *Myocastor coypus*	1929	Fur
Mink *Mustela vison*	1929	Fur
Bennett's wallaby *Macropus rufogriseus bennetti*	1939 or 1940	Ornament
Reindeer *Rangifer tarandus*	1952	Herding, ornament
Himalayan (Hodgson's) porcupine *Hystrix hodgsoni*	1969	Ornament
Crested porcupine *Hystrix cristata*	1972	Ornament
Mongolian gerbil *Meriones unguiculatus*	1973	Ornament

occurs as an incidental consequence of human activities. There are, of course, many ways in which people have intentionally and effectively promoted the expansion of particular species (Table 3.2 presents data for some introduced mammals in Britain). This may sometimes be done deliberately to reduce the numbers of a species that have expanded as an unwanted consequence of human actions. Perhaps the best known example of this is biological control using introduced predators and parasites. Thus an Australian insect, the cottony-cushion scale, *Icerya purchasi*, was found in California in 1868. By the mid-1880s it was effectively destroying the citrus industry, but its ravages were quickly controlled by deliberately importing a parasitic fly and a predatory beetle (the Australian ladybird) from Australia. Because of the uncertain ecological effects of synthetic pesticides, such biological control has its attractions but the importation of natural enemies is not without its own risks. For example, the introduction into Jamaica of mongooses to control rats led to the undesirable decimation of many native birds and small land mammals.

One further calculated method of increasing the numbers of wild animals, of conserving them, and of gaining an economic return from them, is game cropping.

In some circumstances, because they are better adapted to, and utilize more components of, the environment, wild ungulates provide an alternative means of land use to domestic stock. The exploitation of the saiga antelope, *Saiga tatarica*, in the Soviet Union is the most successful story of this kind (Edington and Edington, 1977). Hunting, much overdone because of the imagined medical properties of its horns, had led to its near demise, but this was banned in the 1920s and a system of controlled cropping was instituted. Under this regime (which produces appreciable quantities of meat and leather) total numbers have risen from about 1000 to over 2 million in the 1970s, and in the process the herds have reoccupied most of their original range. Since the break up of the Soviet Union, however, previously successful conservation schemes have suffered from a shortage of funding and high levels of poverty and unemployment have encouraged poaching.

Causes of animal contractions and declines: pollution

The extreme effect of human interference with animals is extinction, but before that point is reached humans

Table 3.3 Examples of biological magnification
(a) Enrichment factors for the trace element compositions
of shellfish compared with the marine environment.
Source: Merlini (1971) in King (1975: 303, table 8.8)

Element	Enrichment factor		
	Scallops	Oysters	Mussels
Ag	23,002	18,700	330
Cd	260,000	318,000	100,000
Cr	200,000	60,000	320,000
Cu	3000	13,700	3000
Fe	291,500	68,200	196,000
Mn	55,000	4000	13,500
Mo	90	30	60
Ni	12,000	4000	14,000
Pb	5300	3300	4000
V	4500	1500	2500
Zn	28,000	110,300	9100

(b) Mean methyl mercury concentrations in organisms
from a contaminated salt marsh. Source: Gardner et al.
(1978) in Bryan (1979)

Organism	Parts per million
Sediments	< 0.001
Spartina	< 0.001–0.002
Echinoderms	0.01
Annelids	0.13
Bivalves	0.15–0.26
Gastropods	0.25
Crustaceans	0.28
Fish muscle	1.04
Fish liver	1.57
Mammal muscle	2.2
Bird muscle	3.0
Mammal liver	4.3
Bird liver	8.2

(c) The concentration of dichlorodiphenyltrichloroethane
(DDT) in the food chain. Source: King (1975: 301, table 8.7)

Source	Parts per million
River water	0.000003
Estuary water	0.00005
Zooplankton	0.04
Shrimps	0.16
Insects – Diptera	0.30
Minnows	0.50
Fundulus	1.24
Needlefish	2.00
Tern	2.80–5.17
Cormorant	26.40
Immature gull	75.50

may cause major contractions in both animal numbers
and animal distribution. This decline may be brought
about partly by intentional killings for subsistence and
commercial purposes, but much wildlife decline occurs
indirectly (Doughty, 1974), for example, through pollution (Holdgate, 1979; Waldichuk, 1979), habitat change
or loss, and competition from invaders.

Particular concern has been expressed about the
role of certain pesticides in creating undesirable and
unexpected changes. The classic case of this concerns
dichlorodiphenyltrichloroethane (DDT) and related
substances. These were introduced on a worldwide
basis after the Second World War and proved highly
effective in the control of insects such as malarial mosquitoes. However, evidence accumulated that DDT was
persistent, capable of wide dispersal, and reached high
levels of concentration in certain animals at the top
trophic levels. The tendency for DDT to become more
concentrated as one moves up the food chain is illustrated further by the data in Table 3.3(c). Levels of
DDT in river and estuary waters may be low, but the
zooplankton and shrimps contain higher levels, the
fish that feed on them higher levels still, while fish-
eating birds have the highest levels of all.

Dichlorodiphenyltrichloroethane has a major effect
on sea birds. Their eggshells become thinned to the
extent that reproduction fails in fish-eating birds. Similar correlations between eggshell thinning and DDT
residue concentrations have been demonstrated for
various raptorial land-birds. The bald eagle (*Haliaeetus
leucocephalus*), peregrine falcon (Figure 3.8) (*Falco
peregrinus*), and osprey (*Pandion haliaetus*) all showed

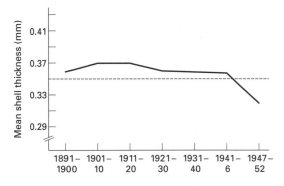

Figure 3.8 Changing thickness of peregrine
falcon eggshells versus time, associated with
dichlorodiphenyltrichloroethane (DDT) use since
the Second World War (modified after Hickey and
Anderson, 1968).

decreases in eggshell thickness and population decline from 1947 to 1967, a decline which correlated with DDT usage and subsequent reproductive and metabolic effects upon the birds (Johnston, 1974).

Another example serves to make the same point about the 'magnification' effects associated with pesticides (Southwick, 1976: 45–6). At Clear Lake in California, periodic appearances of large numbers of gnats caused problems for tourists and residents. An insecticide, dichlorodiphenyldichloroethane (DDD), was applied at a low concentration (up to 0.05 ppm) and killed 99% of the larvae. Following the application of DDD, western grebes (*Alchmorphorus occidentalis*) were found dead, and tissue analysis showed concentrations of DDD in them of 1600 ppm, representing a concentration of 32,000 times the application rate. The DDD had accumulated in the insect-eating fish at levels of 40–2500 ppm and the grebes feeding on the fish received lethal doses of the pesticide. There are also records elsewhere of trout reproduction ceasing when DDT levels build up (Chesters and Konrad, 1971).

However, appreciation of the undesirable side effects of DDD and DDT brought about by ecological concentration has led to severe controls of their use. For example, DDT reached a peak in terms of utilization in the USA in 1959 (35×10^6 kg) and by 1971 was down to 8.1×10^6 kg. The monitoring of birds in Florida over the same period indicated a parallel decline in the concentration of DDT and its metabolites (DDD and dichlorodiphenyldichloroethylene – DDE) in their fat deposits (Johnston, 1974). Other substances are also capable of concentration. For example, heavy metals and methyl mercury may build up in marine organisms, and filter feeders such as shellfish have a strong tendency to concentrate the metals from very dilute solutions, as is clear in Table 3.3(a and b).

It is possible, however, that the importance of biological accumulation and magnification has been overstated in some textbooks. G. W. Bryan (1979) reviewed the situation and noted that, although the absorption of pollutants from foods is often the most important route for bioaccumulation, and transfer along food chains certainly occurs, this does not automatically mean that predators at high trophic levels will always contain the highest levels. He wrote (p. 497):

although, for a number of contaminants, concentrations in individual predators sometimes exceed those of their prey,

when the situation overall is considered only the more persistent organochlorine pesticides, such as DDT and its metabolites and methyl mercury, show appreciable signs of being biologically magnified as a result of food-chain transfer.

Oil pollution is a serious problem for marine and coastal fauna and flora (Figure 3.9), although some of it derives from natural seeps (Blumer, 1972; Landes, 1973). Sea birds are especially vulnerable since oil clogs their feathers, while the ingestion of oil when birds attempt to preen themselves leads to enteritis and other complaints. Local bird populations may be seriously diminished although probably no extinction has resulted (Bourne, 1970). Fortunately, although oil is toxic it becomes less so with time, and the oil spilt from the *Torrey Canyon* in March 1967 was almost biologically inert when it was stranded on the Cornish beaches (Cowell, 1976). There are 'natural' oil-degrading organisms in nature. Much of the damage caused to marine life as a result of this particular disaster was caused not by the oil itself but by the use of 2.5 million gallons of detergents to disperse it (Smith, 1968).

It is possible that oil spills pose a particular risk in cold areas, for **biodegradation** processes achieved by microbes appear to be slow. This is probably because of a combination of low temperatures and limited availability of nitrogen, phosphorus, and oxygen. The last of these is a constraint because, compared with temperate ecosystems, arctic tundra and coastal ecosystems are relatively stagnant – ice dampens re-aeration due to wave action in marine ecosystems, while standing water in tundra soils limits inputs of oxygen to them. Detailed reviews are provided in Engelhardt (1985).

There is far less information available on the effects of oil spills in freshwater environments, although they undoubtedly occur. Freshwater bodies have certain characteristics which, compared with marine environments, tend to modify the effects of oil spills. The most important of these are the smaller and shallower dimensions of most rivers, streams, and lakes, which means that dilution and spreading effects are not as vigorous and significant in reducing surface slicks as would be the case at sea. In addition, the confining dimensions of ponds and lakes are likely to cause organisms to be subjected to prolonged exposure to dissolved and dispersed hydrocarbons. Information on this problem is summarized in Vandermeulen and Hrudey (1987).

Figure 3.9 Oil spills are generally perceived as a major cause of pollution on coastlines. In January 1993 the tanker *Braer* ran aground on the Shetland Isles (Scotland) and large quantities of oil were liberated, causing some distress and mortality among sea birds.

Figure 3.10 Corals reefs are wonderfully attractive, diverse and productive ecosystems that are subjected to a whole series of threats. At Sharks Bay, on the Red Sea in Egypt, tourists are encouraged not to damage the coral reef offshore by avoiding trampling on the coral.

One particular type of aquatic ecosystem where pollution is an increasingly serious problem is the coral reef (Figure 3.10) (see Kuhlmann, 1988). Coral reefs are important because they are among the most diverse, productive, and beautiful communities in the world. Although they only cover an area of less than 0.2% of the world's ocean beds, up to one-quarter of all marine species and one-fifth of known marine fish species live in coral reef ecosystems (World Resources Institute, 1998: 253). They also provide coastal protection, opportunities for recreation, and potential sources of substances such as drugs. Accelerated sedimentation resulting from poor land management (Nowlis et al., 1997), together with dredging, is probably responsible for more damage to reef communities than all the other forms of human insult combined, for suspended sediments restrict the light penetration necessary for coral growth. Also, the soft, shifting sediments

may not favor colonization by reef organisms. Sewage is the second worst form of stress to which coral reefs are exposed, for oxygen-consuming substances in sewage result in reduced levels of oxygen in the water of lagoons. The detrimental effects of oxygen starvation are compounded by the fact that sewage may cause nutrient enrichment to stimulate algal growth, which in turn can overwhelm coral. Poor land management can also cause salinity levels to be reduced below the level of tolerance of reef communities as a consequence of accelerated runoff of freshwater from catchments draining into lagoons. One of the reasons why all these stresses may be especially serious for reefs, is that corals live and grow for several decades or more, so that it can take a long time for them to recover from damage.

An interesting example of the role of sedimentation in damaging reef health is provided by geochemical studies of long-lived corals from Australia's Great Barrier Reef (McCulloch et al., 2003). These showed that since 1870 and the start of European settlement in its catchment the Burdekin River has carried five to ten times more sediment to the reef than it did previously.

Since the late nineteenth century it has been clear that industrial air pollution has had an adverse effect on domestic animals, but there is less information about its influence on wildlife. Nonetheless, there is some evidence that it does affect wild animals (Newman, 1979). Arsenic emissions from silver foundries are known to have killed deer and wild rabbits in Germany; sulfur emissions from a pulp mill in Canada are known to have killed many song-birds; industrial fluorosis has been found in deer living in the USA and Canada; asbestosis has been found in free-living baboons and rodents in the vicinity of asbestos mines in South Africa; and oxidants from air pollution are recognized causes of blindness in big-horn sheep in the San Bernadino Mountains near Los Angeles in California.

In Britain considerable concern has arisen over the effects of lead poisoning on wildlife, particularly on wildfowl. Poisoning occurs in swans and mallards, for example, when they feed and seek grit, since they ingest spent shotgun pellets or discarded anglers' weights in the process. Such pellets are eroded in the bird's gizzard and the absorbed lead causes a variety of adverse physiological effects that can result in death: damage to the nervous system, muscular paralysis, anemia, and liver and kidney damage (Mudge, 1983).

Soil erosion, by increasing stream **turbidity** (another type of pollution), may adversely affect fish habitats (Ritchie, 1972). The reduction of light penetration inhibits photosynthesis, which in turn leads to a decline in food and a decline in carrying capacity. Decomposition of the organic matter, which is frequently deposited with sediment, uses dissolved oxygen, thereby reducing the oxygen supply around the fish; sediment restricts the emergence of fry from eggs; and turbidity reduces the ability of fish to find food (though conversely it may also allow young fish to escape predators).

Turbidity has also been increased by mining waste, by construction (Barton, 1977), and by dredging. In the case of some of the Cornish rivers in china-clay mining areas, river turbidities reach 5000 mg L^{-1} and trout are not present in streams so affected. Indeed, the non-toxic turbidity tolerances of river fish are much less than that figure. Alabaster (1972), reviewing data from a wide range of sources, believes that in the absence of other pollution fisheries are unlikely to be harmed at chemically inert suspended sediment concentrations of less than 25 mg L^{-1}, that there should be good or moderate fisheries at 25–80 mg L^{-1}, that good fisheries are unlikely at 80–400 mg L^{-1} and that at best only poor fisheries would exist at more than 400 mg L^{-1}.

Habitat change and animal decline

Land use and land cover changes, including the conversion of natural ecosystems to cropland (Ramankutty and Foley, 1999), have so modified habitats that animal declines have ensued. Elsewhere ancient and diverse anthropogenic landscapes have themselves been modified by modern agricultural intensification. Good reviews of changes in biodiversity are provided by Heywood and Watson (1995) and Chapin et al. (2001). Certainly, many species have lost a very large part of their natural ranges as a result of human activities, and a study of historical and present distributions of 173 declining mammal species from six continents has indicated that these species have collectively lost over 50% of their historic range area (Ceballos and Ehilich, 2002).

One of the most important habitat changes in Britain has been the removal of many of the hedgerows that form the patchwork quilt so characteristic of large tracts of the country (Sturrock and Cathie, 1980). Their removal has taken place for a variety of reasons: to create larger fields, so that larger machinery can be employed and so that it spends less time turning; because as farms are amalgamated boundary hedges may become redundant; because as farms become more specialized (e.g., just arable) hedges may not be needed to control stock or to provide areas for lambing and calving; and because drainage improvements may be more efficiently executed if hedges are absent. In 1962 there were almost 1 million kilometers of hedge in Britain, probably covering the order of 180,000 hectares,

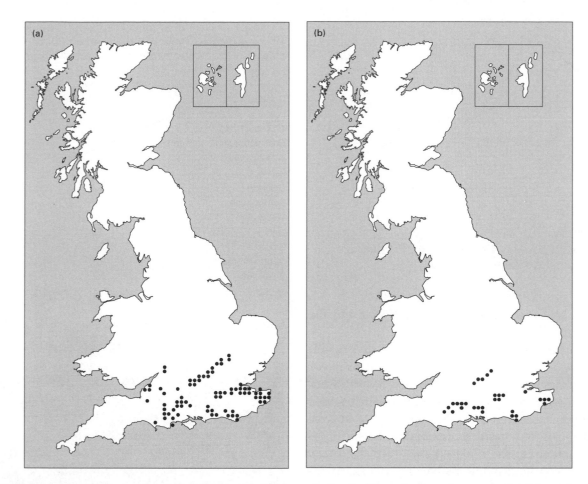

Figure 3.11 Reduction in the range of the silver-spotted skipper butterfly (*Hesperia comma*) as a result of the reduction in the availability of the chalk downland and limestone pastures upon which it depends. (a) Distribution prior to 1960 (after Nature Conservancy Council, 1977, figure 3.8). (b) Distribution in the late 1990s (from data provided by National Biodiversity Network).

an area greater than that of the national nature reserves. Bird numbers and diversity are likely to be severely reduced if the removal of hedgerows continues, for most British birds are essentially of woodland type and, since woods only cover about 5% of the land surface of Britain, depend very much on hedgerows (Moore et al., 1967).

Other types of habitat have also been disappearing at a fast rate with the agricultural changes of recent years (Nature Conservancy Council, 1977). For example, the botanical diversity of old grasslands is often reduced by replacing the lands with grass leys or by treating them with selective herbicides and fertilizers; this can mean that the habitat does not contain some of the basic requirements essential for many species. One can illustrate this by reference to the larva of the common blue butterfly (*Polyommatus icarus*) which feeds upon bird-foot trefoil (*Lotus corniculatus*). This plant disappears when pasture is ploughed and converted into a ley, or when it is treated with a selective herbicide. Once the plant has gone the butterfly vanishes too because it is not adapted to feeding on the plants grown in the ley of improved pasture. Figure 3.11 illustrates the way in which the distribution of the silver-spotted skipper butterfly (*Hesperia comma*) has contracted with the diminished availability of chalk downland and limestone pastures upon which it depends. Likewise, the large blue butterfly (*Maculinea arion*) has decreased in Britain. Its larvae live solely on the wild thyme (*Thymus druccei*), a plant that thrives on close-cropped grassland. Since the decimation of the rabbit by myxomatosis, conditions for the thyme

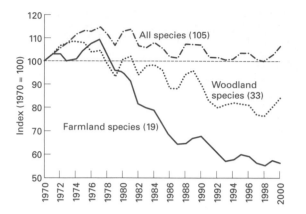

Figure 3.12 Population of wild birds in the UK, 1970–2000 (based on data from UK Department for Environment Food and Rural Affairs (Defra), 2003).

Table 3.4 Decline in farmland birds in Britain between 1972 and 1996 as shown by the Common Birds Census. Source: http://www.jncc.gov.uk/species/Birds/c0c_98.htm (accessed 15 October 2003)

Species	% decline
Corn bunting	74
Grey partridge	78
Tree sparrow	87
Lapwing	42
Bullfinch	62
Song thrush	66
Turtle dove	62
Linnet	40
Skylark	60
Spotted flycatcher	78
Blackbird	33

have been less favorable so that both thyme and the large blue butterfly have declined.

The replacement of natural oak-dominated woodlands in Britain by conifer plantations, another major land-use change of recent decades, also has its implications for wildlife. It has been estimated that where this change has taken place the number of species of birds found has been approximately halved (Figure 3.12). Likewise the replacement of upland sheep walks with conifer plantations has led to a sharp decline in raven numbers in southern Scotland and northern England. The raven (*Corvus corax*) obtains much of its carrion from open sheep country (Marquiss et al., 1978). Other birds that have suffered from planting moorland areas with forest plantations are miscellaneous types of wader, golden eagles, peregrine falcons, and buzzards (Grove, 1983).

Agricultural intensification on British farms also appears to have had a deleterious effect on species of farmland bird, many of which have shown a downward trend in numbers over the past three decades (Table 3.4). This is because of habitat changes – lack of fallows, less mixed farming, new crops, modern farm management, biocide use, hedgerow removal, etc. (Fuller et al., 1991).

On a global basis the loss of **wetland** habitats (Figure 3.13) (marshes, bogs, swamps, fens, mires, etc.) is a cause of considerable concern (Maltby, 1986; Williams, 1990). In all, wetlands cover about 6% of Earth's surface (not far short of the total under tropical rain forest), and so they are far from being trivial, even though

Figure 3.13 The Okavango Swamps in Botswana are an example of an important wetland habitat that is home to many species of plant and animal. Were the waters that drain into it to be diverted, an ecological and aesthetic tragedy would result.

they tend to occur in relatively small patches. However, they also account for about one-quarter of Earth's total net primary productivity, have a very diverse fauna and flora, and provide crucial wintering, breeding, and refuge areas for wildlife. According to some sources, the world may have lost half its wetlands since 1900, and the USA alone has lost 54% of its original wetland area, primarily because of agricultural developments. There are, however, other threats, including drainage, dredging, filling, peat removal, pollution, and **channelization**.

In Europe and the USA 90% of riverine flood plains have been gravely and extensively modified as habitats and in ecological terms have been described as 'functionally extinct' (Tockner and Stanford, 2002).

Other causes of animal decline

One could list many other indirect causes of wildlife decline. Vehicle speed, noise, and mobility upset remote and sensitive wildlife populations, and this problem has intensified with the rapid development in the use of off-road recreation vehicles. The effects of roads are becoming ever more pervasive as a result of increasing traffic levels, vehicle speed, and road width, and a new subdiscipline of road ecology has emerged (Sherwood et al., 2002; Forman et al., 2003). A fine-meshed road network is a highly effective cause of habitat isolation, acting as a series of barriers to movement, particularly of small, cover-loving animals. As Oxley et al. (1974) have put it, 'a four-lane divided highway is as effective a barrier to the dispersal of small forest mammals as a body of fresh water twice as wide.' This barrier effect results from a variety of causes (Mader, 1984): roads interrupt microclimatic conditions; they are a broad band of emissions and disturbances; they are zones of instability due to cutting and spraying, etc.; they provide little cover against predators; and they subject animals to death and injury by moving wheels. Leisure activities in general may create problems for some species (Speight, 1973). A survey of the breeding status of the little tern (*Sterna albifrons*) in Britain gave a number of instances of breeding failure by the species, apparently caused by the presence of fishermen and bathers on nesting beaches. The presence of even a few people inhibited the birds from returning to their nests. In like manner, species building floating nests on inland waters, such as great crested grebes (*Podiceps cristatus*), are very prone to disturbance by water-skiers and powerboats.

New types of construction can cause problems. As Doughty (1974) has remarked '"wirescapes", tall buildings, and towers can become to birds what dams, locks, canals and irrigation ditches are to the passage of fish.' The construction of canals can cause changes in aquatic communities. Likewise turbines associated with hydroelectric schemes can cause numerous fish deaths, either directly through ingesting them, or through

gas-bubble disease. This resembles the 'bends' in divers and is produced if a fish takes in water supersaturated with gases. The excess gas may come out of solution as bubbles, which can lodge in various parts of the fish's body causing injury or death. Supersaturation of water with gas can be produced in turbines or when a spillway plunges into a deep basin (Baxter, 1977).

It might also be thought that human use of fire could be directly detrimental to animals, though the evidence is not conclusive. Many forest animals appear to be able to adapt to fire, and fire also tends to maintain habitat diversity. Thus, after a general review of the available ecological literature Bendell (1974) found that fires did not seem to produce as much change in birds or small mammals as one might have expected. Some of his data are presented in Table 3.5. The amount of change is fairly small, although some increase in species diversities and animal densities is evident. Indeed Vogl (1977) has pointed to the diversity of fauna associated with fire-affected ecosystems (p. 281):

Birds and mammals usually do not panic or show fear in the presence of fire, and are even attracted to fires and smoking or burned landscapes . . . The greatest arrays of higher animal species, and the largest numbers per unit area, are associated with fire-dependent, fire-maintained, and fire-initiated ecosystems.

Purposeful hunting, particularly when it involves modern firearms, means that the distribution of many animal species has become smaller very rapidly (Figure 3.14). This is clearly illustrated by a study of the North American bison (Figure 3.15), which on the eve of European colonization still had a population of some 60 million in spite of the presence of a small Indian population. By 1850 only a few dozen examples of bison still survived, though conscious protective and conservation measures have saved it from extinction.

Animals native to remote ocean islands have been especially vulnerable since many have no flight instinct. They also provided a convenient source of provisions for seafarers. Thus the last example of the dodo (*Raphus cucullatus*) was slaughtered on Mauritius in 1681 and the last Steller's sea cow (*Hydrodamalis gigas*) was killed on Bering Island in 1768 (Illies, 1974: 96). The land-birds, which are such a feature of coral atoll ecosystems, rapidly become extinct when cats, dogs, and rats are introduced. The endemic flightless rail of Wake Island, *Rallus wakensis*, became extinct during

Table 3.5 Environmental effects of burning
(a) Change in number of species of breeding birds and small mammals after fire

	Foraging zone	Before fire	After fire	Gained (%)	Lost (%)
Birds	Grassland and shrub	48	v62	38	8
	Tree trunk	25	26	20	16
	Tree	63	58	10	17
	Totals	136	146	21	14
Small mammals	Grassland and shrub	42	45	17	10
	Forest	16	14	13	25
	Totals	58	59	16	14

(b) Change in density and trend of population of breeding birds and small mammals after fire

	Foraging zone	Density (%)			Trend (%)		
		Increase	Decrease	No change	Increase	Decrease	No change
Birds	Grassland and shrub	50	9	41	24	10	66
	Tree trunk	28	16	56	4	8	88
	Tree	24	19	57	6	6	88
	Totals	35	15	50	12	8	80
Small mammals	Grassland and shrub	24	13	63	20	5	75
	Forest	23	42	35	0	11	89
	Totals	23	25	52	14	1	80

Figure 3.14 A major cause of animal decline is trapping. Here we see a sample of the many snares that have been found in a game reserve in Swaziland, southern Africa.

the prolonged siege of the island in the Second World War, and the flightless rail of Laysan, *Porzanula palmeri*, is also now extinct. Seabirds are more numerous on the atolls, but many colonies of terns, noddies, and boobies have been drastically reduced by vegetation

clearance and by introduced rats and cats in recent times (Stoddart, 1968). Other atoll birds, including the albatross, have been culled because of the threat they pose to aircraft using military airfields on the islands. Some birds have declined because they are large, vulnerable, and edible (Feare, 1978).

The fish resources of the oceans are still exploited by hunting and gathering techniques, and about 95 million tonnes of fish are caught in the world each year, a more than fivefold increase since 1950. The world catch may have peaked, however, for some major fisheries have collapsed as a result of unsustainable exploitation (Hilborn et al., 2003). Chronic overfishing is a particular problem in shallow marginal seas (e.g., the North Sea) and in the vicinity of coral reefs, where the removal of herbivorous fish can enable algal bloom to develop catastrophically (Hughes, 1994). Populations of the large varieties of whales, those that have been most heavily exploited, have been cut to a tiny fraction of their former sizes, and the hunting, especially by the Russians and the Japanese, still goes on, although since the 1980s most nations have banned whaling.

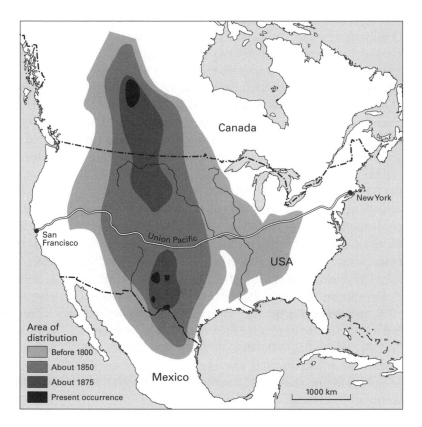

Figure 3.15 The former and present distribution of the bison in North America (after Ziswiler, in Illies, 1974, figure 3.1).

It is extremely difficult to estimate whale stocks before whaling began and therefore to quantify the magnitude of hunting effects that have taken place. However, genetic techniques have recently been used to establish pre-whaling stocks (Roman and Palumbi, 2003). These techniques have indicated that previous estimates of natural stocking levels were too low and that current populations are a fraction of past numbers. Pre-whaling populations in the North Atlantic of humpback, fin, and minke whales were perhaps 865,000 in total, whereas current populations are 215,000. With bans on whaling that now exist, a number of large whale species have shown signs of recovery, although the status of some baleen whale populations continues to cause concern. In addition even if commercial whaling was banned in perpetuity, whales would face other threats of mankind's making, including acoustic disturbance from seismic exploration, sonar, and motor-vessel traffic.

The introduction of a new animal species can cause the decline of another, whether by predation, competition, or by hybridization. This last mechanism has been found to be a major problem with fish in California, where many species have been introduced by humans. Hybridization results when fish are transferred from one basin to a neighboring basin, since closely related species of the type likely to hybridize usually exist in adjacent basins. One species can eliminate another closely related to it through genetic swamping (Moyle, 1976).

Animal extinctions in prehistoric times

Some workers, notably Martin (1967, 1974), believe that the human role in animal extinctions goes back to the Stone Age and to the Late Pleistocene. They believe that extinction closely follows the chronology of the spread of prehistoric civilization and the development of big-game hunting. They would also maintain that there are no known continents or islands in which accelerated extinction definitely pre-dates the arrival of human settlements. An alternative interpretation, however, is that the Late Pleistocene extinctions of big mammals were caused by the rapid and substantial changes of climate at the termination of the last glacial (Martin and Wright, 1967; Martin and Klein, 1984). Martin (1982) argues that the global pattern of extinctions of the large land mammals follows the footsteps of Paleolithic settlements. He suggests that Africa and parts of southern Asia were affected first, with substantial losses at the end of the Acheulean, around 200,000 years ago. Europe and northern Asia were affected between 20,000 and 10,000 years ago, while North and South America were stripped of large herbivores between 12,000 and 10,000 years ago. Megafaunal extinctions in Australia may have occurred around 41,000 years ago, shortly after human arrivals (Roberts et al., 2001). Extinctions continued into the Holocene on oceanic islands and in the Galapagos Islands, for example, virtually all extinctions took place after the first human contact in AD 1535 (Steadman et al., 1991). Likewise the complete deforestation of Easter Island in the Pacific between 1200 and 800 years BP was an ecological disaster that led to the demise of much of the native flora and fauna and also precipitated the decline of the megalithic civilization that had erected the famous statues on the islands (Flenley et al., 1991). As Burney (1993: 536) has written:

For millennia after the late-Pleistocene extinctions, these seemingly fragile and ungainly ecosystems apparently thrived throughout the world. This is one of the soundest pieces of evidence against purely climatic explanations for the passing of Pleistocene faunas. In Europe, for instance, the last members of the elephant family survived climatic warming not on the vast Eurasian land mass, but on small islands in the Mediterranean.

The dates of the major episodes of Pleistocene extinction are crucial to the discussion. Martin (1967: 111–15) writes:

Radiocarbon dates, pollen profiles associated with extinct animal remains, and new stratigraphic and archaeological evidence show that, depending on the region involved, late-Pleistocene extinction occurred either after, during or somewhat before worldwide climatic cooling of the last maximum . . . of glaciation . . . Outside continental Africa and South East Asia, massive extinction is unknown before the earliest known arrival of prehistoric man. In the case of Africa, massive extinction coincides with the final development of Acheulean hunting cultures which are widespread throughout the continent . . . The thought that prehistoric hunters ten to fifteen thousand years ago (and in Africa over forty thousand years ago) exterminated far more large

animals than has modern man with modern weapons and advanced technology is certainly provocative and perhaps even deeply disturbing.

The nature of the human impact on animal extinctions can conveniently be classified into three types (Marshall, 1984): the 'blitzkrieg effect', which involves rapid deployment of human populations with big-game hunting technology so that there is very rapid demise of animal populations; the 'innovation effect', whereby long-established human population groups adopt new hunting technologies and erase fauna that have already been stressed by climatic changes; and the 'attrition effect', whereby extinction takes place relatively slowly after a long history of human activity because of loss of habitat and competition for resources.

The arguments that have been used in favor of an anthropogenic interpretation can be summarized as follows. First, massive extinction in North America seems to coincide in time with the arrival of humans in sufficient quantity and with sufficient technological skill in making suitable artifacts (the bifacial Clovis blades) to be able to kill large numbers of animals (Krantz, 1970). Second, in Europe the efficiency of Upper Paleolithic hunters is attested by such sites as Solutré in France, where a late Perigordian level is estimated to contain the remains of over 100,000 horses. Third, many beasts unfamiliar with people are remarkably tame and stupid in their presence, and it would have taken them a considerable time to learn to flee or seek concealment at the sight or scent of human life. Fourth, humans, in addition to hunting animals to death, may also have competed with them for particular food or water supplies or may have introduced predators, such as the dingo in Australia, which then competed with native fauna (Johnson and Wroe, 2003). Fifth, Coe (1981, 1982) has suggested that the supposed preferential extinction of the larger mammals could also lend support to the role of human actions. He argues that while large body size has certain definite advantages, especially in terms of avoiding predation and being able to cover vast areas of savanna in search of food, large body size also means that these herbivores are required to feed almost continuously to sustain a large body mass. Furthermore, as the size and generation time of a mammal increases, the rate at which they turn over their biomass decreases (Figure 3.16).

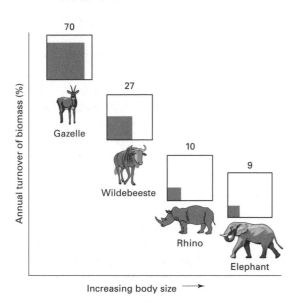

Figure 3.16 Decreasing rate of population biomass turnover with increasing body size in warm- and cold-blooded vertebrates, with some mammal examples illustrated (from Coe, 1982).

Hence a population of Thompson's gazelle will turn over up to 70% of their biomass each year, a wildebeest over 27%, but a rhino only 10% and the elephant just 9%. The significance of this is that, since large mammals can only turn over a small percentage of their population biomass each year, the rate of slaughter that such a population can sustain in the face of even a primitive hunter is very low indeed.

Certain objections have been leveled against the climatic change model and these tend to support the anthropogenic model. It has been suggested, for instance, that changes in climatic zones are generally sufficiently gradual for beasts to be able to follow the shifting vegetation and climatic zones of their choice. Similar environments are available in North America today as were present, in different locations and in different proportions, during Late Pleistocene times. Second, it can be argued that the climatic changes associated with the multiple glaciations, interglacials, pluvials, and interpluvials do not seem to have caused the same striking degree of elimination as those in the Late Pleistocene. A third difficulty with the climatic-cause theory is that animals such as the mammoth occupied a broad range of habitats from arctic to tropical latitudes, so it is unlikely all would perish as a result of a climatic change (Martin, 1982).

This is not to say, however, that the climatic hypothesis is without foundation or support. The migration of animals in response to rapid climatic change could be halted by geographic barriers such as high mountain ranges or seas. The relatively rich state of the African big mammalian fauna is due, according to this point of view, to the fact that the African biota is not, or was not, greatly restricted by an insuperable geographical barrier. Another way in which climatic change could cause extinction is through its influence on disease transmission. It has been suggested that during glacials animals would be split into discrete groups cut off by ice sheets but that, as the ice melted (before 11,000 years BP in many areas), contacts between groups would once again be made enabling diseases to which immunity might have been lost during isolation to spread rapidly. Large mammals, because of their low reproduction rates, would recover their numbers only slowly, and it was large mammals (according to Martin) that were the main sufferers in the Late Pleistocene extinctions.

The detailed dating of the European megafauna's demise lends further credence to the climatic model (Reed, 1970). The Eurasiatic **boreal** mammals, such as mammoth, woolly rhinoceros, musk ox, and steppe bison, were associated with and adapted to the cold steppe, which was the dominant environment in northern Europe during the glacial phases of the last glaciation. Each of these forms, especially the mammoth and the steppe bison, had been hunted by humans for tens of thousands of years, yet managed to survive through the last glacial. They appear to have disappeared, according to Reed, within the space of a few hundred years when warm conditions associated with the Alleröd Interstadial led to the restriction and near disappearance of their habitat.

Grayson (1977) has added to the doubts expressed about the anthropogenic overkill hypothesis. He suggests that the overkill theory, because it states that the end of the North American Pleistocene was marked by extraordinary high rates of extinction of mammalian genera, requires terminal Pleistocene mammalian generic extinctions to have been relatively greater than generic extinctions within other classes of vertebrates at this time. When he examined the extinction of birds he found that an almost exactly comparable proportion became extinct at the end of the Pleistocene as one finds for the megafauna. Moreover, as the radiocarbon dates for early societies in countries such as Australia are pushed back, it becomes increasingly clear that humans and several species of megafauna were living together for quite long periods, thereby undermining the notion of rapid overkill (Gillespie et al., 1978).

Further arguments can be marshalled against the view that humans as predators played a critical role in the Late Pleistocene extinctions in North America (Butzer, 1972: 509–10). There are, for example, relatively few Paleo-Indian sites over an immense area, and the majority of these have a very limited cultural inventory. In addition there is no clear evidence that Paleo-Indian subsistence was necessarily based, in the main part, on big-game hunting; if it was, only two genera were hunted intensively: mammoth and bison. A final point that militates against the argument that humans were primarily responsible for the waves of extinction is the survival of many big-game species well into the nineteenth century, despite a much larger and more efficient Indian population.

Thus the human role in the great Late Pleistocene extinction is still a matter of debate. The problem is complicated because certain major cultural changes in human societies may have occurred in response to climatic change. The cultural changes may have assisted in the extinction process, and increasing numbers of technologically competent humans may have delivered the final coup de grace to isolated remnants already doomed by rapid post-glacial environmental changes (Guilday, 1967). In this context it is worth noting that Haynes (1991) has suggested that at around 11,000 years ago conditions were dry in the interior of the USA and that Clovis hunters may as a result have found large game animals easier prey when concentrated around waterholes and under stress.

Actual extinction may have occurred after or concurrently with a dwarfing in the size of animals, and this too is a subject of controversy (Guthrie, 2003). On the one hand, there are those who believe that the dwarfing could result from a reduction in food availability brought about by climatic deterioration. On the other, it can be maintained that this phenomenon derives from the fact that small animals, being more adept at hiding and being a less attractive target for a hunter, are more likely to survive human predators, so that a genetic selection towards reduced body size takes place (Marshall, 1984).

Although the debate about the cause of this extinction spasm has now persisted for a long time, there are still great uncertainties. This is particularly true with regard to the chronologies of extinction and of human colonization (Brook and Bowman, 2002). This is particularly so in the context of North America, as noted by Grayson (1988). It is probable that the Late Pleistocene extinctions themselves had major ecological consequences. As Birks (1986: 49) has put it:

The ecological effects of rapid extinction of over 75% of the New World's large herbivores . . . must have been profound, for example on seed dispersal, browsing, grazing, trampling and tree regeneration . . . large grazers and browsers such as bison, mammoth and woolly rhinoceros may have been important in delaying or even inhibiting tree growth.

Modern-day extinctions

As the size of human populations has increased and technology has developed, humans have been responsible for the extinction of many species of animals and a reduction in biodiversity – 'the only truly irreversible global environmental change the Earth faces today' (Dirzo and Raven, 2003). Indeed, it has often been maintained that there has been a close correlation between the curve of population growth since the mid-seventeenth century and the curve of the number of species that have become extinct (Figure 3.17a, b). However, although it is likely that European expansion overseas during that time has led to many extinctions, the dramatic increase in the rate of extinctions over the past few centuries that is shown in the figure may in part be a consequence of a dramatic increase in the documentation of natural phenomena.

A more recent attempt to look at the historical trends in extinction rates (Figure 3.17c) for 20-year intervals from the year 1600 shows a rather different pattern, with a decline in rate for 1960–1990. This may partly be because of a time lag in the recording of the most recent extinction events, but it may also be the result of successful conservation efforts and the spread of the areal extent of protected areas. What is clear, however, is just what a large proportion of total extinctions have been recorded from islands.

When considering modern-day extinctions the role of human agency is much less controversial. Nonetheless,

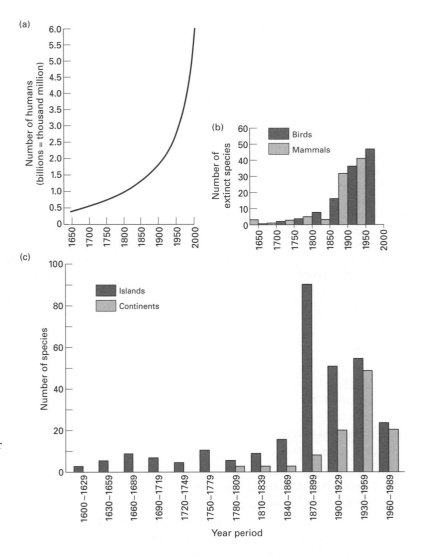

Figure 3.17 Time series of animal extinctions since the seventeenth century, in relation to human population increase (a). Figure (b) shows an early attempt to trace the number of extinctions in birds and mammals (after Ziswiler in National Academy of Sciences, 1972, figure 3.2), while figure (c) is a more recent attempt and displays a decreasing rate of animal extinctions in recent decades (after World Conservation Monitoring Centre, 1992, figure 16.5).

some modern extinctions of species are natural. Extinction is a biological reality: it is part of the process of evolution. In any period, including the present, there are naturally doomed species, which are bound to disappear, either through overspecialization or an incapacity to adapt themselves to climatic change and the competition of others, or because of natural cataclysms such as earthquakes, volcanic eruptions, and floods. Fisher et al. (1969) believe that probably one-quarter of the species of birds and mammals that have become extinct since 1600 may have died out naturally. In spite of this, however, the rate of animal extinctions brought about by humans in the past 400 years is of a very high order when compared with the norm of geologic time. In 1600 there were approximately 4226 living species of mammals: since then 36 (0.85%) have

become extinct; and at least 120 of them (2.84%) are presently believed to be in some (or great) danger of extinction. A similar picture applies to birds. In 1600 there were about 8684 living species; since then 94 (1.09%) have become extinct, and at least 187 of them (2.16%) are presently, or have very lately been, in danger of extinction (Fisher et al., 1969).

Some beasts appear to be more prone to extinction than others, and a distinction is now often drawn between *r*-selected species and *k*-selected species. These are two ends of a spectrum. The former have a high rate of increase, short gestation periods, quick maturation, and the advantage that they have the ability either to react quickly to new environmental opportunities or to make use of transient habitats (such as seasonal ponds). The life duration of individuals tends to be short,

Table 3.6 Characteristics of species affecting survival

Endangered	Safe
Large size	Small size
Predator	Grazer, scavenger, insectivore
Narrow habitat tolerance	Wide habitat tolerance
Valuable fur, oil, hide, etc.	Not a source of natural products
Restricted distribution	Broad distribution
Lives largely in international waters	
Migrates across international boundaries	Lives largely in one country
Reproduction in one or two vast aggregates	Reproduction by solitary pairs or in many small aggregates
Long gestation period	Short gestation period
Small litters	Big litters and quick maturation
Behavioral idiosyncrasies that are nonadaptive today	Adaptive
Intolerance to the presence of humans	Tolerance of humans
Dangerous to humans, livestock, etc.	Perceived as harmless

populations tend to be unstable, and the species may overexploit their environment to their eventual detriment. Many pests come into this category. At the other end of the spectrum the *k*-selection species are those which tend to be endangered or to become extinct. Their prime characteristics are that they are better adapted to physical changes in their environment (such as seasonal fluctuations in temperature and moisture) and live in a relatively stable environment (Miller and Botkin, 1974). These species tend to have much greater longevity, longer generation times, fewer offspring, but a higher probability of survival of young and adults. They have traded a high rate of increase and the ability to exploit transient environments for the ability to maintain more stable populations with low rates of increase, but correspondingly low rates of mortality and a closer adjustment to the long-term capacity of the environment to support their population.

Table 3.6, which is modified after Ehrenfield (1972), attempts to bring together some of the characteristics of species that affect their survival in the human world. This is also a theme discussed in Jeffries (1997).

There are at least nine ecological or life history traits that have been proposed as factors which determine the sensitivity of an animal species to a reduction and fragmentation in habitat (World Conservation Monitoring Centre, 1992: 193):

Rarity
Several studies have found that the abundance of a species prior to habitat fragmentation is a significant predictor of extinction . . . This is only to be expected, since fewer individuals of a rare species than a common species are likely to occur in habitat fragments, and the mechanisms of extinction mean that small populations are inherently more likely to become extinct than large.

Dispersal ability
If animals are capable of migrating between fragments or between 'mainland' areas and fragments, the effects of small population size may be partly or even greatly mitigated by the arrival of 'rescuers'. Species that are good dispersers may therefore be less prone to extinction in fragmented habitats than poor dispersers.

Degree of specialization
Ecological specialists often exploit resources which are patchily distributed in space and time, and therefore tend to be rare. Specialists may also be vulnerable to successional changes in fragments and to the collapse of coevolved mutualisms or food webs.

Niche location
Species adapted to, or able to tolerate, conditions at the interface between different types of habitats may be less affected by fragmentation than others. For example, forest edge species may actually benefit from habitat fragmentation.

Population variability
Species with relatively stable populations are less vulnerable than species with pronounced population fluctuations, since they are less likely to decline below some critical threshold from which recovery becomes unlikely.

Trophic status
Animals occupying high trophic levels usually have small populations: e.g., insectivores are far fewer in number than their insect prey and, as noted above, rarer species are more vulnerable to extinction.

Adult survival rate
Species with naturally low adult survival rates may be more likely to become extinct.

Longevity
Long-lived animals are less vulnerable to extinction than short-lived.

Intrinsic rate of population increase
Populations which can expand rapidly are more likely to recover after population declines than those which cannot.

Possibly one of the most fundamental ways in which humans are causing extinction is by reducing the area of natural habitat available to a species (Turner, 1996) and by fragmenting it (Fahrig, 2003). Habitat fragmentation divides once continuous, large populations into many smaller ones, which can be more or less isolated. Small population size and strong isolation of populations are associated with various negative consequences, which include susceptibility to natural disasters, genetic drift, and inbreeding (Lienert, 2004). Even wildlife reserves tend to be small 'islands' in an inhospitable sea of artificially modified vegetation or urban sprawl.

We know from many of the classic studies in true island biogeography that the number of species living at a particular location is related to area (see Figure 3.18

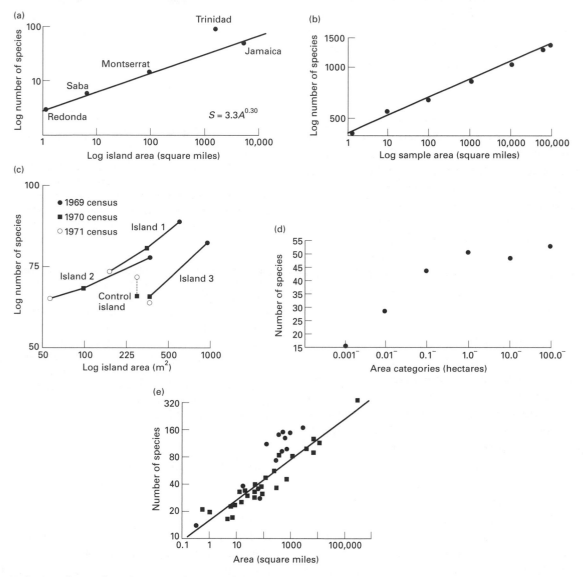

Figure 3.18 Some relationships between the size of 'islands' and numbers of species (after Gorman, 1979, figures 3.1, 3.2, 8.1 and 8.2). (a) The number of amphibian and reptile species living on West Indian islands of various sizes. Trinidad, joined to South America 10,000 years ago, lies well above the species–area curve on other islands. (b) The area–species curve for the number of species of flowering plants found in a sample of area of England. (c) The effect on the number of arthropod species of reducing the size of mangrove islands. Islands 1 and 2 were reduced after both the 1969 and the 1970 census. Island 3 was reduced only after the 1969 census. The control island was not reduced, the changes in species number being attributable to random fluctuation. (d) The number of species of bird living in British woods of various size categories. (e) The number of species of land birds living in the lowland rainforest on small islands in New Guinea, plotted as a function of island area. The squares represent islands which have not had a land connection to New Guinea and whose avifaunas are at equilibrium. The regression line is fitted through these points. The circles represent former land-bridge islands connected to New Guinea some 10,000 years ago. Note that the large ones have more species than one would expect for their size.

for some examples); islands support fewer species than do similar areas of mainland, and small islands have fewer species than do large ones. Thus it may well follow that if humans destroy the greater part of a vast belt of natural forest, leaving just a small reserve, initially it will be 'supersaturated' with species, containing more than is appropriate to its area when at equilibrium (Gorman, 1979). Since the population sizes of the species living in the forest will now be greatly reduced, the extinction rate will increase and the number of species will decline towards equilibrium. For this reason it may be a sound principle to make reserves as large as possible; a larger reserve will support more species at equilibrium by allowing the existence of larger populations with lower extinction rates. Several small reserves will plainly be better than no reserves at all, but they will tend to hold fewer species at equilibrium than will a single reserve of the same area (Figure 3.19). If it is necessary to have several small reserves, they should be placed as close to each other as possible so that each may act as a source area for the others. Connectivity of reserves is an important issue, and if it can be achieved their equilibrial number of species will be raised due to increased immigration rates.

There are situations when small reserves may have advantages over a single large reserve: they will be less prone to total decimation by some natural catastrophe such as fire; they may allow the preservation of a range of rare and scattered habitats; and they may allow the survival of a group of competitors, one of which would exclude the others from a single reserve.

Reduction in area leads to reduction in numbers, and this in turn can lead to genetic impoverishment through inbreeding (Frankel, 1984). The effect on reproductive performance appears to be particularly marked. Inbreeding degeneration is, however, not the only effect of small population size for, in the longer term, the depletion of genetic variance is more serious since it reduces the capacity for adaptive change. Space is therefore an important consideration, especially for those animals that require large expanses of territory. For example, the population density of the wolf is about one adult per 20 km^2, and it has been calculated that for a viable population to exist one might need 600 individuals ranging over an area of 12,000 km^2. The significance of this is apparent when one realizes that most nature reserves are small: 93% of the world's

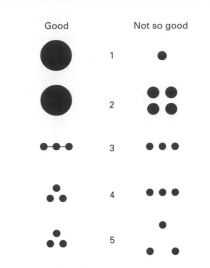

Figure 3.19 A set of general design rules for nature reserves based on theories of island biogeography. The designs on the left are preferable to those on the right because they should enjoy lower rates of species extinction. (1) A large reserve will hold larger populations with lower probabilities of going extinct than a small reserve. (2) A single reserve is preferable to a series of smaller reserves of equal total area, since these will support only small populations with relatively high probabilities of going extinct. (3–5) If reserves must be fragmented, then they should be connected by corridors of similar vegetation or be placed equidistant and as near to each other as possible. In this way immigration rates between the fragments will be increased, thereby maximizing the chances of extinct populations being replaced from elsewhere in the reserve complex (after Gorman, 1979, figure 8.4).

national parks and reserves have an area less than 5000 km^2, and 78% less than 1000 km^2.

Equally, range loss, the shrinking of the geographical area in which a given species is found, often marks the start of a downward spiral towards extinction. Such a contraction in range may result from habitat loss or from such processes as hunting and capture. Particular concern has been expressed in this context about the pressures on primates, notably in Southeast Asia. Of 44 species, 33 have lost at least half their natural range in the region. In two cases, those of the Javan leaf monkey and the Javan and gray gibbons, the loss of range is no less than 96%. Recent figures produced by the International Union for the Conservation of Nature and the United Nations Environment Program for wildlife habitat loss show the severity of the problem

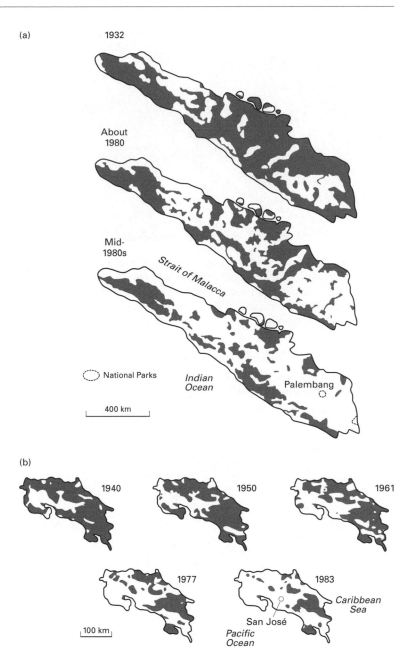

Figure 3.20 Progressive habitat fragmentation in the rainforest environments of (a) Sumatra and (b) Costa Rica (modified after Whitten et al., 1987 and Terborgh, 1992).

(see World Resources Institute, 1988, tables 6.4 and 6.5). In the Indomalayan countries 68% of the original wild-life habitat has been lost, and the comparable figure for tropical Africa is 65%. In these regions, only Brunei and Zambia have lost less than 30% of their original habitat, while at the other end of the spectrum Bangladesh, the most densely populated large country in the world, has suffered a loss of 94%. Figure 3.20 shows the fragmentation and reduction in area of rainforest within two areas: Sumatra and Costa Rica.

Although habitat change and destruction is clearly a major cause of extinction in the modern era, a re-markably important cause is the introduction of com-petitive species (Figure 3.21). When new species are deliberately or accidentally introduced to an area, they can cause the extermination of local fauna by preying on them or out-competing them for food and space. As we have seen elsewhere island species have proved to be especially vulnerable. The World Conservation Monitoring Centre, in its analysis of the known causes

Figure 3.21 The Fynbos of the Cape Region of South Africa is a biodiversity hot spot that has a great diversity of flora and fauna. It is a habitat that is under a number of threats, including the invasion of competitive species from Australia, which smother the native heathland.

of animal extinctions since AD 1600, believed that 39% were caused by species introduction, 36% by habitat destruction, 23% by hunting, and 2% for other reasons (World Resources Institute, 1996: 149).

There are certainly some particularly important environments in terms of their species diversity (Myers, 1990) and such biodiversity 'hot spots' (Figure 3.22) need to be made priorities for conservation. They include coral reefs, tropical forests (which support well over half the planet's species on only about 6% of its land area), and some of the Mediterranean climate ecosystems (including the extraordinarily diverse Fynbos shrublands of the Cape region of South Africa). Some environments are crucial because of their high levels of species diversity or endemic species; others are crucial because their loss would have consequences elsewhere. This applies, for example, to wetlands, which provide habitats for migratory birds and produce the nutrients for many fisheries. Myers et al. (2000) has argued that as many as 44% of all species of vascular plants and 35% of all species in four groups of vertebrate animals are confined to just 25 hot spots that comprise a mere 1.4% of Earth's land surface. These are major locations for conservation.

Finally, future climate change could act as a major cause of extinctions in the coming decades and could be a major threat to biodiversity (Thomas et al., 2004).

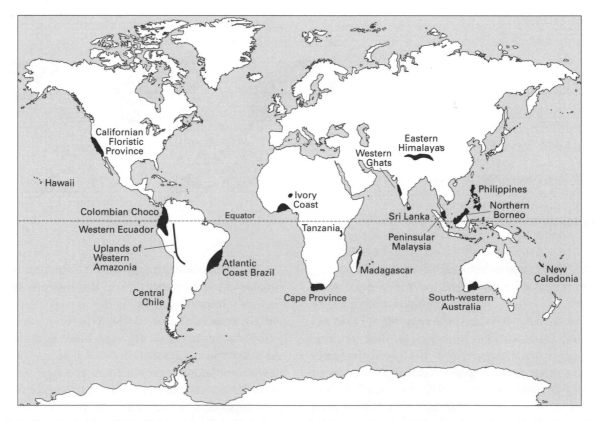

Figure 3.22 Forest and heathland hot-spot areas. Hot spots are habitats with many species found nowhere else and in greatest danger of extinction from human activity (after Wilson, 1992: 262–3)

As we will see in Chapters 8 to 13, many habitats will change markedly, and as a result many nature reserves will be in the wrong place for the species they are meant to protect. For instance, high altitude habitats at the tops of mountains may simply disappear.

Points for review

What do you understand by plant and animal domestication?

What are the ecological effects of invasive animals?

What are the main factors that lead to declines in animal populations?

Why did some animals become extinct in prehistoric times?

Guide to reading

Clutton-Brock, J., 1987, *A natural history of domesticated mammals*. London: British Museum and Cambridge: Cambridge University Press. A global survey of the domestication of a wide range of animals from asses to zebu cattle.

Drake, J. A., et al. (eds), 1989, *Biological invasions. A global perspective*. Chichester: Wiley. An edited paper of essays that considers both animal and plant invaders.

Martin, P. S. and Klein, R. G., 1984, *Pleistocene extinctions*. Tucson: University of Arizona Press. An enormous survey of whether or not late Pleistocene extinctions were caused by humans.

Wilson, E. O., 1992, *The diversity of life*. Harvard: Belknap Press. A beautifully written and highly readable discussion of biodiversity.

4 THE HUMAN IMPACT ON THE SOIL

Introduction

Soil is material composed of mineral particles, voids filled by gases or solutes, and organic remains that overlies the bedrock or parent sediment and supports the growth of roots. Humans live close to and depend on the soil. It is one of the thinnest and most vulnerable human resources and is one upon which, both deliberately and inadvertently, humans have had major impacts. Moreover, such impacts can occur with great rapidity in response to land-use change, new technologies, or waves of colonization (see Russell and Isbell, 1986, for a review in the context of Australia).

Natural soil is the product of a whole range of factors and the classic expression of this is that of Jenny (1941):

$$s = f(cl, o, r, p, t \ldots)$$

where s denotes any soil property, cl is the regional climate, o the biota, r the topography, p the parent material, t the time (or period of soil formation), and the dots represent additional, unspecified factors. In reality soils are the product of highly complex interactions of many interdependent variables, and the soils themselves are not merely a passive and dependent factor in the environment. Nonetheless, following Jenny's subdivision of the classic factors of soil formation, one can see more clearly the effects humans have had on soil, be they detrimental or beneficial. These can be summarized as follows (adapted from the work of Bidwell and Hole, 1965):

1 *Parent material*
   Beneficial: adding mineral fertilizers; accumulating shells and bones; accumulating ash locally; removing excess amounts of substances such as salts.
   Detrimental: removing through harvest more plants and animal nutrients than are replaced; adding materials in amounts toxic to plants or animals; altering soil constituents in a way to depress plant growth.
2 *Topography*
   Beneficial: checking erosion through surface roughening, land forming and structure building; raising land level by accumulation of material; land leveling.
   Detrimental: causing subsidence by drainage of wetlands and by mining; accelerating erosion; excavating.

3 *Climate*
Beneficial: adding water by irrigation; rainmaking by seeding clouds; removing water by drainage; diverting winds, etc.
Detrimental: subjecting soil to excessive insolation, to extended frost action, to wind, etc.

4 *Organisms*
Beneficial: introducing and controlling populations of plants and animals; adding organic matter including 'night-soil'; loosening soil by plowing to admit more oxygen; fallowing; removing pathogenic organisms, e.g., by controlled burning.
Detrimental: removing plants and animals; reducing organic content of soil through burning, plowing, overgrazing, harvesting, etc.; adding or fostering pathogenic organisms; adding radioactive substances.

5 *Time*
Beneficial: rejuvenating the soil by adding fresh parent material or through exposure of local parent material by soil erosion; reclaiming land from under water.
Detrimental: degrading the soil by accelerated removal of nutrients from soil and vegetation cover; burying soil under solid fill or water.

Space precludes, however, that we can follow all these aspects of anthropogenic soil modification or, to use the terminology of Yaalon and Yaron (1966), of *metapedogenesis*. We will therefore concentrate on certain highly important changes which humans have brought about, especially chemical changes (such as salinization and lateritization), various structural changes (such as compaction), some hydrological changes (including the effects of drainage and the factors leading to peat-bog development), and, perhaps most important of all, soil erosion.

Salinity: natural sources

Increasing salinity has a whole series of consequences that include a reduction in the availability of potable water (for humans and/or their stock), deterioration in soil structure, reduction in crop yields, and decay of engineering structures and cultural treasures. It is, therefore, a major environmental issue.

Many semi-arid and arid areas are, however, naturally salty. By definition they are areas of substantial water deficit where evapotranspiration exceeds precipitation. Thus, whereas in humid areas there is suf-

ficient water to percolate through the soil and to leach soluble materials from the soil and the rocks into the rivers and hence into the sea, in deserts this is not the case. Salts therefore tend to accumulate. This tendency is exacerbated by the fact that many desert areas are characterized by closed drainage basins, which act as terminal evaporative sumps for rivers.

The amount of natural salinity varies according to numerous factors, one of which is the source of salts. Some of the salts are brought into the deserts by rivers. A second source of salts is the atmosphere – a source that in the past has often been accorded insufficient importance. Rainfall, coastal fogs, and **dust storms** all transport significant quantities of soluble salts. Further soluble salts may be derived from the weathering and solution of bedrock. In the Middle East, for example, there are extensive salt domes and evaporite beds within the bedrock, which create locally high groundwater and surface-water salinity levels. In other areas, such as the Rift Valley of East Africa, volcanic rocks may provide a large source of sodium carbonate to groundwater, while elsewhere the rocks in which groundwater occurs may contain salt because they are themselves ancient desert sediments. Even in the absence of such localized sources of highly saline groundwater it needs to be remembered that over a period of time most rocks will provide soluble products to groundwater, and in a closed hydrological system such salts will eventually accumulate to significant levels.

A further source of salinity may be marine transgressions. At times of higher sea levels, it has sometimes been proposed (see, e.g., Godbole, 1972) that salts would have been laid down by the sea. Likewise in coastal areas, salts in groundwater aquifers may be contaminated by contact with seawater.

Human agency and increased salinity

Human activities cause enhanced or secondary salinization in drylands in a variety of ways (Goudie and Viles, 1997). In Table 4.1 these mechanisms are grouped into five main classes: irrigation salinity; dryland salinity; urban salinity; salinity brought about by interbasin water transfers; and coastal zone salinity.

Human-induced salinization affects about 77 million hectares on a global basis, of which 48 million hectares are in susceptible drylands (Middleton and Thomas, 1997) (Table 4.2).

Table 4.1 Enhanced salinization

Type	Cause
Irrigation salinity	Rise in groundwater
	Evaporation of water from fields
	Evaporation of water from canals and reservoirs
	Waterlogging produced by seepage losses
Dryland salinity	Vegetation clearance
Urban salinity	Water importation and irrigation
	Faulty drains and sewers
Interbasin water transfers	Mineralization of lake waters
	Deflation of salts from desiccating lakes
Coastal zone salinity	Overpumping
	Reduced freshwater recharge
	Sea-level rise
	Ground subsidence

Table 4.2 Global extent of human-induced salinization in the susceptible drylands (million ha). Source: GLASOD; Middleton and Thomas (1997, table 4.17)

Continent	Light	Moderate	Strong	Extreme	Total
Africa	3.3	1.9	0.6	–	5.8
Asia	10.7	8.1	16.2	0.4	35.4
South America	0.9	0.1	–	–	1.0
North America	0.3	1.2	0.3	–	1.8
Europe	0.8	1.7	0.5	–	3.0
Australasia	–	0.5	–	0.4	0.9
Global total	16.0	13.5	17.6	0.8	47.9

Irrigation salinity

In recent decades there has been a rapid and substantial spread of irrigation across the world (Table 4.3). The irrigated area in 1900 amounted to less than 50 million hectares. By 2000 the total area amounted to five times that figure. During the 1950s the irrigated area was increasing at over 4% annually, though this figure has now dropped to only about 1%. This spread of irrigation has brought about a great deal of salinization and waterlogging (Figures 4.1 and 4.2).

The amount of salinized irrigated land varies from area to area (Table 4.4), but in general ranges between 10 and 50% of the total. However, there is a considerable range in these values according to the source of

Table 4.3 Estimates of the increasing area of irrigated land on a global basis. Source: Goudie and Viles (1997, table 1.3)

Year	Irrigated area (10^6 ha)
1900	44–48
1930	80
1950	94
1955	120
1960	168
1980	211
1990 (estimate)	240

Figure 4.1 An irrigated field in southern Morocco. The application of large amounts of irrigation of water causes groundwater levels to rise and high air temperatures lead to rapid evaporation of the water. This leads to the eventual build up of salts in the soil.

Figure 4.2 The extension of irrigation in the Indus valley of Pakistan by means of large canals has caused widespread salination of the soils. Waterlogging is also prevalent. The white efflorescence of salt in the fields has been termed 'a satanic mockery of snow'.

Table 4.4 Salinization of irrigated cropland. Source: FAQ data as summarized in World Resources Institute (1988, table 19.3)

Country	Percentage of irrigated lands affected by salinization
Algeria	10–15
Australia	15–20
China	15
Colombia	20
Cyprus	25
Egypt	30–40
Greece	7
India	27
Iran	< 30
Iraq	50
Israel	13
Jordan	16
Pakistan	< 40
Peru	12
Portugal	10–15
Senegal	10–15
Sri Lanka	13
Spain	10–15
Sudan	< 20
Syria	30–35
USA	20–25

the data (compare Table 4.5) and this may in part reflect differences in the definition of the terms 'salinization' and 'waterlogging' (see Thomas and Middleton, 1993).

Irrigation causes secondary salinization in a variety of ways (Rhoades, 1990). First, the application of irrigation water to the soil leads to a rise in the water table so that it may become near enough to the ground surface for capillary rise and subsequent evaporative concentration to take place. When groundwater comes within 3 m of the surface in clay soils, and even less for silty and sandy soils, capillary forces bring moisture to the surface where evaporation occurs. There is plenty of evidence that irrigation does indeed lead to rapid and substantial rises in the position of the water table. Rates typically range between 0.2 and 3 m per year.

Second, many irrigation schemes, being in areas of high temperatures and rates of evaporation, suffer from the fact that the water applied over the soil surface is readily concentrated in terms of any dissolved salts it may contain. This is especially true for crops with a high water demand (e.g., rice) or in areas where, for one reason or another, farmers are profligate in their application of water.

Third, the construction of large dams and barrages creates extensive water bodies from which further evaporation can take place, once again leading to the concentration of dissolved solutes.

Fourth, notably in sandy soils with high permeability, water seeps both laterally and downwards from irrigation canals so that waterlogging may occur. Many irrigation canals are not lined, with the consequence that substantial water losses can result.

Table 4.5 Global estimate of secondary salinization in the world's irrigated lands. Source: Ghassemi et al. (1995, table 18). Reproduced by permission of CAB International and the University of New South Wales Press

Country	Cropped area (Mha)	Irrigated area (Mha)	Share of irrigated to cropped area (%)	Salt-affected land in irrigated area (%)	Share of salt-affected to irrigated land (%)
China	96.97	44.83	46.2	6.70	15.0
India	168.99	42.10	24.9	7.00	16.6
CIS	232.57	20.48	8.8	3.70	18.1
USA	189.91	18.10	9.5	4.16	23.0
Pakistan	20.76	16.08	77.5	4.22	26.2
Iran	14.83	5.74	38.7	1.72	30.0
Thailand	20.05	4.00	19.9	0.40	10.0
Egypt	2.69	2.69	100.00	0.88	33.0
Australia	47.11	1.83	3.9	0.16	8.7
Argentina	35.75	1.72	4.8	0.58	33.7
South Africa	13.17	1.13	8.6	0.10	8.9
Subtotal	852.80	158.70	18.8	29.62	20.0
World	1473.70	227.11	15.4	45.4	20.0

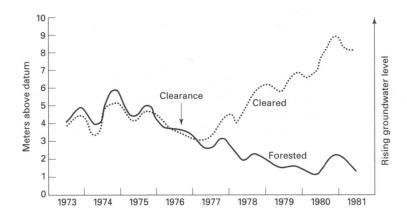

Figure 4.3 Comparison of hydrographs recorded from the boreholes in Wights (——) and Salmon (------) catchments in Western Australia. Both catchments were forested until late in 1976 when Wights was cleared (modified after Peck, 1983, figure 1).

Dryland salinity

A prime cause of dryland salinity extension is vegetation clearance (Peck and Halton, 2003). The removal of native forest, by reducing interception and evapotranspirational losses, allows a greater penetration of rainfall into deeper soil layers which causes groundwater levels to rise, thereby creating conditions for the seepage of sometimes saline water into low-lying areas. This is a particularly serious problem in the wheat belt of Western Australia and in some of the prairies of North America. In the case of the former area it is the clearance of *Eucalyptus* forest that has led to the increased rate of groundwater recharge and to the spreading salinity of streams and bottomlands. Salt 'scalds' have developed. The speed and extent of groundwater rise following such forest clearance is shown in Figure 4.3. Until late 1976 both the Wights and Salmon catchments were forested. Then the Wights was cleared. Before 1976 both catchments showed a similar pattern of groundwater fluctuation, but after that date there was a marked divergence of 5.7 m (Peck, 1983). The process can be reversed by afforestation (Bari and Schofield, 1992). Revegetation policies could also provide increased carbon sinks and so could provide synergistic value (Pittock and Wratt, 2001: 603).

Groundwater levels have increased some tens of meters since clearance of the natural vegetation began. They have increased by up to 30 m since the 1880s in southeastern Australia and by about 20 m in parts of southwest Australia. In some of the upland areas of New South Wales, groundwater levels have increased by up to 60 m over the past 70 to 80 years (http://www.agso.gov.au/information/structure/egg/mb/salinity2.html)

While in New South Wales, the area of land affected by dryland salinity is currently reported to be about 120,000 hectares. However, if current land-use trends and groundwater rise continue, this figure has the potential to increase to as much as 7.5 million hectares by 2050 (http://nccnsw.org.au/veg/context/salinity_fs.html). In Western Australia there is already an estimated 1.8 million hectares of farmland that is salt-affected. This area could double in the next 15 to 20 years, and then double again (http://www.agric.wa.gov.au/progserv/natural/trees/salinity/salwa.html) before reaching equilibrium. In all, some 6.1 million hectares have the potential to be affected by dryland salinity.

Dryland salinity is also a major problem on the Canadian prairies. In Alberta, approximately 0.65 million hectares are affected by secondary salinity, with an average crop yield reduction of 25%. In Saskatchewan 1.3 million hectares are affected, and in Manitoba 0.24 million hectares (http://www.agric.gov.ab.ca/sustain/soil/salinity/).

Rising water tables resulting from land-use changes are now being identified in other areas. For example, there has been a marked rise in the water table of the Continental Terminal in southwest Niger (Leduc et al., 2001). The rates have been between 0.01 and 0.45 million per year. The reason for this is the replacement of natural woodland savanna with millet fields and associated fallows. This has promoted increased surface runoff, which concentrates in temporary endoreic ponds and then infiltrates to the water table.

Urban salinity

Recent decades have seen a great growth in urban areas in drylands. This is, for example, the case with some of the Gulf States of the Middle East. The city of Abu Dhabi had a population of around 8000 in 1960. By 1984 it had reached 243,000, a 30-fold increase, and by 1995, 928,000, a 116-fold increase.

Urbanization can cause a rise in groundwater levels by affecting the amount of moisture lost by evapotranspiration. Many elements of urbanization, and in particular the spread of impermeable surfaces (roads, buildings, car parks, etc.), interrupt the soil evaporation process so that groundwater levels in *sabkha* (salt plain) areas along the coast of the Arabian Gulf rise at a rate of 40 cm per year until a new equilibrium condition is attained; the total rise from this cause may be 1–2 m (Shehata and Lotfi, 1993). This can require the construction of horizontal drains.

Urbanization can lead to other changes in groundwater conditions that can aggravate salinization. In some large desert cities the importation of water, its usage, wastage, and leakage can produce the ingredients to feed this phenomenon. This has, for example, been identified as a problem in Cairo and its immediate environs (Hawass, 1993). The very rapid expansion of Cairo's population has outstripped the development of an adequate municipal infrastructure. In particular, leakage losses from water pipes and sewers have led to a substantial rise in the groundwater level and have subjected many buildings to attack by sulfate- and chloride-rich water. There are other sites in Egypt where urbanization and associated changes in groundwater levels have been identified as a major cause of accelerated salt weathering of important monuments. Smith (1986: 510), for example, described the damage to tombs and monuments in Thebes and Luxor in the south of the country.

Interbasin water transfers

A further reason for increases in levels of salinity is the changing state of water bodies caused by interbasin water transfers. The most famous example of this is the shrinkage of the Aral Sea, the increase in its mineralization, and the deflation of saline materials from

Figure 4.4 Dust plumes caused by the deflation of salty sediments from the drying floor of the Aral Sea as revealed by a satellite image (153/Metero-Priroda, 18 May 1975) (modified after Mainguet, 1995, figure 4).

its desiccating floor and their subsequent deposition downwind. Some tens of millions of tonnes of salt are being translocated by dust storms (Figure 4.4) each year (Saiko and Zonn, 2000). The sea itself has had its mineral content increased more than threefold since 1960. Data on recent salinity enhancement in lakes from the USA, Asia, and Australia is provided by Williams (1999) (Table 4.6).

Another important illustration of the effects of interbasin water transfers is the desiccation of the Owens Lake in California. Diversion of water to feed the insatiable demands of Los Angeles has caused the lake to dry out, so that saline dust storms have become

Table 4.6 Enhanced salinity of lake basins. Source: Data in Williams (1999)

Lake	Period	Change (g L⁻¹)
Mono (California)	1941–1992	48 to 90
Pyramid Lake (Nevada)	1933–1980	3.75 to 5.5
Dead Sea	1910–1990s	200 to 300
Aral Sea	1960–1991	10 to 30
Qinghai (China)	1950s–1990s	6 to 12
Corangamite (Australia)	1960–1990s	35 to 50

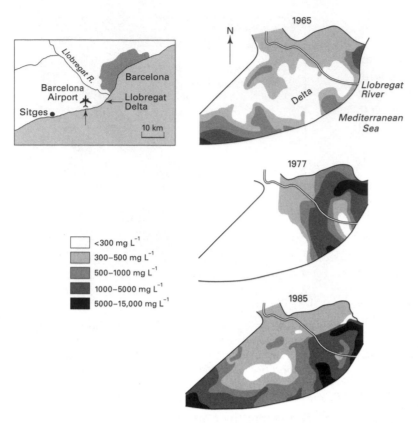

Figure 4.5 Changes in the chloride concentration of the Llobregat Delta aquifer, Barcelona, Spain as a result of seawater incursion caused by the overpumping of groundwater (modified from Custodio et al., 1986).

an increasingly serious issue (Gill, 1996). Sampling of airborne dust in the area has shown that 20–70% of the dust is soluble salts (Tyler et al., 1997).

Coastal zone salinity

Another prime cause of the spread of saline conditions is the incursion of seawater brought about by the overpumping of groundwater. Saltwater displaces less saline groundwater through a mechanism called the Ghyben–Herzberg principle. The problem presents itself on the coastal plain of Israel, in parts of California, on the island of Bahrain, and in some of the coastal aquifers of the United Arab Emirates. A comparable situation has also arisen in the Nile Delta (Kotb, 2000), though here the cause is not necessarily solely groundwater overpumping, but may also be due to changes in water levels and freshwater recharge caused by the construction of the Aswan High Dam. Figure 4.5 shows the way in which chloride concentrations have increased and spread in the Llobregat Delta area of eastern Spain because of the incursion of seawater.

The Ghyben–Herzberg relationship (Figure 4.6) is based on the fact that freshwater has a lower density than saltwater, such that a column of seawater can support a column of freshwater approximately 2.5% higher than itself (or a ratio of about 40:41). So where a body of freshwater has accumulated in a reservoir rock or sediment which is also open to penetration from the sea, it does not simply lie flat on top of the saltwater but forms a lens, with a thickness that is approximately 41 times the elevation of the piezometric surface above sea level. The corollary of this rule is that if the hydrostatic pressure of the freshwater falls as a result of overpumping in a well, then the underlying saltwater will rise by 40 units for every unit by which the freshwater table is lowered. A rise in sea level can cause a comparably dramatic alteration in the balance between fresh and saltwater bodies. This is especially serious on low islands (Figure 4.7) (Broadus, 1990).

An example of concern about salinization following on from sea-level rise is the Shanghai area of China (Chen and Zong, 1999). The area is suffering subsidence as a result of delta sedimentation (at 3 mm per year) and groundwater extraction (now less than 10 mm

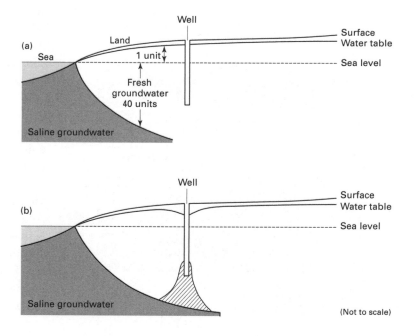

Figure 4.6 (a) The Ghyben–Herzberg relationship between fresh and saline groundwater. (b) The effect of excessive pumping from the well. The diagonal hatching represents the increasing incursion of saline water (after Goudie and Wilkinson, 1977, figure 63).

per year). It is very low-lying and in some counties the groundwater table is at 1.0–1.2 m. Thus a rise in sea level will raise the groundwater level, prolong waterlogging and cause a greater dominance of saline water in the deltaic area.

In any one location the causes of seawater incursion may be complex. A particularly good exemplification of this is provided by the coastal plains of northern Australia. Here Mulrennan and Woodroffe (1998) have assessed the potential role of such factors as sea-level change, rainfall variability, boat erosion of creeks, the activities of feral water buffalo, and sediment compaction.

Consequences of salinity

One consequence of the evaporative concentration of salts, and the pumping of saline waters back into rivers and irrigation canals from tubewells and other sources, is that river waters leading from irrigation areas show higher levels of dissolved salts. These, particularly when they contain nitrates, can make the water undesirable for human consumption.

A further problem is that, as irrigation water is concentrated by evapotranspiration, calcium and magnesium components tend to precipitate as carbonates, leaving sodium ions dominant in the soil solution. The

sodium ions tend to be absorbed by colloidal clay particles, deflocculating them and leaving the resultant structureless soil almost impermeable to water and unfavorable to root development.

The death of vegetation in areas of saline patches, due both to poor soil structure and toxicity, creates bare ground that becomes a focal point for erosion by wind and water. Likewise, deflation from the desiccating surface of the Aral Sea causes large amounts of salt to be blown away in dust storms and to be deposited downwind. Some tens of millions of tonnes are being translocated by such means and their plumes are evident on satellite images.

Probably the most serious impact of salination is on plant growth. This takes place partly through its effect on soil structure, but more significantly through its effects on osmotic pressures and through direct toxicity. When a water solution containing large quantities of dissolved salts comes into contact with a plant cell it causes shrinkage of the protoplasmic lining. The phenomenon is due to the osmotic movement of the water, which passes from the cell towards the more concentrated soil solution. The cell collapses and the plant succumbs. Crop yields fall.

The toxicity effect varies with different plants and different salts. Sodium carbonate, by creating highly alkaline soil conditions, may damage plants by a direct caustic effect, whereas high nitrate may promote

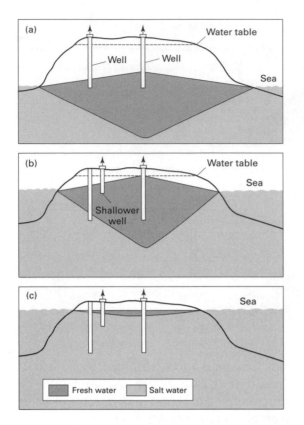

Figure 4.7 Impact of sea-level rise on an island water table. Note: the freshwater table extends below sea level 40 cm for every 1 cm by which it extends above sea level. (a) For islands with substantial elevation a 1-m rise in sea level simply shifts the entire water table up 1 m, and the only problem is that a few wells will have to be replaced with shallower wells (b). For very low islands, however, the water table cannot rise due to runoff, evaporation, and transpiration. A rise in sea level would thus narrow the water table by 40 cm for every 1 cm that the sea level rises (c), effectively eliminating groundwater supplies for the lowest islands (modified after Broadus, 1990).

undesirable vegetative growth in grapes or sugar beets at the expense of sugar content. Boron is injurious to many crop plants at solution concentrations of more than 1 or 2 ppm.

Anthropogenic increases in salinity are not new. Jacobsen and Adams (1958) have shown that they were a problem in Mesopotamian agriculture after about 2400 BC. Individual fields, which in 2400 BC were registered as salt-free, can be seen in the records of ancient temple surveyors to have developed conditions of sporadic salinity by 2100 BC. Further evidence is provided by crop choice, for the onset of salinization strongly favors the adoption of crops that are most salt-tolerant. Counts of grain impressions in excavated pottery from sites in southern Iraq dated at about 3500 BC suggest that at that time the proportions of wheat and barley were nearly equal. A little more than 1000 years later the less salt-tolerant wheat accounted for less than 20% of the crop, and by about 2100 BC it accounted for less than 2% of the crop. By 1700 BC the cultivation of wheat had been abandoned completely in the southern part of the alluvial plain. These changes in crop choice were accompanied by serious declines in yield, which can also probably be attributed to salinity. At 2400 BC the yield was 2537 L per hectare, by 2100 BC it was 1460, and by 1700 BC it was down to 897 L. It seems likely that this played an important part in the break-up of Sumerian civilization, although the evidence is not conclusive. Moreover, the Sumarians appear to have understood the problem and to have had coping strategies (Powell, 1985).

Reclamation of salt-affected lands

Because of the extent and seriousness of salinity, be the causes natural or anthropogenic, various reclamation techniques have been initiated. These can be divided into three main types: eradication, conversion, and control.

Eradication predominantly involves the removal of salt either by improved drainage or by the addition of quantities of freshwater to leach the salt out of the soil. Both solutions involve considerable expense and pose severe technological problems in areas of low relief and limited freshwater availability. Improved drainage can either be provided by open drains or by the use of tubewells (as at Mohenjo Daro, Pakistan) to reduce groundwater levels and associated salinity and waterlogging. A minor eradication measure, which may have some potential, is the biotic treatment of salinity through the harvesting of salt-accumulating plants such as *Suaeda fruticosa*.

Conversion involves the use of chemical methods to convert harmful salts into less harmful ones. For example, gypsum is frequently added to sodic soils to convert caustic alkali carbonates to soluble sodium sulfate and relatively harmless calcium carbonate:

$$Na_2CO_3 + CaSO_4 \leftrightarrow CaCO_3 + Na_2SO_4 \downarrow \text{leachable}$$

Some of the most effective ways of reducing the salinity hazard involve miscellaneous control measures, such as: less wasteful and lavish application of water through the use of sprinklers rather than traditional irrigation methods; the lining of the canals to reduce seepage; the realignment of canals through less permeable soil; and the use of more salt-tolerant plants. As salinity is a particularly serious threat at the time of germination and for seedlings, various strategies can be adopted during this critical phase of plant growth: plots can be irrigated lightly each day after seeding to prevent salt build-up; major leaching can be carried out just before planting; and areas to be seeded can be bedded in such a way that salts accumulate at the ridge tops, with the seed planted on the slope between the furrow bottom and the ridge top (Carter, 1975). Useful general reviews of methods for controlling soil salinity are given by Rhoades (1990) and Qadir et al. (2000).

Lateritization

In some parts of the tropics there are extensive sheets of a material called **laterite**, an iron and/or aluminum-rich duricrust (see Maignien, 1966, or Macfarlane, 1976). These iron-rich sheets result naturally, either because of a preferential removal of silica during the course of extensive weathering (leading to a *relative* accumulation of the sesquioxides of iron and aluminum), or because of an *absolute* accumulation of these compounds.

One of the properties of laterites is that they harden on exposure to air and through desiccation. Once hardened they are not favorable to plant growth. One particular way in which exposure may take place is by accelerated erosion, while forest removal may so cause a change in microclimate that desiccation of the laterite surface can take place. Indeed, one of the main problems with the removal of humid tropical rain forest is that soil hardening may occur. The phenomenon may occur in some, but by no means in all, so-called tropical soils (Richter and Babbar, 1991). Should hardening occur, it tends to limit the extent of successful soil utilization and severely retards the re-establishment of forest. Although Vine (1968: 90) and Sanchez and Buol (1975) have rightly warned against exaggerating this difficulty in agricultural land use, there are records from many parts of the tropics of accelerated induration brought about by forest removal (Goudie, 1973). In the Cameroons, for example, around 2 m of complete induration can take place in less than a century. In India, foresters have for a long time been worried by the role that plantations of teak (*Tectona grandis*) can play in lateritization. Teak is deciduous, demands light, likes to be well spaced (to avoid crown friction), dislikes competition from undergrowth, and is shallow-rooted. These characteristics mean that teak plantations tend to expose the soil surface to erosive and desiccative forces more than does the native vegetation cover.

One of the main exponents of the role that human agency has played in lateritization in the tropical world has been Gourou (1961: 21–2). Although he may be guilty of exaggerating the extent and significance of laterite, Gourou gives many examples from low latitudes of falling agricultural productivity resulting from the onset of lateritization. It is worth quoting him at length:

On the whole, laterite is hostile to agriculture owing to its sterility and compactness. All tropical countries have not reached the same degree of lateritic 'suicide' but when the evolution has advanced a considerable way, man is placed in very strange conditions . . . Laterite is a pedological leprosy. Man's activities aggravate the dangers of laterite and increase the rate of the process of lateritization. To begin with, erosion when started by negligent removal of the forest simply wears away the friable and relatively fertile soil which would otherwise cover the laterite and support forest or crops . . . The forest checks the formation of the laterite in various ways. The trees supply plenty of organic matter and maintain a good proportion of humus in the soil. The action of capillary attraction is checked by the loosening of the soil; and the bases are retained through the absorbent capacity of humus. The forest slows down evaporation from the soil . . . it reduces percolation and consequently leaching. Lastly, the forest may improve the composition of the soil by fixing atmospheric dust.

Accelerated podzolization and acidification

There is an increasing amount of evidence that the introduction of agriculture, deforestation, and pastoralism to parts of upland western Europe promoted some major changes in soil character: notably an increase in the development of acidic and **podzolized** conditions, associated with the development of peat bogs. Climatic changes may have played a role, as could progressive leaching of last glacial drifts during the passage of the Holocene. But the association in time and space of human activities with soil deterioration has become increasingly clear (Evans et al., 1975).

Replacing the natural forest vegetation with cultivation and pasture, human societies set in train various related processes, especially on base-poor materials. First, the destruction of deep-rooting trees curtailed the enrichment of the surface of the soil by bases brought up from the deeper layers. Second, the use of fire to clear forest may have released nutrients in the form of readily soluble salts, some of which were inevitably lost in drainage, especially in soils poor in colloids (Dimbleby, 1974). Third, the taking of crops and animal products depleted the soil reserves to an extent probably greater than that arising from any of the manuring practices of prehistoric settlements. Fourth, as the soil degraded, the vegetation which invaded – especially bracken and heather – itself tended to produce a more acidic humus type of soil than the original mixed deciduous forest, and so continued the process.

Various workers now attribute much of the podzolization in upland Britain to such processes, and Dimbleby has concluded that 'although a few soils have been podzols since the Atlantic period [in the mid-Holocene], the majority are secondary, having arisen as a result of man's assault on the landscape, particularly in the Bronze Age' (see also Bridges, 1978).

The development of podzols and their associated ironpans (Cunningham et al., 2001), by impeding downward percolating waters, may have accelerated the formation of peats, which tend to develop where there is waterlogging through impeded drainage. Many peat bogs in highland Britain appear to coincide broadly in age with the first major land-clearance episodes (Merryfield and Moore, 1971; Moore, 1973). Another fact that would have contributed to their development is that when a forest canopy is removed (as by deforestation) the transpiration demand of the vegetation is reduced, less rainfall is intercepted, so that the supply of groundwater is increased, aggravating any waterlogging.

However, the role of natural processes must not be totally forgotten, and Ball's assessment would seem judicious (1975: 26):

It seems to be on balance that the highland trends in soil formation due to climate, geology and relief have been clearly running in the direction of leaching, acidity, podzolization, **gleying** and peat formation. For the British highlands generally, man has only intervened to hasten or slow the rate of these trends, rather than being in a position to alter the whole trend from one pedogenetic trend to another.

Moreover, it would be plainly misleading to stress only the deleterious effects of human actions on European soils. Traditional agricultural systems have often employed laborious techniques to augment soil fertility and to reduce such properties as undesirable acidity. In Britain, for example, the addition of chalk to light sandy land goes back at least to Roman times and the marl pits from which the chalk was dug are a striking feature of the Norfolk landscape, where Prince (1962) has identified at least 27,000 hollows. Similarly, in The Netherlands, Germany, and Belgium there are soils which for centuries (certainly more than 1000 years) have been built up (often over 50 cm) and fertilized with a mixture of manure, sods, litter, or sand. Such soils are called Plaggen soils (Pope, 1970). Plaggen soils also occur in Ireland, where the addition of sea-sand to peat was carried out in pre-Christian times. Likewise, before European settlement in New Zealand, the Maoris used thousands of tonnes of gravel and sand, carried in flax baskets, to improve soil structure (Cumberland, 1961).

Another type of soil that owes much to human influence is the category called 'paddy soil'. Long-continued irrigation, leveling, and manuring of terraced land in China and elsewhere have changed the nature of the pre-existing soils in the area. Among the most important modifications that have been recognized (Gong, 1983; Zhang and Gong, 2003) are an increase in organic matter, and increase in base saturation, and the translocation and reduction of iron and manganese.

One serious type of soil acidification is that which produces acid sulfate soils. As Dent and Pons (1995: 263) wrote:

Acid sulphate soils are the nastiest soils in the world. They generate sulphuric acid that brings their pH as low as 2 and leaks into drainage and floodwaters. In this acid environment, aluminium and other toxic elements kill vegetation and aquatic life or, in sub-lethal doses, render many species stunted and sickly. Generations of people depending on these soils have been impoverished and, probably, poisoned by their drinking water.

The reason for the development of such extremely acid soils is that originally they accumulated as sediments under severely reducing conditions in environments such as tidal (e.g., mangrove) swamps, or brackish lakes. Large amounts of sulfitic mud accumulated. When such materials are drained reduction is replaced by oxidation, and sulfuric acid is produced. Infamous examples are known from the drained polders of The Netherlands and from the drained coastal swamps of southeast Asia.

Some soils are currently being acidified by air pollution and the deposition of acid precipitation (Grieve, 2001) (see also Chapter 7). Many soils have a resistance to acidification because of their buffering capacity, which enables them to neutralize acidity. However, this resistance very much depends on soil type and situation, and soils which have a low buffering capacity because of their low calcium content (as, e.g., on granite), and which are subjected to high levels of precipitation, may build up high levels of acidity.

The concept of **critical loads** has been developed. They are defined as exposures below which significant harmful effects on sensitive elements of the environment do not occur according to currently available knowledge. The critical load for sensitive forest soils on gneiss, granite or other slow-weathering rocks is often less than 3 kg of sulfur per hectare per year. In some of the more polluted parts of central Europe, the rates of sulfur deposition may be between 20 and 100 kg per hectare per year (Ågren and Elvingson, 1996).

The immediate impact of high levels of acid input to soils is to increase the exchange between hydrogen (H^+) ions and the nutrient cations, such as potassium (K^+), magnesium (Mg^{2+}), and calcium (Ca^{2+}). As a result of this exchange, such cations can be quickly leached from the soil, along with the sulfate from the acid input. This leaching leads to nutrient deficiency. Acidification also leads to a change in the rate at which dead organic matter is broken down by soil microbes. It can also render some ions, such as aluminum, more mobile and this has been implicated in the phenomenon of forest decline (see Chapter 2).

Soil carbon

One important consequence of land use and land cover changes is changes in the carbon content of soils. This has potential significance in terms of carbon release to the atmosphere and hence for the enhanced greenhouse effect (Lal et al., 1995). Soil organic carbon includes plant, animal and microbial residues.

Under prolonged cultivation there is strong evidence for loss of soil carbon. Agricultural practices that contribute to this are deforestation and biomass burning, drainage of wetlands, plowing and other forms of soil disturbance, and removal of crop residues. Soil carbon is depleted by oxidation or mineralization due to breakdown of aggregates, leading to exposure of carbon, leaching, and translocation of dissolved organic carbon or particulate organic carbon, and accelerated erosion by runoff or wind (Lal, 2002). Losses of soil organic carbon of as much as 50% in surface soils (20 cm) have been observed after cultivation for 30–50 years. Reductions average around 30% of the original amount in the top 100 cm (Post and Kwon, 2000).

Much depends, however, on the nature of land management. In recent decades, for example, the carbon stock in agricultural land in the USA may have increased because of the adoption of conservation tillage practices on cropland and a reduction in the use of bare fallow (Eve et al., 2002). The use of nitrogenous fertilizers may also explain some of the increase (Buyanovsky and Wagner, 1998).

Conversely, reafforestation and conversion of cropland to grassland can cause substantial gains in soil organic carbon. In general, after afforestation there may be a period when soil organic carbon declines, because of low rates of litter fall and continuing decomposition of residues from the preceding agricultural phase. As the forest cover develops, inputs of carbon exceed outputs. The rates of building of carbon vary, and tend to be greater in cool and humid regions and under hardwoods and softwoods rather than under *Eucalyptus* or *Pinus radiata* (Paul et al., 2002).

Soil structure alteration

One of the most important features of a soil, in terms of both its suitability for plant growth and its inherent erodibility, is its structure. There are many ways in which humans can alter this, especially by compacting it with agricultural machinery (Horn et al., 2000), by the use of recreation vehicles, by changing its chemical character through irrigation, and by trampling

Table 4.7 Change to soil properties resulting from the passage of 100 motorcycles in New Zealand. Source: Crozier et al. (1978, table 1)

Soil property	Total number of sites	Number of sites with significant change*	Mean percentage at significant sites	Number of sites with significant increase	Number of sites with significant decrease	Main direction of change	Mean percentage change in direction
Infiltration capacity	16	16 (100%)	84.3	3	13	Decrease	78.1
Bearing capacity	21	10 (48%)	22.8	2	8	Decrease	18.6
Soil moisture	20	14 (70%)	15.5	5	9	Decrease	16.7
Dry bulk density	19	11 (58%)	13.6	9	2	Decrease	13.3

*Change is significant when greater than: 10% for infiltration capacity; 10% for bearing capacity; 5% for soil moisture; 5% for bulk density.

(Grieve, 2001). Soil compaction, which involves the compression of a mass of soil into a smaller volume, tends to increase the resistance of soil to penetration by roots and emerging seedlings, and limits oxygen and carbon dioxide exchange between the root zone and the atmosphere. Moreover, it reduces the rate of water infiltration into the soil, which may change the soil moisture status and accelerate surface runoff and soil erosion (Chancellor, 1977). For example, the effects of the passage of vehicles on some soil structural properties are shown in Table 4.7. Excessive use of heavy agricultural machinery is perhaps the major cause of soil compaction, and most procedures in the cropping cycle, from tillage and seedbed preparation, through drilling, weeding and agrochemical application to harvesting, are now largely mechanized, particularly in the developed world. Most notable of all is the reduction that is caused in soil **infiltration capacity**, which may explain why vehicle movements can often lead to gully development. Whether one is dealing with primitive sledges (as in Swaziland), or with the latest recreational toys of leisured Californian adolescents, the effects may be comparable.

Grazing is another activity that can damage soil structure through trampling and compaction. Heavily grazed lands tend to have considerably lower infiltration capacities than those found in ungrazed lands. Trimble and Mendel (1995) have drawn particular attention to the capability that cows have to cause soil compaction. Given their large mass, their small hoof area, and the stress that may be imposed on the ground when they are scrambling up a slope, they are probably remarkably effective in compacting soils. The removal of vegetation cover and associated litter also changes infiltration capacity, since cover protects the soil from packing by raindrops and provides organic matter for binding soil particles together in open aggregates. Soil fauna that live on the organic matter assist this process by churning together the organic material and mineral particles. Dunne and Leopold have ranked the relative influence of different land-use types on infiltration (1978, table 6.2, after US Soil Conservation Service):

Highest infiltration	Woods, good
	Meadows
	Woods, fair
	Pasture, good
	Woods, poor
	Pasture, fair
	Small grains, good rotation
	Small grains, poor rotation
	Legumes after row crops
	Pasture, poor
	Row crops, good rotation (more than one quarter in hay or sod)
	Row crops, poor rotation (one quarter or less in hay or sod)
Lowest infiltration	Fallow

In general, experiments show that reafforestation improves soil structure, especially the pore volume of the soils (see, e.g., Challinor, 1968). Plowing is also known to produce a compacted layer at the base of the zone of plowing (Baver et al., 1972). This layer has been termed the 'plow sole'. The normal action of the plow is to leave behind a loose surface layer and a dense subsoil where the soil aggregates have been pressed together by the sole of the plow. The compacting action can be especially injurious when the depth of plowing is both constant and long term, and when heavy machinery is used on wet ground (Greenland and Lal, 1977).

On the other hand, for many centuries farmers have achieved improvements in soil structure by deliberate practice, particularly with a view to developing the all-important crumb structure. In pre-Roman times people in Britain and France added lime to heavy clay soils, while the agricultural improvers of the eighteenth and nineteenth centuries improved the structure of sandy heath soils by adding clay, and of clay soils by adding calcium carbonate (marl). They also compacted such soils and added binding organic matter by breeding sheep and feeding them with turnips and other fodder plants (Russell, 1961).

In the modern era attempts have been made to reduce soil crusting by applying municipal and animal wastes to farm land, by adding chemicals such as phosphoric acid, and by adopting a cultivation system of the no-tillage type. The last practice is based on the idea that the use of herbicides has eliminated much of the need for tillage and cultivation in row crops; seeds are planted directly into the soil without plowing, and weeds are controlled by the herbicides. With this method less bare soil is exposed and heavy farm machinery is less likely to create soil compaction problems (see Carlson, 1978, for some of these methods).

Soil structures may also be modified to increase water runoff, particularly in arid zones where the runoff obtained can augment the meager water supply for crops, livestock, industrial and urban reservoirs, and groundwater recharge projects. In the Negev farming was practiced in this way, especially in the Nabatean and the Romano-Byzantine periods (about 300 BC to AD 630), and attempts were made to induce runoff by clearing the surface gravel of the soil and heaping it into thousands of mounds. This exposed the finer silty soil beneath, facilitating soil crust formation by raindrop impact, decreasing infiltration capacity and reducing surface roughness, so that runoff increased (Evenari et al., 1971). Today a greater range of techniques are available for the same end: soils can be smoothed and compacted by heavy machinery, soil crusting can be promoted by dispersion of soil colloids with sodium salts, the permeability of the soil surface can be reduced by applying water-repellent materials, and soil pores can be filled with binders (Hillel, 1971).

Soil drainage and its impact

Soil drainage 'has been a gradual process and the environmental changes to which it has led have, by reason of that gradualness, often passed unnoticed' (Green, 1978: 171). To be sure, the most spectacular feats of drainage – arterial drainage – involving the construction of veritable rivers and large dike systems, as seen in The Netherlands and the Fenlands of eastern England, have received attention. However, more widespread than arterial drainage, and sometimes independent of it, is the drainage of individual fields, either by surface ditching or by underdrainage with tile pipes and the like. Green (1978) has attempted to map the areas of drained agricultural land. In Finland, Denmark, Great Britain, The Netherlands, and Hungary, the majority of agricultural land is drained (Figure 4.8).

In Britain underdrainage was promoted by government grants and reached a peak of about one million hectares per year in the 1970s in England and Wales. More recently government subsidies have been cut and the uncertain economic future of farming has led to a reduction in farm expenditure. Both tendencies have led to a reduction in the rate of field drainage, which by 1900 was only 40,000 hectares per year (Robinson, 1990).

The drainage conditions of the soil have also frequently been altered by the development of ridge and furrow patterns created by plowing. Such patterns are a characteristic feature of many of the heavy soils of lowland England where large areas, especially of the Midland lowlands, are striped by long, narrow ridges of soil, lying more or less parallel to each other and usually arranged in blocks of approximately rectangular shape. They were formed by plowing with heavy plows pulled by teams of oxen; the precise mechanism has been described with clarity by Coones and Patten (1986: 154).

Soil drainage has been one of the most successful ways in which communities have striven to increase agricultural productivity; it was practiced by Etruscans, Greeks and Romans (Smith, 1976). Large areas of marshland and floodplain have been drained to human advantage. By leading water away, the water table is lowered and stabilized, providing greater depth for the root zone. Moreover, well-drained soils warm up

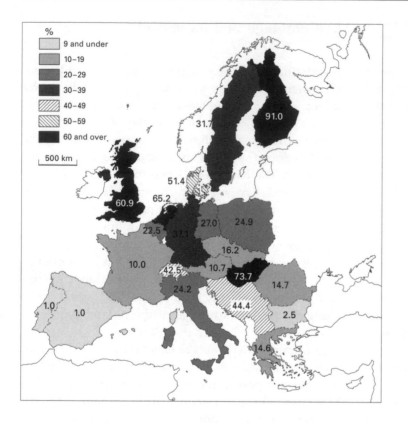

Figure 4.8 Percentage of drained agricultural land in Europe. There are no data for the blank areas (after Green, 1978, figure 1).

earlier in the spring and thus permit earlier planting and germination of crops. Farming is easier if the soil is not too wet, since the damage to crops by winter freezing may be minimized, undesirable salts carried away, and the general physical condition of the soil improved. In addition, drained land may have certain inherent virtues: tending to be flat it is less prone to erosion and more amenable to mechanical cultivation. It will also be less prone to drought risk than certain other types of land (Karnes, 1971). Paradoxically, by reducing the area of saturated ground, drainage can alleviate flood risk in some situations by limiting the extent of a drainage basin that generates saturation excess overland flow.

Conversely, soil drainage can have quite undesirable or unplanned effects. For example, some drainage systems, by raising drainage densities, can increase flood risk by reducing the distance over which unconcentrated overland flow (which is relatively slow) has to travel before reaching a channel (where flow is relatively fast). In central Wales, for instance, the establishment of drainage ditches in peaty areas to enable afforestation has tended to increase flood peaks in the rivers Wye and Severn (Howe et al., 1966).

Drainage can also cause long-term damage to soil quality. A fall in water level in organic soils can lead to the oxidation and eventual disappearance of peaty materials, which in the early stages of post-drainage use may be highly productive. This has occurred in the English Fenland, and in the Everglades of Florida, where drainage of peat soils has led to a subsidence in the soil of 32 mm per year (Stephens, 1956).

The soil climate of neighboring areas may also be modified when the water tables of drained land are lowered. This has been known to create problems for forestry in areas of marginal water availability.

Soil moisture content can also determine the degree to which soils are subjected to expansion and contraction effects, which in turn may affect engineering structures in areas with expansive soils (Holtz, 1983). Soils containing sodium montmorillonite type clays, when drained or planted with large trees, may dry out and cause foundation problems (Driscoll, 1983).

Some of the most contentious effects of drainage are those associated with the reduction of wetland wildlife habitats. In Britain this has become an important political issue, especially in the context of the Somerset Levels and the Halvergate Marshes.

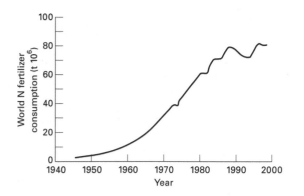

Figure 4.9 Global consumption of nitrogen fertilizer over the past five decades.

Soil fertilization

The chemistry of soils has been changed deliberately by the introduction of chemical fertilizers. The employment of chemical fertilizers on a large scale is little more than 150 years old. In the early nineteenth century nitrates were first imported from Chile, and sulfate of ammonia was produced only after the 1820s as a by-product of coal gas manufacture. In 1843 the first fertilizer factory was established at Deptford Creek (London), but for a long time superphosphates were the only manufactured fertilizers in use. In the twentieth century synthetic fertilizers, particularly nitrates, were developed, notably by Scandinavian countries that used their vast resources of waterpower. Potassic fertilizers came into use much later than the phosphatic and nitrogenous; nineteenth-century farmers hardly knew them (Russell, 1961). The rise in nitrogen fertilizer consumption for the world as a whole between 1950 and 1990 is shown in Figure 4.9. Global fertilizer consumption in 2000 was around 141 million tonnes of nutrient, of which around 61% was nitrogen, 23% phosphate, and 16% potash. The use of synthetic fertilizers has greatly increased agricultural productivity in many parts of the world, and remarkable increases in yields have been achieved. It is also true that in some circumstances proper fertilizer use can help minimize erosion by ensuring an ample supply of roots and plant residues, particularly on infertile or partially degraded soils (Bockman et al., 1990).

On the other hand, the increasing use of synthetic fertilizers can create environmental problems such as water pollution, while their substitution for more traditional fertilizers may accelerate soil-structure deterioration and soil erosion. One effect has been the increase in the water repellency of some surface soil materials. This in turn can reduce soil infiltration rates and a consequential increase in erosion by overland flow (Conacher and Conacher, 1995: 39). Fertilizers can also promote soil acidity and may lead to deficiencies or toxic excesses of major nutrients and trace elements. They may also contain impurities, such as fluoride, lead, cadmium, zinc, and uranium. Some of these heavy metals can inhibit water uptake and plant growth. They may also become concentrated in food crops, which can have important implications for human health.

Fires and soil quality

The importance and antiquity of fire as an agency through which the environment is transformed requires that some attention be given to the effects of fire on soil characteristics.

Fire has often been used intentionally to change soil properties, and both the release of nutrients by fire and the value of ash have long been recognized, notably by those involved in shifting agriculture based on slash-and-burn techniques. Following cultivation, the loss of nutrients by leaching and erosion is very rapid (Nye and Greenland, 1964), and this is why after only a few years the shifting cultivators have to move on to new plots. Fire rapidly alters the amount, form, and distribution of plant nutrients in ecosystems, and, compared with normal biological decay of plant remains, burning rapidly releases some nutrients into a plant-available form. Indeed, the amounts of P, Mg, K, and Ca released by burning forest and scrub vegetation are high in relation to both the total and available quantities of these elements in soils (Raison, 1979). In particular phosphate loss can be very detrimental to soil fertility (Thomas et al., 2000). In forests, burning often causes the pH of the soil to rise by three units or more, creating alkaline conditions where formerly there was acidity. Burning also leads to some direct nutrient loss by volatilization and convective transfer of ash, or by loss of ash to water erosion or wind deflation. The removal of the forest causes soil temperatures to increase because of the absence of shade, so that humus is often lost at a faster rate than it is formed (Grigg, 1970).

Table 4.8 The beneficial properties of humus. Source: Swift and Sanchez (1984, table 2)

Property		Explanation	Effect
Chemical	Mineralization	Decomposition of humus yields CO_2, NH_4^+, NO_3^-, PO_4^{3-}, and SO_4^{2-}	A source of nutrient elements for plant growth
	Cation exchange	Humus has negatively charged surfaces which bind cations such as Ca^+ and K^+	Improved cation exchange capacity (CEC) of soil. From 20 to 70% of the CEC of some soils (e.g., Mollisols) is attributable to humus
	Buffer action	Humus exhibits buffering in slightly acid, neutral and alkaline ranges	Helps to maintain a uniform pH in the soil
	Acts as a matrix for biochemical action in soil	Binds other organic molecules electrostatically or by covalent bonds	Affects bioactivity persistence and biodegradability of pesticides
	Chelation	Forms stable complexes with Cu^{2+}, Mn^{2+}, Zn^{2+}, and other polyvalent cations	May enhance the availability of micronutrients to higher plants
Physical	Water retention	Organic matter can hold up to 20 times its weight in water	Helps prevent drying and shrinking. May significantly improve moisture-retaining properties of sandy soils
	Combination with clay	Cements soil particles into structural units called aggregates	Improves aeration. Stabilizes structure. Increases permeability
	Color	The typical dark color of many soils is caused by organic matter	May facilitate warming

Soil erosion: general considerations

Loss of soil humus, whether as a result of fire, drainage, deforestation or plowing, is an especially serious manifestation of human alteration of soil. As Table 4.8 indicates, humus has many beneficial effects on both the chemical and the physical properties of soil. Its removal by human activity can be a potent contributory cause of soil erosion.

The scale of accelerated soil erosion that has been achieved by human activities has been well summarized by Myers (1988: 6):

Since the development of agriculture some 12,000 years ago, soil erosion is said by some to have ruined 4.3 million km² of agricultural lands, or an area equivalent to rather more than one-third of today's crop-lands . . . the amount of agricultural land now being lost through soil erosion, in conjunction with other forms of degradation, can already be put at a minimum of 200,000 km² per year.

That soil erosion is a major and serious aspect of the human role in environmental change is not to be doubted (Sauer, 1938). There is a long history of weighty books and papers on the subject (see, e.g., Marsh, 1864; Bennett, 1938; Jacks and Whyte, 1939; Morgan, 1995). Although many techniques have been developed to

reduce the intensity of the problem (see Hudson, 1987) it appears to remain intractable. As L. J. Carter (1977: 409) has reported of the USA:

Although nearly $15 billion has been spent on soil conservation since the mid-1930s, the erosion of croplands by wind and water . . . remains one of the biggest, most pervasive environmental problems the nation faces. The problem's surprising persistence apparently can be attributed at least in part to the fact that, in the calculation of many farmers, the hope of maximizing short-term crop yields and profits has taken precedence over the longer term advantages of conserving the soil. For even where the loss of topsoil has begun to reduce the land's natural fertility and productivity, the effect is often masked by the positive response to heavy application of fertilizer and pesticides, which keep crop yields relatively high.

Although construction, urbanization, war, mining, and other such activities are often significant in accelerating the problem, the prime causes of soil erosion are deforestation and agriculture (Figure 4.10). Pimentel (1976) estimated that in the USA soil erosion on agricultural land operates at an average rate of about 30 tonnes per hectare per year, which is approximately eight times quicker than topsoil is formed. He calculated that water runoff delivers around 4 billion tonnes

Figure 4.10 Soil erosion near Baringo in Kenya has exposed the roots of a tree, thereby indicating the speed at which soil can be lost.

of soil each year to the rivers of the 48 contiguous states, and that three-quarters of this comes from agricultural land. He estimated that another billion tonnes of soil is eroded by the wind, a process that created the Dust Bowl of the 1930s. More recently, Pimentel et al. (1995) have argued that in the USA about 90% of cropland is losing soil above the sustainable rate, that about 54% of pasture land is overgrazed and subject to high rates of erosion, and that erosion costs about $44 billion each year. They argue that on a global basis soil erosion costs the world about $400 billion each year. However, as Trimble and Crosson (2000) and Boardman (1998) point out, determination of general rates of soil erosion is fraught with uncertainties.

One serious consequence of accelerated erosion is the sedimentation that takes place in reservoirs, shortening their lives and reducing their capacity. Many small reservoirs, especially in semi-arid areas and in areas with erodible sediments in their catchments such as the loess lands of China, appear to have an expected life of only 30 years or even less (see, e.g., Rapp et al., 1972). Soil erosion also has serious implications for soil productivity. A reduction in soil thickness reduces available water capacity and the depth through which root development can occur. The water-holding properties of the soil may be lessened as a result of the preferential removal of organic material and fine sediment. Hardpans and duricrusts may become exposed at the surface, and provide a barrier to root penetration. Furthermore, splash erosion may cause soil compaction

and crusting, both of which may be unfavorable to germination and seedling establishment. Erosion also removes nutrients preferentially from the soil. Some damage may be caused by associated excessive sedimentation, while wind erosion may lead to the direct sandblasting of crops. Finally, extreme erosion may lead to wholesale removal of both seeds and fertilizer. Stocking (1984) provides a useful review of these problems.

Soil erosion associated with deforestation and agriculture

Forests protect the underlying soil from the direct effects of rainfall, generating what is generally an environment in which erosion rates tend to be low. The canopy plays an important role by shortening the fall of raindrops, decreasing their velocity and thus reducing kinetic energy. There are some examples of certain types (e.g., beech) in certain environments (e.g., maritime temperate) creating large raindrops, but in general most canopies reduce the erosion effects of rainfalls. Possibly more important than the canopy in reducing erosion rates in forest is the presence of humus in forest soils (Trimble, 1988). This both absorbs the impact of raindrops and has an extremely high permeability. Thus forest soils have high infiltration capacities. Another reason that forest soils have an ability to transmit large quantities of water through their fabrics is that they have many macropores produced by roots and their rich soil fauna. Forest soils are also well aggregated, making them resistant to both wetting and water drop impact. This superior degree of aggradation is a result of the presence of considerable organic material, which is an important cementing agent in the formation of large water-stable aggregates. Furthermore, earthworms also help to produce large aggregates. Finally, deep-rooted trees help to stabilize steep slopes by increasing the total shear strength of the soils.

It is therefore to be expected that with the removal of forest, for agriculture or for other reasons, rates of soil loss will rise (Figure 4.11) and mass movements will increase in magnitude and frequency. The rates of erosion that result will be particularly high if the ground is left bare; under crops the increase will be less marked. Furthermore, the method of plowing, the

Figure 4.11 The removal of vegetation in Swaziland creates spectacular gully systems, which in southern Africa are called *dongas*. The smelting of local iron ores in the early nineteenth century required the use of a great deal of firewood, which may have contributed to the formation of this example.

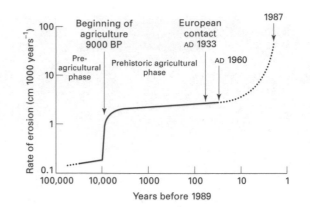

Figure 4.12 Rates of erosion in Papua New Guinea in the Holocene derived from rates of sedimentation in Kuk Swamp (after Hughes et al., 1991, figure 5, with modifications).

time of planting, the nature of the crop, and size of the fields, will all have an influence on the severity of erosion.

It is seldom that we have reliable records of rates of erosion over a sufficiently long time-span to show just how much human activities have accelerated these effects, and it is important to try and isolate the role of human impacts from climatic changes (Wilby et al., 1997). Recently, however, techniques have been developed which enable rates of erosion on slopes to be gauged over a lengthy time-span by means of dendrochronological techniques that date the time of root exposure for suitable species of tree. In Colorado, USA, Carrara and Carroll (1979) found that rates over the past 100 years have been about 1.8 mm per year, whereas in the previous 300 years rates were between 0.2 and 0.5 mm per year, indicating an acceleration of about sixfold. This great jump has been attributed to the introduction of large numbers of cattle to the area about a century ago.

Another way of obtaining long-term rates of soil erosion is to look at rates of sedimentation on continental shelves and on lake floors. The former method was employed by Milliman et al. (1987) to evaluate sediment removal down the Yellow River in China during the Holocene. They found that, because of accelerated erosion, rates of sediment accumulation on the shelf over the past 2300 years have been ten

times higher than those for the rest of the Holocene (i.e., since around 10,000 years BP).

Another good example of using long-term sedimentation rates to infer long-term rates of erosion is provided by Hughes et al.'s (1991) study of the Kuk Swamp in Papua New Guinea (Figure 4.12). They identify low rates of erosion until 9000 years BP, when, with the onset of the first phase of forest clearance, erosion rates increased from 0.15 cm per 1000 years to about 1.2 cm per 1000 years. Rates remained relatively stable until the past few decades when, following European contact, the extension of anthropogenic grasslands, subsistence gardens, and coffee plantations has produced a rate that is very markedly higher: 34 cm per 1000 years.

A further good long-term study of the response rates of erosion to land use changes is provided by a study undertaken on the North Island of New Zealand by Page and Trustrum (1997). During the past 2000 years of human settlement their catchment has undergone a change from indigenous forest to fern/scrub following Polynesian settlement (*c.* 560 years BP) and then a change to pasture following European settlement (AD 1878). Sedimentation rates under European pastoral land use are between five and six times the rates that occurred under fern/scrub and between eight and seventeen times the rate under indigenous forest. In a broadly comparable study, Sheffield et al. (1995) looked at rates of infilling of an estuary fed by a sheepland catchment in another part of New Zealand. In pre-Polynesian times rates of sedimentation were 0.1 mm per year, during Polynesian times the rate

Table 4.9 Runoff and erosion under various covers of vegetation in parts of Africa. Source: After Charreau (table 5.5, p. 153) in Greenland and Lal (1977)

Locality	Average annual rainfall (mm)	Slope (%)	Annual runoff* (%)			Erosion* (t ha^{-1} year^{-1})		
			A	B	C	A	B	C
Ouagadougou (Burkina Faso)	850	0.5	2.5	2–32	40–60	0.1	0.6–0.8	10–20
Sefa (Senegal)	1300	1.2	1.0	21.2	39.5	0.2	7.3	21.3
Bouake (Ivory Coast)	1200	4.0	0.3	0.1–26	15–30	0.1	1–26	18–30
Abidjan (Ivory Coast)	2100	7.0	0.1	0.5–20	38	0.03	0.1–90	108–170
Mpwapwa† (Tanzania)	c. 570	6.0	0.4	26.0	50.4	0	78	146

*A = forest or ungrazed thicket; B = crop; C = barren soil.
†From Rapp et al. (1972: 259, figure 5).

climbed to 0.3 mm per year, and since European land clearance in the 1880s the rate has shot up to 11 mm per year (see also Nichol et al., 2000). A good case study of the effect of European settlement on soil erosion rates in neighboring Australia is given by Olley and Wasson (2003).

Rates of sediment accumulation on floodplains also give an indication of historical rates of soil erosion. This is a topic that has been well reviewed by Knox (2002) and which is discussed further in Chapter 6.

In a more general sense there are plainly huge difficulties in estimating erosion rates in pre-human times, but in a recent analysis McLennan (1993) has estimated that the pre-human suspended sediment discharge from the continents was about 12.6×10^{15} grams per year, which is about 60% of the present figure.

Table 4.9, which is based on data from tropical Africa, shows the comparative rates of erosion for three main types of land use: trees, crops, and barren soil. It is very evident from these data that under crops, but more especially when ground is left bare or under fallow, soil erosion rates are greatly magnified. At the same time, and causally related, the percentage of rainfall that becomes runoff is increased.

In some cases the erosion produced by forest removal will be in the form of widespread surface stripping. In other cases the erosion will occur as more spectacular forms of mass movement, such as mudflows, landslides, and debris avalanches. Some detailed data on debris-avalanche production in North American catchments as a result of deforestation and forest road construction are presented in Table 4.10. They illustrate the substantial effects created by clear-cutting and by the construction of logging roads. It is indeed

probable that a large proportion of the erosion associated with forestry operations is caused by road construction, and care needs to be exercised to minimize these effects. The digging of drainage ditches in upland pastures and peat moors to permit tree-planting in central Wales has also been found to cause accelerated erosion (Clarke and McCulloch, 1979), while the elevated sediment loads can cause reservoir pollution (Burt et al., 1983).

In general, the greater the deforested proportion of a river basin the higher the sediment yield per unit area will be. In the USA the rate of sediment yield appears to double for every 20% loss in forest cover.

Soil erosion resulting from deforestation and agricultural practice is often thought to be especially serious in tropical areas or semi-arid areas (see Moore, 1979, for a good case study), but it is also a problem in the UK (Figure 4.13), in mainland Europe (Fuller, 2003), and in Russia (Sidorchuk and Golosov, 2003). Measurements by Morgan (1977) on sandy soils in the English East Midlands near Bedford indicate that rates of soil loss under bare soil on steep slopes can reach 17.69 tonnes per hectare per year, compared with 2.39 under grass and nothing under woodland (Table 4.11), and subsequent studies have demonstrated that water-induced soil erosion is a substantial problem, in spite of the relatively low erosivity of British rainfall. Walling and Quine (1991: 123) have identified the following farming practices as contributing to this developing problem.

1 Plowing up of steep slopes that were formerly under grass, in order to increase the area of arable cultivation.

Table 4.10 Debris-avalanche erosion in forest, clear-cut, and roaded areas. Source: after Swanston and Swanson, 1976, table 4

Site	Area type	Period of records (years)	Area (%)	(km²)	Number of slides	Debris-avalanche erosion (m³ km⁻² yr⁻¹)	Rate of debris-avalanche erosion relative to forested areas
Stequaleho Creek,	Forest	84	79	19.3	25	71.8	× 1.0
Olympic Peninsula	Clear-cut	6	18	4.4	0	0	0
	Road	6	3	0.7	83	11,825	× 165
	Total			24.4	108		
Alder Creek, western	Forest	25	70.5	12.3	7	45.3	× 1.0
Cascade Range,	Clear-cut	15	26.0	4.5	18	117.1	× 2.6
Oregon	Road	15	3.5	0.6	75	15,565	× 344
	Total			17.4	100		
Selected drainages,	Forest	32	88.9	246.1	29	11.2	× 1.0
Coast Mountains,	Clear-cut	32	9.5	26.4	18	24.5	× 2.2
southwest British	Road	32	1.5	4.2	11	282.5	× 25.2
Columbia	Total	–	–	276.7	58	–	–
H. J. Andrews	Forest	25	77.5	49.8	31	35.9	× 1.0
Experimental Forest,	Clear-cut	25	19.3	12.4	30	132.2	× 3.7
western Cascade	Road	25	3.2	2.0	69	1772	× 49
Range, Oregon	Total	–	–	64.2	130	–	–

Figure 4.13 Soil erosion on a bare field in Oxfordshire, central England.

Table 4.11 Annual rates of soil loss (tonnes per hectare) under different land-use types in eastern England. Source: From Morgan (1977)

Plot type	Slope location	Splash	Overland flow	Rill	Total
Bare soil	Top	0.33	6.67	0.10	7.10
	Middle	0.82	16.48	0.39	17.69
	Lower	0.62	14.34	0.06	15.02
Bare soil	Top	0.60	1.11	–	1.71
	Middle	0.43	7.78	–	8.21
	Lower	0.37	3.01	–	3.38
Grass	Top	0.09	0.09	–	0.18
	Middle	0.09	0.57	–	0.68
	Lower	0.12	0.05	–	0.17
Woodland	Top	–	–	–	0.00
	Middle	–	0.012	–	0.012
	Lower	–	0.008	–	0.008

2 Use of larger and heavier agricultural machinery, which has a tendency to increase soil compaction.
3 Removal of hedgerows and the associated increase in field size. Larger fields cause an increase in slope length with a concomitant increase in erosion risk.
4 Declining levels of organic matter resulting from intensive cultivation and reliance on chemical fertilizers, which in turn lead to reduced aggregate stability.

5 Availability of more powerful machinery, which permits cultivation in the direction of maximum slope rather than along the contour. Rills often develop along tractor and implement wheel tracks and along drill lines.

6 Use of powered harrows in seedbed preparation and the rolling of fields after drilling.

7 Widespread introduction of autumn-sown cereals to replace spring-sown cereals. Because of their longer growing season, winter cereals produce greater yields and are therefore more profitable. The change means that seedbeds are exposed with little vegetation cover throughout the period of winter rainfall.

Water is not the only active process creating accelerated erosion in eastern England, though it is important (Evans and Northcliff, 1978). Ever since the 1920s dust storms have been recorded in the Fenlands, the Brecklands, East Yorkshire (Radley and Sims, 1967), and Lincolnshire (see, e.g., Arber, 1946), and they seem to be occurring with increasing frequency. The storms result from changing agricultural practices, including the substitution of artificial fertilizers for farmyard manure, a reduction in the process of 'claying', whereby clay was added to the peat to stabilize it, the removal of hedgerows to facilitate the use of bigger farm machinery, and, perhaps most importantly, the increased cultivation of sugar beet. This crop requires a fine tilth and tends to leave the soil relatively bare in early summer compared with other crops (Pollard and Miller, 1968).

Nevertheless, possibly the most famous case of soil erosion by deflation was the Dust Bowl of the 1930s in the USA (Figure 4.14). In part this was caused by a series of hot, dry years which depleted the vegetation cover and made the soils dry enough to be susceptible to wind erosion. The effects of this drought were gravely exacerbated by years of overgrazing and unsatisfactory farming techniques. However, perhaps the prime cause of the event was the rapid expansion of wheat cultivation in the Great Plains. The number of cultivated hectares doubled during the First World War as tractors (for the first time) rolled out on to the plains by the thousands. In Kansas alone wheat hectarage increased from under 2 million hectares in 1910 to almost 5 million in 1919. After the war wheat cultivation continued apace, helped by the development of the combine harvester and government assistance. The farmer, busy sowing wheat and reaping gold, could foresee no end to his land of milk and honey, but the years of favorable climate were not to last, and over large areas the tough sod which exasperated the earlier homesteaders had given way to friable soils of high

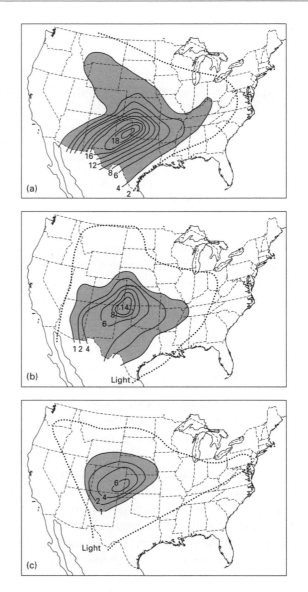

Figure 4.14 The concentration of dust storms (number of days per month) in the USA in 1939, illustrating the extreme localization over the High Plains of Texas, Colorado, Oklahoma, and Kansas: (a) March, (b) April, and (c) May (after Goudie, 1983).

erosion potential. Drought, acting on damaged soils, created the 'black blizzards' that have been so graphically described by Coffey (1978).

Dust storms are still a serious problem in various parts of the USA; the Dust Bowl was not solely a feature of the 1930s (Figure 4.15). Thus, for example, in the San Joaquin Valley area of California in 1977 a dust storm caused extensive damage and erosion over an area of about 2000 km². More than 25 million tonnes

Figure 4.15 In the High Plains near Lubbock in Texas the effect of soil erosion and drifting was very evident in 1977. Note the vast fields and absence of windbreaks.

of soil were stripped from grazing land within a 24-hour period. While the combination of drought and a very high wind provided the predisposing natural conditions for the stripping to occur, overgrazing and the general lack of windbreaks in the agricultural land played a more significant role. In addition, broad areas of land had recently been stripped of vegetation, leveled or plowed up prior to planting. Other quantitatively less important factors included stripping of vegetation for urban expansion, extensive denudation of land in the vicinity of oilfields, and local denudation of land by vehicular recreation (Wilshire et al., 1981). One interesting observation made in the months after the dust storm was that in subsequent rainstorms runoff occurred at an accelerated rate from those areas that had been stripped by the wind, exacerbating problems of flooding and initiating numerous gullies. Elsewhere in California dust yield has been considerably increased by mining operations in dry lakebeds (Wilshire, 1980) and by disturbance of playas (Gill, 1996).

A comparable acceleration of dust storm activity occurred in the former Soviet Union. After the 'Virgin Lands' program of agricultural expansion in the 1950s, dust storm frequencies in the southern Omsk region increased on average by a factor of 2.5 and locally by factors of 5 to 6. Data on trends elsewhere are evaluated by Goudie (1983: 520) and Goudie and Middleton (1992). A good review of wind erosion of agricultural land is provided by Warren (2002).

Soil erosion produced by fire

Many fires are started by humans, either deliberately or non-deliberately, and because fires remove vegetation

Table 4.12 Recent examples of fire-induced soil erosion

Source	Location
Wilson (1999)	Tasmania
Cerda (1998)	Mediterranean region
Moody and Martin (2001)	Colorado Front Range, USA
Dragovich and Morris (2002)	Eastern Australia
Pierson et al. (2002)	Idaho, USA

Table 4.13 Soil erosion associated with *Calluna* (heather) burning on the North Yorkshire Moors. Source: Data in Imeson (1971)

Condition of Calluna or ground surface	Mean rate of litter accumulation (+) or erosion (−) (mm year^{-1})	Number of observations
Calluna 30–40 cm high. Complete canopy	+3.81	60
Calluna 20–30 cm high. Complete canopy	+0.25	20
Calluna 15–20 cm high. 40–100% cover	−0.74	20
Calluna 5–15 cm high. 10–100% cover	−6.4	20
Bare ground. Surface of burnt *Calluna*	−9.5	19
Bare ground. Surface of peaty or mineral subsoil	−45.3	25

and expose the ground they tend to increase rates of soil erosion (see Table 4.12).

The burning of forests, for example, can, especially in the first years after the fire event, lead to high rates of soil loss. Burnt forests often have rates a whole order of magnitude higher than those of protected areas. Comparably large changes in soil erosion rates have been observed to result from the burning of heather in the Yorkshire moors in northern England (see Table 4.13), and the effects of burning may be felt for the six years or more that may be required to regenerate the heather (*Calluna*). In the Australian Alps fire in experimental catchments has been found to lead to a greatly increased flow in the streams, together with a marked surge in the delivery of suspended load. Combining the two effects of increased flow rate and sediment yield, it was found that, after fire, the total sediment load was increased 1000 times (Pereira, 1973). Likewise, watershed experiments in the chaparral scrub of Arizona, involving denudation by a destructive fire, indicated that whereas erosion losses before the fire were only 43 tonnes per square kilometer per year, after the fire they were between 50,000 and 150,000 tonnes per square kilometer per year. The causes of the marked erosion associated with chaparral burning are particularly interesting. There is normally a distinctive 'non-wettable' layer in the soils supporting chaparral. This layer, composed of soil particles coated by hydrophobic substances leached from the shrubs or their litter, is normally associated with the upper part of the soil profile (Mooney and Parsons, 1973), and builds up through time in the unburned chaparral. The high temperatures that accompany chaparral fires cause these hydrophobic substances to be distilled so that they condense on lower soil layers. This process results in a shallow layer of wettable soil overlying a non-wettable layer. Such a condition, especially on steep slopes, can result in severe surface erosion (DeBano, 2000; Shakesby et al., 2000; Letey, 2001).

In chaparral terrain it is possible to envisage a fire-induced sediment cycle (Graf, 1988: 243). It starts with a fire that destroys the scrub and the root net, and changes surface soil properties in the way already discussed. After the fire, a precipitation event of low magnitude (with a return interval of around one or two years) is sufficient to induce extensive sheet and rill erosion, which removes enough soil to retard vegetation recovery. Eventually, a larger precipitation event occurs (with a return interval of around five to ten years) and, because of limited vegetation cover, produces severe debris slides. Slowly the vegetation cover re-establishes itself, and erosion rates diminish. However, in due course enough vegetation grows to create a fire hazard, and the whole process starts again.

Soil erosion associated with construction and urbanization

There are now a number of studies which illustrate clearly that urbanization can create significant changes in erosion rates.

The highest rates of erosion are produced in the construction phase, when there is a large amount of exposed ground and much disturbance produced by vehicle movements and excavations. Wolman and Schick (1967) and Wolman (1967) have shown that the

equivalent of many decades of natural or even agricultural erosion may take place during a single year in areas cleared for construction. In Maryland they found that sediment yields during construction reached 55,000 tonnes per square kilometer per year, while in the same area rates under forest were around 80–200 tonnes per square kilometer per year and those under farm 400 tonnes per square kilometer per year. New road cuttings in Georgia were found to have sediment yields up to 20,000–50,000 tonnes per square kilometer per year. Likewise, in Devon, England, Walling and Gregory (1970) found that suspended sediment concentrations in streams draining construction areas were two to ten times (occasionally up to 100 times) higher than those in undisturbed areas. In Virginia, USA, Vice et al. (1969) noted equally high rates of erosion during construction and reported that they were ten times those from agricultural land, 200 times those from grassland, and 2000 times those from forest in the same area.

However, construction does not go on forever, and once the disturbance ceases, roads are surfaced, and gardens and lawns are cultivated. The rates of erosion fall dramatically and may be of the same order as those under natural or pre-agricultural conditions (Table 4.14). Moreover, even during the construction phase several techniques can be used to reduce sediment removal, including the excavation of settling ponds, the seeding and mulching of bare surfaces, and the erection of rock dams and straw bales (Reed, 1980).

Attempts at soil conservation

Because of the adverse effects of accelerated erosion a whole array of techniques has now been widely adopted to conserve soil resources (Hudson, 1987). Some of the techniques such as hillslope terracing may be of some antiquity, and traditional techniques have both a wide range of types and also many virtues (see Critchley et al., 1994; Reij et al., 1996). The following are some of the main ways in which soil cover may be conserved.

1 Revegetation:
   • deliberate planting;
   • suppression of fire, grazing, etc., to allow regeneration.
2 Measures to stop stream bank erosion.

Table 4.14 Rates of erosion associated with construction and urbanization

Location	Land use	Source	Rate (t km^{-2} yr^{-1})
Maryland, USA	Forest	Wolman (1967)	39
	Agriculture		116–309
	Construction		38,610
	Urban		19–39
Virginia, USA	Forest	Vice et al. (1969)	9
	Grassland		94
	Cultivation		1876
	Construction		18,764
Detroit, USA	General non-urban	Thompson (1970)	642
	Construction		17,000
	Urban		741
Maryland, USA	Rural	Fox (1976)	22
	Construction		37
	Urban		337
Maryland, USA	Forest and grassland	Yorke and Herb (1978)	7–45
	Cultivated land		150–960
	Construction		1600–22,400
	Urban		830
Wisconsin, USA	Agricultural	Daniel et al. (1979)	< 1
	Construction		19.2
Tama New Town, Japan	Construction	Kadomura (1983)	c. 40,000
Okinawa, Japan	Construction	Kadomura (1983)	25,000–125,000

3 Measures to stop gully enlargement:
  • planting of trailing plants, etc.;
  • weirs, dams, gabions, etc.
4 Crop management:
  • maintaining cover at critical times of year;
  • rotation;
  • cover crops.
5 Slope runoff control:
  • terracing;
  • deep tillage and application of humus;
  • transverse hillside ditches to interrupt runoff;
  • contour plowing;
  • preservation of vegetation strips (to limit field width).
6 Prevention of erosion from point sources such as roads and feedlots:
  • intelligent geomorphic location;
  • channeling of drainage water to nonsusceptible areas;
  • covering of banks, cuttings, etc., with vegetation.

7 Suppression of wind erosion:
  • soil moisture preservation;
  • increase in surface roughness through plowing up clods or by planting windbreaks.

An alternative way of classifying soil conservation is provided by Morgan (1995). He identifies three main types of measure: agronomic, soil management, and mechanical. The effect of these in relation to the main detachment and transport phases of erosion are shown in Table 4.15.

There are some parts of the world where terraces (a mechanical measure) are one of the most prominent components of the landscape. This applies to many wine-growing areas, to some arid zone regions (such as southwest USA, Yemen, and Peru), and to a wide selection of localities in the more humid tropics (Luzon, Java, Sumatra, Assam, Ceylon, Uganda, Cameroons, the Andes, etc.). However, much traditional terracing only had soil erosion control as a secondary motive.

Table 4.15 Effect† of various soil conservation practices on the detachment (D) and transport (T) phases of erosion. Source: Morgan (1995, table 7.1)

Practice	Rainsplash		Runoff		Wind	
	D	T	D	T	D	T
Agronomic measures:						
(a) covering soil surface	*	*	*	*	*	*
(b) increasing surface roughness	–	–	*	*	*	*
(c) increasing surface depression storage	+	+	*	*	–	–
(d) increasing infiltration	–	–	+	*	–	–
Soil management:						
(a) fertilizers, manures	+	+	+	*	+	*
(b) subsoiling, drainage	–	–	+	*	–	–
Mechanical measures:						
(a) contouring, ridging	–	+	+	*	+	*
(b) terraces	–	+	+	*	–	–
(c) shelterbelts	–	–	–	–	*	*
(d) waterways	–	–	–	–	*	–

†– no control; + moderate control; *strong control.

More often these were constructed to provide level planting surfaces, to provide deeper soil, and to manage the flow of water (Doolittle, 2000). In areas subject to wind erosion other strategies may be necessary. Since soil only blows when it is dry, anything which conserves soil moisture is beneficial. Another approach to wind erosion is to slow down the wind by physical barriers, either in the form of an increased roughness of the soil surface brought about by careful plowing, or by planted vegetative barriers, such as windbreaks and shelterbelts.

Some attempts at soil conservation have been particularly successful. For example, in Wisconsin, a study by Trimble and Lund (1982) showed that in the Coon Creek Basin erosion rates declined fourfold between the 1930s and the 1970s. One of the main reasons for this was the progressive adoption of contour-strip plowing (Figure 4.16).

Attempts at soil conservation have not always been without their drawbacks. For example, the establishment of ground cover in dry areas to limit erosion may so reduce soil moisture because of accelerated

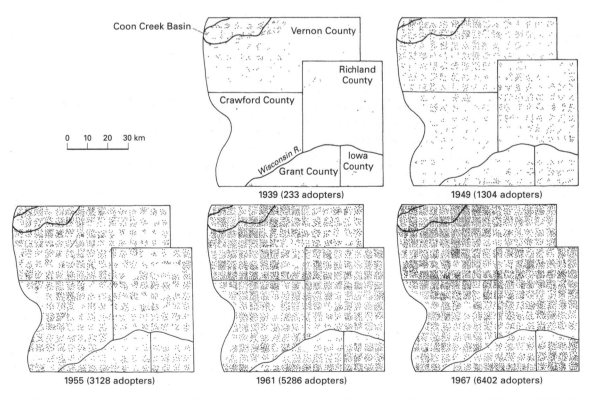

Figure 4.16 The spread of contour-strip soil conservation methods in Wisconsin, USA, between 1939 and 1967. One dot represents one adopter (after H. E. Johansen, in Trimble and Lund, 1982, figure 22).

evapotranspiration that the growth of the main crop is adversely affected. On a wider scale major afforestation schemes can cause substantial runoff depletion in river catchments. Likewise, the provision of mulching is sometimes detrimental: in cool climates, reduced soil temperature shortens the growing season, while in wet areas, higher soil moisture may induce gleying and anaerobic conditions (Morgan, 1979: 60). Some terrace schemes have also had their shortcomings. They have been known to hold back so much water on hillsides that the soils have become saturated and landsliding has been induced. Similarly strip cropping, because it involves the farming of small areas, is incompatible with highly mechanized agricultural systems, and insect infestation and weed control are additional problems which it has posed.

Soil conservation measures may not always be appropriate, for, paradoxically, soil erosion can be a useful phenomenon. Sanchez and Buol (1975), for instance, have pointed out that in recent volcanic areas soil erosion has enabled removal of the more weathered base-depleted material from the soil surface, exposing the more fertile, less weathered, base-rich material beneath. Likewise, in the Nochixtland area of southern Mexico, soil erosion has been utilized by local farmers to *produce* agricultural land. Severe gullies have cut into steep valley-side slopes, and since the Spanish Conquest an average depth of 5 m has been stripped from the entire surface area. The local Mixtec farmers, far from seeing this high rate of erosion as a hazard to be feared, have directed the flow of the eroded material to feed their fields with fertile soil and to extend their land. Over the past 1000 years (see Whyte, 1977), the Mixtec cultivators have managed to use gully erosion to double the width of the main valley floors with flights of terraces. Judicious use of the phenomenon of gully erosion has enabled them to convert poor hilltop fields into the rich alluvial farmland below.

Nonetheless it is undoubtedly true that manipulation of the soil is one of the most significant ways in which humans change the environment, and one in which they have had some of the most detrimental effects. Soil deterioration has led to many cases of what W. C. Lowdermilk once termed 'regional suicide', and the overall situation is at least as bleak as it was in the post-Dust Bowl years when Jacks commented: 'The organization of civilized societies is founded upon the measures taken to wrest control of the soil from wild Nature, and not until complete control has passed into human hands can a stable super-structure of what we call civilization be erected on the land' (Jacks and Whyte, 1939: 17).

Points for review

What are the causes of accelerated salinization?

Do humans contribute to lateritization and podzolization?

How does the removal of forest cover affect soil erosion?

What effects do fires have on soils?

How might you seek to reduce rates of soil erosion by (a) water, and (b) wind?

How serious a risk is soil erosion to agricultural sustainability?

Guide to reading

Conacher, A. J. and Sala, M. (eds), 1998, *Land degradation in Mediterranean environments of the world.* Chichester: Wiley. An edited discussion of land degradation processes in Mediterranean environments.

Fullen, M. A. and Catt, J. A., 2004, *Soil management: problems and solutions.* A review of some of the problems involved in the management of soils in the face of human impacts.

Holliday, V. T., 2004, *Soils in archaeological research.* New York: Oxford University Press. This book includes a treatment of long-term human impacts on soils.

Johnson, D. L. and Lewis, L. A., 1995, *Land degradation: creation and destruction.* Oxford: Blackwell. A broadly based study of intentional and unintentional causes of many aspects of land degradation.

McTainsh, G. and Boughton, W. C., 1993, *Land degradation processes in Australia.* Melbourne: Longman Cheshire. An Australian perspective on soil modification.

Morgan, R. P. C., 1995, *Soil erosion and conservation* (2nd edn). Harlow: Longman. A revised edition of a fundamental work.

Russell, J. S. and Isbell, F. R. (eds), 1986, *Australian soils: the human impact.* St. Lucia: University of Queensland Press. A detailed collection of studies on land degradation in Australia.

5 THE HUMAN IMPACT ON THE WATERS

Introduction

Because water is so important to human affairs, humans have sought to control water resources in a whole variety of ways. Also because water is such an important part of so many natural and human systems, its quantity and quality have undergone major changes as a consequence of human activities. We can quote Gleick (1993: 3):

... we must now acknowledge that many of our efforts to harness water have been inadequate or misdirected ... Rivers, lakes, and groundwater aquifers are increasingly contaminated with biological and chemical wastes. Vast numbers of people lack clean drinking water and rudimentary sanitation services. Millions of people die every year from water-related diseases such as malaria, typhoid, and cholera. Massive water developments have destroyed many of the world's most productive wetlands and other aquatic habitats.

In recent decades human demand for freshwater has increased rapidly. Global water use has more than tripled since 1950, and annual irretrievable water losses increased about sevenfold in the twentieth century (Table 5.1a).

Deliberate modification of rivers

Although there are many ways in which humans influence water quantity and quality in rivers and streams – for example, by direct channel manipulation, modification of basin characteristics, urbanization and pollution – the first of these is of particularly great importance (Mrowka, 1974). Indeed, there are a great variety of methods of direct channel manipulation and many of them have a long history. Perhaps the most widespread of these is the construction of dams and reservoirs (Figure 5.1). The first recorded dam was constructed in Egypt some 5000 years ago, but since that time the adoption of this technique has spread variously to improve agriculture, to prevent floods, to generate power, or to provide a reliable source of water.

There are some 75,000 dams in the USA. Most are small, but the bulk of the storage of water is associated

Table 5.1 Major changes in the hydrological environment
(a) Irretrievable water losses (km³ year⁻¹)

Users	1900	1940	1950	1960	1970	1980	1990	2000
Agriculture	409	679	859	1180	1400	1730	2050	2500
Industry	3.5	9.7	14.5	24.9	38.0	61.9	88.5	117
Municipal supply	4.0	9.0	14	20.3	29.2	41.1	52.4	64.5
Reservoirs	0.3	3.7	6.5	23.0	66.0	120	170	220
Total	417	701	894	1250	1540	1950	2360	2900

(b) Number of large dams (> 15 m high) constructed or under construction, 1950–86. Source: data provided by United Nations Environment Program (UNEP) and World Commission on Dams

Continent	1950	1982	1986	Under construction (31 December 1986)
Africa	133	665	763	58
Asia	1562	22,789	22,389	613
Australasia/Oceania	151	448	492	25
Europe	1323	3961	4114	222
North and Central America	2099	7303	6595	39
South America			884	69
Total	5268	35,166	36,327*	1036

*The figure by the end of the twentieth century was c. 45,000.

Figure 5.1 The Kariba Dam on the Zambezi River between Zambia and Zimbabwe. Such large dams can provide protection against floods and water shortages, and generate a great deal of electricity. However, they can have a whole suite of environmental consequences.

with a relatively limited number of structures. Those dams creating reservoirs of more than 1.2×10^9 m³ (1×10^6 acre-feet) account for only 3% of the total number of structures, but they account for 63% of the total storage. In all the dams are capable of storing a volume of water almost equaling one year's runoff and they store around 5000 m³ (4 acre-feet) of water per person. The decade of the 1960s saw the greatest spate of dam construction in American history (18,833 dams were built then). Since the 1980s, however, there have been only relatively minor increases in storage. The dam building era is over, but the environmental effects remain and the physical integrity of many rivers has been damaged (Graf, 2001).

The construction of large dams increased markedly, especially between 1945 and the early 1970s (Beaumont, 1978). Engineers have now built more than 45,000 large dams around the world and, as Table 5.1b shows, such large dams (i.e., more than 15 m high) are still being constructed at an appreciable rate, especially in Asia. In the late 1980s some 45 very large dams (more than 150 m high) were under construction. Indeed, one of the most striking features of dams and reservoirs is that they have become increasingly large (Beckinsale, 1969). Thus in the 1930s the Hoover or Boulder Dam in the USA (221 m high) was by far the tallest in the world and it impounded the largest reservoir, Lake Mead, which contained 38×10^9 m³ of water. By the 1980s it was exceeded in height by at least 18 others,

Table 5.2 Peak flow reduction downstream from selected British reservoirs. Source: after Petts and Lewin (1979: 82, table 1)

Reservoir	Catchment inundated (%)	Peak flow reduction (%)
Avon, Dartmoor	1.38	16
Fernworth, Dartmoor	2.80	28
Meldon, Dartmoor	1.30	9
Vyrnwy, mid-Wales	6.13	69
Sutton Bingham, Somerset	1.90	35
Blagdon, Mendip	6.84	51
Stocks, Forest of Bowland	3.70	70
Daer, Southern Uplands	4.33	56
Camps, Southern Uplands	3.13	41
Catcleugh, Cheviots	2.72	71
Ladybower, Peak District	1.60	42
Chew Magna, Mendips	8.33	73

Table 5.3 The ratios of post- to pre-dam discharges for flood magnitudes of selected frequency. Source: after Petts and Lewin (1979: 84, table 2)

Reservoir	Recurrence interval (years)			
	1.5	2.3	5.0	10.0
Avon, River Avon	0.90	0.89	0.93	1.02
Stocks, River Hodder	0.83	0.86	0.84	0.95
Sutton Bingham, River Yeo	0.52	0.61	0.69	0.79

and some of these impounded reservoirs with four times the volume of Lake Mead. The massive new Three Gorges Dam in China is a major cause of current environmental concern, as is the damming of the Narmada River in India.

Large dams are capable of causing almost total regulation of the streams they impound but, in general, the degree to which peak flows are reduced depends on the size of the dam and the impounded lake in relation to catchment characteristics. In Britain, as Table 5.2 shows, peak flow reduction downstream from selected reservoirs varies considerably, with some tendency for the greatest degree of reduction to occur in those catchments where the reservoirs cover the largest percentage of the area. When considering the magnitude of floods of different recurrence intervals before and after dam construction, it is clear that dams have much less effect on rare events of high magnitude (Petts and Lewin, 1979), and this is brought out in Table 5.3.

Nonetheless, most dams achieve their aim: to regulate river discharge. They are also highly successful in fulfilling the needs of surrounding communities: millions of people depend upon them for survival, welfare, and employment.

However, dams may have a whole series of environmental consequences that may or may not have been anticipated (Figure 5.2) (World Commission on Dams, 2000). Some of these are dealt with in greater detail elsewhere, such as subsidence (p. 167), earthquake triggering (p. 193), the transmission and expansion in the range of organisms, inhibition of fish migration (p. 55), the build-up of soil salinity (p. 96), changes in groundwater levels creating slope instability (p. 176) and water-logging (p. 97). Several of these processes may in turn affect the viability of the scheme for which the dam was created.

A particularly important consequence of impounding a reservoir behind a dam is the reduction in the sediment load of the river downstream. A clear demonstration of this effect has been given for the South Saskatchewan River in Canada (Table 5.4) by Rasid (1979). During the pre-dam period, typified by 1962, the total annual suspended loads at Saskatoon and Lemsford Ferry were remarkably similar. As soon as the reservoir began to fill late in 1963, however, some of the suspended sediment began to be trapped, and the transitional period was marked by a progressive reduction in the proportion of sediment which reached Saskatoon. In the four years after the dam was fully operational the mean annual sediment load at Saskatoon was only 9% that at Lemsford Ferry.

Even more dramatic are the data for the Colorado River in the USA (Figure 5.3). Prior to 1930 it carried around 125–150 million tonnes of suspended sediment per year to its delta at the head of the Gulf of California. Following a series of dams the Colorado now discharges neither sediment nor water to the sea (Schwarz et al., 1991). There have also been marked changes in the amount of sediment passing along the Missouri and Mississippi rivers over the period 1938 to 1982. Downstream sediment loads have been reduced by about half over that period (Figure 5.4a). Meade (1996) has attempted to compare the situation in the 1980s with that which existed before humans started to interfere will those rivers (c. AD 1700) (Figure 5.4b).

Sediment retention is also well illustrated by the Nile (Table 5.5), both before and after the construction

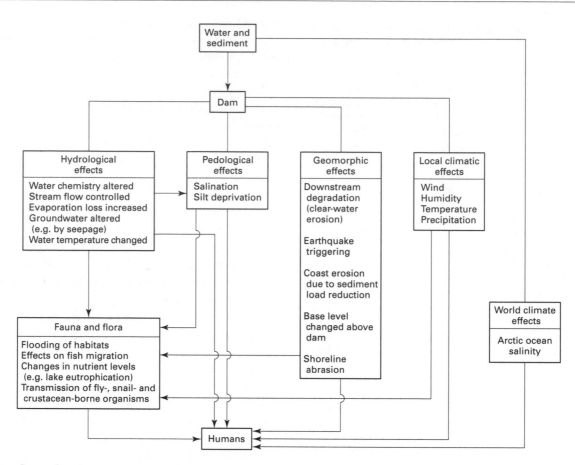

Figure 5.2 Generalized representation of the possible effects of dam construction on human life and various components of the environment.

Table 5.4 Total yearly suspended load (thousand imperial tons) of the South Saskatchewan River at Lemsford Ferry and Saskatoon, 1962–70. Source: Rasid (1979, table 1)

Period of record	Year	Lemsford Ferry	Saskatoon	Difference at Saskatoon (%)
Pre-dam	1962	1813	1873	+3
Transitional	1963	4892	4478	−8
	1964	7711	4146	−46
	1965	9732	2721	−72
	1966	5228	1675	−68
	Mean	6891	3255	−53
Post-dam	1967	12,619	446	−96
	1968	2661	101	−96
	1969	10,562	2146	−80
	1970	5643	118	−98
	Mean	7871	703	−91

of the great Aswan High Dam. Until its construction the late summer and autumn period of high flow was characterized by high silt concentrations, but since it has been finished the silt load is rendered lower throughout the year and the seasonal peak is removed. Petts (1985, table XVIII) indicates that the Nile now only transports 8% of its natural load below the Aswan High Dam, although this figure seems to be exceptionally low. Other rivers for which data are available carry between 8 and 50% of their natural suspended loads below dams.

Sediment removal in turn has various possible consequences, including a reduction in flood-deposited nutrients on fields, less nutrients for fish in the southeast Mediterranean Sea, accelerated erosion of the Nile Delta, and accelerated riverbed erosion since less sediment is available to cause bed aggradation. The

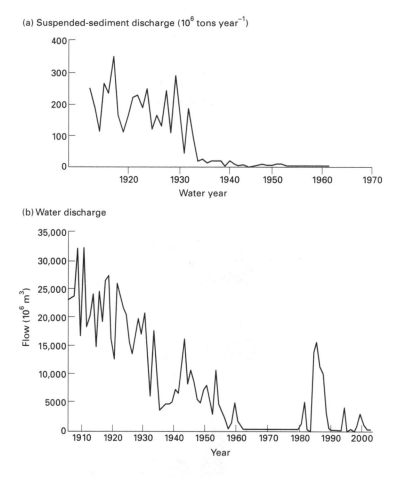

Figure 5.3 Historical (a) sediment and (b) water discharge trends for the Colorado River, USA (after the US Geological Survey, in Schwarz et al., 1991).

last process is often called 'clear-water erosion' (see Beckinsale, 1972), and in the case of the Hoover Dam it affected the river channel of the Colorado for 150 km downstream by causing incision. In turn, such channel incision may initiate headward erosion in tributaries and may cause the lowering of groundwater tables and the undermining of bridge piers and other structures downstream of the dam. On the other hand, in regions such as northern China, where modern dams trap silt, the incision of the river channel downstream may alleviate the strain on **levees** and lessen the expense of levee strengthening or heightening.

However, clear-water erosion does not always follow from silt retention in reservoirs. There are examples of rivers where, before impoundment, floods carried away the sediment brought into the main stream by steep tributaries. Reduction of the peak discharge after the completion of the dam leaves some rivers unable to scour away the sediment that accumulates as large fans of sand or gravel below each tributary mouth

(Dunne and Leopold, 1978). The bed of the main stream is raised and if water-intakes, or other structures, lie alongside the river they can be threatened again by flooding or channel shifting across the accumulating wedge of sediment. Rates of **aggradation** of a meter a year have been observed, and tens of kilometers of channel have been affected by sedimentation. One of the best-documented cases of aggradation concerns the Colorado River below Glen Canyon Dam in the USA. Since dam closure the extremes of river flow have been largely eliminated so that the 10 years' recurrence interval flow has been reduced to less than one-third. The main channel flow is no longer capable of removing sediment provided by flash-flooding tributaries, and deposits up to 2.6 m thick have accumulated within the upper Grand Canyon (Petts, 1985: 133).

Some landscapes in the world are dominated by dams, canals, and reservoirs. Probably the most striking example of this (Figure 5.5) is the 'tank' landscape of southeast India where myriads of little streams

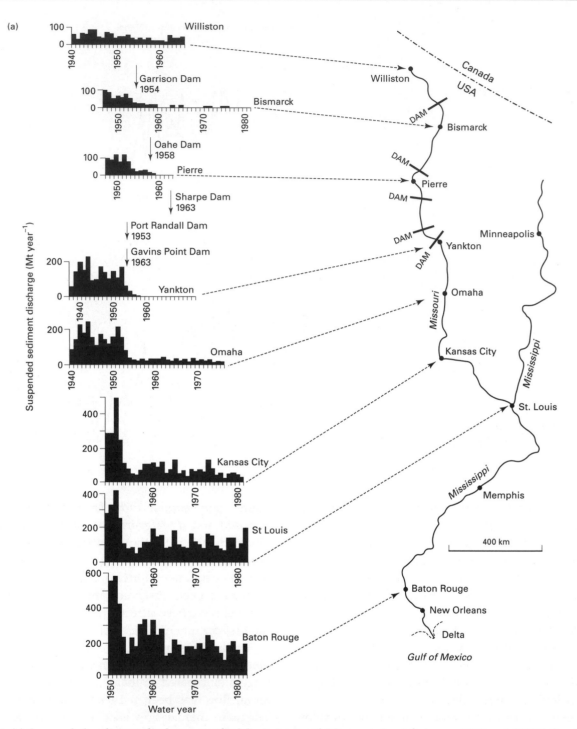

Figure 5.4 (a) Suspended sediment discharge on the Mississippi and Missouri rivers between 1939 and 1982 (after Meade and Parker, 1985, with modifications). (b) Long-term average discharges of suspended sediment in the lower Mississippi River *c.* 1700 and *c.* 1980.

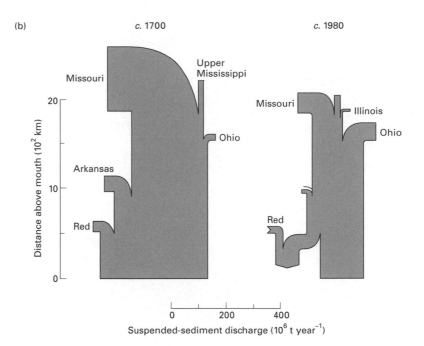

Figure 5.4 (*cont'd*)

Table 5.5 Silt concentrations (in parts per million) in the Nile at Gaafra before and after the construction of the Aswan High Dam. Source: Abul-Atta (1978: 199)

Month	Before (averages for the period 1958–63)	After	Ratio
January	64	44	1.5
February	50	47	1.1
March	45	45	1
April	42	50	0.8
May	43	51	0.8
June	85	49	1.7
July	674	48	14.0
August	2702	45	60
September	2422	41	59.1
October	925	43	21.5
November	124	48	2.58
December	71	47	1.63

and areas of overland flow have been dammed by small earth structures to give what Spate (Spate and Learmonth, 1967: 778) has likened to 'a surface of vast overlapping fish-scales'. In the northern part of the sub-continent, in Sind, the landscape changes wrought by hydrology are no less striking, with the mighty snow-fed Indus being controlled by large embankments (*bunds*) and interrupted by great barrages. Its waters are distributed over thousands of square kilometers

by a canal network (Figure 5.6) that has evolved over the past 4000 years (Figure 5.7). Another landscape where equally far-reaching changes have been wrought is The Netherlands. Coates (1976) has calculated that, before 1860, reclamation of that country from the sea, in the extension of drainage lines, involved the movement of 1000×10^6 m^3 of material. The area is dominated by human constructions: canals, rivers, drains, and lakes.

Another direct means of river manipulation is channelization. This involves the construction of embankments, dikes, levees, and floodwalls to confine floodwaters; and improving the ability of channels to transmit floods by enlarging their capacity through straightening, widening, deepening, or smoothing (Table 5.6).

Some of the great rivers of the world are now lined by extensive embankment systems such as those that run for more than 1000 km alongside the Nile, 700 km along the Hwang Ho, 1400 km by the Red River in Vietnam, and over 4500 km in the Mississippi Valley (Ward, 1978). Like dams, embankments and related structures often fulfil their purpose but they may also create some environmental problems and have some disadvantages. For example, they reduce natural storage for floodwaters, both by preventing water from spilling on to much of the floodplain, and by stopping bank storage in cases where impermeable floodwalls

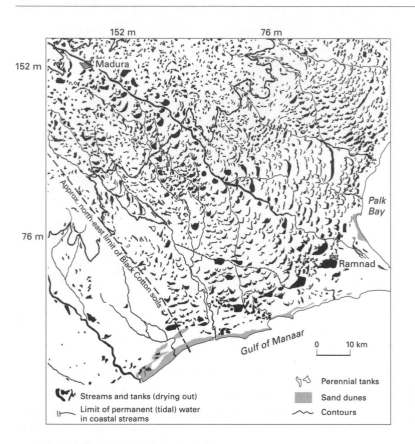

Figure 5.5 The Madurai–Ramanathapuram tank country in south India (after Spate and Learmonth, 1967, figure 25.12).

Figure 5.6 A large irrigation canal taking water across the Indus Plain from the Sukkur Barrage in Sind, Pakistan.

are used. Likewise the flow of water in tributaries may be constrained. In addition, embankments may occasionally exacerbate the flood problem they were designed to reduce by preventing floodwaters downstream of a breach from draining back into the channel once the peak has passed.

Channel improvement, designed to improve water flow, may also have unforeseen or undesirable effects. For example, the more rapid movement of water along improved channel sections can aggravate flood peaks further downstream and cause excessive erosion. The lowering of water tables in the 'improved' reach may cause overdrainage of adjacent agricultural land so that sluices must be constructed in the channel to maintain levels. On the other hand, lined channels may obstruct soil water movement (interflow) and shallow groundwater and so cause surface saturation. Brookes (1985) and Gregory (1985) provide useful reviews on the impact of channelization.

Channelization may also have miscellaneous effects on fauna through the increased velocities of water flow, reductions in the extent of shelter in the channel bed, and by reduced nutrient inputs due to the destruction of overhanging bank vegetation (see Keller, 1976). In the case of large swamps, such as those of the Sudd in Sudan or the Okavango in Botswana, the channelization of rivers could completely transform the whole character of the swamp environment. Figure 5.8 illus-

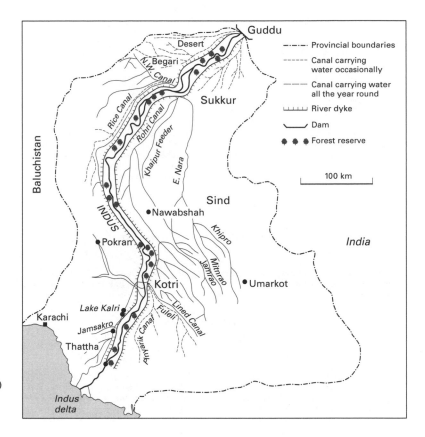

Figure 5.7 The irrigated areas in Sind (Pakistan) along the Indus Valley (after Manshard, 1974, figure 5.7).

Table 5.6 Selected terminologies for the methods of river channelization in the USA and UK. Source: Brookes (1985, table 1)

American term	British equivalent	Method involved
Widening Deepening	Resectioning Resectioning	Increase of channel capacity by manipulating width and/or depth variable
Straightening	Realigning	Increasing velocity of flow by steepening the gradient
Diking	Embanking	Raising of channel banks to confine floodwaters
Bank stabilization	Bank protection	Methods to control bank erosion, e.g., gabions and concrete structures
Clearing and snagging	Pioneer tree clearance Control of aquatic plants Dredging of sediments Urban clearing	Removal of obstructions from a watercourse, thereby decreasing the resistance and increasing the velocity of flow

trates some of the differences between natural and artificial channels.

Another type of channel modification is produced by the construction of bypass and diversion channels to carry excess floodwater or to enable irrigation to take place. Such channels may be as old as irrigation itself. They may contribute to the salinity problems encountered in many irrigated areas.

Deliberate modification of a river regime can also be achieved by long-distance interbasin water transfers (Shiklomanov, 1985), transfers necessitated by the unequal spatial distribution of water resources, and by the increasing rates of water consumption. The total volume of water in the various transfer systems in operation and under construction on a global scale is about 300 km^3 per year, with the largest countries in terms of volume of transfers being Canada, the Commonwealth of Independent States, the USA, and India.

In future decades it is likely that many even greater schemes will be constructed (see Figure 5.9); route lengths of some hundreds of kilometers will be common, and the water balances of many rivers and lakes

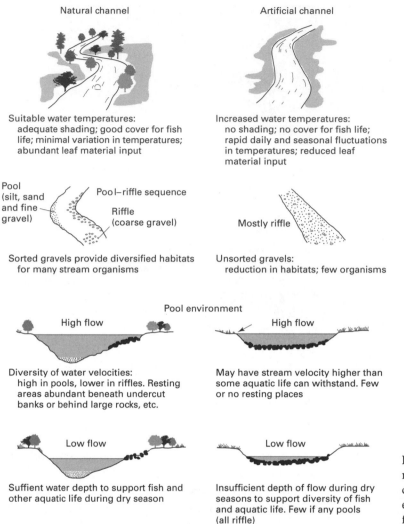

Natural channel

Suitable water temperatures: adequate shading; good cover for fish life; minimal variation in temperatures; abundant leaf material input

Pool (silt, sand and fine gravel)   Pool–riffle sequence
Riffle (coarse gravel)

Sorted gravels provide diversified habitats for many stream organisms

Pool environment
High flow

Diversity of water velocities: high in pools, lower in riffles. Resting areas abundant beneath undercut banks or behind large rocks, etc.

Low flow

Suffient water depth to support fish and other aquatic life during dry season

Artificial channel

Increased water temperatures: no shading; no cover for fish life; rapid daily and seasonal fluctuations in temperatures; reduced leaf material input

Mostly riffle

Unsorted gravels: reduction in habitats; few organisms

High flow

May have stream velocity higher than some aquatic life can withstand. Few or no resting places

Low flow

Insufficient depth of flow during dry seasons to support diversity of fish and aquatic life. Few if any pools (all riffle)

Figure 5.8 Comparison of the natural channel morphology and hydrology with that of a channelized stream, suggesting some possible ecological consequences (after Keller, 1976, figure 4).

will be transformed. This is already happening in the CIS (Figure 5.10), where the operation of various anthropogenic activities of this type have caused runoff in the most intensely cultivated central and southern areas to decrease by 30–50% compared with normal natural runoff. At the same time, inflows into the Caspian and Aral Seas have declined sharply. The level of the Aral has fallen and its area decreased.

When one turns to the coastal portions of rivers, to estuaries, the possible effects of another human impact, dredging, can be as complex as the effects of dams and reservoirs upstream (La Roe, 1977). Dredging and filling are certainly widespread and often desirable. Dredging may be performed to create and maintain canals, navigation channels, turning basins, harbors, and marinas; to lay pipelines; and to obtain a source of material for fill or construction. Filling is the deposition of dredged materials to create new land. There are miscellaneous ecological effects of such actions. In the first place, filling directly disrupts habitats. Second, the generation of large quantities of suspended silt tends physically to smother bottom-dwelling plants and animals; tends to smother fish by clogging their gills; reduces photosynthesis through the effects of turbidity; and tends to lead to eutrophication by an increased nutrient release. Likewise, the destruction of marshes, mangroves, and sea grasses by dredge and fill can result in the loss of these natural purifying systems. The removal of vegetation may also cause erosion. Moreover, as silt deposits stirred up by dredging accumulate elsewhere in the estuary they tend to create a 'false bottom'. Characterized by

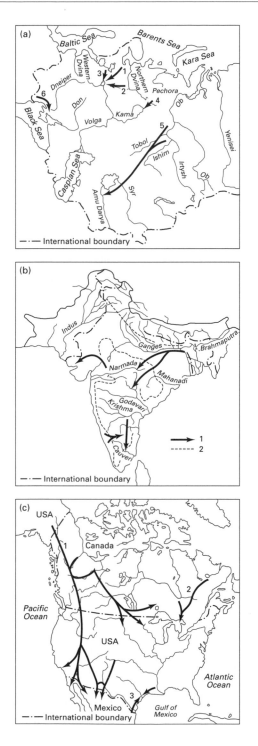

shifting, unstable sediments, the dredged bottom, fill deposits, or spoil areas are slowly – if at all – recolonized by fauna and flora. Furthermore, dredging tends to change the configuration of currents, the rate of freshwater drainage and may provide avenues for saltwater intrusion.

Urbanization and its effects on river flow

The process of urbanization has a considerable hydrological impact, in terms of controlling rates of erosion and the delivery of pollutants to rivers, and in terms of influencing the nature of runoff and other hydrological characteristics (Hollis, 1988). An attempt to generalize some of these impacts using a historical model of urbanization is summarized usefully by Savini and Kammerer (1961) and reproduced here in Table 5.7.

One of the most important effects is the way in which urbanization affects flood runoff. Research both in the USA and in Britain has shown that, because urbanization produces extended impermeable surfaces of bitumen, tarmac, tiles, and concrete, there is a tendency for flood runoff to increase in comparison with rural sites. City drainage densities may be greater than those in natural conditions (Graf, 1977) and the installation of sewers and storm drains accelerates runoff, as illustrated in Figure 5.11a. The greater the area that is sewered, the greater is the discharge for a particular recurrence level (Figure 5.11b). Peak discharges are higher and occur sooner after runoff starts in basins that have been affected by urbanization and the installation of sewers. Some runoff may be generated in

Figure 5.9 (*left*) Some major schemes proposed for large-scale interbasin water transfers: (a) projected water transfer systems in the Commonwealth of Independent States: 1, from the Onega River and in future from Onega Bay; 2, from the Sukhona and Northern Dvina rivers; 3, from the Svir River and Lake Onega; 4, from the Pechora River; 5, from the Ob River; 6, from the Danube delta. (b) Projected systems for water transfers in India: 1, scheme of the national water network; 2, scheme of the Grand Water Garland. (c) Some major projects for water transfers in North America: 1, North American Water and Power Alliance (NAWAPA); 2, Grand Canal; 3, Texas River basins (after Shiklomanov, 1985, figs 12.6, 12.9 and 12.11, in *Facets of hydrology II*, ed. J. C. Rodda, by permission of John Wiley and Sons Ltd).

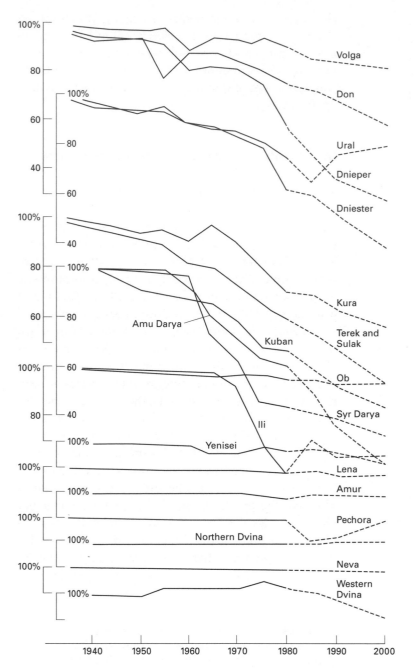

Figure 5.10 Changes in annual runoff in the Commonwealth of Independent States due to human activity during 1936–2000 (from Shiklomanov, 1985, figure 12.7, in *Facets of hydrology II*, ed. J. C. Rodda, by permission of John Wiley and Sons Ltd).

urban areas because low vegetation densities mean that evapotranspiration is limited.

However, in many cases the effect of urbanization is greater on small floods; as the size of the flood and its recurrence interval increase, so the effect of urbanization diminishes (Martens, 1968; Hollis, 1975). A probable explanation for this is that, during a severe and prolonged storm event, a nonurbanized catchment may become so saturated and its channel network so extended that it begins to behave hydrologically as if it were an impervious catchment with a dense surface-water drain network. Under these conditions, a rural catchment produces floods of a type and size similar to those of its urban counterpart. Moreover, a further mechanism probably operates in the same direction, for in an urban catchment it seems probable that some throttling of flow occurs in surface-water drains during intense storms, tending to attenuate the

Table 5.7 Stages of urban growth and their miscellaneous hydrological impacts. Source: modified after Savini and Kammerer (1961)

Stage	Activity	Impact
Transition from pre-urban stage	Removal of trees or vegetation	Decrease in transpiration and increase in storm flow
	Construction of scattered houses with limited sewerage and water facilities	
	Drilling of wells	Some lowering of water table
	Construction of septic tanks, etc.	Some increase in soil moisture and perhaps some contamination
Transition from early-urban to middle-urban stage	Bulldozing of land	Accelerated land erosion
	Mass construction of houses, etc.	Decreased infiltration
	Discontinued use and abandonment of some shallower wells	Rise in water table
	Diversion of nearby streams for public supply	Decrease in runoff between points of diversion of disposal
	Untreated or inadequately treated sewerage into streams and wells	Pollution of streams and wells
Transition from middle-urban to late-urban stage	Urbanization of area completed by addition of more buildings	Reduced infiltration and lowered water table, higher flood peaks and lower low flows
	Larger quantities of untreated waste into local streams	Increased pollution
	Abandonment of remaining shallow wells because of pollution	Rise in water table
	Increase in population requiring establishment of new water supply and distribution systems	Increase in local stream flow if supply is from outside basin
	Channels of streams restricted at least in part to artificial channels and tunnels	Higher stage for a given flow (therefore increased flood damage), changes in channel geometry and sediment load
	Construction of sanitary drainage system and treatment plant for sewerage	Removal of additional water from area
	Improvement of storm drainage system	
	Drilling of deeper, large-capacity industrial wells	Lowered water pressure, some subsidence, saltwater encroachment
	Increased use of water for air-conditioning	Overloading of sewers and other drainage facilities
	Drilling of recharge wells	
	Waste-water reclamation and utilization	Raising of water pressure surface
		Recharge of groundwater aquifers: more efficient use of water resources

very highest discharges. Thus, Hollis believes, while the size of small frequent floods is increased many times by urbanization, large, rare floods (the ones likely to cause extreme damage) are not significantly affected by the construction of suburban areas within a catchment area (Figure 5.11c). Hollis's findings may not be universally applicable. For example, K. V. Wilson (1967) working in Jackson, Mississippi, found that the 50-year flood for an urbanized catchment was three times higher than that of a rural one.

Nonetheless, a whole series of techniques has been developed in an attempt to reduce and delay urban storm runoff (Table 5.8). Of particular current interest are sustainable urban drainage systems (SUDS). These tackle urban surface runoff problems at source using

features such as soakaways, permeable pavements, grassed swales or vegetated filter strips, infiltration trenches, ponds (detention and retention basins), and wetlands to attenuate flood peak flows.

Vegetation modification and its effect on river flow

As we have already noted in Chapter 1, one of the first major indications that humans could inadvertently adversely affect the environment was the observation that deforestation could create torrents and floods. The deforestation that gives rise to such flows can be produced both by felling and by fire (Scott, 1997).

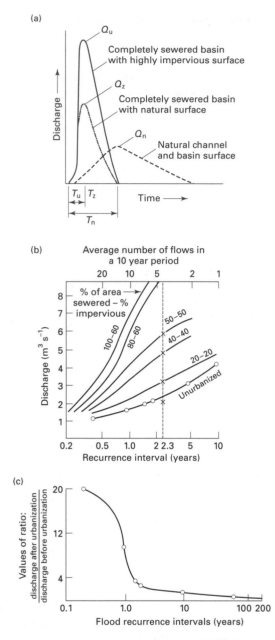

Figure 5.11 Some hydrological consequences of urbanization. (a) Effect of urban development on flood hydrographs. Peak discharges (Q) are higher and occur sooner after runoff starts (T) in basins that have been developed or sewered (after Fox, 1976, figure 3). (b) Flood frequency curves for a 1 mile^2 basin in various states of urbanization (after US Geological Survey, in Viessman et al., 1977, figure 11.33). (c) Effects of flood magnitude with paving 20% of a basin (after Hollis, 1975).

The first experimental study, in which a planned land-use change was executed to enable observation of the effects of stream flow, began at Wagon Wheel Gap, Colorado, in 1910. Here stream flow from two similar watersheds of about 80 hectares each were compared for eight years. One valley was then clear-felled and the records were continued. After the clear-felling the annual water yield was 17% above that predicted from the flows of the unchanged control valley. Peak flows are also increased. Studies on two small basins in the Australian Alps (Wallace's Creek, 41 km^2; Yarrango Billy River, 224 km^2), which were burned over, showed that rainstorms, which from previous records would have been expected to give rise to flows of 6080 m^3 per second, produced a peak of 370 m^3 per second (a five or sixfold increase). Likewise catchment experiments in Arizona have shown that when chaparral scrub is burned there is a tenfold increase in water yield.

Experiments with tropical catchments have shown typical maximum and mean annual stream-flow increases of 400–450 mm per year on clearance with increases in water yield of up to 6 mm per year for each percentage reduction in forest area above a 15% change in cover characteristics (Anderson and Spencer, 1991: 49). Additional data for the effects of land-use change on annual runoff levels are presented for tropical areas in Table 5.9.

As vegetation regenerates after a forest has been cut or burned, so stream flow tends to revert to normal, although the process may take some decades. This is illustrated in Figure 5.12 which shows the dramatic effects produced on the Coweeta catchments in North Carolina by two spasms of clear-felling, together with the gradual return to normality in between.

The substitution of one forest type for another may also affect stream flow. This can again be exemplified from the Coweeta catchments, where two experimental catchments were converted from a mature deciduous hardwood forest cover to a cover of pine (*Pinus strobus*). Fifteen years after the conversion, annual stream flow was found to be reduced by about 20% (Swank and Douglass, 1974). The reason for this notable change is that the interception and subsequent evaporation of rainfall is greater for pine than it is for hardwoods during the dormant season.

Fears have also been expressed that the replacement of tall natural forests by eucalyptus will produce a

Table 5.8 Measures for reducing and delaying urban storm runoff, including various types of sustainable urban drainage systems (SUDS). Source: after US Department of Agriculture, Soil Conservation Service (1972) in Viessman et al. (1977: 569)

Area	Reducing runoff	Delaying runoff
Large flat roof	Cistern storage Rooftop gardens Pool storage or fountain storage Sod roof cover	Ponding on roof by constricted drainpipes Increasing roof roughness: 　rippled roof 　graveled roof
Car parks	Porous pavement: 　gravel car parks 　porous or punctured asphalt Concrete vaults and cisterns beneath car parks in high value areas Vegetated ponding areas around car parks Gravel trenches	Grassy strips on car parks Grassed waterways draining car parks Ponding and detention measures for impervious areas: 　rippled pavement 　depressions 　basins
Residential	Cisterns for individual homes or groups of homes Gravel drives (porous) Contoured landscape Groundwater recharge: 　perforated pipe 　gravel (sand) 　trench 　porous pipe 　dry wells Vegetated depressions	Reservoir or detention basin Planting a high-delaying grass (high roughness) Grassy gutters or channels Increased length of travel of runoff by means of 　gutters, diversions and so on
General	Gravel alleys Porous pavements	Gravel alleys

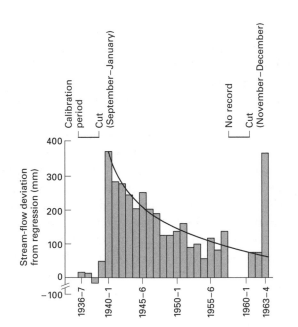

Figure 5.12 The increase of water yield after clear-felling a forest: a unique confirmation from the Coweeta catchment in North Carolina, USA.

decline in stream flow. However, most current research does not support this contention, for transpiration rates from eucalyptus are similar to those from other tree species (except in situations with a shallow groundwater table) while their interception losses are, if anything, generally rather less than those from other tree species of similar height and planting density (Bruijnzeel, 1990).

The reasons why the removal of a forest cover and its replacement with pasture, crops or bare ground have such important effects on stream flow are many. A mature forest probably has a higher rainfall interception rate, a tendency to reduce rates of overland flow, and probably generates soils with a higher infiltration capacity and better general structure. All these factors will tend to produce both a reduction in overall runoff levels and less extreme flood peaks. However, with careful management the replacement of forest by other land-use types need not be detrimental in terms of either sediment loss or flood generation. In Kenya, for example, the tea plantations with shade

Table 5.9 Impact of water use by invading alien plants on the mean annual runoff (MAR) in primary catchment areas of South Africa and Lesotho. Source: Le Maitre et al. (2000)

River system	MAR ($10^6\,m^3$)	Condensed invaded area (ha)	Incremental water use ($10^6\,m^3$)	Water use (% of MAR)	Reduction in rainfall equivalents (mm)
Limpopo	2381.82	122,457	190.38	7.99	155
Olifants	2904.10	217,855	290.44	10.00	133
Vaal	4567.37	64,632	190.53	4.17	295
Orange	7147.76	141,012	141.40	1.98	100
Olifants, Sout and Doring	1008.35	37,623	35.52	3.52	94
Namaqualand coast	25.01	46,618	22.76	91.00	49
W Cape and Agulhas coast	2056.75	384,636	646.50	31.43	168
Breede and Riversdale coast	2088.35	84,398	181.63	8.70	215
Gouritz	670.63	59,399	74.79	11.15	126
South Cape coast	1297.30	52,993	134.46	10.36	254
Gamtoos	494.71	34,289	96.53	19.51	282
Port Elizabeth Coast, Swartkops and Coega	150.04	11,358	40.18	26.78	354
Sundays	279.89	3964	8.34	2.98	210
Bushmans and Alexandria coast	172.92	22,894	73.08	42.26	319
Gt Fish	520.72	6980	21.12	4.06	303
Border Coast	578.91	12,483	55.58	9.60	445
Great Kei	1042.35	30,694	138.22	13.26	450
Former Transkei	7383.76	68,493	217.38	2.94	317
S KwaZulu-Natal	3121.20	46,442	126.37	4.05	272
Tugela	3990.88	62,151	104.67	2.62	168
North KwaZulu-Natal	4741.74	100,574	229.86	4.85	229
Komati to Nwanedzi	2871.4	124,494	283.26	9.86	228
RSA	49,495.96	1,736,438	3303.00	6.67	190

trees, protective grass covers, and carefully designed culverts were found to be 'a hydrologically effective substitute for natural forest' (Pereira, 1973: 127).

In many studies the runoff from clean-tilled land tends to be greater than that from areas under a dense crop cover, but tilling the soil surface does not always increase runoff (Gregory and Walling, 1973: 345). There are reports from the CIS suggesting that the reverse may be the case, and that autumn plowing can decrease surface runoff, presumably because of its effect on surface retention and on soil structure.

Grazing practices also influence runoff, for heavy grazing can both compact the soil and cause vegetation removal (Trimble and Mendel, 1995). In general it tends to lead to an increase in runoff.

In some parts of the world, such as southern Africa, the spread of invasive exotic plants (see p. 53) may cause greater loss of water then the native vegetation and so cause reduction in stream flow. Some calculations of this are presented in Table 5.9. The incremental water use of alien plants is estimated to be $3300 \times$

$10^6\,m^3$ per year, equivalent to a 190 mm reduction in rainfall, and equivalent to almost three-quarters of the virgin mean annual runoff of the huge Vaal River (Le Maitre et al., 2000).

Changes in riverbank vegetation may have a particularly strong influence on river flow. In the southwest USA, for instance, many streams are lined by the salt cedar (*Tamarix pentandra*). With roots either in the water table or freely supplied by the capillary fringe, these shrubs have full potential transpiration opportunity. The removal of such vegetation can cause large increases in stream flow. It is interesting to note that the salt cedar itself is an alien, native to Eurasia, which was introduced by humans. It has spread explosively in the southwest USA, increasing from about 4000 hectares in 1920 to almost 400,000 hectares in the early 1960s (Harris, 1966). In the Upper Rio Grande valley in New Mexico, the salt cedar was introduced to try and combat anthropogenic soil erosion, but it spread so explosively that it came to consume approximately 45% of the area's total available water (Hay, 1973).

Reforestation of abandoned farmlands reverses the effects of deforestation: increased interception and evapotranspiration can cause a decline in water yield. In parts of the eastern USA, farm abandonment and recolonization of the land by pines, spruce, and cedar has been occurring throughout the twentieth century, and this has reduced stream flow by important amounts at a time when water supplies for some eastern cities were becoming critically short (Dunne and Leopold, 1978).

A process that is often associated with afforestation is peat drainage. The hydrological effects of this are the subject of controversy, since there are cases of both increased and decreased flood peaks after drainage (Holden et al., 2004). It has been suggested that differences in peat type alone might account for the different effects. Thus it is conceivable that the drainage of a *Sphagnum* catchment would lead to increased flooding since *Sphagnum* compacts with drainage, reducing its storage volume and its permeability. On the other hand, in the case of non-*Sphagnum* peat there would be relatively less change in structure, but there would be a reduction in moisture content and an increase in storage capacity, thereby tending to reduce flood flows. The nature of the peat is, however, but one feature to consider (Robinson, 1979). The intensity of the drainage works (depth, spacing, etc.) may also be important. In any case, there may be two (sometimes conflicting) processes operating as a result of peat drainage: the increased drainage network will facilitate rapid runoff, and the drier soil conditions will provide greater storage for rainfall. Which of these two tendencies is dominant will depend on local catchment conditions.

However, the impact of land drainage upon downstream flood incidence has long been a source of controversy. Much depends on the scale of study, the nature of land management, and the character of the soil that has been drained. After a detailed review of experience in the UK, Robinson (1990) found that the drainage of heavy clay soils that are prone to prolonged surface saturation in their undrained state generally led to a reduction of large and medium flow peaks. He attributed this to the fact that their natural response is 'flashy' (with limited soil water storage available), whereas their change to more permeable soils, which are less prone to such surface saturation, improves the speed of subsurface flow, thereby tending to increase peak flow levels.

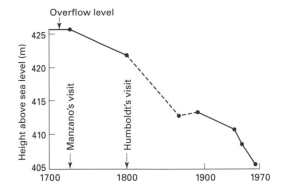

Figure 5.13 Variations in the level of Lake Valencia, Venezuela, to 1968 (after Böckh, 1973, figure 18.2).

The human impact on lake levels

One of the results of human modification of river regimes is that lake levels have suffered some change, although it is not always possible to distinguish between the part played by humans and that played by natural climatic changes.

A lake basin for which there are particularly long records of change is the Valencia Basin in Venezuela (Böckh, 1973). It was the declining level of the waters in this lake that so struck the great German geographer von Humboldt in 1800. He recorded its level as being about 422 m above sea level, and some previous observations on its level were made by Manzano in 1727, which established it as being at 426 m. The 1968 level was about 405 m, representing a fall of no less than 21 m in about 240 years (Figure 5.13). Humboldt believed that the cause of the declining level was the deforestation brought about by humans, and this has been supported by Böckh (1973), who points also to the abstraction of water for irrigation. This remarkable fall in level meant that the lake ceased to have an overflow into the River Orinoco. It has as a consequence become subject to a build-up in salinity, and is now eight times more saline than it was 250 years ago.

Even the world's largest lake, the Caspian, has been modified by human activities. The most important change was the fall of 3 m in its level between 1929 and the late 1970s (see Figure 5.14). This decline was undoubtedly partly the product of climatic change (Micklin, 1972), for winter precipitation in the northern Volga Basin, the chief flow-generating area of the Caspian, was generally below normal for that period

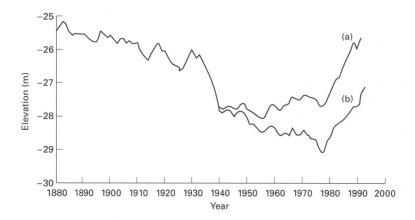

Figure 5.14 Annual fluctuations in the level of the Caspian Sea, for the period 1880–1993. Curve (a) shows the changes in level which would have occurred but for anthropogenic influences, while curve (b) shows the actual observed levels (modified from World Meteorological Organization, 1995, figure 15.3).

because of a reduction in the number of moist cyclones penetrating into the Volga Basin from the Atlantic. Nonetheless, human actions have contributed to this fall, particularly since the 1950s because of reservoir formation, irrigation, municipal and industrial withdrawals, and agricultural practices. In addition to the fall in level, salinity in the northern Caspian has increased by 30% since the early 1930s. A secondary effect of the changes in level has been a decline in fish numbers due to the disappearance of the shallows. These are biologically the most productive zones of the lake, providing a food base for the more valuable types of fish and also serving as spawning grounds for some species. There are plans to divert some water from northward-flowing rivers in Siberia towards the Volga to correct the decline in Caspian levels, but the possible climatic impacts of such action have caused some concern. In any event, an amelioration of climate since the late 1970s has caused some recovery in the level of the lake.

Perhaps the most severe change to a major inland sea is that taking place in the Aral Sea of the CIS (Figure 5.15) (Saiko and Zonn, 2000). Between 1960 and 1990, largely because of diversions of river flow, the Aral Sea lost more than 40% of its area and about 60% of its volume, and its level fell by more than 14 m (Kotlyakov, 1991). By 2002 its level had fallen another 6 m. This has lowered the artesian water table over a band 80–170 km in width, has exposed 24,000 km² of former lake bed to desiccation, and has created salty surfaces from which salts are deflated to be transported in dust storms, to the detriment of soil quality. The mineral content of what remains has increased almost threefold over the same period. It is probably the most

dire ecological tragedy to have afflicted the CIS, and as with the Caspian's decline much of the blame rests with excessive use of water which would otherwise replenish the sea.

Water abstraction from the Jordan River has caused a decline in the level of the Dead Sea. During the past four decades increasing amounts of water have been diverted from surface and groundwater sources in its catchment. Under current conditions, on average, the annual inflow from all sources to the Dead Sea has been one-half to one-quarter that of the inflow prior to development. The water level has fallen about 20 m over that period (Figure 5.16). Under natural conditions freshwater from the Jordan constantly fed the less salty layer of the sea, which occupied roughly the top 40 m of the 320-m-deep body of water. Because the amount of water entering the lake via the Jordan was more or less equal to the quantity lost by evaporation, the lake maintained its stable, stratified state, with less salty water resting on the waters of high salinity. However, with the recent human-induced diminution in Jordan discharge, the sea's upper layer has receded because of the intense levels of evaporation. Its salinity has approached that of the older and deeper waters. It has now been established that as a consequence the layered structure has collapsed, creating a situation where there is increased precipitation of salts. Moreover, now that circulating waters carry oxygen to the bottom, the characteristic hydrogen-sulfide smell has largely disappeared (Maugh, 1979).

From time to time there have been proposals for major augmentation of lake volumes, either by means of river diversions or by allowing the ingress of seawater through tunnels or canals. Among such plans have

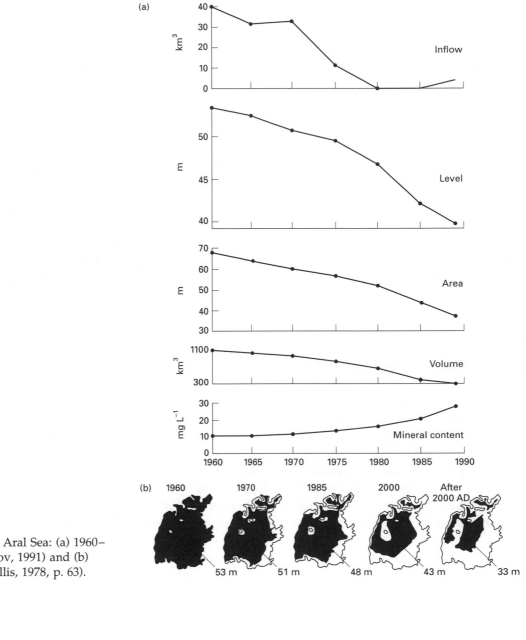

Figure 5.15 Changes in the Aral Sea: (a) 1960–1989 (from data in Kotlyakov, 1991) and (b) 1960 to after 2000 (after Hollis, 1978, p. 63).

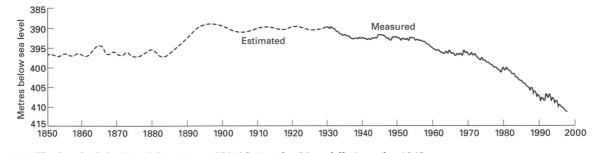

Figure 5.16 The level of the Dead Sea since 1850. Notice the 20-m fall since the 1960s.

been those to flood the salt lakes of the Kalahari by transferring water from the rivers of central Africa such as the Zambezi; the scheme to transfer Mediterranean water to the Dead Sea and to the Quattara Depression; and the Zaire–Chad scheme. Perhaps the most ambitious idea has been the so-called 'Atlantropa' project, whereby the Mediterranean would be empoldered, a dam built at the Straits of Gibraltar, and water fed into the new lake from the Zaire River (Cathcart, 1983).

Changes in groundwater conditions

In many parts of the world humans obtain water supplies by pumping from groundwater (see, e.g., Drennan, 1979). This has two main effects: the reduction in the levels of water tables and the replacement, in coastal areas, of freshwater by saltwater. Environmental consequences of these two phenomena include ground subsidence and soil salinization. Increasing population levels and the adoption of new exploitation techniques (e.g., the replacement of irrigation methods involving animal or human power by electric and diesel pumps) have increased these problems.

Some of the reductions in groundwater levels that have been caused by abstraction are considerable. Figure 5.17 shows the rapid increase in the number of wells tapping groundwater in the London area from 1850 until after the Second World War, while Figure 5.18 illustrates the widespread and substantial changes in groundwater conditions that resulted. The **piezometric** surface in the confined chalk **aquifer** has fallen by more than 60 m over hundreds of square kilometers. Likewise, beneath Chicago, Illinois, pumping since the late nineteenth century has lowered the piezometric head by some 200 m. The drawdown that has taken place in the Great Artesian Basin of Australia locally exceeds 80–100 m (Lloyd, 1986).

The reductions in water levels that are taking place in Nebraska and the High Plains of Texas are some of the most serious, and threaten the long-term viability of irrigated agriculture in that area. Before irrigation development started in the 1930s, the High Plains groundwater system was in a state of dynamic equilibrium, with long-term recharge equal to long-term discharge. However, the groundwater is now being mined at a rapid rate to supply center-pivot (Figure 5.19) and other schemes. In a matter of only 50 years or less, the water level declined by 30 to 50 m in a large area to the north of Lubbock. Since the 1980s the rate of decline has slowed from an average of *c.* 10 cm per year to just over 5 cm per year. This is the result of a combination of higher rainfall over the period together with reduced groundwater withdrawals for irrigation.

Another example of excessive 'mining' of a finite resource is the exploitation of groundwater resources in the oil-rich kingdom of Saudi Arabia. Most of Saudi Arabia is desert, so climatic conditions are not favorable for rapid large-scale recharge of aquifers. Also, much of the groundwater that lies beneath the desert is a fossil resource, created during more humid conditions

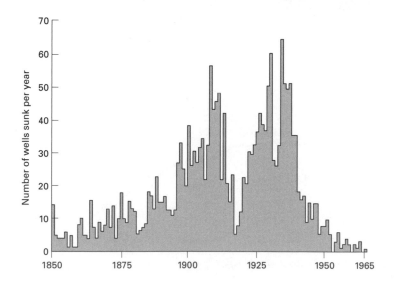

Figure 5.17 Construction of wells tapping the confined aquifer below London, 1850–1965.

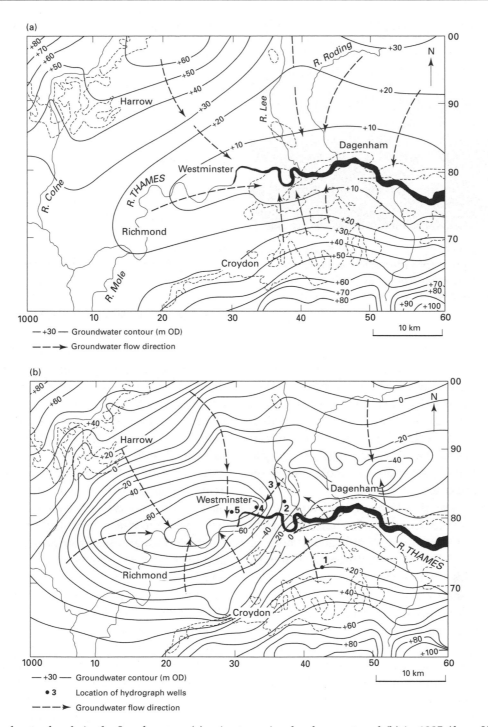

Figure 5.18 Groundwater levels in the London area (a) prior to major development and (b) in 1985 (from Wilkinson and Brassington, 1991, figures 4.5, 4.6, 4.7).

– pluvials – that existed in the Late Pleistocene, between 15,000 and 30,000 years ago. In spite of these inherently unfavorable circumstances, Saudi Arabia's demand for water is growing inexorably as its economy develops. In 1980 the annual demand was 2.4×10^9 m^3. By 1990 it had reached 12×10^9 m^3 (a fivefold increase in just a decade), and it is expected to reach 20×10^9 m^3 by 2010. Only a very small part of the demand can

Figure 5.19 In the High Plains of the USA, fields are irrigated by center-pivot irrigation schemes which use groundwater. Groundwater levels have fallen rapidly in many areas because of the adoption of this type of irrigation technology.

Table 5.10 Possible effects of a rising groundwater level. Source: modified after Wilkinson and Brassington (1991, table 4.3: 42)

Increase in spring and river flows
Re-emergence of 'dry springs'
Surface water flooding
Pollution of surface waters and spread of underground pollution
Flooding of basements
Increased leakage into tunnels
Reduction of slope and retaining wall stability
Reduction in bearing capacity of foundations and piles
Increased hydrostatic uplift and swelling pressures on foundations and structures
Swelling of clays
Chemical attack on foundations

be met by runoff; over three-quarters of the supply is obtained from predominantly nonrenewable groundwater resources. The drawdown on its aquifers is thus enormous. It has been calculated that by 2010 the deep aquifers will contain 42% less water than in 1985. Much of the water is used ineffectively and inefficiently in the agricultural sector (Al-Ibrahim, 1991), to irrigate crops that could easily be grown in more humid regions and then imported.

However, there are situations where humans deliberately endeavor to increase the natural supply of groundwater by attempting artificial recharge of groundwater basins (Peters, 1998). Where the materials containing the aquifer are permeable (as in some alluvial fans, coastal sand dunes, or glacial deposits) the technique of water spreading is much used. In relatively flat areas river water may be diverted to spread evenly over the ground so that infiltration takes place. Alternative water-spreading methods may involve releasing water into basins which are formed either by natural processes (such as the playas of the High Plains in the USA), or by excavation, or by the construction of dikes or small dams. On alluvial plains water can also be encouraged to percolate down to the water table by distributing it to a series of ditches or furrows. In some situations natural channel infiltration can be promoted by building small check dams down a stream course. In irrigated areas surplus water can be spread by irrigating with excess water during the

dormant season. When artificial groundwater recharge is required in sediments with impermeable layers such water-spreading techniques are not effective and the appropriate method may then be to pump water into deep pits or into wells. This last technique is used on the coastal plain of Israel, both to replenish the groundwater reservoirs when surplus irrigation water is available, and to attempt to diminish the problems associated with saltwater intrusion.

In some industrial areas, reductions in industrial activity have caused a recent reduction in groundwater abstraction, and as a consequence groundwater levels have begun to rise, a trend that is exacerbated by considerable leakage losses from ancient, deteriorating pipe and sewer systems. In cities such as London, Liverpool, and Birmingham an upward trend has already been identified (Brassington and Rushton, 1987; Price and Reed, 1989). In London, because of a 46% reduction in groundwater abstraction, the water table in the Cretaceous Chalk and Tertiary beds has risen by as much as 20 m. Such a rise has numerous implications, which are listed in Table 5.10. The trend in groundwater levels beneath Trafalgar Square is shown in Figure 5.20.

There is, finally, a series of ways in which unintentionally changes in groundwater conditions can occur in urban areas. As we have clearly seen, in cities surface runoff is increased by the presence of impermeable surfaces. One consequence of this would be that less water went to recharge groundwater. However, there is an alternative point of view, namely that groundwater recharge can be accelerated in urban

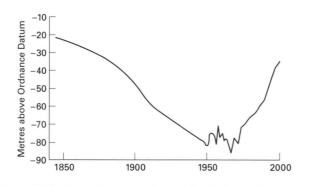

Figure 5.20 Changing groundwater levels beneath Trafalgar Square, central London, showing the sharp decline until the 1950s and the 1960s, and the substantial rise since then (from Environment Agency (UK) data).

areas because of leaking water mains, sewers, septic tanks, and soakaways (Figure 5.21). In cities in arid areas there is often no adequate provision for storm runoff, and the (rare) increased runoff from impermeable surfaces will infiltrate into the permeable surroundings. In some cities recharge may result from over-irrigation of parks and gardens. Indeed, where the climate is dry, or where large supplies of water are imported, or where pipes and drains are poorly maintained, groundwater recharge in urban areas is likely to exceed that in rural areas.

Water pollution

Water pollution is not new, but is frequently undesirable: it causes disease transmission through infection; it may poison humans and animals; it may create objectionable odors and unsightliness; it may be the cause of the unsatisfactory quality even of treated water; it may cause the eutrophication of water bodies; and it may affect economic activities such as shellfish culture.

The causes and forms of water pollution created by humans are many and can be classified into groups as follows (after Strandberg, 1971):

1 sewage and other oxygen-demanding wastes;
2 infectious agents;
3 organic chemicals;
4 other chemical and mineral substances;
5 sediments (turbidity);
6 radioactive substances;
7 heat (thermal pollution).

Moreover, many human activities can contribute to changes in water quality, including agriculture, fire, urbanization, industry, mining, irrigation, and many others. Of these, agriculture is probably the most important. Some pollutants merely have local effects,

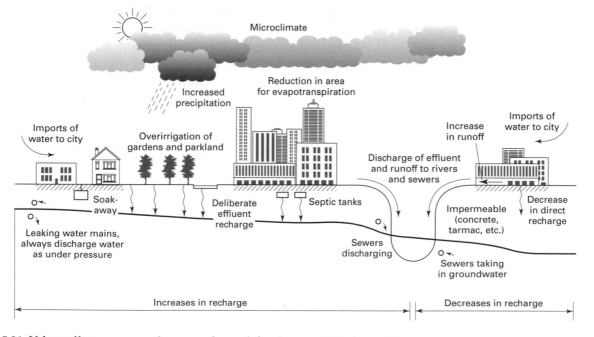

Figure 5.21 Urban effects on groundwater recharge (after Lerner, 1990, figure 2).

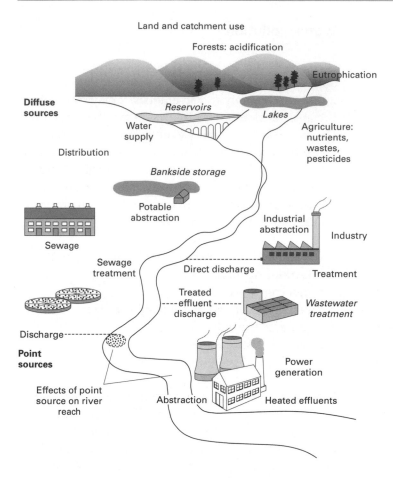

Land and catchment use

Forests: acidification

Eutrophication

Diffuse sources

Reservoirs

Lakes

Water supply

Distribution

Agriculture: nutrients, wastes, pesticides

Bankside storage

Potable abstraction

Industrial abstraction

Industry

Sewage

Sewage treatment

Direct discharge

Treatment

Treated effluent discharge

Wastewater treatment

Discharge

Point sources

Power generation

Effects of point source on river reach

Abstraction

Heated effluents

Figure 5.22 Diffuse and point sources of pollution into river systems (after Newson, 1992, figure 7.7).

while others, such as acid rain or DDT (dichlorodi-phenyltrichloroethane), may have continental or even planetary implications.

It is also possible to categorize water pollutants according to whether or not they are derived from 'point' or 'nonpoint' (also called 'diffuse') sources (Figure 5.22). Municipal and industrial wastes tend to fall into the former category because they are emitted from one specific and identifiable place (e.g., a sewage pipe or industrial outfall). Pollutants from nonpoint sources include agricultural wastes, many of which enter rivers in a diffuse manner as chemicals percolate into groundwater or are washed off into fields, as well as some mining pollutants, uncollected sewage, and some urban storm-water runoff.

It is plainly not a simple matter to try to estimate the global figure for the extent of water pollution caused by humans. For one thing we know too little about the natural long-term levels of dissolved materials in the world's rivers. Nonetheless, Meybeck (1979)

calculated that about 500 million tonnes of dissolved salts reach the oceans each year as a result of human activity. These inputs have increased by more than 30% the natural values for sodium, chloride, and sulfate, and have created an overall global augmentation of river mineralization by about 12%. Likewise, Peierls et al. (1991) have demonstrated that the quantity of nitrates in world rivers now appears to be closely correlated to human population density. Using published data for 42 major world rivers they found a highly significant correlation between annual nitrate concentration and human population density that explained 76% of the variation in nitrate concentration for the 42 rivers. They maintain that 'human activity clearly dominates nitrate export from land.' Meybeck (2001a) argued that on a global scale nutrient inputs to the oceans by rivers have already increased 2.2 times for nitrate and four times for ammonia.

Nitrate trends in most rivers in Europe and North America reveal a marked increase since the 1950s. This

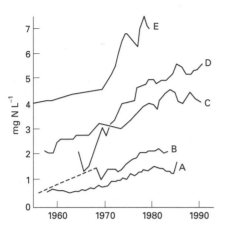

Figure 5.23 Recent trends of nitrate concentrations in some rivers: A, Mississippi at mouth; B, Danube at Budapest; C, Rhine at the Dutch–German border; D, Seine at mouth; E, Thames at mouth (from Meybeck, 2001b, figure 17.6).

can be attributed to the growth in use of nitrate fertilizers (Meybeck, 2001b). Trends for the Mississippi, Danube, Rhine, Seine, and Thames are shown in Figure 5.23.

There are three main classes of chemical pollutant that deserve particular attention: nitrates and phosphates; metals; and synthetic and industrial organic pollutants.

Nitrates and phosphates, trends in the concentration of which are reviewed by Heathwaite et al. (1996), are an important cause of a process called eutrophication. Nitrates normally occur in drainage waters and are derived from soil nitrogen, from nitrogen-rich glacial deposits, and from atmospheric deposition. Anthropogenic sources include synthetic fertilizers, sewage, and animal wastes from feedlots. Land-use changes (e.g., logging) can also increase nitrate inputs to streams. Phosphate levels are also rising in some parts of the world. Major sources include detergents, fertilizers and human wastes.

Metals, such as nitrates and phosphates, occur naturally in soil and water. However, as the human use of metals has burgeoned, so has the amount of water pollution they cause. In addition, some metal ions reach river waters because they become more quickly mobilized as a result of acid rain. Aluminum is a notable example of this. From a human point of view, the metals of greatest concern are probably lead, mercury, arsenic, and cadmium, all of which have adverse

health effects. Other metals can be toxic to aquatic life, and these include copper, silver, selenium, zinc, and chromium.

The anthropogenic sources of metal pollution include the industrial processing of ores and minerals, the use of metals, the leaching of metals from garbage and solid waste dumps, and animal and human excretions. Nriagu and Pacyna (1988) estimated the global anthropogenic inputs of trace metals into aquatic systems (including the oceans), and concluded that the sources producing the greatest quantities were, in descending order, the following (the metals produced by each source are listed in parentheses):

- domestic wastewater effluents (arsenic, chromium, copper, manganese, nickel);
- coal-burning power stations (arsenic, mercury, selenium);
- nonferrous metal smelters (cadmium, nickel, lead, selenium);
- iron and steel plants (chromium, molybdenum, antimony, zinc);
- the dumping of sewage sludge (arsenic, manganese, lead).

However, in some parts of the world metal pollution may be derived from other sources. There is increasing evidence, for example, that in the western USA water derived from the drainage of irrigated lands may contain high concentrations of toxic or potentially toxic trace elements such as arsenic, boron, chromium, molybdenum, and selenium. These can cause human health problems and poison fish and wildlife in desert wetlands (Lently, 1994).

Synthetic and industrial organic pollutants have been manufactured and released in very large quantities since the 1960s. The dispersal of these substances into watercourses has resulted in widespread environmental contamination. There are many tens of thousands of synthetic organic compounds currently in use, and many are thought to be hazardous to life, even at quite low concentrations – concentrations possibly lower than those that can be measured routinely by commonly available analytical methods. Among these pollutants are synthetic organic pesticides, including chlorinated hydrocarbon insecticides (e.g., DDT). Some of these can reach harmful concentrations as a result of biological magnification in the food chain. Other important

Table 5.11 Characteristics of lakes experiencing 'cultural eutrophication'. Source: modified after Mannion (1992, table 11.4)

Biological factors	Physical factors
Primary productivity: usually much higher than in unpolluted water and is manifest as extensive algal blooms *Diversity of primary producers*: initially green algae increase, but blue-green algae rapidly become dominant and produce toxins. Similarly, macrophytes (e.g., reed maces) respond well initially but due to increased turbidity and anoxia (see below) they decline in diversity as eutrophication proceeds *Higher trophic level productivity*: overall decrease in response to factors given in this table *Higher trophic level diversity*: decreases due to factors given in this table. The species of macro- and microinvertebrates that tolerate more extreme conditions increase in numbers. Fish are also adversely affected and populations are dominated by surface-dwelling coarse fish such as pike and perch	*Mean depth of water body*: as infill occurs the depth decreases *Volume of hypolimnion*: varies *Turbidity*: this increases, as sediment input increases, and restricts the depth of light penetration that can become a limiting factor for photosynthesis. It is also increased if boating is a significant activity

organic pollutants include PCBs (polychlorinated biphenyls), which have been used extensively in the electrical industry as dielectrics in large transformers and capacitors; PAHs (polycyclic aromatic hydrocarbons), which result from the incomplete burning of fossil fuels; various organic solvents used in industrial and domestic processes; phthalates, which are plasticizers used, for example, in the production of polyvinyl chloride resins; and DBPs (1,4-diphenyl-1,3-butadienes), which are a range of disinfection by-products. The long-term health effects of cumulative exposure to such substances are difficult to quantify. However, some work suggests that they may be implicated in the development of birth defects and certain types of cancer.

Chemical pollution by agriculture and other activities

Agriculture may be one, if not the most, important cause of pollution, either by the production of sediments or by the generation of chemical wastes. With regard to the latter, it has been suggested that denitrification processes in the environment are incapable of keeping pace with the rate at which atmospheric nitrogen is being mobilized through industrial fixation processes and being introduced into the biosphere in the form of commercial fertilizers (Manners, 1978). Nitrogen, with phosphorus, tends to regulate the growth of aquatic plants and therefore the eutrophication of inland waters. Excess nitrates can also cause health hazards to humans and animals.

Eutrophication is the enrichment of waters by nutrients (Ryding and Rast, 1989). The process occurs naturally during, for example, the slow aging of lakes, but it can be accelerated both by runoff from fertilized agricultural land and by the discharge of domestic sewage and industrial effluents (Lund, 1972). This process, often called 'cultural eutrophication', commonly leads to excessive growths of algae, followed in some cases by a serious depletion of dissolved oxygen as the algae decay after death. Oxygen levels may become too low to support fish life, resulting in fish kills. Changes in diatom assemblages can also occur (see especially Battarbee, 1977). The nature of these changes is summarized in Table 5.11.

As agriculture has, in the developed world, become of an increasingly specialized and intensive nature, so the pollution impact has increased. The traditional mixed farm tended to be a more or less closed system that generated relatively few external impacts. This was because crop residues were fed to livestock or incorporated in the soil; and manure was returned to the land in amounts that could be absorbed and utilized. Many farms have become more specialized, with the separation of crop and livestock activities; large numbers of stock may be kept on feedlots, silage may be produced in large silos, and synthetic fertilizers may be applied to fields in large quantities (Conway and Pretty, 1991).

Although there are considerable fluctuations from year to year, the trend in nitrate levels in English rivers is clearly apparent (Figure 5.24b) (Royal Society Study Group, 1983). By 1980 they were 50 to 400%

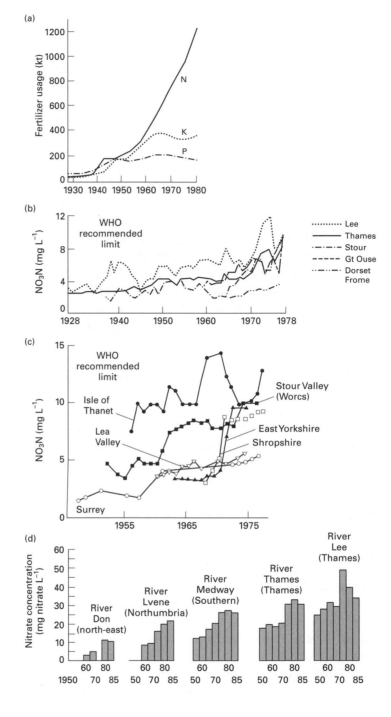

Figure 5.24 Nitrates in surface and groundwaters in the UK: (a) the trends in annual fertilizer usage in the UK during the period 1928–1980; (b) trends in mean annual nitrate concentration in five rivers for which long-term data are available – WHO, World Health Organization; (c) nitrate concentrations in selected public water supply abstraction boreholes in the Cretaceous Chalk and Triassic sandstone aquifers of the UK (Royal Society Study Group, 1983, figures 4, 18, 31); (d) changing nitrate concentrations in five UK rivers. The averages are five-year means (from Department of the Environment statistics, in Conway and Pretty, 1991, figure 4.8).

higher than they were 20 years before. The River Thames, which provides the major supply of London's water, increased its mean annual NO_3 concentration from around 11 mg L^{-1} in 1928 to 35 mg L^{-1} in the 1980s. Trends in nitrogen and phosphorus concentration in lowland regions of North America are not as great as those reported in lowland regions of Britain. This probably reflects the higher population density and greater intensity of land use in Britain (Heathwaite et al., 1996).

Also causing concern is the trend in nitrate concentrations in British groundwaters. Investigations have revealed a large quantity of nitrate in the unsaturated zone of the principal aquifers (mainly Cretaceous Chalk and Triassic sandstone), and this is slowly moving down towards the main groundwater body. The slow transit time means that in many water supply wells increased nitrate concentrations will not occur for 20 to 30 years, but they will then be above acceptable levels for human health (Figure 5.24c).

However, even if an increase in nitrate levels is evident this may not necessarily be because of the application of fertilizers (Burt and Haycock, 1992). Some nitrate pollution may be derived from organic wastes. There has also been some anxiety in Britain over a possible decline in the amount of organic matter present in the soil, which could limit its ability to assimilate nitrogen. Moreover, the pattern of tillage may have affected the liberation of nitrogen via mineralization of organic matter. The increased area and depth of modern plowing accelerates the decay of residues and may change the pattern of water movement in the soil. Finally, tile drainage has also expanded in area very greatly in recent decades in England. This has affected the movement of water through the soil and hence the degree of leaching of nitrates and other materials (Edwards, 1975).

Given the uncertainties surrounding the question of nitrate pollution, and bearing in mind the major advances in agricultural productivity that they have permitted, attempts at controlling their use have not been received with complete favor. Indeed, as Viets (1971) has pointed out, if fertilizer use were curtailed there would be less vegetation cover and increased erosion and sediment delivery to rivers. This, he suggests, would necessitate an increase in land hectarage to maintain production levels which would also cause greater erosion; at the same time the increased plowing would lead to greater nitrate loss from grasslands. Nevertheless some farmers are inefficient and wasteful in their use of fertilizers and there is scope for economy (Cooke, 1977).

In addition to the influence of nitrate fertilizers, the confining of animals on feedlots results in tremendously concentrated sources of nutrients and pollutants. Animal wastes in the USA are estimated to be as much as 1.6 billion tonnes per year, with 50% of this amount originating from feedlots. It needs to be remembered that 1000 head of beef produce the equivalent organic load of 6000 people (Sanders, 1972). The number of cattle fed on feedlots in the USA has increased rapidly, partly because per capita consumption

of meat has moved sharply upward, and partly because the confinement of cattle on small lots, on which they are fed a controlled selection of feeds, leads to the greatest possible weight gains with the least possible cost and time (Bussing, 1972). Cattle feedlot runoff is a high-strength organic waste, high in oxygen-demanding material, and has caused many fish kills in rivers in states such as Kansas. In North Carolina and Virginia industrial style hog and poultry operations have caused large waste spills, especially during severe storms, and these too have caused massive fish kills.

Pesticides are another source of chemical pollution brought about by agriculture. There is now a tremendous range of pesticides and they differ greatly in their mode of action, in the length of time they remain in the biosphere, and in their toxicity. Much of the most adverse criticism of pesticides has been directed against the chlorinated hydrocarbon group of insecticides, which includes DDT and Dieldrin. These insecticides are toxic not only to the target organism but to other insects too (that is, they are nonspecific). They are also highly persistent. Appreciable quantities of the original application may survive in the environment in unaltered form for years. This can have two rather severe effects: global dispersal and the 'biological magnification' of these substances in food chains (Manners, 1978). This last problem is well illustrated for DDT in Figure 5.25.

Changes in surface-water chemistry may be produced by the washing out of acids that pollute the air in precipitation. This is particularly clear in the heavily industrialized area of northwestern Europe, where these is a marked zonation of acid rain with pH values that often fall below 4 or 5. A broadly comparable situation exists in North America (Johnson, 1979). Very considerable fears have been expressed about the possible effects that acid deposition may have on aquatic ecosystems, and particularly on fish populations. It is now generally accepted that decades of increasing lake and river acidification caused fish kills and stock depletion. Fishless lakes now occur in areas such as the Adirondacks in the northeast USA. Species of fish vary in their tolerance to low pH, with rainbow trout being largely intolerant to values of pH below 6.0, while salmon, brown and brook trout are somewhat less sensitive. Values of 4.0–3.5 are lethal to salmonids, while values of 3.5 or less are lethal to most fish. Numerous

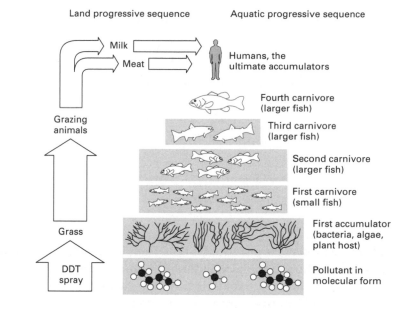

Figure 5.25 Biological concentration occurs when relatively indestructible substances (dichlorodiphenyltrichloroethane (DDT), for example) are ingested by lesser organisms at the base of the food pyramid. An estimated 1000 kg of plant plankton are needed to produce 100 kg of animal plankton. These in turn are consumed by 10 kg of fish, the amount needed by one person to gain 1 kg. The ultimate consumer (man or woman) then takes in the DDT taken in by the 1000 kg of the lesser creatures at the base of the food pyramid, when he or she ingests enough fish to gain 1 kg (after Strandberg, 1971, figure 2).

lacustrine molluscs and crustaceans are not found even at weakly acid pH values of 5.8–6.0, and crayfish rapidly lose their ability to recalcify after molting as the pH drops from 6.0 to 5.5 (Committee on the Atmosphere and the Biosphere, 1981).

Fish declines and deaths in areas afflicted by acid rain are not solely caused by water acidity per se. It has become evident that under conditions of high acidity certain metal ions, including aluminum, are readily mobilized, leading to their increasing concentration in freshwaters. Some of these metal ions are highly toxic and many mass mortalities of fish have been attributed to aluminum poisoning rather than high acidity alone. The aluminum adversely affects the operation of fish gills, causing mucus to collect in large quantities; this eventually inhibits the ability of the fish to take in necessary oxygen and salts. In addition, aluminum's presence in water can reduce the amount of available phosphates, an essential food for phytoplankton and other aquatic plants. This decreases the available food for fish higher in the food chain, leading to population decline (Park, 1987).

Fortunately, however, during the 1980s and 1990s rates of acid deposition decreased over large portions of North America and Europe (Stoddart et al., 1999; Cooper and Jenkins, 2003), and some recovery in streams and lakes has resulted. Reversal of acidification, while not universal, has been noted, for example,

in the English Lake District (Tipping et al., 2000), in Scandinavia (Fölster and Wilander, 2002), in central Europe (Vesely et al., 2002), and in the English Pennines (Evans and Jenkins, 2000). In Europe, recovery has been strongest in the Czech Republic and Slovakia, moderate in Scandinavia and the UK, and weakest in Germany (Evans et al., 2001). The period required for recovery may in some areas be considerable. In the Adirondacks of the eastern USA recovery may take many decades because, on the one hand, they have been exposed to acid deposition for long periods of time so that they have been relatively depleted of substances that can neutralize acids, and on the other their soils are thin and so can offer less material to neutralize the acid from precipitation (US General Accounting Office, 2000).

Mining can also create serious chemical pollution. In the western USA, the forest service estimates that between 20,000 and 50,000 mines are currently generating acid on forest service lands, having an impact upon 8000–16,000 km of stream. In the eastern USA more than 7000 km of stream are affected by acid drainage from coal mines (US Environmental Protection Agency, 1994).

In Britain, coal seams and mudstones in the coal measures contain pyrite and marcasite (ferrous sulfide). In the course of mining the water table is lowered, air gains access to these minerals and they are oxidized.

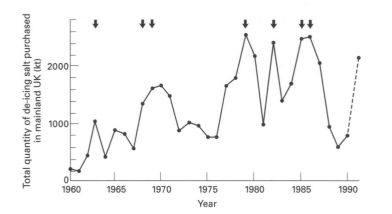

Figure 5.26 Estimates of the total quantity of de-icing salt purchased annually in mainland Britain during the period 1960–1991. Arrows highlight severe winters. Data provided by ICI (in Dobson, 1991, figure 1.1).

Sulfuric acid may be produced. Should mining cease, as is the situation in the coalfields of Britain, pumping may cease, groundwater may rise through the worked mine, and acid derived from it may discharge into more and more surface waters.

As a consequence, mine drainage waters may be very acid and have pH values as low as 2 or 3. They have high sulfate and iron concentrations, they may contain toxic metals such as arsenic and cadmium, and, because of the reaction of the acid on clay and silicates in the rocks, they may also contain appreciable amounts of calcium, magnesium, aluminum, and manganese. Following reactions with sediments or mixing with alkaline river waters, or when chemical or bacterial oxidation of ferrous compounds occurs, the iron may precipitate as ferric hydroxide. This may discolor the water and leave unsightly deposits (Rodda et al., 1976: 299). Some of the reactions involved in the development of what is called acid rock drainage (ARD) or acid mine drainage (AMD) have been set down by Down and Stocks (1977: 110–11).

1 Oxidation of the sulfide (usually FeS_2) in a wet environment producing ferrous sulfate and sulfuric acid

$$2FeS_2 + 2H_2O + 7O_2 \rightarrow 2FeSO_4 + 2H_2SO_4$$

2 Ferrous sulfate, in the presence of sulfuric acid and oxygen, can oxidize to produce ferric sulfate, especially in the presence of the bacterium *Thiobacillus ferro-oxidans*

$$4FeSO_4 + 2H_2SO_4 + O_2 \rightarrow 2Fe_2(SO_4)_3 + 2H_2O$$

3 The ferric iron so produced combines with the hydroxyl $(OH)^-$ ions of water to form ferric hydroxide, which is insoluble in acid and precipitates

$$Fe_2(SO_4)_3 + 6H_2O \rightarrow 2Fe(OH)_3 + 3H_2SO_4$$

The nature of acid drainage from mines and the remedial measures that can be adopted to deal with it are discussed by Robb and Robinson (1995).

Rock salt (sodium chloride) has been used in increasing quantities since the Second World War for minimizing the dangers to motorists and pedestrians from icy road and pavements. With the rise in the number of vehicles there has been a tendency throughout Europe and North America for a corresponding increase in the use of salt for de-icing purposes (see, e.g., Howard and Beck, 1993). Data for the UK between 1960 and 1991 are shown in Figure 5.26 and demonstrate that while there is considerable interannual variability (related to weather severity) there has also been a general upward trend so that in the mid-1970s to end-1980s more than two million tonnes of de-icing salt were being purchased each year (Dobson, 1991). Data on de-icing salt application in North America are given by Scott and Wylie (1980). They indicate that the total use of de-icing salts in the USA increased at a nearly exponential rate between the 1940s and the 1970s, increasing from about 200,000 tonnes in 1940 to approximately 9,000,000 tonnes in 1970. This represented a doubling time of about five years. Runoff from salted roads can cause ponds and lakes to become salinized, concrete to decay, and susceptible trees to die.

Similarly, storm-water runoff from urban areas may contain large amounts of contaminants, derived from

Table 5.12 Comparison of contaminant profiles for urban surface runoff and raw domestic sewage, based on surveys throughout the USA. Source: Burke (1972, table 7.36.1)

Constituent*	Urban surface runoff	Raw domestic sewage
Suspended solids	250–300	150–250
BOD	10–250	300–350
Nutrients:		
total nitrogen	0.5–5.0	25–85
total phosphorus	0.5–5.0	2–15
Coliform bacteria (MPN 100 mL^{-1})	10^4–10^6	10^6 or greater
Chlorides	20–200	15–75
Miscellaneous substances:		
oil and grease	Yes	Yes
heavy metals	10–100 times sewage concentration	Traces
pesticides	Yes	Seldom
other toxins	Potential exists	Seldom

*All concentrations are expressed in mg L^{-1} unless stated otherwise: BOD, biochemical oxygen demand; MPN, most probable number.

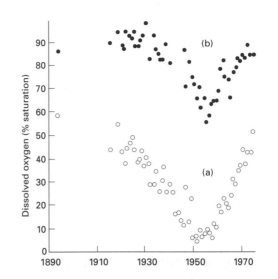

Figure 5.27 The average dissolved oxygen content of the River Thames at half-tide in the July–September quarter since 1890: (a) 79 km below Teddington Weir; (b) 95 km below Teddington Weir (after Gameson and Wheeler, 1977, figure 4).

litter, garbage, car-washings, horticultural treatments, vehicle drippings, industry, construction, animal droppings, and the chemicals used for snow and ice clearance. Comparison of contaminant profiles for urban runoff and raw domestic sewage, based on surveys throughout the USA, indicates the importance of pollution from urban runoff sources (Table 5.12). In New York City some half million dogs leave up to 20,000 tonnes of pollutant feces and up to 3.8×10^6 L of urine in the city streets each year, all of which is flushed by gutters to storm-water sewers. Taking Britain as a whole, dogs deposit 1000 tonnes of excrement and three million gallons of urine on the streets every day (Ponting, 1991).

Nonetheless, it would be unfair to create the impression that water pollution is either a totally insoluble problem or that levels of pollution must inevitably rise. It is true that some components of the hydrological system – for example, lakes, because they may act as sumps – will prove relatively intractable to improvement, but in many rivers and estuaries striking developments have occurred in water quality in recent decades. This is clear in the case of the River Thames, which suffered a serious decline in its quality and in its fish as London's pollution and industries expanded.

However, after about 1950, because of more stringent controls on effluent discharge, the downward trend in quality was reversed (see Figure 5.27), and many fish, long absent from the Thames, are now returning (Gameson and Wheeler, 1977). Likewise, a long-term study of the levels of many heavy metals in seawater off the British Isles suggests that levels are generally stable and that concentrations do not seem to be increasing (Preston, 1973).

A similar picture emerges if one considers the trend in the state of Lake Washington near Seattle, USA. Studies there in 1933, 1950, and 1952 showed steady increases in the nutrient content and in associated growth of phytoplankton (algae). These resulted from the increasing levels of untreated sewage pumped into the lake from the Seattle area. A sewage diversion project, started in 1963 and completed in 1968 (see Figure 5.28), resulted in a significant decrease in the sewage input to the lake and was reflected in a rapid decline in phosphates, plankton, and turbidity, and a less marked decline in nitrate levels. There also appears to have been some improvement in pollution levels in the Great Lakes since the mid-1970s (Figure 5.29), as indicated by phosphorus loadings, and DDT and PCB levels.

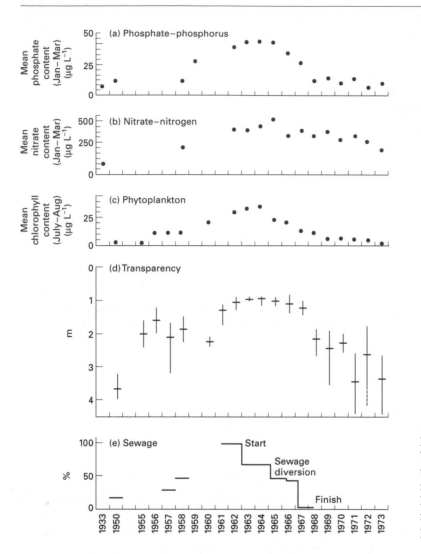

Figure 5.28 Changes in the state of Lake Washington, USA, associated with levels of untreated sewage from 1933 to 1973. The relative amount of treated sewage entering the lake is shown as a percentage of the maximum rated capacity of the treatment plants, 76×10^6 L per day (after Edmonson, 1975, figure 3).

One very important way of reducing the input of pollutants to stream courses is to provide a buffer zone of vegetation between them and the fields from which the pollution comes. Alternatively, bans can be instituted to prevent fertilizer applications in particularly sensitive zones with respect to pollutants such as nitrates.

Deforestation and its effects on water quality

The removal of tree cover not only may affect river flow and stream-water temperature, it may also cause changes in stream-water chemistry. The reason for this is that in many forests large quantities of nutrients are cycled through the vegetation, and in some cases (notably in humid tropical rain forest) the trees are a great store of the nutrients. If the trees are destroyed the cycle of nutrients is broken and a major store is disrupted. The nutrients thus released may become available for crops, and shifting cultivators, for example, may utilize this fact and gain good crop yields in the first year after burning and felling forest. Some of the nutrients may, however, be leached out of the soils, and appear as dissolved load in streams. Large increases in dissolved load may have undesirable effects, including eutrophication (Hutchinson, 1973), soil salination and a deterioration in public water supply (Conacher, 1979).

A classic but, it must be added, extreme exemplification of this is the experiment that was carried out in

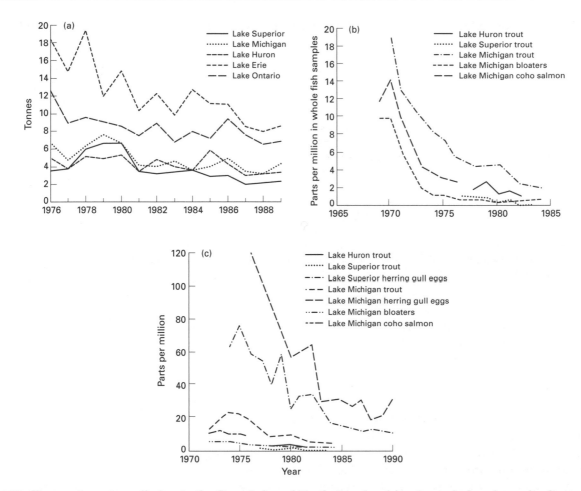

Figure 5.29 Changes in water pollution in the Great Lakes of North America: (a) estimated phosphorus loadings; (b) dichlorodiphenyltrichloroethane (DDT) levels; (c) polychlorinated biphenyl (PCB) levels (after Council on Environmental Quality, 17th (1986) and 22nd (1992) Annual Reports).

the Hubbard Brook catchments in New Hampshire, USA (Bormann et al., 1968). These workers monitored the effects of 'savage' treatment of the forest on the chemical budgets of the streams. The treatment was to fell the trees, leave them in place, kill lesser vegetation and prevent regrowth by the application of herbicide. The effects were dramatic: the total dissolved inorganic material exported from the basin was about 15 times larger after treatment, and, in particular, there was a very substantial increase in stream nitrate levels. Their concentration increased by an average of 50 times.

It has been argued, however, that the Hubbard Brook results are atypical (Sopper, 1975) because of the severity of the treatment. Most other forest clear-cutting experiments in the USA (Table 5.13) indicate that

nitrate-nitrogen additions to stream water are not so greatly increased. Under conventional clear-cutting the trees are harvested rather than left, all saleable material is removed, and encouragement is given to the establishment of a new stand of trees. Aubertin and Patric (1974), using such conventional clear-cutting methods rather than the drastic measures employed in Hubbard Brook, found that clear-cutting in their catchments in West Virginia had a negligible effect on stream temperatures (they left a forest strip along the stream), pH, and the concentration of most dissolved solids.

Even if clear-cutting is practiced, there is some evidence of a rapid return to steady-state nutrient cycling because of quick regeneration. As Marks and Bormann (1972) have put it:

Table 5.13 Nitrate-nitrogen losses from control and disturbed forest ecosystems. Source: modified after Vitousek et al. (1979)

Site	Nature of disturbance	Nitrate-nitrogen loss (kg ha⁻¹ year⁻¹)	
		Control	Disturbed
Hubbard Brook	Clear-cutting without vegetation removal, herbicide inhibition of re-growth	2.0	97
Gale River (New Hampshire)	Commercial clear-cutting	2.0	38
Fernow (West Virginia)	Commercial clear-cutting	0.6	3.0
Coweeta (North Carolina)	Complex	0.05	7.3*
H. J. Andrews Forest (Oregon)	Clear-cutting with slash-burning	0.08	0.26
Alsea River (Oregon)	Clear-cutting with slash-burning	3.9	15.4
Mean		1.44	26.83

*This value represents the second year of recovery after a long-term disturbance. All other results for disturbed ecosystems reflect the first year after disturbance.

Because terrestrial plant communities have always been subjected to various forms of natural disturbances, such as wind storms, fires, and insect outbreaks, it is only reasonable to consider recovery from disturbance as a normal part of community maintenance and repair.

Likewise Hewlett et al. (1984) found no evidence in the Piedmont region of the USA that clear-cutting would create such extreme nitrate loss that soil fertility would be impaired or water eutrophication caused. They point out that rapid growth of vegetation minimizes nutrient losses from the ecosystem by three main mechanisms. It channels water from runoff to evapotranspiration, thereby reducing erosion and nutrient loss; it reduces the rates of decomposition of organic matter through moderation of the microclimate so that the supply of soluble ions for loss in drainage water is reduced; and it causes the simultaneous incorporation of nutrients into the rapidly developing biomass so that they are not lost from the system.

In Britain, a comparison of water chemistry in forested and clear-cut areas in the Plynlimon area of mid-Wales has revealed certain trends (Figure 5.30). The most notable of these is the increase in nitrate levels, which impacts on catchment acidification processes by causing a decrease in pH and releasing toxic metals such as aluminum. Also important is the increase in dissolved organic carbon, which causes discoloration of stream water (Institute of Hydrology, 1991).

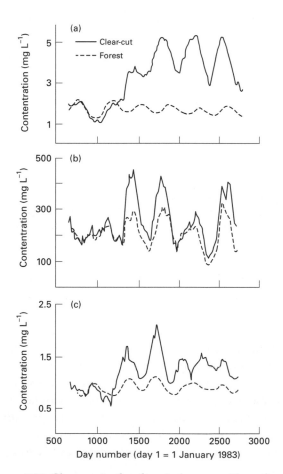

Figure 5.30 Changes in the chemical composition of stream water from clear-cut and forested catchments in mid-Wales, 1983–1990: (a) nitrate (NO₃); (b) aluminum (Al); (c) dissolved organic carbon (DOC) (modified after Institute of Hydrology, 1991, figure 20).

Thermal pollution

The pollution of water by increasing its temperature is called thermal pollution. Many fauna are affected by temperature, so this environmental impact has some significance (Langford, 1990).

In industrial countries probably the main source of thermal pollution is from condenser cooling water released from electricity generating stations. Water discharged from power stations has been heated some 6–9°C, but usually has a temperature of less than 30°C. The extent to which water affects river temperature depends very much on the state of flow. For example, below the Ironbridge power station in England, the Severn River undergoes a temperature increase of only 0.5°C during floods, compared with an 8°C increase at times of low flow (Rodda et al., 1976).

Over the past few decades both the increase in the capacity of individual electricity power-generating units and the improvements in their thermal efficiency have led to a diminution in the heat rejected in relation to the amount of cooling water per unit of production. Economic optimization of the generating plant has cut down the flow of cooling water per unit of electricity, although it has raised its temperature. This increased temperature rise of the cooling water is more than offset by the reduction achieved in the volume of water utilized. Nonetheless, the expansion of generating capacity has meant that the total quantity of heat discharged has increased, even though it is much less than would be expected on a simple proportional basis.

Thermal pollution of streams may also follow from urbanization (Pluhowski, 1970). This results from various sources: changes in the temperature regime of streams brought about by reservoirs according to their size, their depth, and the season; changes produced by the urban heat island effect; changes in the configuration of urban channels (e.g., their width/depth ratio); changes in the degree of shading of the channel, either by covering it over or by removing natural vegetation cover; changes in the volume of storm runoff; and changes in the groundwater contribution. Pluhowski found that the basic effects of cities on river-water temperatures on Long Island, New York, was to raise temperatures in summer (by as much as 5–8°C) and to lower them in winter (by as much as 1.5–3.9°C).

Reservoir construction can also affect stream-water temperatures. Crisp (1977) found, for example, that as a result of the construction of a reservoir at Cow Green (Upper Teesdale, northern England) the temperature of the river downstream was modified. He noted a reduction in the amplitude of annual water-temperature fluctuations, and a delay in the spring rise of water temperature by 20–50 days, and in the autumn fall by up to 20 days. In rural areas human activities can also cause significant modifications in river-water temperature. Deforestation is especially important (Lynch et al., 1984). Swift and Messer (1971), for example, examined stream temperature measurements during six forest-cutting experiments on small basins in the southern Appalachians, USA. They found that with complete cutting, because of shade removal, the maximum stream temperatures in summer went up from 19 to 23°C. They believed that such temperature increases were detrimental to temperature-sensitive fish such as trout (see Figure 5.31). On the other hand, the temporary shutdown of power plants may create severe cold-shock kills of fish in discharge-receiving waters in winter (Clark, 1977). Moreover, an increase in water temperature causes a decrease in the solubility of oxygen that is needed for the oxidation of biodegradable wastes. At the same time, the rate of oxidation is accelerated, imposing a faster oxygen demand on the smaller supply and thereby depleting the oxygen content of the water still further. Temperature also affects the lower organisms, such as plankton and crustaceans. In general, the more elevated the temperature is, the less desirable the types of algae in water. In cooler waters diatoms are the predominant phytoplankton in water that is not heavily eutrophic; with the same nutrient levels, green algae begin to become dominant at higher temperatures and diatoms decline; at the highest water temperatures, blue-green algae thrive and often develop into heavy blooms. One further ecological consequence of thermal pollution is that the spawning and migration of many fish are triggered by temperature and this behavior can be disrupted by thermal change.

Pollution with suspended sediments

Probably the most important effect that humans have had on water quality is the contribution to levels of

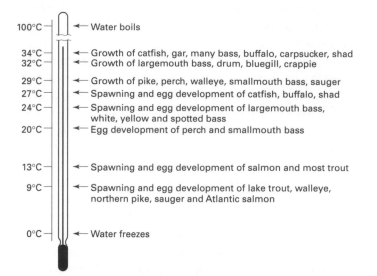

Figure 5.31 Maximum temperatures for the spawning and growth of fish. Heated waste water may be up to 5 to 10°C warmer than receiving waters and consequently the local fish populations cannot reproduce or grow properly. (After J. C. Giddings, *Chemistry, man and environmental change*, Harper and Row; from *Encounter with the Earth* by Leo F. Laporte, figure 13-2. Copyright © by permission of Harper & Row Publishers, Inc.)

suspended sediments in streams. This is a theme that is intimately tied up with that of soil erosion, on the one hand, and with channel manipulation by activities such as reservoir construction, on the other. The clearance of forest, the introduction of plowing and grazing by domestic animals, the construction of buildings, and the introduction of spoil materials into rivers by mineral extraction industries have all led to very substantial increases in levels of stream turbidity. Frequently sediment levels are a whole order of magnitude higher than they would have been under natural conditions. However, the introduction of soil conservation measures, or a reduction in the intensity of land use, or the construction of reservoirs, can cause a relative (and sometimes absolute) reduction in sediment loads. All three of these factors have contributed to the observed reduction in turbidity levels monitored in the rivers of the Piedmont region of Georgia in the USA.

Marine pollution

It is not within the scope of this book to discuss human impacts on the oceans to any great extent. However, it is worth making a few points on this subject, which has been well reviewed by Clark (1997). At first sight, as Jickells et al. (1991: 313) point out, two contradictory thoughts may cross our minds on this issue:

The first is the observation of ocean explorers, such as Thor Heyerdahl, of lumps of tar, flotsam and jetsam, and other products of human society thousands of kilometers from inhabited land. An alternative, vaguer feeling is that given the vastness of the oceans (more than 1000 billion billion liters of water!), how can man have significantly polluted them?

What is the answer to this conundrum? Jickells et al. (p. 330) draw a clear distinction between the open oceans and regional seas and in part come up with an answer:

The physical and chemical environment of the open oceans has not been greatly affected by events over the past 300 years, principally because of their large diluting capacity . . . Material that floats and is therefore not diluted, such as tar balls and litter, can be shown to have increased in amount and to have changed in character over the past 300 years.

In contrast to the open oceans, regional areas in close proximity to large concentrations of people show evidence of increasing concentrations of various substances that are almost certainly linked to human activities. Thus the partially enclosed North Sea and Baltic show increases in phosphate concentrations as a result of discharges of sewage and agriculture. The same is true of the more enclosed Black Sea (Mee, 1992).

This can cause accelerated eutrophication – often called 'cultural eutrophication' – which can lead to excessive growths of algae. Coastal and estuary water are sometimes affected by algal foam and scum, often called 'red tides'. Some of these blooms are so toxic that consumers of seafood that have been exposed to them can be affected by diarrhea, sometimes fatally. A thorough analysis of eutrophication in Europe's coastal waters is provided by the European Environment Agency (2001).

The nature of red tides has recently been discussed by Anderson (1994), who points out that these blooms, produced by certain types of phytoplankton (tiny pigmented plants), can grow in such abundance that they change the color of the seawater not only red but also brown or even green. They may be sufficiently toxic to kill marine animals such as fish and seals. Long-term studies at the local and regional level in many parts of the world suggest that these so-called red tides are increasing in extent and frequency as coastal pollution worsens and nutrient enrichment occurs more often.

Eutrophication also has adverse effects on coral reefs, as explained by Weber (1993: 49):

Initially, coral productivity increases with rising nutrient supplies. At the same time, however, corals are losing their key advantage over other organisms: their symbiotic self-sufficiency in nutrient poor seas. As eutrophication progresses, algae start to win out over corals for newly opened spaces on the reef because they grow more rapidly than corals when fertilized. The normally clear waters cloud as phytoplankton begin to multiply, reducing the intensity of the sunlight reaching the corals, further lowering their ability to compete. At a certain point, nutrients in the surrounding waters begin to overfertilize the corals' own zooxanthellae, which multiply to toxic levels inside the polyps. Eutrophication may also lead to black band and white band disease, two deadly coral disorders thought to be caused by algal infections. Through these stages of eutrophication, the health and diversity of reefs declines, potentially leading to death.

Likewise it is clear that pollution in the open ocean is, as yet, of limited biological significance. GESAMP (1990), an authoritative review of the state of the marine environment for the United Nations Environment Program, reported (p. 1):

The open sea is still relatively clean. Low levels of lead, synthetic organic compounds and artificial radionuclides, though widely detectable, are biologically insignificant. Oil slicks and litter are common along sea-lanes, but are, at present, of minor consequence to communities of organisms living in open-ocean waters.

On coastal waters it reported (p. 1):

The rate of introduction of nutrients, chiefly nitrates but sometimes also phosphates, is increasing, and areas of eutrophication are expanding, along with enhanced frequency and scale of unusual plankton blooms and excessive seaweed growth.

The two major sources of nutrients to coastal waters are sewage disposal and agricultural run-off from fertilizer-treated fields and from intensive stock raising.

Attention is also drawn to the presence of synthetic organic compounds – chlorinated hydrocarbons, which build up in the fatty tissues of top predators such as seals that dwell in coastal waters. Levels of contamination are decreasing in northern temperate areas but rising in tropical and subtropical areas due to continued use of chlorinated pesticides there.

The world's oceans have been greatly contaminated by oil. The effects of this pollution on animals have been widely studied. The sources of pollution include tanker collisions with other ships, the explosion of individual tankers because of the build-up of gas levels, the wrecking of tankers on coasts through navigational or mechanical failure, seepage from offshore oil installations, and the flushing of tanker holds. There is no doubt that the great bulk of the oil and related materials polluting the oceans results from human action (Blumer, 1972), although humans should not be attributed with all the blame, for natural seepages are reasonably common (Landes, 1973). Paradoxically, human actions mean that many natural seepages have diminished as wells have drawn down the levels of hydrocarbons in the oil-bearing rocks. Likewise, the flow of asphalt that once poured from the Trinidad Pitch Lake into the Gulf of Paria has also ceased because mining of the asphalt has lowered the lake below its outlet.

Of the various serious causes of oil pollution, a more important mechanism than the much-publicized role of tanker accidents is the discharge of the ballast water taken into empty tankers to provide stability on the return voyage to the loading terminal (Pitt, 1979). It is in this field, however, that technical developments have led to most amelioration. For example, the quantity of oil in tanker ballast water has been drastically reduced by the LOT (load on top) system, in which the ballast water is allowed to settle so that the oil rises to the surface. The tank is then drained until only the surface oil in the tank remains, and this forms part of the new cargo. In the future large tankers will be fitted with separate ballast tanks not used for oil cargoes, called segregated or clean ballast systems. Given such advances it appears likely than an increasing proportion of oil pollution will be derived from accidents involving tankers of ever increasing size, such as the spillage of 300×10^6 L that came from the wreck of the *Amoco Cadiz* in 1978. Even here, however, some progress has been made. Since 1980 there has been a reduction in the number of major oil spills, partly as a consequence of diminished long-distance oil transport to western Europe.

Points for review

What are the ecological effects of dams?

How do humans modify river channels?

To what extent do land cover changes lead to changes in river regimes?

Why should we be concerned about changes in groundwater levels?

Assess the causes and trends in nitrate pollution of water.

Guide to reading

Clark, R. B., 2001, *Marine pollution* (5th edn). Oxford: Clarendon Press. The standard work on this important topic.

Downs, P. W. and Gregory, K. J., 2004, *River channel management*. London: Arnold. A comprehensive treatment of river channels, human impacts, and river system management techniques.

Gleick, P. H. (ed.), 1993, *Water in crisis: a guide to the world's freshwater resources*. New York: Oxford University Press. A compendium of information on trends in water quality and consumption.

Kabat, P. and 9 others (eds), 2004, *Vegetation, water, humans and climate*. Berlin: Springer. A high-level research collection that investigates in particular the effects of vegetation change on climate and hydrological processes.

Petts, G. E., 1985, *Impounded rivers: perspectives for ecological management*. Chichester: Wiley. A treatment of the many consequences of dam construction.

6 HUMAN AGENCY IN GEOMORPHOLOGY

Introduction

The human role in creating landforms and modifying the operation of geomorphologic processes such as weathering, erosion, and deposition is a theme of great importance, though one that, particularly in the Western world, has not received the attention it deserves.

The range of the human impact on both forms and processes is considerable. For example, in Table 6.1 there is a list of some anthropogenic landforms together with some of the causes of their creation. There are very few spheres of human activity which do not, even indirectly, create landforms (Haigh, 1978). It is, however, useful to recognize that some features are produced by *direct* anthropogenic processes. These tend to be more obvious in their form and origin and are frequently created deliberately and knowingly. They include landforms produced by constructional activity (such as tipping), excavation, travel, hydrologic interference, and farming. Hillsides have been terraced in many parts of the world for many centuries, notably in the arid and semi-arid highlands of the New World (Donkin, 1979; Denevan, 2001), but examples are

also known from southern England where in Roman and Medieval times *strip lynchets* have been produced by plowing on steep slopes (Figure 6.1).

Landforms produced by *indirect* anthropogenic processes are often less easy to recognize, not least because

Table 6.1 Some anthropogenic landforms

Feature	Cause
Pits and ponds	Mining, marling
Broads	Peat extraction
Spoil heaps	Mining
Terracing, lynchets	Agriculture
Ridge and furrow	Agriculture
Cuttings and sunken lanes	Transport
Embankments	Transport, river and coast management
Dikes	River and coast management
Mounds	Defense, memorials
Craters	War, *qanat* construction
City mounds (*tells*)	Human occupation
Canals	Transport, irrigation
Reservoirs	Water management
Subsidence depressions	Mineral and water extraction
Moats	Defense
Banks along roads	Noise abatement

Figure 6.1 Strip lynchets (terraces) in Dorset, southern England, produced by plowing on steep slopes.

they tend to involve, not the operation of a new process or processes, but the acceleration of natural processes. They are the result of environmental changes brought about inadvertently by human technology. Nonetheless, it is probably this indirect and inadvertent modification of process and form that is the most crucial aspect of **anthropogeomorphology**. By removing natural vegetation cover – through the agency of cutting, burning, and grazing (see Trimble and Mendel, 1995) – humans have accelerated erosion and sedimentation. Sometimes the results will be obvious, for example when major gully systems rapidly develop; other results may have less immediate effect on landforms but are, nevertheless, of great importance. By other indirect means humans may create subsidence features and hazards, trigger off mass movements such as landslides, and even influence the operation of phenomena such as earthquakes.

Finally there are situations where, through a lack of understanding of the operation of processes and the links between different processes and phenomena, humans may deliberately and directly alter landforms and processes and thereby set in train a series of events which were not anticipated or desired. There are, for example, many records of attempts to reduce coast erosion by expensive engineering solutions, which, far from solving erosion problems, only exacerbated them.

Landforms produced by excavation

Of the landforms produced by direct anthropogenic processes those resulting from excavation are widespread,

and may have some antiquity. For example, Neolithic peoples in the Breckland of East Anglia in England used antler picks and other means to dig a remarkable cluster of deep pits in the chalk. The purpose of this was to obtain good-quality nonfrost-shattered flint to make stone tools. In many parts of Britain chalk has also been excavated to provide marl for improving acidic, light, sandy soils, and Prince (1962, 1964) made a meticulous study of the 27,000 pits and ponds in Norfolk that have resulted mainly from this activity – an activity particularly prevalent in the eighteenth century. It is often difficult in individual cases to decide whether the depressions are the results of human intervention, for solutional and periglacial depressions are often evident in the same area, but pits caused by human action do tend to have some distinctive features: irregular shape, a track leading into them, proximity to roads, etc. (see the debate between Prince (1979) and Sperling et al. (1979)).

Difficulties of identifying the true origin of excavational features were also encountered in explaining the Broads, a group of 25 freshwater lakes in the county of Norfolk. They are of sufficient area, and depth, for early workers to have precluded a human origin. Later it was proposed instead that they were natural features caused by uneven alluviation and siltation of river valleys which were flooded by the rapidly rising sea level of the Holocene (Flandrian) transgression. It was postulated by Jennings (1952) that the Broads were initiated as a series of discontinuous natural lakes, formed beyond the limits of a thick estuarine clay wedge laid down in Romano-British times by a transgression of the sea over earlier valley peats. The waters of the Broads were thought to have been impounded in natural peaty hollows between the flanges of the clay and the marginal valley slopes, or in tributary valleys whose mouths were blocked by the clay.

It is now clear, however, that the Broads are the result of human work (Lambert et al., 1970). Some of them have rectilinear and steep boundaries, most of them are not mentioned in early topographic books, and archival records indicate that peat cutting (*turbary*) was widely practiced in the area. On these and other grounds it is believed that peat-diggers, before AD 1300, excavated 25.5×10^6 m^3 of peat and so created the depressions in which the lakes have formed. The flooding may have been aided by sea-level change. Comparably extensive peat excavation was also carried on in The Netherlands, notably in the fifteenth century.

Figure 6.2 A limestone pavement at Hutton Roof Crags in northwest England. Such bare rock surfaces may result in part from accelerated erosion induced by the first farmers in prehistoric times. Many of them are now being damaged by quarrying and removal of stone for garden ornamentation.

Figure 6.3 The Rössing uranium mine near Swakopmund in Namibia, southern Africa. The excavation of such mines involves the movement of prodigious amounts of material.

Other excavational features result from war, especially craters caused by bomb or shell impact. Regrettably, human power to create such forms is increasing. It has been calculated (Westing and Pfeiffer, 1972) that between 1965 and 1971, 26 million craters, covering an area of 171,000 hectares, were produced by bombing in Indo-China. This represents a total displacement of no less than 2.6×10^9 m^3 of earth, a figure much greater than calculated as being involved in the peaceable creation of The Netherlands.

Some excavation is undertaken on a large scale for purely aesthetic reasons, when Nature offends the eye (Prince, 1959), while in many countries where land is scarce, whole hills are leveled and extensive areas stripped to provide fill for harbor reclamation. One of the most spectacular examples of this kind was the deliberate removal of steep-sided hills in the center of Brazil's Rio de Janeiro, for housing development.

An excavational activity of a rather specialized kind is the removal of limestone pavements (Figure 6.2) – areas of exposed limestone in northern England which were stripped by glaciers and then molded into bizarre shapes by solutional activity – for ornamental rock gardens. These pavements which consist of arid, bare rock surfaces (clints) bounded by deep, humid fissures (grikes) have both an aesthetic and a biological significance (Ward, 1979).

One of the most important causes of excavation is still mineral extraction (Figure 6.3), producing open-

pit mines, strip mines, quarries for structural materials, borrow pits along roads, and similar features (Doerr and Guernsely, 1956). Of these, 'without question, the environmental devastation produced by strip mining exceeds in quantity and intensity all of the other varied forms of man-made land destruction' (Strahler and Strahler, 1973: 284). This form of mining is a particular environmental problem in the states of Pennsylvania, Ohio, West Virginia, Kentucky, and Illinois. Oil shales, a potential source of oil that at present is relatively untapped, can be exploited by open-pit mining, by traditional room and pillar mining, and by underground *in situ* pyrolysis. Vast reserves exist (as in Canada) but the amount of excavation required will probably be about three times the amount of oil produced (on a volume basis), suggesting that the extent of both the excavation and subsequent dumping of overburden and waste will be considerable (Routson et al., 1979). Some of the waste, produced by the retorting of the shale to release the oil, contains soluble salts and potentially harmful trace elements, which limit the speed of ground reclamation (Petersen, 1981).

An early attempt to provide a general picture of the importance of excavation in the creation of the landscape of Britain was given by Sherlock (1922). He estimated (see Table 6.2) that up until the time in which he wrote, human society had excavated around 31×10^9 m^3 of material in the pursuit of its economic activities. That figure must now be a gross underestimate, partly because Sherlock himself was not in a position to appreciate the anthropogenic role in creating features such as the Norfolk Broads, and partly

Table 6.2 Total excavation of material in Great Britain until 1922. Source: Sherlock (1922: 86)

Activity	Approximate volume (m³)
Mines	15,147,000,000
Quarries and pits	11,920,000,000
Railways	2,331,000,000
Manchester Ship Canal	41,154,000
Other canals	153,800,000
Road cuttings	480,000,000
Docks and harbors	77,000,000
Foundations of buildings and street excavation	385,000,000
Total	30,534,954,000

Table 6.3 Humans as earthmovers. Source: Hooke (1994)
(a) Deliberate human earthmoving actions in the USA, omitting the indirect effects of actions such as deforestation and cultivation

Activity	(10⁹ tonnes year⁻¹)
Excavation for housing and other construction	0.8
Mining	3.8
Road work	3.0
Total USA	7.6
World total (roughly four times the USA total)	30.0

(b) Estimated world totals due to natural earthmoving processes

Activity	(10^9 tonnes year^{-1})
River transport	
(a) to oceans and lakes	14
(b) short-distance transport within river basins	40
Tectonic forces lifting continents	14
Volcanic activity elevating sea floor	30
Glacial transport	4.3
Wind transport	1.0

because, since his time, the rate of excavation has greatly accelerated. Sherlock's 1922 study covers a period when earth-moving equipment was still ill-developed. Nonetheless, on the basis of his calculations, he was able to state that 'at the present time, in a densely peopled country such as England, Man is many times more powerful, as an agent of denudation, than all the atmospheric denuding forces combined' (p. 333). The most notable change since Sherlock wrote has taken place in the production of aggregates for concrete. Demand for these materials in the UK grew from 20 million tonnes per annum in 1900 to 202 million tonnes in 2001, a tenfold increase.

For the world as a whole, the annual movement of soil and rock resulting from mineral extraction may be as high as 3000 billion tonnes (Holdgate et al., 1982: 186). By comparison it has been estimated that the amount of sediment carried into the ocean by the world's rivers each year amounts to 24 billion tonnes per year (Judson, 1968). More recently, Hooke (1994) has tried to produce some data on the significance of deliberate human earthmoving actions in the USA and globally and these are shown in Table 6.3.

Landforms produced by construction and dumping

The process of constructing mounds and embankments and the creation of dry land where none previously existed is longstanding. In Britain the mound at Silbury Hill dates back to prehistoric times, and the pyramids

of Central America, Egypt, and the Far East are even more spectacular early feats of landform creation. Likewise in the Americas Native Indians, prior to the arrival of Europeans, created large numbers of mounds of different shapes and sizes for temples, burials, settlement, and effigies (Denevan, 1992: 377). In the same way, hydrologic management has involved, over many centuries, the construction of massive banks and walls – the ultimate result being the present-day landscape of The Netherlands. Transport developments have also required the creation of large constructional landmarks, but probably the most important features are those resulting from the dumping of waste materials, especially those derived from mining (see Figure 6.4). It has been calculated that there are at least 2000 million tonnes of shale lying in pit heaps in the coalfields of Britain (Richardson, 1976). In the Middle East and other areas of long-continued human urban settlement the accumulated debris of life has gradually raised the level of the land surface, and occupation mounds (*tells*) are a fertile source of information to the archaeologist. Today, with the technical ability to build that humans have, even estuaries may be converted from

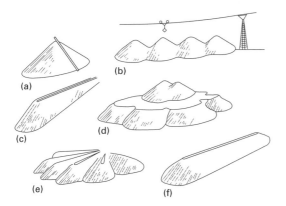

Figure 6.4 Some shapes produced by shale tipping: (a) conical, resulting from MacClaine tipping; (b) multiple cones tipped from aerial ropeways; (c) high fan-ridge by tramway tipping over slopes; (d) high plateau mounds topped with cones; (e) low multiple fan ridges by tramway tipping; (f) lower ridge by tramway tipping (after Haigh, 1978, figure 2.1).

ecologically productive environments into suburban sprawl by the processes of dredging and filling. Indeed, one of the striking features of the distribution of the world's population is the tendency for large human concentrations to occur near vast expanses of water. Many of these cities have extended out on to land that has been reclaimed from the sea (e.g., Hong Kong, Figure 6.5), thereby providing valuable sites for development, but sometimes causing the loss of rich fishing grounds and ecologically valuable wetlands (Hudson, 1979).

The ocean floors are also being affected because of the vast bulk of waste material that humankind is creating. Disposal of solid waste by coastal cities is now sufficiently large to modify shorelines, and it covers adjacent ocean bottoms with characteristic deposits on a scale large enough to be geologically significant. This has been brought out dramatically by Gross (1972: 3174), who undertook a quantitative comparison of the amount of solid wastes dumped into the Atlantic by humans in the New York metropolitan region with the amount of sediment brought into the ocean by rivers:

The discharge of waste solids exceeds the suspended sediment load of any single river along the U.S. Atlantic coast. Indeed, the discharge of wastes from the New York metropolitan region is comparable to the estimated suspended-sediment yield (6.1 megatons per year) of all rivers along the Atlantic coast between Maine and Cape Hatteras, North Carolina.

Not only are the rates of sedimentation high, but the anthropogenic sediments tend to contain abnormally high contents of such substances as carbon and heavy metals (Goldberg et al., 1978).

Many of the features created by excavation in one generation are filled in by another, since phenomena such as water-filled hollows produced by mineral extraction are often both wasteful of land and also suitable locations for the receipt of waste. The same applies to natural hollows such as **karstic** or ground-ice depressions. Watson (1976), for example, has mapped

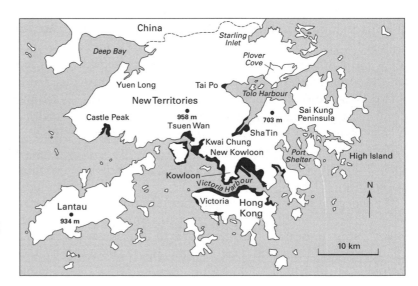

Figure 6.5 Map of Hong Kong showing main urban reclamation areas shaded black (reprinted from Hudson, 1979, figure 1, in *Reclamation review*, 2, 3–16, permission of Pergamon Press Ltd © Pergamon Press Ltd).

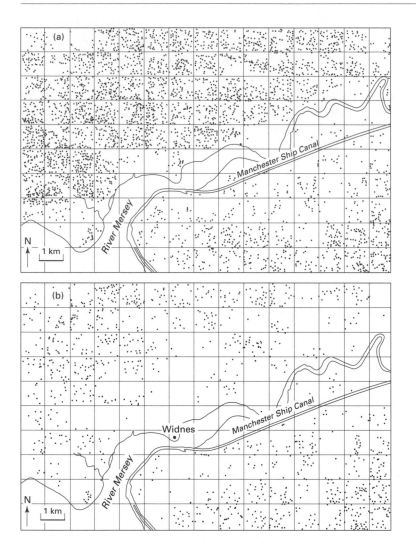

Figure 6.6 The distribution of pits and ponds in a portion of northwestern England: (a) in the mid-nineteenth century; (b) in the mid-twentieth century (after Watson, 1976).

the distribution of hollows, which were largely created by marl diggers in the lowlands of southwest Lancashire and northwest Cheshire, as they were represented on mid-nineteenth-century topographic maps (Figure 6.6a). When this distribution is compared with the present-day distribution for the same area (Figure 6.6b), it is evident that a very substantial proportion of the holes has been infilled and obliterated by humans, with only 2114 out of 5380 remaining. Hole densities have fallen from 121 to 47 km^{-2}.

At the present time, large quantities of waste are sent to landfill sites. In 1995, member states of the European Union landfilled more than 80% of their waste, but under the EU Landfill Directive this figure should be reduced substantially in coming decades.

Accelerated sedimentation

An inevitable consequence of the accelerated erosion produced by human activities has been accelerated sedimentation (see, e.g., Komar et al.'s 2004 study of sedimentation in Tillamook Bay, Oregon). This has been heightened by the deliberate addition of sediments to stream channels as a result of the need to dispose of mining and other wastes.

In a classic study, G. K. Gilbert (1917) demonstrated that hydraulic mining in the Sierra Nevada mountains of California led to the addition of vast quantities of sediments into the river valleys draining the range. This in itself raised their bed levels, changed their channel configurations and caused the flooding of lands that had previously been immune. Of even greater significance was the fact that the rivers transported vast quantities of debris into the estuarine bays of the San Francisco system, and caused extensive shoaling which in turn diminished the tidal prism of the bay. Gilbert calculated the volume of shoaling produced by hydraulic mining since the discovery of gold to be 846×10^6 m^3.

Comparably serious sedimentation of bays and estuaries has also been caused by human activity on the eastern coast of America. As Gottschalk (1945: 219) wrote:

Both historical and geologic evidence indicates that the pre-agricultural rate of silting of eastern tidal estuaries was low. The history of sedimentation of ports in the Chesapeake Bay area is an epic of the effects of uncontrolled erosion since the beginning of the wholesale land clearing and cultivation more than three centuries ago.

He has calculated that at the head of the Chesapeake Bay, 65×10^6 m^3 of sediment were deposited between 1846 and 1938. The average depth of water over an area of 83 km^2 was reduced by 0.76 m. New land comprising 318 hectares was added to the state of Maryland and, as Gottschalk remarked, 'the Susquehanna River is repeating the history of the Tigris and Euphrates'. Much of the material entrained by erosive processes on upper slopes as a result of agriculture in Maryland, however, was not evacuated as far as the coast. Costa (1975) has suggested, on the basis of the study of sedimentation, that only about one-third of the eroded material left the river valleys. The remainder accumulated on floodplains as alluvium and colluvium at rates of up to 1.6 cm per year. Similarly, Happ (1944), working in Wisconsin, carried out an intensive augering survey of floodplain soil and established that, since the development of agriculture, floodplain aggradation had proceeded at a rate of approximately 0.85 cm per year. He noted that channel and floodplain aggradation had caused the flooding of low alluvial terraces to be more frequent, more extensive and deeper. The rate of sedimentation has since declined (Trimble, 1976), because of less intensive land use and the institution of effective erosion control measures on farmland (see also Magilligan, 1985).

Such valley sedimentation is by no means restricted to the newly settled terrains of North America. There

is increasing evidence to suggest that silty valley fills in Germany, France, and Britain, many of them dating back to the Bronze Age and the Iron Age, are the result of accelerated slope erosion produced by the activities of early farmers (Bell, 1982). Indeed, in recent years, various studies have been undertaken with a view to assessing the importance of changes in sedimentation rate caused by humans at different times in the Holocene in Britain. Among the formative events that have been identified are: initial land clearance by Mesolithic and Neolithic peoples; agricultural intensification and sedentarization in the late Bronze Age; the widespread adoption of the iron plow in the early Iron Age; settlement by the Vikings; and the introduction of sheep farming (Table 6.4).

A core from Llangorse Lake in the Brecon Beacons of Wales (Jones et al., 1985) provides excellent long-term data on changing sedimentation rates:

Period (years BP)	Sedimentation rate (cm 100 years^{-1})
9000–7500	3.57
7500–5000	1.0
5000–2800	13.2
2800–AD 1840	14.1
c. AD 1840–present	59.0

The thirteenfold increase in rates after 5000 years BP seems to have occurred rapidly and is attributed to initial forest clearance. The second dramatic increase of more than fourfold took place in the past 150 years and is a result of agricultural intensification.

In the past two centuries rates of sedimentation in lake basins have changed in different ways in different basins according to the differing nature of economic activities in catchments. Some data from various sources are listed for comparison in Table 6.5. In the case of the Loe Pool in Cornwall (southwest England) rates of sedimentation were high while mining industry was active, but fell dramatically when mining was curtailed. In the case of Seeswood Pool in Warwickshire, a dominantly agricultural catchment area in central England, the highest rates have occurred since 1978 in response to various land management changes, such as larger fields, continuous cropping, and increased dairy herd size. In other catchments, pre-afforestation plowing may have caused sufficient disturbance to cause accelerated sedimentation. For example, Battarbee et al. (1985b) looked at sediment cores in the Galloway area of southwest Scotland and found that in Loch

Table 6.4 Accelerated sedimentation in Britain in prehistoric and historic times

Location	Source	Evidence and date
Howgill Fells	Harvey et al. (1981)	Debris cone production following tenth century AD introduction of sheep farming
Upper Thames Basin	Robinson and Lambrek (1984)	River alluviation in the late Bronze Age and early Iron Age
Lake District	Pennington (1981)	Accelerated lake sedimentation at 5000 years BP as a result of Neolithic agriculture
Mid-Wales	Macklin and Lewin (1986)	Floodplain sedimentation (Capel Bangor unit) on Breidol as a result of early Iron Age sedentary agriculture
Brecon Beacons	Jones et al. (1985)	Lake sedimentation increase after 5000 years BP at Llangorse due to forest clearance
Weald	Burrin (1985)	Valley alluviation from Neolithic onwards until early Iron Age
Bowland Fells	Harvey and Renwick (1987)	Valley terraces at 5000–2000 years BP (Bronze or Iron Age settlement) and after 1000 years BP (Viking settlement)
Southern England	Bell (1982)	Fills in dry valleys: Bronze and Iron Age
Callaly Moor (Northumberland)	Macklin et al. (1991)	Valley fill sediments of late Neolithic to Bronze Age

Table 6.5 Data on rates of erosion and sedimentation in lakes in the last two centuries

Location	Dates	Activity	Rates of erosion in catchment (R. Cober) as determined from lake sedimentation rates ($t\ km^{-2}\ year^{-1}$)
Loe Pool (Cornwall)	1860–1920	Mining and agriculture	174
(from O'Sullivan et al., 1982)	1930–6	Intensive mining and agriculture	421
	1937–8	Intensive mining and agriculture	361
	1938–81	Agriculture	12
Seeswood Pool (Warwickshire)	1765–1853		7.0
(from Foster et al., 1986)	1854–80		12.2
	1881–1902		8.1
	1903–19		9.6
	1920–5		21.6
	1926–33		16.1
	1934–47		12.7
	1948–64		12.0
	1965–72		13.9
	1973–7		18.3
	1978–82		36.2

Grannoch the introduction of plowing in the catchment caused an increase in sedimentation from 0.2 cm per year to 2.2 cm per year.

The work of Binford et al. (1987) on the lakes of the Peten region of northern Guatemala (Central America), an area of tropical lowland dry forest, is also instructive with respect to early agricultural colonization. Combining studies of archaeology and lake sediment stratigraphy, they were able to reconstruct the diverse environmental consequences of the growth of Mayan civilization (Figure 6.7). This civilization showed a dramatic growth after 3000 years BP, but collapsed in the ninth century AD. The hypotheses put forward to explain this collapse include warfare, disease, earthquakes, and soil degradation. The population has remained relatively low ever since, and after the first European contact (AD 1525) the region was virtually depopulated. The period of Mayan success saw a marked

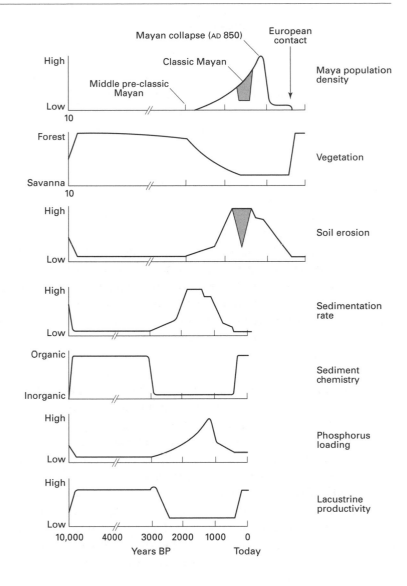

Figure 6.7 The human and environmental history of the Peten Lakes, Guatemala. The shaded areas indicate a phase of local population decline (modified after Binford et al., 1987).

reduction in vegetation cover, an increase in lake sedimentation rates and in catchment soil erosion, an increased supply of inorganic silts and clays to the lakes, a pulse of phosphorus derived from human wastes, and a decrease in lacustrine productivity caused by high levels of turbidity.

Sedimentation has severe economic implications because of the role it plays in reducing the effective lifetime of reservoirs. In the tropics capacity depletion through sedimentation is commonly around 2% per year (Myers, 1988: 14), so that the expected useful life of the Paute hydroelectric project in Ecuador (cost US$ 600 million) is 32 years, that of the Mangla Dam in Pakistan (cost US$ 600 million) is 57 years, and that of the Tarbela Dam, also in Pakistan, is just 40 years.

Ground subsidence

Not all ground subsidence is caused by humans. For example, limestone solutional processes can, in the absence of humankind, create a situation where a cavern collapses to produce a surface depression or sinkhole, and permafrost will sometimes melt to produce a **thermokarst** depression without human intervention. Nonetheless, ground subsidence can be caused or accelerated by humans in a variety of ways: by the transfer of subterranean fluids (such as oil, gas, and water); by the removal of solids through underground mining or by dissolving solids and removing them in solution (e.g., sulfur and salt); by the disruption of permafrost; and by the compaction or reduction of sediments because of drainage and irrigation (Johnson, 1991; Barends et al., 1995).

Some of the most dangerous and dramatic collapses have occurred in limestone areas because of the dewatering of limestone caused by mining activities. In the Far West Rand of South Africa, gold mining has required the abstraction of water to such a degree that the local water table has been lowered by more than 300 m. The fall of the water table caused miscellaneous clays and other materials filling the roofs of large caves to dry out and shrink so that they collapsed into the underlying void. One collapse created a depression 30 m deep and 55 m across, killing 29 people. In Alabama in the southern USA, water-level decline consequent upon pumping has had equally serious consequences in a limestone terrain; and Newton (1976) has estimated that since 1900 about 4000 induced sinkholes or related features have been formed, while fewer than 50 natural sinkholes have been reported over the same interval. Sinkholes that may result from such human activity are also found in Georgia, Florida, Tennessee, Pennsylvania, and Missouri. In 1904, near Tampa (Florida), overpumping of an artesian aquifer caused 64 new sinkholes to form within a one-month period.

In some limestone areas, however, a reverse process can operate. The application of water to overburden above the limestone may render it more plastic so that the likelihood of collapse is increased. This has occurred beneath reservoirs, such as the May Reservoir in central Turkey, and as a result of the application of wastewater and sewerage to the land surface. Williams (1993) provides a survey of the diverse effects of human activities on limestone terrains.

The process can be accelerated by the direct solution of susceptible rocks. For example, collapses have occurred in gypsum bedrock because of solution brought about by the construction of a reservoir. In 1893 the MacMillan Dam was built on the Pecos River in New Mexico, but within 12 years the whole river flowed through caves which had developed since construction. Both the San Fernando and Rattlesnake Dams in California suffer severe leakage for similar reasons.

Subsidence produced by oil abstraction is an increasing problem in some parts of the world. The classic area is Los Angeles, where 9.3 m of subsidence occurred as a result of exploitation of the Wilmington oilfield between 1928 and 1971. The Inglewood oilfield displayed 2.9 m of subsidence between 1917 and 1963. Some coastal flooding problems occurred at Long Beach because of this process. Similar subsidence has been recorded from the Lake Maracaibo field in Venezuela (Prokopovich, 1972) and from some Russian fields (Nikonov, 1977).

A more widespread problem is posed by groundwater abstraction for industrial, domestic, and agricultural purposes. Table 6.6 presents some data for such subsidences from various parts of the world. The ratios of subsidence to water level decline are strongly dependent on the nature of the sediment composing the aquifer. Ratios range from 1:7 for Mexico City, to 1:80 for the Pecos in Texas, and to less than 1:400 for London, England (Rosepiler and Reilinger, 1977). The extent of subsidence that has taken place in the USA as a result of groundwater abstraction has recently been assessed by Chi and Reilinger (1984). In Japan subsidence has also now emerged as a major problem (Nakano and Matsuda, 1976). In 1960 only 35.2 km^2 of the Tokyo lowland was below sea level, but continuing subsidence meant that by 1974 this had increased to 67.6 km^2, exposing a total of 1.5 million people to major flood hazard. Shanghai in China is another low-lying Asian city that has suffered from subsidence, with as much as 2 to 3 m since 1921 (Chai et al., 2004).

The subsidence caused by mining (which led to court cases in England as early as the fifteenth century as a consequence of associated damage to property) is perhaps the most familiar, although its importance varies according to such factors as the thickness of seam removed, its depth, the width of working, the degree of filling with solid waste after extraction, the geologic structure, and the method of working adopted (Wallwork, 1974). In general terms, however, the vertical displacement by subsidence is less than the thickness of the seam being worked, and decreases with an increase in the depth of mining. This is because the overlying strata collapse, fragment, and fracture, so that the mass of rock fills a greater space than it did when naturally compacted. Consequently the surface expression of deep-seated subsidence may be equal to little more than one-third of the thickness of the material removed. Subsidence associated with coal mining may disrupt surface drainage and the resultant depressions then become permanently flooded.

Coal-mining regions are not the only areas where subsidence problems are serious. In Cheshire, northwest England, rock salt is extracted from two major

Table 6.6 Ground subsidence. Source: data in Cooke and Doornkamp (1974); Prokopovich (1972); Rosepiler and Reilinger (1977); Holzer (1979); Nutalaya and Ran (1981); Johnson (1991)

Cause	Location	Amount (m)	Date	Rate (mm year^{-1})
Ground subsidence produced by oil and gas abstraction	Azerbaydzhan, USSR	2.5	1912–62	50
	Atravopol, USSR	1.5	1956–62	125
	Wilmington, USA	9.3	1928–71	216
Ground subsidence produced by groundwater abstraction	Inglewood, USA	2.9	1917–63	63
	Maracaibo, Venezuela	5.03	1929–90	84
	London, England	0.06–0.08	1865–1931	0.91–1.21
	Savannah, Georgia (USA)	0.1	1918–55	2.7
	Mexico City	7.5	–	250–300
	Houston, Galveston, Texas	1.52	1943–64	60–76
	Central Valley, California	8.53	–	–
	Tokyo, Japan	4	1892–1972	500
	Osaka, Japan	>2.8	1935–72	76
	Niigata, Japan	>1.5	–	–
	Pecos, Texas	0.2	1935–66	6.5
	South-central Arizona	2.9	1934–77	96
	Bangkok, Thailand	0.5	–	100
	Shanghai, China	2.62	1921–65	60

seams, each about 30 m in thickness. Moreover, these seams occur at no great depth – the uppermost being at about 70 m below the surface. A further factor to be considered is that the rock salt is highly soluble in water, so the flooding of mines may cause additional collapse. These three conditions – thick seams, shallow depth and high solubility – have produced optimum conditions for subsidence and many subsidence lakes called 'flashes' have developed (Wallwork, 1956). Some of these are illustrated in Figure 6.8.

Some subsidence is created by a process called hydrocompaction, which is explained thus. Moisture-deficient, unconsolidated, low-density sediments tend to have sufficient dry strength to support considerable effective stresses without compacting. However, when such sediments, which may include alluvial fans or loess, are thoroughly wetted for the first time (e.g., by percolating irrigation water) the intergranular strength of the deposits is diminished, rapid compaction takes place, and ground surface subsidence follows. Unequal subsidence can create problems for irrigation schemes.

Land drainage can promote subsidence of a different type, notably in areas of organic soils. The lowering of the water table makes peat susceptible to oxidation and deflation so that its volume decreases. One of the

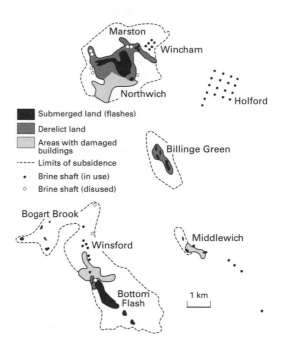

Figure 6.8 Subsidence in the salt area of mid-Cheshire, England, in 1954 (after Wallwork, 1956, figure 3).

longest records of this process, and one of the clearest demonstrations of its efficacy, has been provided by the measurements at Holme Fen Post in the English Fenlands. Approximately 3.8 m of subsidence occurred

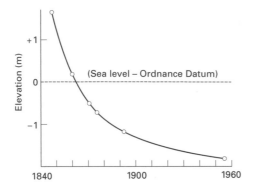

Figure 6.9 The subsidence of the English Fenlands peat at Holme Fen Post from 1842 to 1960 following drainage (from data in Fillenham, 1963).

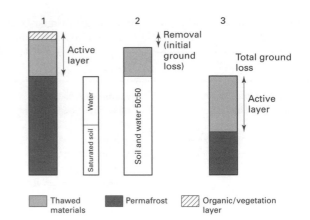

Figure 6.10 Diagram illustrating how the disturbance of high ice-content terrain can lead to permanent ground subsidence. 1–3 indicate stages before, immediately after, and subsequent to disturbance (after Mackay in French, 1976, figure 6.1).

between 1848 and 1957 (Fillenham, 1963), with the fastest rate occurring soon after drainage had been initiated (Figure 6.9). The present rate averages about 1.4 cm per year (Richardson and Smith, 1977). At its maximum natural extent the peat of the English Fenland covered around 1750 km^2. Now only about one-quarter (430 km^2) remains.

A further type of subsidence, sometimes associated with earthquake activity, results from the effects on Earth's crust of large masses of water impounded behind reservoirs. As we shall see later, seismic effects can be generated in areas with susceptible fault systems and this may account for earthquakes recorded at Koyna (India) and elsewhere. This process whereby a mass of water causes coastal depression is called hydro-isostasy.

In tundra regions ground subsidence is associated with thermokarst development, thermokarst being irregular, hummocky terrain produced by the melting of ground ice, permafrost. The development of thermokarst is due primarily to the disruption of the thermal equilibrium of the permafrost and an increase in the depth of the active layer. This is illustrated in Figure 6.10. Following French (1976: 106), consider an undisturbed tundra soil with an active layer of 45 cm. Assume also that the soil beneath 45 cm is supersaturated permafrost and yields on a volume basis upon thawing 50% excess water and 50% saturated soil. If the top 15 cm were removed, the equilibrium thickness of the active layer, under the bare ground conditions, might increase to 60 cm. As only 30 cm of the

original active layer remains, 60 cm of the permafrost must thaw before the active layer can thicken to 60 cm, since 30 cm of supernatant water will be released. Thus, the surface subsides 30 cm because of thermal melting associated with the degrading permafrost, to give an overall depression of 45 cm.

Thus the key process involved in thermokarst subsidence is the state of the active layer and its thermal relationships. When, for example, surface vegetation is cleared for agricultural or constructional purposes the depth of thaw will tend to increase. The movement of tracked vehicles has been particularly harmful to surface vegetation and deep channels may soon result from permafrost degradation. Similar effects may be produced by the siting of heated buildings on permafrost, and by the laying of oil, sewer, and water pipes in or on the active layer (Ferrians et al., 1969; Lawson, 1986).

Thus subsidence is a diverse but significant aspect of the part humans play as geomorphologic agents. The damage caused on a worldwide basis can be measured in billions of dollars each year (Coates, 1983), and among the effects are broken dams, cracked buildings (Figure 6.11), offset roads and railways, fractured well casings, deformed canals and ditches, bridges that need releveling, saline encroachment, and increased flood damage.

Figure 6.11 The city of Mexico has subsided by many meters as a result of groundwater abstraction. Many of the ancient buildings in the city center have been severely damaged, and have been cracked and deformed.

Arroyo trenching, gullies, and peat haggs

In the southwestern USA many broad valleys and plains became deeply incised with valley-bottom gullies (*arroyos*) over a short period between 1865 and 1915, with the 1880s being especially important (Cooke and Reeves, 1976). This cutting had a rapid and detrimental effect on the flat, fertile, and easily irrigated valley floors, which are the most desirable sites for settlement and economic activity in a harsh environment. The causes of the phenomenon have been the subject of prolonged debate (Elliott et al., 1999; Gonzalez, 2001).

Many students of this phenomenon have believed that thoughtless human actions caused the entrenchment, and the apparent coincidence of white settlement and arroyo development tended to give credence to this viewpoint. The range of actions that could have been culpable is large: timber-felling, overgrazing, cutting grass for hay in valley bottoms, compaction along well-traveled routes, channeling of runoff from trails and railways, disruption of valley-bottom sods by animals' feet, and the invasion of grasslands by miscellaneous types of scrub.

On the other hand, study of the long-term history of the valley fills shows that there have been repeated phases of aggradation and incision and that some of these took place before the influence of humans could have been a significant factor (Waters and Haynes, 2001). This has prompted debate as to whether the arrival of white communities was in fact responsible for this particularly severe phase of environmental degradation. Huntington (1914), for example, argued that valley filling would be a consequence of a climatic shift to more arid conditions. These, he believed, would cause a reduction in vegetation, which in turn would promote rapid removal of soil from devegetated mountain slopes during storms, and overload streams with sediment. With a return to humid conditions vegetation would be re-established, sediment yields would be reduced, and entrenchment of valley fills would take place. Bryan (1928) put forward a contradictory climatic explanation. He argued that a slight move towards drier conditions, by depleting vegetation cover and reducing soil infiltration capacity, would produce significant increases in storm runoff, which would erode valleys. Another climatic interpretation was advanced by Leopold (1951), involving a change in rainfall intensity rather than quantity. He indicated that a reduced frequency of low-intensity rains would weaken the vegetation cover, while an increased frequency of heavy rains at the same time would increase the incidence of erosion. Support for this contention comes from the work of Balling and Wells (1990) in New Mexico. They attributed early twentieth-century arroyo trenching to a run of years with intense and erosive rainfall characteristics that succeeded a phase of drought conditions in which the productive ability of the vegetation had declined. It is also possible, as Schumm et al. (1984) have pointed out, that arroyo incision could result from neither climatic change nor

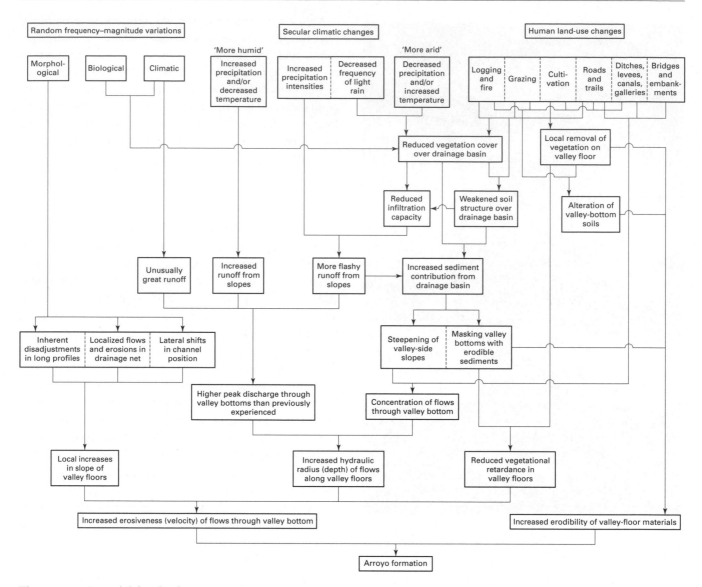

Figure 6.12 A model for the formation of arroyos (gullies) in southwestern USA (after Cooke and Reeves, 1976, figure 1.2).

human influence. It could be the result of some natural geomorphologic threshold being crossed. Under this argument, conditions of valley-floor stability decrease slowly over time until some triggering event initiates incision of the previously 'stable' reach.

It is therefore clear that the possible mechanisms that can lead to alternations of cut-and-fill of valley sediments are extremely complex, and that any attribution of all arroyos in all areas to human activities may be a serious oversimplification of the problem

(Figure 6.12). In addition, it is possible that natural environmental changes, such as changes in rainfall characteristics, have operated at the same time and in the same direction as human actions.

In the Mediterranean lands there have also been controversies surrounding the age and causes of alternating phases of aggradation and erosion in valley bottoms. Vita-Finzi (1969) suggested that at some stage during historical times many of the steams in the Mediterranean area, which had hitherto been engaged

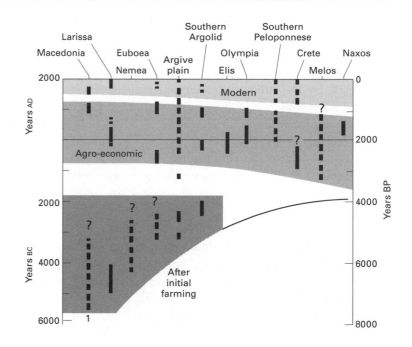

Figure 6.13 Chronology of Holocene alluviation in Greece and the Aegean. Broken bars are dated uncertainly or represent intermittent deposition (from various sources in Van Andel et al., 1990, figure 10).

primarily in downcutting, began to build up their beds. Renewed downcutting, still seemingly in operation today, has since incised the channels into the alluvial fill. He proposed that the reversal of the downcutting trend in the Middle Ages was both ubiquitous and confined in time, and that some universal and time-specific agency was required to explain it. He believed that devegetation by humans was not a medieval innovation and that some other mechanism was required. A solution he gave to account for the phenomenon was precipitation change during the climatic fluctuation known as the **Little Ice Age** (AD 1550–1850). This was not an interpretation that found favor with Butzer (1974). He reported that his investigations showed plenty of post-Classical and pre-1500 alluviation (which could not therefore be ascribed to the Little Ice Age), and he doubted whether Vita-Finzi's dating was precise enough to warrant a 1550–1850 date. Instead, he suggested that humans were responsible for multiple phases of accelerated erosion from slopes, and accelerated sedimentation in valley bottoms, from as early as the middle of the first millennium BC.

Butzer's interpretation has found favor with van Andel et al. (1990) who have detected an intermittent and complex record of cut-and-fill episodes during the late Holocene in various parts of Greece (Figure 6.13). They believe that this evidence is compatible with a

model of the control of timing and intensity of landscape destabilization by local economic and political conditions. This is a view shared in the context of the Algarve in Portugal by Chester and James (1991).

A further location with spectacular gullies, locally called *lavaka*, is Madagascar. Here too there have been debates about cultural versus natural causation. Proponents of cultural causes have argued that since humans arrived on the island in the last two thousand years there has been excessive cattle grazing, removal of forest for charcoal and for slash-and-burn cultivation, devastating winter (dry season) burning of grasslands, and erosion along tracks and trails. However, the situation is more complex than that and the *lavaka* are polygenetic. Tectonism and natural climatic acidification may be at least as important, and given the climatic and soil types of the island many *lavaka* are a natural part of the landscape's evolution. Some of them also clearly predate primary (i.e., uncut) rain forest. The many factors, natural and cultural, involved in *lavaka* development are well reviewed by Wells and Andriamihaja (1993).

Another example of drainage incision that demonstrates the problem of disentangling the human from the natural causes of erosion is provided by the eroding peat bogs of highland Britain (Bragg and Tallis, 2001). Over many areas, including the Pennines of northern

England and the Brecon Beacons of Wales, blanket peats are being severely eroded to produce pool and hummock topography, areas of bare peat, and incised gullies (haggs). Many rivers draining such areas are discolored by the presence of eroded peat particles, and sediment yields of organic material are appreciable (Labadz et al., 1991).

Some of the observed peat erosion may be an essentially natural process, for the high water content and low cohesion of undrained peat masses make them inherently unstable. Moreover, the instability must normally become more pronounced as peat continues to accumulate, leading to bog slides and bursts round margins of expanded peat blankets. Conway (1954) suggested that an inevitable end-point of peat build-up on high-altitude, flat or convex surfaces is that a considerable depth of unconsolidated and poorly humidified peat overlies denser and well-humidified peat, so adding to the instability. Once a bog burst or slide occurs, this leads to the formation of drainage gullies which extend back into the peat mass, slumping-off marginal peat downslope, and leading to the drawing off of water from the pools of the hummock and hollow topography of the watershed.

Tallis (1985) believes that there have been two main phases of erosion in the Pennines. The first, initiated 1000–1200 years ago, may have been caused by natural instability of the type outlined above. However, there has been a second stage of erosion, initiated 200–300 years ago, in which miscellaneous human activities appear to have been important. Radley (1962) suggested that among the pressures that had caused erosion were heavy sheep grazing, regular burning, peat cutting, the digging of boundary ditches, the incision of packhorse tracks, and military maneuvers during the First World War. Other causes may include footpath erosion (Wishart and Warburton, 2001) and severe air pollution (Tallis, 1965), the latter causing the loss of a very important peat forming moss, *Sphagnum*. In South Wales, there is some evidence that the blanket peats have degenerated as a result of contamination by particulate pollution (soot, etc.) during the Industrial Revolution (Chambers et al., 1979). On the other hand, in Scotland lake-core studies indicate that severe peat erosion was initiated between AD 1500 and 1700, prior to air pollution associated with industrial growth, and Stevenson et al. (1990) suggest that this erosion initiation may have been caused either by the adverse climatic conditions of the Little Ice Age or by an increasing intensity of burning as land use pressures increased.

Accelerated weathering and the tufa decline

Although fewer data are available and the effects are generally less immediately obvious, there is some evidence that human activities have produced changes in the nature and rate of weathering (Winkler, 1970). The prime cause of this is probably air pollution. It is clear that, as a result of increased emissions of sulfur dioxide through the burning of fossil fuels, there are higher levels of sulfuric acid in rain over many industrial areas. This in itself may react with stones and cause their decay. Chemical reactions involving sulfur dioxide can also generate salts such as calcium sulfate and magnesium sulfate, which may be effective in causing the physical breakdown of rock through the mechanism of salt weathering.

Similarly, atmospheric carbon-dioxide levels have been rising steadily because of the burning of fossil fuels, and deforestation. Carbon dioxide may combine with water, especially at lower temperatures, to produce weak carbonic acid, which can dissolve limestone, marbles, and dolomites. Weathering can also be accelerated by changes in groundwater levels resulting from irrigation. This can be illustrated by considering the Indus Plain in Pakistan (Goudie, 1977), where irrigation has caused the water table to be raised by about 6 m since 1922. This has produced increased evaporation and salinization. The salts that are precipitated by evaporation above the capillary fringe include sodium sulfate, a very effective cause of stone decay. Indeed buildings, such as the great archaeological site of Mohenjo-Daro, are decaying at a catastrophic rate (Figure 6.14).

In other cases accelerated weathering has been achieved by moving stone from one environment to another. Cleopatra's Needle, an Egyptian obelisk in New York City, is an example of rapid weathering of stone in an inhospitable environment. Originally erected on the Nile opposite Cairo about 1500 BC, it was toppled in about 500 BC by Persian invaders, and lay partially buried in Nile sediments until, in 1880, it was moved to New York. It immediately began to mechanism(s) that might be involved. If the late Holocene reduction in tufa deposition is a reality, then it is necessary to consider a whole range of possible mechanisms, both natural and anthropogenic (Nicod, 1986: 71–80; Table 6.7). As yet the case for an anthropogenic role is not proven (Goudie et al., 1993).

Accelerated mass movements

There are many examples of **mass movements** being triggered by human actions (Selby, 1979). For instance, landslides can be created either by undercutting or by overloading (Figure 6.15). When a road is constructed, material derived from undercutting the upper hillside may be cast on to the lower hillslope as a relatively loose fill to widen the road bed. Storm water is then often diverted from the road on to the loose fill.

Because of the hazards presented by both natural and accelerated mass movements, humans have developed a whole series of techniques to attempt to control them. Such methods, many of which are widely used by engineers, are listed in Table 6.8. These techniques are increasingly necessary, as human capacity to change a hillside and to make it more prone to slope failure has been transformed by engineering development. Excavations are going deeper, buildings and other structures are larger, and many sites that are at best marginally suitable for engineering projects are now being used because of increasing pressure on land. This applies especially to some of the expanding urban areas in the humid parts of low latitudes – Hong Kong, Kuala Lumpur, Rio de Janeiro, and many others. It is very seldom that human agency deliberately accelerates mass movements; most are accidentally

Figure 6.14 The ancient city of Mohenjo-Daro in Pakistan was excavated in the 1920s. Irrigation has been introduced into the area, causing groundwater levels to be raised. This has brought salt into the bricks of the ancient city, producing severe disintegration.

suffer from scaling and the inscriptions were largely obliterated within ten years because of the penetration of moisture, which enabled frost-wedging and hydration of salts to occur.

During the 1970s and 1980s an increasing body of isotopic dates became available for deposits of tufa (secondary freshwater deposits of limestone, also known as travertine). Some of these dates suggest that over large parts of Europe, from Britain to the Mediterranean basin and from Spain to Poland, rates of tufa formation were high in the early and mid-Holocene, but declined markedly thereafter (Weisrock, 1986: 165–7). Vaudour (1986) maintains that since around 3000 years BP 'the impact of man on the environment has liberated their disappearance', but he gives no clear indication of either the basis of this point of view or of the precise

Table 6.7 Some possible mechanisms to account for the alleged Holocene tufa decline

Climatic/natural	Anthropogenic
Discharge reduction following rainfall decline leading to less turbulence	Discharge reduction due to overpumping, diversions, etc.
Degassing leads to less deposition	Increased flood scour and runoff variability of channels due to deforestation, urbanization, ditching, etc.
Increased rainfall causing more flood scour	Channel shifting due to deforestation of floodplains leads to tufa erosion
Decreasing temperature leads to less evaporation and more CO_2 solubility	Reduced CO_2 flux in system after deforestation
Progressive Holocene peat development and soil podzol development through time leads to more acidic surface waters	Introduction of domestic stock causes breakdown of fragile tufa structures
	Deforestation = less fallen trees to act as foci for tufa barrages
	Increased stream turbidity following deforestation reduces algal productivity

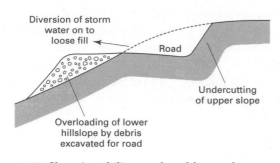

Figure 6.15 Slope instability produced by road construction.

Table 6.8 Methods used to control mass movements on slopes. Source: after R. F. Baker and H. E. Marshall, in Dunne and Leopold (1978, table 15.16)

Type of movement	Method of control
Falls	Flattening the slope
	Benching the slope
	Drainage
	Reinforcement of rock walls using anchor bolts and grouting with cement
	Covering of walls with steel mesh
Slides and flows	Grading or benching to flatten slope
	Drainage of surface water with ditches
	Sealing surface cracks to prevent infiltration
	Subsurface drainage
	Rock and earth buttresses at foot
	Retaining walls at foot
	Pilings through the potential slide mass

caused, the exception possibly being the deliberate triggering of a threatening snow avalanche (Perla, 1978).

The forces producing slope instability and landsliding can be divided usefully into disturbing forces and resisting properties (Cooke and Doornkamp, 1990: 113–14). The factors leading to an increase in shear stress (disturbing forces) can be listed as follows (modified after Cooke and Doornkamp, 1990: 113):

1 removal of lateral or underlying support
  • undercutting by water (e.g., river, waves), or glacier ice
  • weathering of weaker strata at the toe of a slope
  • washing out of granular material by seepage erosion
  • *human cuts and excavations, drainage of lakes or reservoirs*

2 increased disturbing forces
  • natural accumulations of water, snow, talus
  • *pressure caused by human activity* (e.g., stockpiles of ore, tip-heaps, rubbish dumps, or buildings)
3 transitory earth stresses
  • earthquakes
  • *continual passing of heavy traffic*
4 increased internal pressure
  • build-up of pore-water pressures (e.g., in joints and cracks, especially in the tension crack zone at the rear or the slide)

Some of the factors are natural, while others (italicized) are affected by humans. Factors leading to a decrease in the shearing resistance of materials making up a slope can also be summarized (also modified after Cooke and Doornkamp, 1990: 113).

1 Materials:
  • beds which decrease in shear strength if water content increases (clays, shale, mica, schist, talc, serpentine) (e.g., *when local water-table is artificially increased in height by reservoir construction*), or as a result of stress release (vertical and/or horizontal) following slope formation;
  • low internal cohesion (e.g., consolidated clays, sands, porous organic matter);
  • in bedrock – faults, bedding planes, joints, foliation in schists, cleavage, brecciated zones, and pre-existing shears.
2 Weathering changes:
  • weathering reduces effective cohesion, and to a lesser extent the angle of shearing resistance;
  • absorption of water leading to changes in the fabric of clays (e.g., loss of bonds between particles or the formation of fissures).
3 Pore-water pressure increase:
  • high groundwater table as a result of increased precipitation, or *as a result of human interference* (e.g., *dam construction*) (see 1 above).

Once again the italics show that there are a variety of ways in which humans can play a role.

Some mass movements are created by humans piling up waste soil and rock into unstable accumulations that fail spontaneously. At Aberfan, in South Wales, a major disaster occurred when a coal-waste tip 180 m high began to move as an earth flow. The tip had been constructed not only as a steep slope but also upon a spring line. This made an unstable configuration, which eventually destroyed a school and claimed over 150 lives. In Hong Kong, where a large proportion of the population is forced to occupy steep slopes developed on deeply weathered granites and other rocks, mass movements are a severe problem, and So (1971) has shown that many of the landslides and washouts (70% of those in the great storm of June 1966, for example) were associated with road sections and slopes artificially modified through construction and cultivation.

In southeastern France humans have accelerated landslide activity by building excavations for roads and by loading slopes with construction material (Julian and Anthony, 1996). The undercutting and removal of the trees on slopes for the construction of roads and paths has also led to landsliding in the Himalayas (Barnard et al., 2001). The arrival of European settlers in New Zealand, particularly since the 1840s, had a profound effect on landslide activity, as they cleared the forest and converted it to pasture (Glade, 2003).

One of the most serious mass movements partly caused by human activity was that which caused the Vaiont Dam disaster in Italy in 1963, in which 2600 people were killed (Kiersch, 1965). Heavy antecedent rainfall and the presence of young, highly folded sedimentary rocks provided the necessary conditions for a slip to take place, but it was the construction of the Vaiont Dam itself which changed the local groundwater conditions sufficiently to affect the stability of a rock mass on the margins of the reservoir: 240×10^6 m^3 of ground slipped, causing a rise in water level which overtopped the dam and caused flooding and loss of life downstream. Comparable slope instability resulted when the Franklin D. Roosevelt lake was impounded by the Columbia River in the USA (Coates, 1977), but the effects were, happily, less serious.

It is evident from what has been said about the predisposing causes of the slope failure *triggered* by the Vaiont Dam that human agency was only able to have such an impact because the natural conditions were broadly favorable. Exactly the same lesson can be learnt from the accelerated landsliding in southern Italy. Nossin (1972) has demonstrated how, in Calabria, road construction has triggered off (and been hindered by) landsliding, but he has also stressed that the area is fundamentally susceptible to such mass movement

activity because of geologic conditions. It is an area where recent rapid uplift has caused downcutting by rivers and the undercutting of slopes by erosion. It is also an area of incoherent metamorphic rocks, with frequent faulting. Further, water is often trapped by Tertiary clay layers, providing further stimulus to movement.

Although the examples of accelerated mass movements that have been given here are essentially associated with the effects of modern construction projects, more long-established activities, including deforestation and agriculture, are also highly important. For example, Innes (1983) has demonstrated, on the basis of lichenometric dating of debris-flow deposits in the Scottish Highlands, that most of the flows have developed in the past 250 years, and he suggests that intensive burning and grazing may be responsible.

Fire, whether natural or man-induced, can be a major cause of slope instability and debris flow generation by removing or reducing protective vegetation, by increasing peak stream flows, and by leading to larger soil moisture contents and soil-water pore pressures (because of reduced interception of rainfall and decreased moisture loss by transpiration) (Wondzell and King, 2003). Examples of fire-related debris flow generation are known from many sites in the USA, including Colorado (Cannon et al., 2001b), New Mexico (Cannon et al., 2001a), the Rocky Mountains and the Pacific North West (Wondzell and King, 2003).

However, as with so many environmental changes of that nature in the past, there are considerable difficulties in being certain about causation. This has been well expressed by Ballantyne (1991: 84):

Although there is growing evidence for Late Holocene erosion in upland Britain, the causes of this remain elusive. A few studies have presented evidence linking erosion to vegetation degradation and destruction due to human influence, but the validity of climatic deterioration as a cause of erosion remains unsubstantiated. This uncertainty stems from a tendency to link erosion with particular causes only through assumed coincidence in timing, a procedure fraught with difficulty because of imprecision in the dating of both putative causes and erosional effects. Indeed, in many reported instances, it is impossible to refute the possibility that the timing of erosional events or episodes may be linked to high magnitude storms of random occurrence, and bears little relation to either of the casual hypotheses outlined above. . . .

Deliberate modification of channels

Both for purposes of navigation and flood control humans have deliberately straightened many river channels. Indeed, the elimination of meanders contributes to flood control in two ways. First, it eliminates some overbank floods on the outside of curves, against which the swiftest current is thrown and where the water surface rises highest. Second, and more importantly, the resultant shortened course increases both the gradient and the flow velocity, and the floodwaters erode and deepen the channel, thereby increasing its flood capacity.

It was for this reason that a program of channel cutoffs was initiated along the Mississippi in the early 1930s. By 1940 it had lowered flood stages by as much as 4 m at Arkansas City, Arkansas. By 1950 the length of the river between Memphis, Tennessee and Baton Rouge, Louisiana (600 km down the valley) had been reduced by no less than 270 km as a result of 16 cutoffs.

Some landscapes have become dominated by artificial channels, normally once again because of the need for flood alleviation and drainage. This is especially evident in an area such as the English Fenlands where straight constructed channels contrast with the sinuous courses of original rivers such as the Great Ouse.

Nondeliberate river-channel changes

There are thus many examples of the intentional modification of river-channel geometry by humans – by the construction of embankments, by channelization, and by other such processes. The complexity and diversity of causes of stream channel change is brought out in the context of Australia (Table 6.9) (Rutherford, 2000), and more generally by Downs and Gregory (2004). Major changes in the configuration of channels can be achieved accidentally (Table 6.10), either because of human-induced changes in stream discharge or in their sediment load: both parameters affect channel capacity (Park, 1977). The causes of observed cases of riverbed degradation are varied and complex and result from a variety of natural and human changes (Table 6.11). A useful distinction can be drawn between degradation that proceeds downstream, and that which

Table 6.9 Human impacts on Australian stream channel morphology

Channel incision by changes to resistance of valley flow (drains, cattle tracks)
Enlargement due to catchment clearing and grazing
Channel enlargement by sand and gravel extraction
Erosion by boats
Scour downstream from dams
Channelization and river training
Acceleration of meander migration rates by removal of riparian vegetation
Channel avulsion because of clearing of floodplain vegetation
Sedimentation resulting from mining
Channel contraction below dams
Channel invasion and narrowing by exotic vegetation

Table 6.10 Causes of riverbed degradation. Source: after Galay (1983)

Type	Primary cause	Contributory cause
Downstream progressing	Decrease of bed-material discharge	Dam construction
		Excavation of bed material
		Diversion of bed material
		Change in land use
		Storage of bed material
	Increased water discharge	Diversion of flow
		Rare floods
	Decrease in bed-material size	River emerging from lake
	Other	Thawing of permafrost
Upstream progressing	Lower base level	Drop in lake level
		Drop in level of main river
		Excavation of bed material
	Decrease in river length	Cutoffs
		Channelization
		Stream capture
		Horizontal shift of base level
	Removal of control point	Natural erosion
		Removal of dam

proceeds upstream, but in both cases the complexity of causes is evident.

Deliberate channel straightening causes various types of sequential channel adjustment both within and downstream from straightened reaches, and the types of adjustment vary according to such influences as stream gradient and sediment characteristics. Brookes (1987) recognized five types of change *within* the straightened reaches (types W1 to W5) and two

Table 6.11 Accidental channel changes

Phenomenon	Cause
Channel incision	Clear-water erosion below dams caused by sediment removal
Channel aggradation	Reduction in peak flows below dams Addition of sediment to streams by mining, agriculture, etc.
Channel enlargement	Increase in discharge level produced by urbanization
Channel diminution	Discharge decrease following water abstraction or flood control Trapping and stabilizing of sediment by artificially introduced plants
Channel planform	Change in nature of sediment load and its composition, together with flow regime

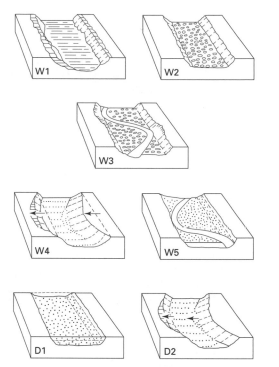

Figure 6.16 Principal types of adjustment in straightened river changes (after Brookes, 1987, figure 4). For an explanation of the different types see text.

types of change downstream (types D1 and D2). They are illustrated in Figure 6.16.

Type W1 is degradation of the channel bed, which results from the fact that straightening increases the slope by providing a shorter channel path. This in turn increases its sediment transport capability.

Type W2 is the development of an armored layer on the channel bed by the more efficient removal of fine materials as a result of the increased sediment transport capability referred to above.

Type W3 is the development of a sinuous thalweg in streams that are not only straightened but which are also widened beyond the width of the natural channel.

Type W4 is the recovery of sinuosity as a result of bank erosion in channels with high slope gradients.

Type W5 is the development of a sinuous course by deposition in streams with a high sediment load and a relatively low valley gradient.

Types D1 and D2 result from deposition downstream as the stream tries to even out its gradient, the deposition occurring as a general raising of the bed level, or as a series of accentuated point bar deposits.

It is now widely recognized that the urbanization of a river basin results in an increase in the peak flood flows in a river. It is also recognized that the morphology of stream channels is related to their discharge characteristics, and especially to the discharge at which bank full flow occurs. As a result of urbanization, the

frequency of discharges which fill the channel will increase, with the effect that the beds and banks of channels in erodible materials will be eroded so as to enlarge the channel (Trimble, 1997b). This in turn will lead to bank caving, possible undermining of structures, and increases in turbidity (Hollis and Luckett, 1976). Trimble (2003) provides a good historical analysis of how the San Diego Creek in Orange County, California, has responded to flow and sediment yield changes related to the spread of both agriculture and urbanization.

Changes in channel morphology also result from discharge diminution and sediment load changes produced by flood-control works and diversions for irrigation (Brandt, 2000). This can be shown for the North Platte and the South Platte in America, where both peak discharge and mean annual discharge have declined to 10–30% of their pre-dam values. The North Platte, 762–1219 m wide in 1890 near the Wyoming Nebraska border, has narrowed to about 60 m at present, while the South Platte River was about 792 m wide, 89 km above its junction with the North Platte

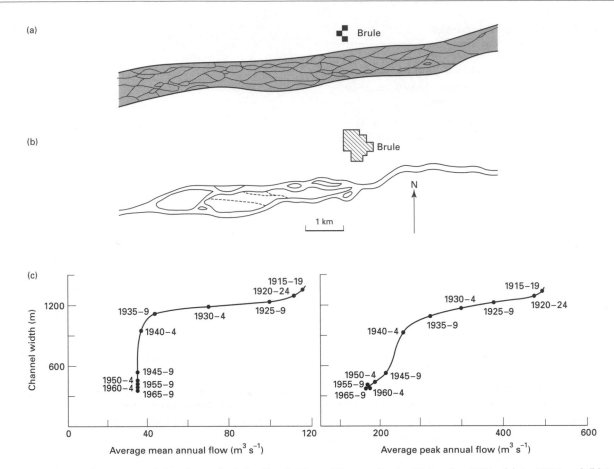

Figure 6.17 The configuration of the channel of the South Platte River at Brule, Nebraska, USA: (a) in 1897 and (b) in 1959. Such changes in channel form result from discharge diminution (c) caused by flood-control works and diversions for irrigation (after Schumm, 1977, figure 5.32 and Williams, 1978).

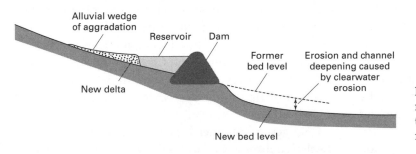

Figure 6.18 Diagrammatic long profile of a river showing the upstream aggradation and the downstream erosion caused by dam and reservoir construction.

in 1897, but had narrowed to about 60 m by 1959 (Schumm, 1977: 161). The tendency of both rivers has been to form one narrow, well-defined channel in place of the previously wide, braided channels, and, in addition, the new channel is generally somewhat more sinuous than the old (Figure 6.17).

Similarly, the building of dams can lead to channel aggradation upstream from the reservoir and channel deepening downstream because of the changes brought about in sediment loads (Figure 6.18). Some data on observed rates of degradation below dams are presented in Table 6.12. They show that the average rate of degradation has been of the order of a few meters over a few decades following closure of the dams. However, over time the rate of degradation seems to become less or to cease altogether, and Leopold et al.

Table 6.12 Riverbed degradation below dams. Source: From data in Galay (1983)

River	Dam	Amount (m)	Length (km)	Time (years)
South Canadian (USA)	Conchos	3.1	30	10
Middle Loup (USA)	Milburn	2.3	8	11
Colorado (USA)	Hoover	7.1	111	14
Colorado (USA)	Davis	6.1	52	30
Red (USA)	Denison	2.0	2.8	3
Cheyenne (USA)	Angostura	1.5	8	16
Saalach (Austria)	Reichenhall	3.1	9	21
South Saskatchewan (Canada)	Diefenbaker	2.4	8	12
Yellow (China)	Samenxia	4.0	68	4

Table 6.13 Channel capacity reduction below reservoirs. Source: modified after Petts (1979, table 1)

River	Dam	Channel capacity loss (%)
Republican, USA	Harlan County	66
Arkansas, USA	John Martin	50
Rio Grande, USA	Elephant Buttre	50
Tone, UK	Clatworthy	54
Meavy, UK	Burrator	73
Nidd, UK	Angram	60
Burn, UK	Burn	34
Derwent, UK	Ladybower	40

(1964: 455) suggest that this can be brought about in several ways. First, because degradation results in a flattening of the channel slope in the vicinity of the dam, the slope may become so flat that the necessary force to transport the available materials is no longer provided by the flow. Second, the reduction of flood peaks by the dam reduces the competence of the transporting stream to carry some of the material on its bed. Thus if the bed contains a mixture of particle size the river may be able to transport the finer sizes but not the larger, and the gradual winnowing of the fine particles will leave an armor of coarser material that prevents further degradation.

The overall effect of the creation of reservoirs by the construction of a dam is to lead to a reduction in downstream channel capacity (see Petts, 1979, for a review). This seems to amount to between about 30 m and 70% (see Table 6.13).

Equally far-reaching changes in channel form are produced by land-use changes and the introduction of soil conservation measures. Figure 6.19 is an idealized representation of how the river basins of Georgia in the USA have been modified through human agency between 1700 (the time of European settlement) and the present. Clearing of the land for cultivation (Figure 6.19b) caused massive slope erosion, which resulted in the transfer of large quantities of sediment into channels and floodplains. The phase of intense erosive land use persisted and was particularly strong during the nineteenth century and the first decades of the twentieth century, but thereafter (Figure 6.19c) conservation measures, reservoir construction, and a reduction in the intensity of agricultural land use led to further channel changes (Trimble, 1974). Streams ceased to carry such a heavy sediment load, they became much less turbid, and incision took place into the floodplain sediments. By means of this active streambed erosion, streams incised themselves into the modern alluvium, lowering their beds by as much as 3–4 m.

In the Platte catchment of southwest Wisconsin a broadly comparable picture of channel change has been documented by Knox (1977). There, as in the Upper Mississippi Valley (Knox, 1987), it is possible to identify stages of channel modification associated with various stages of land use, culminating in decreased overbank sedimentation as a result of better land management in the past half century.

Other significant changes produced in channels include those prompted by accelerated sedimentation associated with changes in the vegetation communities growing along channels. The introduction of salt cedar in the southern USA has caused significant floodplain aggradation. In the case of the Brazos River in Texas, for example, the plants encourage sedimentation by their damming and ponding effect. They clogged channels by invading sand banks and sand bars, and so increased the area subject to flooding. Between 1941 and 1979 the channel width declined from 157 to 67 m, and the amount of aggradation was as much as 5.5 m (Blackburn et al., 1983). Equally, the establishment or re-establishment of riparian forest has been implicated with channel narrowing in southeastern France during the twentieth century (Liébault and Piégay, 2002).

There is, however, a major question about the ways in which different vegetation types affect channel form (Trimble, 2004). Are tree-lined banks more stable than

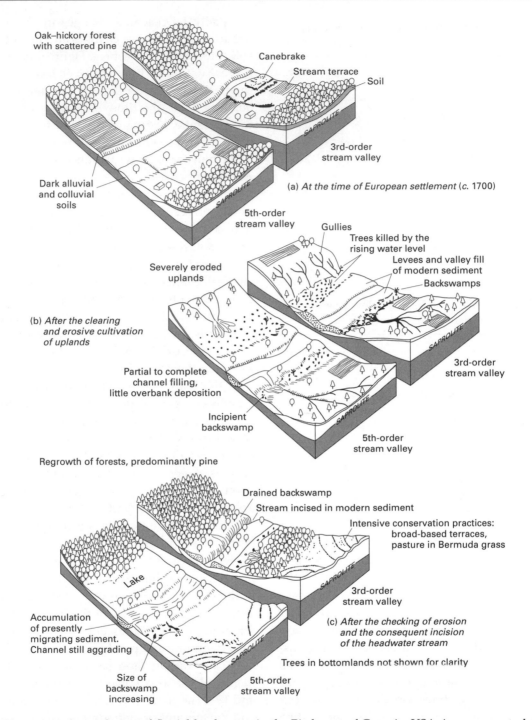

Oak–hickory forest
with scattered pine

Canebrake

Stream terrace

Soil

SAPROLITE

3rd-order
stream valley

Dark alluvial
and colluvial
soils

SAPROLITE

5th-order
stream valley

(a) *At the time of European settlement* (*c.* 1700)

Severely eroded
uplands

Gullies

Trees killed by the
rising water level

Levees and valley fill
of modern sediment

Backswamps

SAPROLITE

3rd-order
stream valley

(b) *After the clearing
and erosive cultivation
of uplands*

Partial to complete
channel filling,
little overbank deposition

Incipient
backswamp

SAPROLITE

5th-order
stream valley

Regrowth of forests, predominantly pine

Drained backswamp

Stream incised in modern sediment

Intensive conservation practices:
broad-based terraces,
pasture in Bermuda grass

Lake

SAPROLITE

3rd-order
stream valley

Accumulation
of presently
migrating sediment.
Channel still aggrading

(c) *After the checking of erosion
and the consequent incision
of the headwater stream*

Trees in bottomlands not shown for clarity

Size of
backswamp
increasing

SAPROLITE

5th-order
stream valley

Figure 6.19 Changes in the evolution of fluvial landscapes in the Piedmont of Georgia, USA, in response to land-use change between 1700 and 1970 (after Trimble, 1974, p. 117, in S. W. Trimble, *Man-induced soil erosion on the southern Piedmont*, Soil Conservation Society of America. © Soil Conservation Society of America).

those flowing through grassland? On the one hand tree roots stabilize banks and their removal might be expected to cause channel widening and shallowing (Brooks and Brierly, 1997). On the other hand, forests produce log-jams that can cause aggradation or concentrate flow on to channel banks, thereby leading to their erosion. These issues are discussed in Trimble (1997b) and Montgomery (1997).

Another organic factor that can modify channel form is the activity of grazing animals. These can break the banks down directly by trampling and can reduce bank resistance by removing protective vegetation and loosening soil (Trimble and Mendel, 1995).

Finally, the addition of sediments to stream channels by mining activity can cause channel aggradation. Mine wastes can clog channel systems (Gilbert, 1917; Lewin et al., 1983). Equally, the mining of aggregates from river beds themselves can lead to channel deepening (Bravard and Petts, 1996: 246–7).

Reactivation and stabilization of sand dunes

To George Perkins Marsh the reactivation and stabilization of sand dunes, especially coastal dunes, was a theme of great importance in his analysis of human transformation of nature. He devoted 54 pages to it:

The preliminary steps, whereby wastes of loose, drifting, barren sands are transformed into wooded knolls and plains, and finally through the accumulation of vegetable mold, into arable ground, constitute a conquest over nature which proceeds agriculture – a geographical revolution – and therefore, an account of the means by which the change has been effected belongs properly to the history of man's influence on the great features of physical geography (1965: 393).

He was fascinated by 'the warfare man wages with the sand hills' and asked (1965: 410) 'in what degree the naked condition of most dunes is to be ascribed to the improvidence and indiscretion of man'.

His analysis showed quite clearly that most of the coastal dunes of Europe and North America had been rendered mobile, and hence a threat to agriculture and settlement, through human action, especially because of grazing and clearing. In Britain the cropping of dune warrens by rabbits was a severe problem, and a most significant event in their long history was the myxomatosis outbreak of the 1950s, which severely reduced the rabbit population and led to dramatic changes in stability and vegetative cover.

Appreciation of the problem of dune reactivation on mid-latitude shorelines, and attempts to overcome it, go back a long way (Kittredge, 1948). For example, the menace of shifting sand following denudation is recognized in a decree of 1539 in Denmark, which imposed a fine on those who destroyed certain species of sand plants on the coast of Jutland. The fixation of coastal sand dunes by planting vegetation was initiated in Japan in the seventeenth century, while attempts at the reafforestation of the spectacular Landes dunes in southwest France began as early at 1717, and came to fruition in the nineteenth century through the plans of the great Bremontier: 81,000 hectares of moving sand had been fixed in the Landes by 1865. In Britain possibly the most impressive example of sand control is provided by the reafforestation of the Culbin Sands in northeast Scotland with conifer plantations (Edlin, 1976).

Human-induced dune instability is not, however, a problem that is restricted to mid-latitude coasts. In inland areas of Europe, clearing, fire, and grazing have affected some of the late Pleistocene dune fields that were created on the arid steppe margins of the great ice sheets, and in eastern England the dunes of the Breckland presented problems on many occasions. There are records of carriages being halted by sand-blocked roads and of one village, Downham, being overwhelmed altogether.

However, it is possibly on the margins of the great subtropical and tropical deserts that some of the strongest fears are being expressed about sand-dune reactivation. This is one of the facets of the process of desertification. The increasing population levels of both humans and their domestic animals, brought about by improvement in health and by the provision of boreholes, has led to an excessive pressure on the limited vegetation resources. As ground cover has been reduced, so dune instability has increased. The problem is not so much that dunes in the desert cores are relentlessly marching on to moister areas, but that the fossil dunes, laid down during the more arid phase peaking around 18,000 years ago, have been reactivated *in situ*.

A wide range of methods (Figure 6.20) is available to attempt to control drifting sand and moving dunes as follows:

1 drifting sand
  • enhancement of deposition of sand by the creation of large ditches, vegetation belts, and barriers and fences

Figure 6.20 Techniques to control dune movement: (a) a sand fence on a dune that threatens part of the town of Walvis Bay in Namibia; (b) a patchwork of palm frond fences being used at Erfoud on the edge of the Sahara in Morocco; (c) vegetation growth on coastal sand dunes at Hout Bay in South Africa being encouraged by irrigation.

- enhancement of transport of sand by aerodynamic streamlining of the surface, change of surface materials, or paneling to direct flow
- reduction of sand supply, by surface treatment, improved vegetation cover, or erection of fences
- deflection of moving sand, by fences, barriers, tree belts, etc.

2 moving dunes
- removal by mechanical excavation
- destruction by reshaping, trenching through dune axis, or surface stabilization of barchan arms
- immobilization by trimming, surface treatment, and fences

In practice most solutions to the problem of dune instability and sand blowing have involved the establishment of a vegetation cover. This is not always easy. Species used to control sand dunes must be able to endure undermining of their roots, burning, abrasion, and often severe deficiencies of soil moisture. Thus the species selected need to have the ability to recover after partial growth in the seedling stages, to promote rapid litter development, and to add nitrogen to the soil through root nodules. During the early stages of growth they may need to be protected by fences, sand traps, and surface mulches. Growth can also be stimulated by the addition of synthetic fertilizers.

In hearts of deserts sand dunes are naturally mobile because of the sparse vegetation cover. Even here, however, humans sometimes attempt to stabilize sand surfaces to protect settlements, pipelines, industrial plant and agricultural land. The use of relatively porous barriers to prevent or divert sand movement has proved comparatively successful, and palm fronds or chicken wire have made adequate stabilizers. Elsewhere surfaces have been strengthened by the application of high-gravity oil or by salt-saturated water (which promotes the development of wind-resistant surface crusts).

In temperate areas coastal dunes have been effectively stabilized by the use of various trees and other plants (Ranwell and Boar, 1986). In Japan *Pinus thumbergii* has been successful, while in the great Culbin Sands plantations of Scotland *P. nigra* and *P. laricio* have been used initially, followed by *P. sylvestris*. Of the smaller shrubs, *Hippophae* has proved highly efficient, sometimes too efficient, at spreading. Its clearance from areas where it is not welcome is difficult precisely

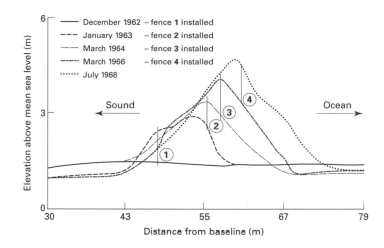

Figure 6.21 Sand accumulation using the method of multiple fences in North Carolina, USA. This raised the dune height approximately 4 m over a period of 6 years (after Savage and Woodhouse, in Goldsmith, 1978, figure 36).

because of some of the properties that make is such an efficient sand stabilizer: vigorous suckering growth and the rapid regrowth of cut stems (Boorman, 1977). Different types of grass have also been employed, especially in the early stages of stabilization. These include two grasses that are moderately tolerant of salt: *Elymus farctus* (sand twitch) and *Leymus arenarius* (lime grass). Another grass that is much used, not least because of its rapid and favorable response to burial by drifting sand, is *Ammophila arenaria* (marram).

Further stabilization of coastal dunes has been achieved by setting up sand fences. These generally consist of slats about 1.0–1.5 m high, and have a porosity of 25–50%. They have proved to be effective in building incipient dunes in most coastal areas. By installing new fences regularly, large dunes can be created with some rapidity (see Figure 6.21). Alternative methods, such as using junk cars on the beaches at Galveston, Texas, have been attempted with little success.

Accelerated coastal erosion

Because of the high concentration of settlements, industries, transport facilities, and recreational developments on coastlines, the pressures placed on coastal landforms are often acute (Nordstrom, 1994) and the consequences of excessive erosion serious. While most areas are subject to some degree of natural erosion and accretion, the balance can be upset by human activity in a whole range of different ways (Table 6.14). However, humans seldom attempt to accelerate coastal

erosion deliberately. More usually, it is an unexpected or unwelcome result of various economic projects. Frequently coast erosion has been accelerated as a result of human efforts to reduce it.

One of the best forms of coastal protection is a good beach. If material is removed from a beach, accelerated cliff retreat may take place. Removal of beach materials may be necessary to secure valuable minerals, including heavy minerals, or to provide aggregates for construction. The classic example of the latter was the mining of 660,000 tonnes of shingle from the beach at Hallsands in Devon, England, in 1887 to provide material for the construction of dockyards at Plymouth. The shingle proved to be undergoing little or no natural replenishment and in consequence the shore level was reduced by about 4 m. The loss of the protective shingle soon resulted in cliff erosion to the extent of 6 m between 1907 and 1957. The village of Hallsands was cruelly attacked by waves and is now in ruins.

Another common cause of beach and cliff erosion at one point is coast protection at another (Figure 6.22). As already stated, a broad beach serves to protect the cliffs behind, and beach formation is often encouraged by the construction of **groynes** and a range of 'hard engineering' structures is available (Figure 6.23). However, these structures sometimes merely displace the erosion (possibly in an even more marked form) further along the coast. This is illustrated in Figure 6.24.

Piers or breakwaters can have similar effects to groynes. This has occurred at various places along the British coast: erosion at Seaford resulted from the Newhaven breakwater, while erosion at Lowestoft resulted

Table 6.14 Mechanisms of human-induced erosion in coastal zones. Source: Hails (1977: 348, table 9.11)

Human-induced erosion zones	Effects
Beach mining for placer deposits (heavy minerals) such as zircon, rutile, ilmenite and monazite	Loss of sand from frontal dunes and beach ridges
Construction of groynes, breakwaters, jetties and other structures	Downdrift erosion
Construction of offshore breakwaters	Reduction in littoral drift
Construction of retaining walls to maintain river entrances	Interruption of littoral drift resulting in downdrift erosion
Construction of sea-walls, revetments, etc.	Wave reflection and accelerated sediment movement
Deforestation	Removal of sand by wind
Fires	Migrating dunes and sand drift after destruction of vegetation
Grazing of sheep and cattle	Initiation of blow-outs and transgressive dunes: sand drift
Off-road recreational vehicles (dune buggies, trail bikes, etc.)	Triggering mechanism for sand drift attendant upon removal of vegetative cover
Reclamation schemes	Changes in coastal configuration and interruption of natural processes, often causing new patterns in sediment transport
Increased recreational needs	Accelerated deterioration, and destruction, of vegetation on dunal areas, promoting erosion by wind and wave action

Figure 6.22 Coastal defense: (a) at Weymouth, southern England; (b) at Arica, northern Chile. The piecemeal emplacement of expensive sea walls and cliff protection structures is often only of short-term effectiveness and can cause accelerated erosion downdrift.

from the pier at Gorleston. Figure 6.25 illustrates the changes in location of beach erosion achieved by the building of jetties or breakwaters at various points. At Madras in southeast India, for example, a 1000-m-long breakwater was constructed in 1875 to create a sheltered harbor on a notoriously inhospitable coast, dominated by sand transport from north to south. On the south side of the breakwater over $1 \times 10^6 \, \text{m}^2$ of new land formed by 1912, but erosion occurred for 5 km north of the breakwater. At Ceara in Brazil, also in 1875, a detached breakwater was erected over a length of 430 m more or less parallel to the shore. It was believed that by using a detached structure littoral drift would be able to move along the coast uninterrupted by the presence of a conventional structure built across the surf zone. This, however, proved to be a fallacy (Komar, 1976), since the removal of the wave action which provided the energy for transporting the littoral sands resulted in their deposition within the protected area.

Figure 6.26 shows the evolution of the coast at West Bay in Dorset, southern England, following the construction of a jetty. As time goes on the beach in the foreground appears to build outwards, while the cliff behind the jetty retreats and so needs to be protected with a sea-wall and with large imported rock armor. Likewise, the construction of some sea-walls, erected to reduce coastal erosion and flooding, has had the opposite effect to the one intended (see Figure 6.27). Given the extent to which artificial structures have

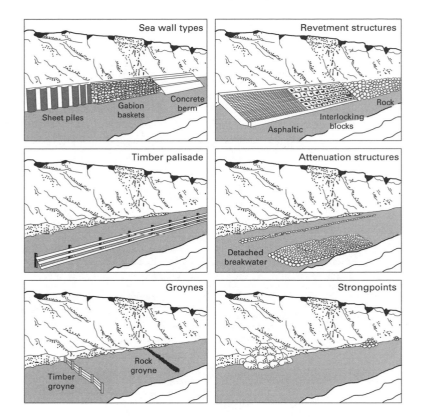

Figure 6.23 A selection of 'hard engineering' structures designed to afford coastal protection (modified from A. H. Brampton, 'Cliff conservation and protection: methods and practices to resolve conflicts', in J. Hooke (ed.), *Coastal and earth science conservation* (Geological Society Publishing House, 1998), figures 3.1, 3.2, 3.4, 3.5, 3.6, and 3.7).

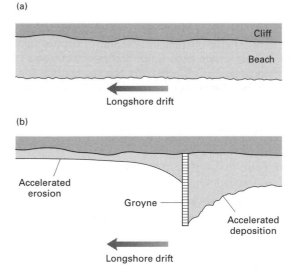

Figure 6.24 Diagrammatic illustration of the effects of groyne construction on sedimentation on a beach.

spread along the coastlines of the world, this is a serious matter (Walker, 1988).

Problems of this type are exacerbated because there is now abundant evidence to suggest that much of the reservoir of sand and shingle that creates beaches is in some respects a relict feature. Much of it was deposited on continental shelves during the maximum of the last glaciation (around 18,000 years BP), when sea level was about 120 m below its present level. It was transported shoreward and incorporated in present-day beaches during the phase of rapidly rising post-glacial sea levels that characterized the Flandrian transgression until about 6000 years BP. Since that time, with the exception of minor oscillations of the order of a few meters, world sea levels have been stable and much less material is, as a consequence, being added to beaches and shingle complexes. Therefore, according to Hails (1977: 322), 'in many areas, there is virtually no offshore supply to be moved onshore, except for small quantities resulting from seasonal changes.' It is because of these problems that many erosion prevention schemes now involve beach nourishment (by the artificial addition of appropriate sediments to build up the beach), or employ miscellaneous sand bypassing techniques (including pumping and dredging) whereby sediments are transferred from the accumulation side of an artificial barrier to the erosional side (King, 1974). Such methods of beach nourishment are reviewed by Bird (1996).

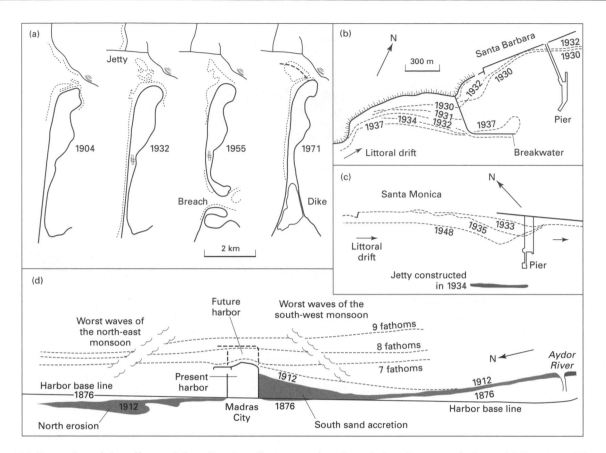

Figure 6.25 Examples of the effects of shoreline installations on beach and shoreline morphology. (a) Erosion of Bayocean Spit, Tillamook Bay, Oregon, after construction of a north jetty in 1914–17. The heavy dashed line shows the position of the new south jetty under construction. (b) The deposition–erosion pattern around the Santa Barbara breakwater in California. (c) Sand deposition in the protected lee of Santa Monica breakwater in California. (d) Madras Harbor, India, showing accretion on updrift side of the harbor and erosion on the downdrift side (after Komar, *Beach processes and sedimentation*, p. 334, © 1976. Reprinted by permission of Prentice-Hall Inc.).

In some areas, however, sediment-laden rivers bring material into the coastal zone, which becomes incorporated into beaches through the mechanism of longshore drift. Thus any change in the sediment load of such rivers may result in a change in the sediment budget of neighboring beaches. When accelerated soil erosion occurs in a river basin the increased sediment load may cause coastal accretion and siltation. But where the sediment load is reduced through action such as the construction of large reservoirs, behind which sediments accumulate, coastal erosion may result. This is believed to be one of the less desirable consequences of the construction of the Aswan Dam on the Nile: parts of its delta have shown recently accelerated recession.

The Nile sediments, on reaching the sea, used to move eastward with the general anticlockwise direction of water movements in that part of the eastern Mediterranean, generating sand bars and dunes that contributed to delta accretion. About a century ago an inverse process was initiated and the delta began to retreat. For example, the Rosetta mouth of the Nile lost about 1.6 km of its length from 1898 to 1954. The imbalance between sedimentation and erosion appears to have started with the Delta Barrages (1861) and culminated with the High Dam itself a century later. In addition, large amounts of sediment are retained in an extremely dense network of irrigation channels and drains that has been developed in the Nile Delta itself (Stanley, 1996). Much of the Egyptian coast is now

Figure 6.26 A jetty was built at West Bay, Dorset, to facilitate entry to the harbor. Top: in 1860 it had had little effect on the coastline. Center: by 1900 sediment accumulation had taken place in the foreground but there was less sediment in front of the cliff behind the town. Bottom: by 1976 the process had gone even further and the cliff had to be protected by a sea-wall. Even this has since been severely damaged by winter storms.

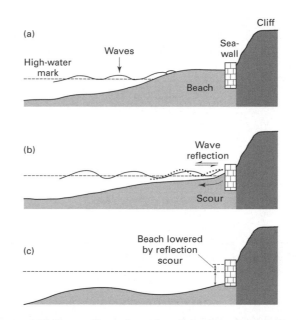

Figure 6.27 Sea-walls and erosion: (a) a broad, high beach prevents storm waves breaking against a sea-wall and will persist, or erode only slowly; but where the waves are reflected by the wall (b) scour is accelerated, and the beach is quickly removed and lowered (c) (modified after Bird, 1979, figure 6.3).

'undernourished' with sediment and, as a result of this overall erosion of the shoreline, the sand bars bordering Lake Manzala and Lake Burullus on the seaward side are eroded and likely to collapse. If this were to happen, the lakes would be converted into marine bays, so that saline water would come into direct contact with low-lying cultivated land and freshwater aquifers.

Likewise in Texas, where over the past century four times as much coastal land has been lost as has been gained, one of the main reasons for this change is believed to be the reduction in the suspended loads of some of the rivers discharging into the Gulf of Mexico (Table 6.15). The four rivers listed carried, in 1961–70, on average only about one-fifth of what they carried in 1931–40. Comparably marked falls in sediment loadings occurred elsewhere in the eastern USA (Figure 6.28). Likewise, in France the once mighty Rhône only carries about 5% of the load it did in the nineteenth century; and in Asia, the Indus discharges less than 20% of the load it did before construction of large barrages over the past half century (Milliman, 1990). On a global basis, large dams may retain 25–30% of the global flux of river sediment (Vörösmarty et al., 2003).

Table 6.15 Suspended loads of Texas rivers discharging into the Gulf of Mexico. Source: modified from Hails (1977, table 9.1) after data from Stout et al. and Curtis et al.

River	Suspended load (million tonnes)		Percent*
	1931–40	1961–70	
Brazos	350	120	30
San Bernard	1	1	100
Colorado	100	11	10
Rio Grande	180	6	3
Total	631	138	20

*1961–70 loads as a percentage of 1931–40 loads.

A good case study of the potential effects of dams on coastal sediment budgets is provided in California by Willis and Griggs (2003). Given that rivers provide the great bulk of beach material (75 to 90%) in the state, the reduction in sediment discharge by dammed rivers can have highly adverse effects. Almost a quarter of the beaches in California are down coast from rivers that have had sediment supplies diminished by one-third or more. Most of those threatened beaches are in southern California where much of the state's tourism and recreation activities are concentrated.

Construction of great levees on the lower Mississippi River since 1717 has also affected the Gulf of Mexico coast. The channelization of the river has in-

creased its velocity, reduced overbank deposition of silt on to swamps, marshes, and estuaries, and changed the salinity conditions of marshland plants (Cronin, 1967). As a result, the coastal marshes and islands have suffered from increased erosion or a reduced rate of development. This has been vividly described by Biglane and Lafleur (1967: 691):

Like a bullet through a rifle barrel, waters of the mighty Mississippi are thrust toward the Gulf between the confines of the flood control levees. Before the day of these man-made structures, these waters poured out over tremendous reaches of the coast . . . Freshwater marshes (salinities averaging 4–6%) were formed by deposited silts and vegetative covers of wire grass . . . As man erected his flood protection devices, these marshes ceased to form as extensively as before.

The changes between 1956 and 1990 are shown in Figure 6.29. However, as with so many examples of environmental change, it is unlikely that just one factor, in this case channelization, is the sole cause of the observed trend. In their study of erosion loss in the Mississippi Delta and neighboring parts of the Louisiana Coast, Walker et al. (1987) suggest that this loss is the result of a variety of complex interactions among a number of physical, chemical, biological, and cultural processes. These processes include, in addition to channelization, worldwide sea-level changes, subsidence resulting from sediment loading by the delta of the underlying crust, changes in the sites of deltaic sedimentation as the delta evolves, catastrophic storm

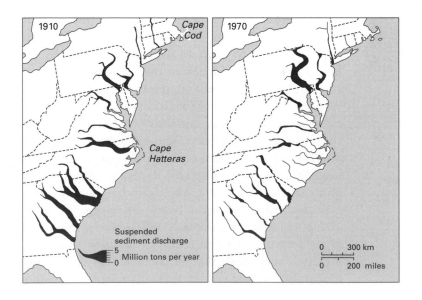

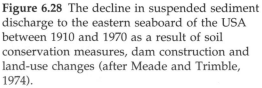

Figure 6.28 The decline in suspended sediment discharge to the eastern seaboard of the USA between 1910 and 1970 as a result of soil conservation measures, dam construction and land-use changes (after Meade and Trimble, 1974).

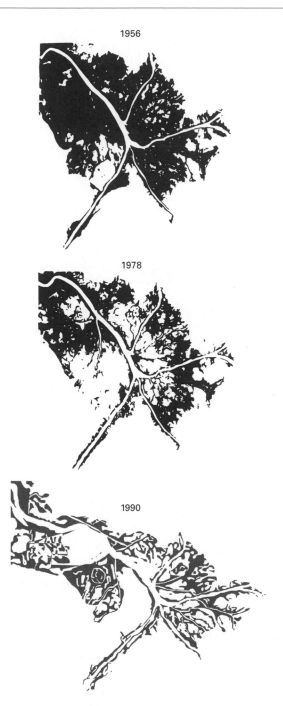

1956

1978

1990

Figure 6.29 Comparison of the outlines of the Mississippi birdsfoot delta from the 1950s to 1990 gives a clear indication of the transformation from marsh to open water. Artificial controls upriver have decreased the amount of sediment carried by the river; artificial levees along much of the lower course have kept flood-borne sediment from replenishing the wetlands; and in the active delta itself rock barriers installed across breaks similarly confine the river. The Gulf of Mexico is intruding as the marshland sinks or is washed away.

surges, and subsidence resulting from subsurface fluid withdrawal.

In some areas anthropogenic vegetation modification creates increased erosion potential. This has been illustrated by Stoddart (1971) for the hurricane-afflicted coast of Belize, Central America. He showed that natural, dense vegetation thickets on low, sand islands (*cays*) acted as a baffle against waves and served as a massive sediment trap for coral blocks, shingle, and sand transported during extreme storms. However, on many islands the natural vegetation had been replaced by coconut plantations. These had an open structure easily penetrated by seawater, they tended to have little or no ground vegetation (thus exposing the cay surface to stripping and channeling), and they had a dense but shallow root net easily undermined by marginal sapping. Thus Stoddart found (p. 191) that 'where the natural vegetation had been replaced by coconuts before the storm (Hurricane Hattie), erosion and beach retreat led to net vertical decreases in height of 3–7 ft; whereas where natural vegetation remained, banking of storm sediment against the vegetation hedge led to a net vertical increase in height of 1–5 ft.'

Other examples of markedly accelerated coastal erosion and flooding result from anthropogenic degradation of dune ridges. Frontal dunes are a natural defense against erosion, and coastal changes may be long-lasting once they are breached. Many of those areas in eastern England which most effectively resisted the great storm and surge of 1952 were those where humans had not intervened to weaken the coastal dune belt.

Not all dune stabilization and creation schemes have proved desirable (Dolan et al., 1973). In North Carolina (see Figure 6.30) the natural barrier-island system along the coastline met the challenge of periodic extreme storms, such as hurricanes, by placing no permanent obstruction in the path of the powerful waves. Under these natural conditions, most of the initial stress of such storms is sustained by relatively broad beaches (Figure 6.30a). Since there is no resistance from impenetrable landforms, water can flow between the dunes (which do not form a continuous line) and across the islands, with the result that wave energy is rapidly exhausted. However, between 1936 and 1940, 1000 km of sand fencing was erected to create an artificial barrier dune along part of the Outer Banks, and 2.5 million trees and various grasses (especially *Ammophila*

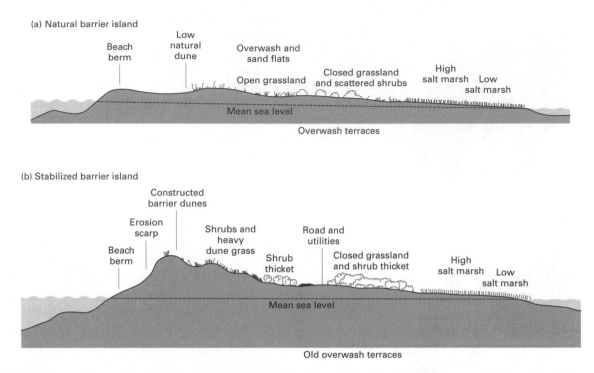

Figure 6.30 Cross-sections of two barrier islands in North Carolina, USA. The upper diagram (a) is typical of the natural systems and the lower (b) illustrates the stabilized systems (after Dolan et al., 1973, figure 4).

breviligulata) were planted to create large artificial dunes. The altered barrier islands (see Figure 6.30b) not only have the artificial barrier-dune system, they also have beaches that are often only 30 m wide, compared with 140 m for the unaltered islands. This beach-narrowing process, combined with the presence of a permanent dune structure, has created a situation in which high wave energy is concentrated in an increasingly restricted run up area, resulting in a steeper beach profile, increased turbulence, and greater erosion. Another problem associated with artificial dune stabilization is the flooding that occurs when northeast storms pile the water of the lagoon, Pamlico Sound, up against the barrier island. In the past, these surge waters simply flowed out between the low, discontinuous dunes and over the beach to the sea, but with the altered dune chain the water cannot drain off readily and vast areas of land are at times submerged.

General treatments of coastal problems and their management are provided by French (1997, 2001) and by Viles and Spencer (1995). What has become apparent in recent years is that there has been an increasing trend towards so-called soft means of coastal protec-

tion, rather than using hard engineering structures such as sea walls or groynes. Beach nourishment, the encouragement of dune formation, and promotion of salt marsh accretion are becoming recognized as being aesthetically pleasing, effective, and economically advantageous.

Changing rates of salt marsh accretion

In Britain in recent decades, the nature of some salt marshes and the rate at which they accrete have been transformed by a major vegetational change, namely the introduction of a salt-marsh plant, *Spartina alterniflora*. This cord-grass appears to have been introduced to Southampton Water in southern England by accident from the east coast of North America, possibly in shipping ballast. The crossing of this species with the native *Spartina maritima* produced an invasive cord-grass of which there were two forms, *Spartina townsendii* and *Spartina anglica*, the latter of which is now the main species. It appeared first at Hythe on Southampton Water in 1870 and then spread rapidly to other salt marshes in Britain: partly because of natural spread and partly because of deliberate planning. *Spartina* now reaches as far north as the Island of Harris in the west of Scotland, and to the Cromarty Firth in the east (Doody, 1984).

The plant has often been effective at excluding other species and also at trapping sediment. Rates of accretion can therefore become very high. Ranwell (1964) gives rates as high as 8–10 cm per year. There is evidence that this has caused progressive silting of estuaries such as those of the Dee (Marker, 1967) and the Severn (Page, 1982).

However, for reasons that are not fully understood, *Spartina* marshes have frequently suffered dieback, which has sometimes led to marsh recession. In the case of Poole Harbour and Beaulieu Estuary, this may date back to the 1920s, but elsewhere on the south coast it has generally been rapid and extensive since about 1960 (Tubbs, 1984). Among the hypotheses that have been put forward to explain dieback are the role of rising sea level, pathogenic fungi, increased wave attack, and the onset of waterlogging and anaerobic conditions on mature marsh. Some support for the latter view comes from the fact that in areas where the introduction has been more recent, for example in Wales and northwest England, the area of *Spartina* still appears to be increasing (Deadman, 1984).

The human impact on seismicity and volcanoes

The seismic and tectonic forces that mold the relief of Earth and cause such hazards to human civilization are two of the fields in which efforts to control natural events have had least success and where least has been attempted. Nonetheless, the fact that humans have been able, inadvertently, to trigger off small earthquakes by nuclear blasts, as in Nevada (Pakiser et al., 1969), by injecting water into deep wells, as in Colorado, by mining, by building reservoirs, and by fluid extraction suggests that in due course it may be possible to 'defuse' earthquakes by relieving tectonic strains gradually in a series of nondestructive, low-intensity earthquakes. One problem, however, is that there is no assurance that an earthquake purposefully triggered by human action will be a small one, or that it will be restricted to a small area. The legal implications are immense.

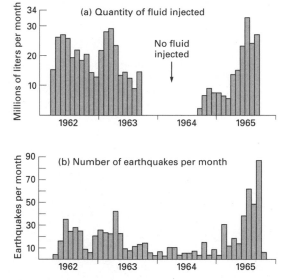

Figure 6.31 Correlation between quantity of waste water pumped into a deep well and the number of earthquakes near Denver, Colorado (after Birkeland and Larson, 1978p. 573).

The demonstration that increasing water pressures could initiate small-scale faulting and seismic activity was unintentionally demonstrated near Denver (Evans, 1966), where nerve-gas waste was being disposed of at a great depth in a well in the hope of avoiding contamination of useful groundwater supplies. The waste was pumped in at high pressures and triggered off a series of earthquakes (see Figure 6.31), the timing of which corresponded very closely to the timing of waste disposal in the well. It is also now thought that the pumping of fluids into the Inglewood Oil Field, Los Angeles, to raise the hydrostatic pressure and increase oil recovery, may have been responsible for triggering the 1963 earthquake which fractured a wall of the Baldwin Hills Reservoir. It appears that increased fluid pressure reduces the frictional force across the contact surface of a fault and allows slippage to occur, thereby causing an earthquake.

In general earthquake triggering has been related to fluid injection, but for reasons that are still obscure there may be some cases where increased seismicity has resulted from fluid abstraction (Segall, 1989).

The significance of these 'accidents' was verified experimentally at an oilfield in Colorado, where variations in seismicity have been produced by deliberately controlled variations in the fluid pressure in a zone that is seismically active (Raleigh et al., 1976).

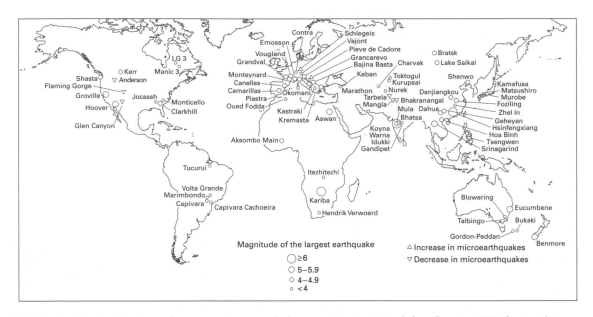

Figure 6.32 Worldwide distribution of reservoir-triggered changes in seismicity (after Gupta, 2002, figure 1).

Perhaps the most important anthropogenically induced seismicity results from the creation of large reservoirs (Talwani, 1997; Guha, 2000; Gupta, 2002). Reservoirs impose stresses of significant magnitude on crustal rocks at depths rarely equaled by any other human construction. With the ever increasing number and size of reservoirs the threat rises. There are at least six cases (Koyna, Jremasta, Hsinfengkiang, Kariba, Hoover, and Marathon) where earthquakes of a magnitude greater than 5, accompanied by a long series of foreshocks and aftershocks, have been related to reservoir impounding. However, as Figure 6.32 shows, there are many more locations where the filling of reservoirs behind dams has led to appreciable levels of seismic activity. Detailed monitoring has shown that earthquake clusters occur in the vicinity of some dams after their reservoirs have been filled, whereas before construction activity was less clustered and less frequent. Similarly, there is evidence from Vaiont (Italy), Lake Mead (USA), Kariba (Central Africa), Koyna (India), and Kremasta (Greece) that there is a linear correlation between the storage level in the reservoir and the logarithm of the frequency of shocks. This is illustrated for Vaiont (Figure 6.33a), Koyna (Figure 6.33b), and Nurek (Figure 6.33c). It is also apparent from Nurek that as the great reservoir has filled, so the depth of the more shallow-seated earthquakes appears to have increased.

One reason why dams induce earthquakes involves the hydro-isostatic pressure exerted by the mass of the water impounded in the reservoir, together with changing water pressures across the contact surfaces of faults. Given that the deepest reservoirs provide surface loads of only 20 bars or so, direct activation by the mass of the impounded water seems an unlikely cause (Bell and Nur, 1978) and the role of changing pore pressure assumes greater importance. Paradoxically, there are some possible examples of reduced seismic activity induced by reservoirs (Milne, 1976). One possible explanation of this is the increased incidence of stable sliding (fault creep) brought about by higher pore-water pressure in the vicinity of the reservoir.

However, the ability to prove an absolutely concrete cause-and-effect relationship between reservoir activity and earthquakes is severely limited by our inability to measure stress below depths of several kilometers, and some examples of induced seismicity may have been built on the false assumption that because an earthquake occurs in proximity to a reservoir it has to be induced by that reservoir (Meade, 1991).

Miscellaneous other human activities appear to affect seismic levels. In Johannesburg, South Africa, for example, gold mining and associated blasting activity have produced tens of thousands of small tremors, and there is a notable reduction in the number that occurs on Sundays, a day of rest. In Staffordshire, England,

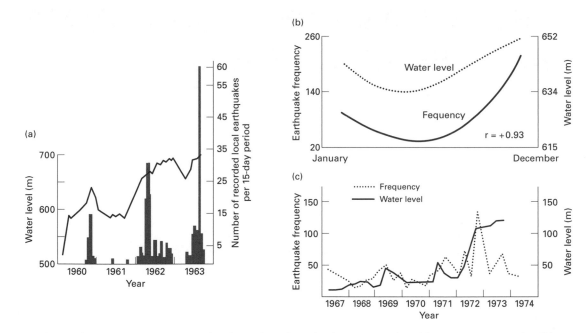

Figure 6.33 Relationships between reservoir levels and earthquake frequencies for: (a) Vaiont Dam, Italy; (b) Koyna, India (these curves show the 3-monthly average of water level and the total number of earthquakes for the same months from 1964 to 1968); (c) The Nurek Dam, Tajikistan (after Judd, 1974 and Tajikistan Academy of Sciences, 1975).

coal mining has caused increased seismic activity and up to 25% of all earthquakes recorded by the British Geological Survey may be related to coal mining. There are also cases where seismicity and faulting can be attributed to fluid extraction, for example, in the oilfields of Texas and California and the gas fields of the Po Valley in Italy and of Uzbekistan (Cypser and Davis, 1998).

When looking at the human impact on volcanic activity human impotence becomes apparent, although some success has been achieved in the control of lava flows. Thus in 1937 and 1947 the US Army attempted to divert lava from the city of Hilo, Hawaii, by bombing threatening flows, while elsewhere, where lava rises in the crater, breaching of the crater wall to direct lava towards uninhabited ground may be possible. In 1973 an attempt was made to halt advance of lava with cold water during the Icelandic eruption of Krikjufell. Using up to 4×10^6 L of pumped waste per hour, the lava was cooled sufficiently to decrease its velocity at the flow front so that the chilled front acted as a dam to divert the still fluid lava behind (Williams and Moore, 1973).

Points for review

What are the causes and consequences of accelerated sedimentation?
How do humans cause land subsidence?
In what ways may humans accelerate mass movements?
Why are many of the world's coastlines eroding?

Guide to reading

Brookes, A., 1988, *Channelised rivers*. Chichester: Wiley. An advanced research monograph with broad scope.

Downs, P. W. and Gregory, K. J., 2004, *River channel management*. Arnold: London. A comprehensive review of river channels and their management.

Goudie, A. S., 1993, Human influence in geomorphology. *Geomorphology*, 7, 37–59. A general review, with a concern for the future, in a major journal.

Nir, D., 1983, *Man, a geomorphological agent: an introduction to anthropic geomorphology*. Jerusalem: Keter. A general survey that was ahead of its time.

7 THE HUMAN IMPACT ON CLIMATE AND THE ATMOSPHERE

World climates

The climate of the world is now known to have fluctuated frequently and extensively in the three or so million years during which humans have inhabited Earth. The bulk of these changes have nothing to do with human intervention. Climate has changed, and is currently changing, because of a wide range of different natural factors which operate over a variety of timescales (Figure 7.1).

No completely acceptable explanation of climatic change has ever been presented, and no one process acting alone can explain all scales of climate change. The complexity of possibly causative factors involved is daunting.

The complexity becomes evident if we follow the pathway of radiation derived from the ultimate force of climate – the sun. First of all, for reasons such as the varying tidal pull being exerted on the sun by the planets, the quality and quantity of solar radiation may change. It has been recognized that radiation from the sun changes both in quantity (through association with familiar phenomena such as **sunspots**, which are dark regions of lower surface temperature on the surface of the sun) and in quality (through changes in the ultraviolet range of the solar spectrum). Cycles of solar activity have been established for the short term by many workers, with 11- to 22-year cycles being noted in particular. Sunspot cycles of 80–90 years have also been postulated. The observations of sunspots in historical times have also given a measure of solar activity and one very striking feature is the near absence of sunspots between AD 1640 and 1710, a period sometimes called the Maunder Minimum. It is perhaps significant that this minimum occurred during some of the more extreme years of the inclement Little Ice Age.

The receipt of such varying radiation at Earth's surface might itself vary because of the presence of fine interstellar matter (nebulae) through which the earth might from time to time pass. This would tend to reduce the receipt of solar radiation. Likewise, the passage of the solar system through a dust lane bordering a spiral arm of the Milky Way galaxy might cause a temporary reduction in receipt of radiation output from the sun.

The receipt of incoming radiation will also be affected by the position and configuration of Earth. Such changes do take place, and there are three main

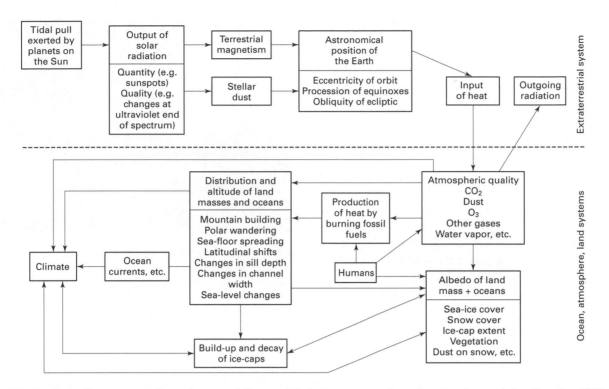

Figure 7.1 A schematic representation of some of the possible influences causing climatic change (after Goudie, 1992, figure 1).

astronomical factors which have been identified as of probable importance, with all three occurring in a cyclic manner. Firstly, Earth's orbit around the sun is not a perfect circle but an ellipse. If the orbit were a perfect circle then the summer and winter parts of the year would be equal in length. With greater eccentricity the length of the seasons will display a greater difference. Over a period of about 96,000 years, the eccentricity of Earth's orbit can 'stretch' by departing much further from a circle and then revert to almost true circularity.

Secondly, changes take place in the 'precession of the equinoxes', which means that the time of year at which the earth is nearest the sun varies. The reason is that Earth wobbles like a child's top and swivels round its axis. This cycle has a periodicity of about 21,000 years.

Thirdly, changes occur, with a periodicity of about 40,000 years, in the 'obliquity of the elliptic' – the angle between the plane of Earth's orbit and the plane of its rotational equator. This movement has been likened to the roll of a ship with a tilt varying from 21°39' to 24°36'. The greater the tilt, the more pronounced is the difference between winter and summer.

These three cycles comprise what is often called the Milankovitch or Orbital Theory of climatic change. They have a temptingly close similarity in their periodicity to the durations of climatic change associated with the many glacials and interglacials of the past 1.6 million years. Indeed, they have been termed the 'pace-maker of the ice ages'.

Once the incoming solar radiation reaches the atmosphere, its passage to the surface of Earth is controlled by the gases, moisture, and particulate matter that are present. Essential importance has been attached to the role of dust clouds emitted from volcanoes. These could increase the backscattering of incoming radiation and thus promote cooling. Volcanic dust veils produced by, for example, the eruption of Krakatoa in the 1880s and by Mount Pinatubo in 1991 caused global cooling for a matter of a few years. However, changing levels of volcanic activity are not the only way in which changes in atmospheric transparency might occur. For example, dust can be emplaced into the atmosphere by the wind erosion of fine-grained sediment and soil, and we know from the extensive deposits of wind-laid silts (loess) of glacial age that during the glacial maxima the atmosphere was probably very dusty, contributing to global cooling.

Carbon dioxide, methane, nitrous oxide, sulfur dioxide, and water vapor can also modify the receipt of solar radiation. Particular attention has focused in recent years on the role of carbon dioxide (CO_2) in the atmosphere. This gas is virtually transparent to incoming solar radiation but absorbs outgoing terrestrial infrared radiation – radiation that would otherwise escape to space and result in heat loss from the lower atmosphere. In general, through the mechanism of this so-called greenhouse effect, low levels of CO_2 in the atmosphere would be expected to lead to cooling, and high levels would be expected to produce a 'heat trap'. The same applies to levels of methane and nitrous oxide, which, molecule for molecule, are even more effective greenhouse gasses than CO_2. Recently it has proved possible to retrieve CO_2 from gas bubbles preserved in layers of ice in deep ice-cores drilled from the polar regions. Analyses of changes in CO_2 concentrations in these cores have provided truly remarkable results and have demonstrated that CO_2 changes and climatic changes have progressed in approximate synchroneity over the past 800,000 years (EPICA, 2004). Thus the last interglacial around 120,000 years ago was a time of high CO_2 levels, and the last glacial maximum around 18,000 years ago of low CO_2 levels, and the early Holocene a time of very rapid rise in CO_2 levels. The reasons for the observed natural change in greenhouse gas concentrations are still the subject of active scientific research.

Once incoming radiation from the sun reaches the surface of Earth it may be absorbed or reflected according to the nature of the surface, and in particular according to whether it is land or water, covered in vegetation or desert, and whether it is mantled by snow.

The effect of the received radiation on climate also depends on the distribution and altitude of landmasses and oceans. These too are subject to change in a wide variety of ways – the plates that comprise Earth's crust are ever moving, mountain belts may grow or subside, and oceans and straits open and close. These processes shift areas into new latitudes, transform the world's wind belts, and modify the climatically very important ocean currents.

In this discussion of causes it is also crucial to consider feedbacks. Such feedbacks are responses to the

Figure 7.2 Air pollution in Cape Town, South Africa. The combustion of fossil fuels, including coal, to generate electricity and to power vehicles, is a major cause not only of local air pollution but also of the increase in greenhouse gas loadings in the atmosphere.

original forcing factors that act either to increase or intensify the original forcing (we call this positive feedback) or to decrease or reverse it (negative feedback). Clouds, ice and snow, and water vapor are three of the most important feedback mechanisms. An example of a positive feedback is the role of snow. Under cold conditions this falls rather than rain, it changes the **albedo** (reflectivity) of the ground surface and causes further cooling of the air above it.

Finally, it may well be that the atmosphere and the oceans possess a degree of internal instability that furnishes a built-in mechanism of change so that some small and random change might, through the operation of positive feedbacks and the passage of thresholds, have extensive and long-term effects. Small triggers might have large consequences.

While humans are at present incapable of modifying some of these natural mechanisms of climate change – the output of solar radiation, the presence of fine interstellar matter, Earth's orbital variations, volcanic eruptions, mountain building, and the overall pattern of land masses and oceans – there are some key areas where humans may be capable of making significant changes to global climates (Figure 7.2). The most important categories of influence are in terms of the chemical composition of the atmosphere and the albedo of Earth's surface, although, as the following list of *possible mechanisms* shows, there are some others that also need to be considered.

1 Gas emissions
  • CO₂ – industrial and agricultural
  • methane
  • chlorofluorocarbons (CFCs)
  • nitrous oxide
  • krypton 85
  • water vapor
  • miscellaneous trace gases
2 Aerosol generation
  • dust, smoke, etc.
3 Thermal pollution
4 Albedo change
  • dust addition to ice caps
  • deforestation
  • overgrazing
5 Extension of irrigation
6 Alteration of ocean currents by constricting straits
7 Diversion of freshwaters into oceans, affecting thermohaline circulation system
8 Impoundment of large reservoirs

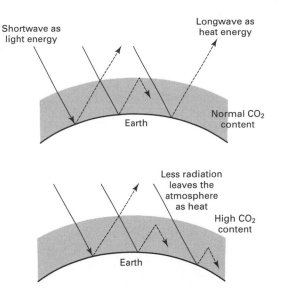

Figure 7.3 The greenhouse effect: shortwave radiation from the sun is absorbed at Earth's surface, which in turn radiates heat at far longer wavelengths because of its temperature of around 280 K, compared with 6000 K for the sun.

The greenhouse gases

Carbon dioxide

The greenhouse effect occurs in the atmosphere because of the presence of certain gases that absorb infrared radiation (Figure 7.3). Light and ultraviolet radiation from the sun are able to penetrate the atmosphere and warm Earth's surface. This energy is re-radiated as infrared radiation, which, because of its longer wavelength, is absorbed by certain substances such as water vapor, carbon dioxide, and other trace gases. This causes the average temperature of Earth's surface and atmosphere to increase. Should the quantities of each substance in the atmosphere be increased, then the greenhouse effect will become enhanced (Hansen et al., 1981).

In reality the term 'greenhouse effect' is something of a misnomer. As Henderson-Sellers and Robinson (1986: 60) explain:

We know that a greenhouse maintains its higher internal temperature largely because the shelter it offers reduces the turbulent transfers of energy away from the surface rather than because of any radiative considerations. Thus while the greenhouse effect remains valid, and vital, for the atmosphere it might be better to think of the physical processes in terms of the 'leaky bucket' analogy. . . . Here an increase in the amount of gas with absorption bands in the infrared part of the spectrum is represented by a decrease in the size of the hole in the bucket. The surface temperature, represented by the depth of the water in the bucket, rises as more absorbing gases enter the atmosphere.

There are a number of ways in which humans have been enhancing the greenhouse effect, the most significant of which is to increase carbon dioxide levels in the atmosphere. This may have started with deforestation in the Holocene (Ruddiman, 2003) but has accelerated in recent centuries. Since the beginning of the industrial revolution humans have been taking stored carbon out of the earth in the form of coal, petroleum, and natural gas, and burning it to make carbon dioxide (CO₂), heat, water vapor, and smaller amounts of sulfur dioxide (SO₂) and other gases. The pre-industrial level of carbon dioxide is a matter of some debate, but may have been as low as 260–70 ppm by volume (Wigley, 1983). The present level is over 370 ppmv, and the upward trend is evident in records from various parts of the world (Figure 7.4). At the present rate it would reach 500 ppmv by the end of the twenty-first century. The prime cause of increased

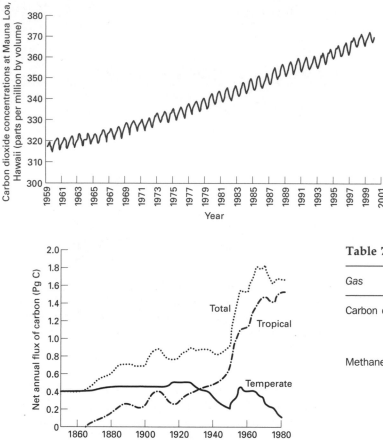

Figure 7.4 Carbon dioxide concentrations at Mauna Loa, Hawaii.

Figure 7.5 The net annual flux of carbon from deforestation in tropical and temperate zones globally, 1850–1980 (after Houghton and Skole, 1990, figure 23.2, and Woodwell, 1992, figure 5.2).

Table 7.1 Sources of four principal greenhouse gases*

Gas	Natural sources	Human-derived sources
Carbon dioxide	Terrestrial biosphere Oceans	Fossil fuel combustion Cement production Land-use modification
Methane	Natural wetlands Termites Oceans and 　freshwater lakes	Fossil fuels (natural gas 　production, coal 　mines, petroleum 　industry, coal 　combustion) Enteric fermentation 　(e.g., cattle) Rice paddies Biomass burning Landfills Animal waste Domestic sewage
Nitrous oxide	Oceans Tropical soils 　(wet forests, 　dry savannas) Temperate soils 　(forests, 　grassland)	Nitrogenous fertilizers Industrial sources Land-use modification 　(biomass burning, 　forest clearing) Cattle and feed lots
Chlorofluorocarbons†	Nil	Rigid and flexible foam Aerosol propellants Teflon polymers Industrial solvents

*Sources listed in order of decreasing magnitude of emission except where otherwise indicated.
†Sources of chlorofluorocarbons not in order of decreasing magnitudes of emission.

carbon dioxide emissions is fossil fuel combustion and cement production (c. 5.5 ± 0.5 Gt C year^{-1} in the 1980s), but with the release of carbon dioxide by changes in tropical land use (primarily deforestation) being a significant factor (c. 1.6 ± 1.0 Gt C year^{-1}). The amount of carbon derived from deforestation has increased greatly from about 0.4 Gt C year^{-1} in 1850 (Figure 7.5) (Woodwell, 1992). The various relationships between land-use change and the build up of greenhouse gases are reviewed by Adger and Brown (1994).

Other gases

In addition to carbon dioxide, it is probable that other gases will contribute to the greenhouse effect (Table 7.1). Individually their effects may be minor, but as a group

Table 7.2 Radiative forcing relative to CO_2 per unit molecule change in the atmosphere. Source: extracted from Houghton et al. (1990: 53, table 2.3)

Gas	Relative radiative forcing	Residence time in atmosphere (years)
CO_2	1	100
CH_4 (methane)	21	10
N_2O (nitrous oxide)	206	100–200
CFC-11	12,400	65
CFC-12	15,800	130

they may be major (Ramanathan, 1988). Indeed, molecule for molecule some of them may be much more effective as greenhouse gases than CO_2, as the data in Table 7.2 show.

One of the more important of the trace gases is methane (CH_4), which has a strong infrared absorption band at 7.66 µm. Ice-core studies and recent direct observations (Figure 7.6b) suggest that until the beginning of the industrial revolution in the eighteenth century background levels were relatively stable at around 600 parts per billion by volume (ppbv) although they may have been increased prior to that by rice farming and other agricultural activities (Ruddiman and Thomsen, 2001). They rose steadily between AD 1700 and 1900, and then increased still more rapidly, attaining levels that averaged 1300 ppbv in the early 1950s and 1600 ppbv by the mid-1980s (Khalil and Rasmussen, 1987) and over 1700 ppbv in the 1990s. This increase of 2.5 times over background levels results primarily from increased rice cultivation in waterlogged paddy fields, the enteric fermentation produced in the growing numbers of flatulent domestic cattle, and the burning of oil and natural gas (Crutzen et al., 1986).

Chlorofluorocarbons (CFCs), despite their relatively trace amounts in the atmosphere, have increased very markedly in terms of their emissions (Figure 7.6c) and their concentrations in recent decades, resulting from their use as refrigerants, foam makers, fire control agents, and propellants in aerosol cans. They have a very strong greenhouse effect even in relatively small amounts. On the other hand, the ozone depletion they have caused in the stratosphere may to some limited extent counteract this effect, for stratospheric ozone depletion results in a decrease in **radiative forcing** (Houghton et al., 1992). Conversely, the build up of

lower-level **tropospheric** ozone can contribute to the greenhouse effect.

Nitrous oxide (N_2O) is also no laughing matter, for it can contribute to the greenhouse effect, primarily by absorption of infrared at the 7.8 and 17 µm bands. Combustion of hydrocarbon fuels, the use of ammonia-based fertilizers, deforestation, and biomass burning are among the processes that could lead to an increase in atmospheric N_2O levels (Figure 7.6a and Table 7.3b). Atmospheric N_2O concentrations have increased from around 275 ppbv in pre-industrial times to 311 ppbv in 1992.

Other trace gases that could play a greenhouse role include bromide compounds, carbon tetrafluoride, carbon tetrachloride, and methyl chloride.

The continued role of greenhouse gases other than CO_2 in changing the climate is already not greatly less important than that of CO_2. If present trends continue, the combined concentrations of atmospheric CO_2 and other greenhouse gases would be radiatively equivalent to a doubling of CO_2 from pre-industrial levels possibly as early as the 2030s. The relative amounts of radiative forcing for different greenhouse gases since pre-industrial times are, according to the Intergovernmental Panel on Climate Change (IPCC, 1996), as follows:

CO_2	1.56 W m^{-2}
CH_4	0.47 W m^{-2}
N_2O	0.14 W m^{-2}
CFCs and HCFCs	0.25 W m^{-2}
tropospheric ozone	0.40 W m^{-2}

Another important feature of the various greenhouse gases is their residence time in the atmosphere. Methane has a residence time of about 10 years, the shortest of all the greenhouse gases. This means that if we could stop the enhanced emissions of that gas, its concentration in the atmosphere should fall to its natural level in a decade. By contrast N_2O (100–200 years) and CO_2 (c. 100 years) have much longer residence times, so that even if we could control their sources immediately it would still take a very long time for them to fall to their natural levels.

Global temperatures have been climbing since the end of the nineteenth century (Figure 7.7) and it is now regarded as highly probable that increased greenhouse gas loadings in the atmosphere have contributed to this.

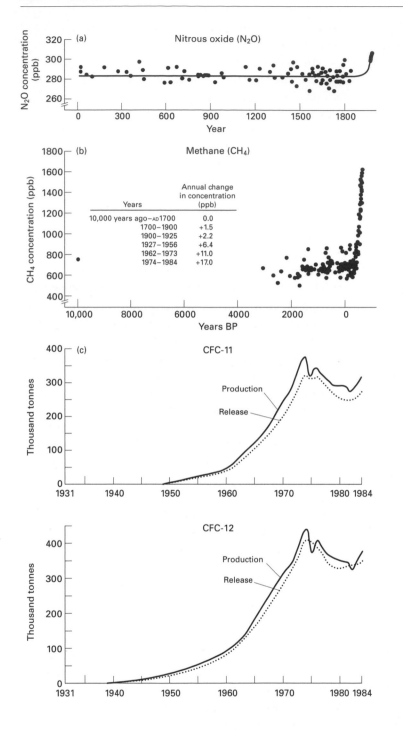

Figure 7.6 The changing concentrations of accessory greenhouse gases in the atmosphere: (a) nitrous oxide – note these remained fairly constant between 23,000 years ago and AD 1850 at approximately 285 parts per billion (after Khalil and Rasmussen, 1987); (b) methane (after Khalil and Rasmussen, 1987); (c) the changing production and release of two CFC gases (CFC-11 and CFC-12) between 1931 and 1992.

Aerosols

Aerosols are finely divided solid or liquid particles dispersed in the atmosphere. In general terms it is believed that they affect climate because they intercept and scatter a small portion of incoming radiation from the sun, thus reducing the energy reaching the ground. This is called the 'direct effect' of aerosols. The direct effect increases with both the number and size of aerosols in the atmosphere. Aerosols also have an 'indirect effect'. This is because they are key elements in cloud formation. The number of cloud droplets that a cloud possesses is determined by the number of aerosols available on to which water vapor can

Table 7.3 (a) Estimates of source strengths and sinks.
Source: data in UNEP (1991)
(a) Methane (CH$_4$)

Sources/sinks	Best estimate (10^6 t year^{-1})	Range (10^6 t year^{-1})
Sources:		
(a) natural wetlands	115	100–200
(b) rice paddies	110	25–170
(c) enteric fermentation (animals)	80	65–100
(d) gas drilling, venting, transmission	45	25–50
(e) biomass burning	40	20–80
(f) termites	40	10–100
(g) landfills	40	20–70
(h) coal mining	35	19–50
(i) oceans	10	5–20
(j) freshwaters	5	1–25
(i) CH$_4$ hydrate destabilization	5	0–100
Sinks:		
(a) removal by soils	30	15–45
(b) reaction with OH	500	400–600
(c) atmospheric increase	44	40–48

(b) Nitrous oxide (N$_2$O)

Sources/sinks	Range (10^6 t year^{-1})
Sources:	
(a) oceans	1.4–2.6
(b) soils (tropical forests)	2.2–3.7
(c) soils (temperate forests)	0.7–1.5
(d) fossil fuel combustion	0.1–0.3
(d) biomass burning	0.02–0.2
(e) fertilizer (including groundwater)	0.01–2.2
Sinks	
(a) removal by soils	Unknown
(b) photolysis in the stratosphere	7–13
(c) atmospheric increase	3–4.5

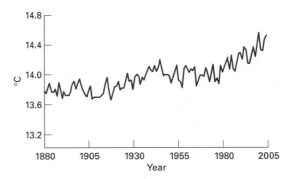

Figure 7.7 Global average temperature at Earth's surface, 1880–2002.

a portion of this back to the land surface, raising surface temperatures. They believe that natural dust from volcanic emissions tends to enter the stratosphere (where backscattering and cooling are the prime consequences), while anthropogenic dust more frequently occurs in the lower levels of the atmosphere, causing thermal blanketing and warming.

There are a variety of ways in which human activities have increased atmospheric aerosol loadings. One of these is industrial emission of smoke and dust particles (Davitya, 1969), though whether this is a matter of only local rather than regional or global significance is a matter for debate.

Industrialization is not, however, the sole source of particles in the atmosphere, nor is a change in temperature the only possible consequence. Bryson and Barreis (1967), for example, argue that intensive agricultural exploitation of desert margins, such as in Rajasthan, India, would create a dust pall in the atmosphere by exposing larger areas of surface materials to deflation in dust storms. This dust pall, they believe, would so change atmospheric temperature that convection, and thus rainfall, would be reduced. Observations on dust levels over the Atlantic during the drought years of the late 1960s and early 1970s in the Sahel suggest that the degraded surfaces of that time led to a great (threefold) increase in atmospheric dust (Prospero and Nees, 1977). There is thus the possibility that human-induced desertification generates dust which could in turn increase the degree of desertification by its effect on rainfall levels. Biomass burning of tropical savannas may also add large numbers of aerosols to the atmosphere. Studies of the links between dust in the atmosphere and temperatures,

condense. Clouds with large numbers of droplets reflect more sunlight back into space and so can also contribute to cooling. Aerosols do not, however, inevitably cause cooling (Weave et al., 1974). So, for example, Idso and Brazel (1978) and Brazel and Idso (1979) point to the two contrasting tendencies of dust: the backscattering effect producing cooling, and the thermalblanketing effect causing warming. The second of these absorbs some of Earth's thermal radiation that would otherwise escape to space, and then re-radiates

Figure 7.8 The Gulf War of 1991 led to the deliberate release and burning of oil in Kuwait. Fears were expressed at the time that smoke palls might have regional and global climate effects. In general, subsequent research has suggested that such fears may have been exaggerated.

precipitation, and clouds is a major area of current research (see, e.g., Tegen et al., 1996; Miller and Tegen, 1998).

The most catastrophic effects of anthropogenic aerosols in the atmosphere could be those resulting from a nuclear exchange between the great powers. Explosion, fire, and wind might generate a great pall of smoke and dust in the atmosphere, which would make the world dark and cold. It has been estimated that if the exchange reached a level of several thousand megatons, a 'nuclear winter' would be as low as −15° to −25°C (Turco et al., 1983), although more recent simulations by Schneider and Thompson (1988) suggest that some previous estimates may have been exaggerated. They suggest that in the Northern Hemisphere maximum average land surface summertime temperature depressions might be of the order of 5–15°C. The concept is discussed by Cotton and Piehlke (1995, chapter 10).

Fears were also expressed that as a result of the severe smoke palls (Figure 7.8) generated by the Gulf War in 1991 there might be severe climate impacts. Studies have suggested that because most of the smoke generated by the oil-well fires stayed in the lower troposphere and only had a short residence time in the air, the effects were local (some cooling) rather than global, and that the operation of the monsoon was not affected to any significant degree (Browning et al., 1991; Bakan et al., 1991). Furthermore, in the event

the emissions of smoke particles were less than some forecasters had predicted, and they were also rather less black (Hobbs and Radke, 1992).

Aircraft, both civil and military, discharge some water vapor into the atmosphere as contrails. At present, the water content of the stratosphere is very low, as is the exchange of air between the lower stratosphere and other regions. Consequently, comparatively modest amounts of water vapor discharge by aircraft could have a significant effect on the natural balance. It is possible that contrails and the development of thin cirrus clouds could lead to warming of the Earth's surface (IPCC, 1999).

Over the world's oceans a major source of aerosols is dimethylsulfide (DMS). This is produced by planktonic algae in seawater and then oxidized in the atmosphere to form sulfate aerosols. Because the albedo of clouds (and thus Earth's radiation budget) is sensitive to cloud-condensation nuclei density, any factor that controls planktonic algae may have an important impact on climate. The production of such plankton could be affected by water pollution in coastal areas or by global warming (Charlson et al., 1987). However, an even more important source of sulfate aerosols is the burning of fossil fuels and the subsequent emission of sulfur dioxide (SO_2) (Charlson et al., 1992), and these types of sulfate aerosol are concentrated over and downwind of major industrial regions. They have probably served to reduce the rate of global warming that has taken place in this century and may help to explain the cessation in global warming that took place in some regions between the 1940s and 1970s. Indeed, climate models that have predicted the amount of increase in global average temperature as a result of the rising concentrations of greenhouse gases have given a greater amount of temperature rise since the last century than has actually occurred. The newer climate models, which include the effect of these aerosols, produce predicted changes that have considerable similarity to the observed patterns of change (Taylor and Penner, 1994).

Vegetation and albedo change

Incoming radiation of all wavelengths is partly absorbed and partly reflected. Albedo is the term used to describe the proportion of energy reflected and

Table 7.4 Albedo values for different land-use types. Source: from miscellaneous data in Pereira (1973) collated by author

Surface type	Location	Albedo (%)
Tall rain forest	Kenya	9
Lake	Israel	11.3
Peat and moss	England and Wales	12
Pine forest	Israel	12.3
Heather moorland	England and Wales	15
Evergreen scrub (maquis)	Israel	15.9
Bamboo forest	Kenya	16
Conifer plantation	England and Wales	16
Citrus orchard	Israel	16.8
Towns	England and Wales	17
Open oak forest	Israel	17.6
Deciduous woodland	England and Wales	18
Tea bushes	Kenya	20
Rough grass hillside	Israel	20.3
Agricultural grassland	England and Wales	24
Desert	Israel	37.3

hence is a measure of the ability of the surface to reflect radiation.

Land-use changes create differences in albedo which have important effects on the energy balance, and hence on the water balance, of an area. Tall rain forest may have an albedo as low as 9%, while the albedo of a desert may be as high as 37% (Table 7.4).

There has been growing interest recently in the possible consequences of deforestation on climate through the effect of albedo change. Ground deprived of vegetation cover as a result of deforestation and overgrazing (as in parts of the Sahel) has a very much higher albedo than ground covered in plants. This could affect temperature levels. Satellite imagery of the Sinai–Negev region of the Middle East shows an enormous difference in image between the relatively dark Negev and the very bright Sinai-Gaza strip area. This line coincides with the 1948–9 armistice line between Israel and Egypt and results from different land-use and population pressures. Otterman (1974) has suggested that this albedo change has produced temperature changes of the order of 5°C.

Charney and others (1975) have argued that the increase in surface albedo, resulting from a decrease in plant cover, would lead to a decrease in the net incoming radiation, and an increase in the radiative cooling of the air. Consequently, they argue, the air would

sink to maintain thermal equilibrium by adiabatic compression, and cumulus convection and its associated rainfall would be suppressed. A positive feedback mechanism would appear at this stage, for the lower rainfall would in turn adversely affect plants and lead to a further decrease in plant cover. However, this view is disputed by Ripley (1976) who suggests that Charney and his colleagues, while considering the impact of vegetation changes on albedo, have completely ignored the effect of vegetation on evapotranspiration. He points out that vegetated surfaces are usually cooler than bare ground since much of the absorbed solar energy is used to evaporate water, and concludes from this that protection from overgrazing and deforestation might, in contrast to Charney's views, be expected to lower surface temperatures and thereby reduce, rather than increase, convection and precipitation.

Removal of humid tropical rain forests has also been seen as a possible mechanism of anthropogenic climatic change through its effect on albedo. Potter et al. (1975) have proposed the following model for such change:

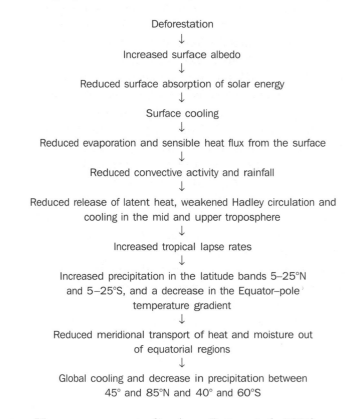

Deforestation
↓
Increased surface albedo
↓
Reduced surface absorption of solar energy
↓
Surface cooling
↓
Reduced evaporation and sensible heat flux from the surface
↓
Reduced convective activity and rainfall
↓
Reduced release of latent heat, weakened Hadley circulation and cooling in the mid and upper troposphere
↓
Increased tropical lapse rates
↓
Increased precipitation in the latitude bands 5–25°N and 5–25°S, and a decrease in the Equator–pole temperature gradient
↓
Reduced meridional transport of heat and moisture out of equatorial regions
↓
Global cooling and decrease in precipitation between 45° and 85°N and 40° and 60°S

However, some studies (e.g., Potter et al., 1981) suggested that globally over the past few thousand years

Table 7.5 Some recent studies of climatic effects of vegetation removal

Source	Location	Subject
Fuller and Ottka (2002)	West Africa	Albedo and desertification
Fu (2003)	East Asia	Reduced atmospheric and soil moisture in East Asian monsoon region
Chase et al. (2000)	Global	Effects on main circulation features
Werth and Avissar (2002)	Amazonia	Reduction of local precipitation, evapotranspiration and cloudiness and also global effects
Berbet and Costa (2003)	Amazonia	Precipitation variability
Reale and Zirmeyer (2000)	Mediterranean Basin	Increased precipitation prior to deforestation in Roman times
Taylor et al. (2002)	Sahel	Rainfall decrease

the climatic effects of albedo changes wrought by humans have been small and probably undetectable. Similarly, Henderson-Sellers and Gornitz (1984) sought to model the possible future effects of albedo changes produced by humans and also predicted that there would be but little alteration brought about by current levels of tropical deforestation. On the other hand, Lean and Warrilow (1989) used a general circulation model (GCM) which showed greater changes than previous models and suggested that Amazon basin deforestation would, through the effects of changes in surface roughness and albedo, lead to reductions in both precipitation and evaporation. Likewise a UK Meteorological Office GCM indicated that the deforestation of both Amazonia and Zaire would by changing surface albedo cause a decrease in precipitation levels (Mylne and Rowntree, 1992). There are now an increasing number of modeling experiments that suggest vegetation removal can have important regional and even global effects (Table 7.5), although there are considerable divergences between different models (Nobre et al., 2004). Nonetheless, most show *decreases* in mean evapotranspiration of from 25.5 to 985.0 mm per year, *increases* in mean surface temperatures of 0.1–3.8°C and *reductions* in regional precipitation. It is even possible that the effects of Amazonian deforestation on precipitation could extend some distance away from there, to the Dakotas and the Midwest Triangle in the USA, as modeling by Werth and Avissar (2002) has shown.

Albedo effects may be especially sensitive in higher latitudes as well. As Betts (2000) has pointed out, in a snowy environment, forests are generally darker than open land, because trees generally remain exposed when cultivated land can become entirely snow-covered. Snow-free foliage is darker than snow. This means that forest has a smaller surface albedo and so may exert a warming influence. Thus land cover changes in the boreal forest zone can have substantial climatic implications.

Forests, irrigation, and climate

The replacement of forest with crops, in addition to leading to a change in surface albedo of the type just discussed, also changes some other factors that may have climatic significance, including surface aerodynamic roughness, leaf and stem areas, and amount of evapotranspiration (Betts, 2003). In particular large amounts of moisture may be transpired by deep rooting plants, which means that moisture is pumped back into the atmosphere, leading to increased levels of precipitation. Conversely, removal of such deep-rooting vegetation could exacerbate drought.

The belief that forests can increase precipitation levels has a long history (Thornthwaite, 1956; Grove, 1997), and it has been the basis of action programs in many lands. For example, the American Timber Culture Act of 1873 was passed in the belief that if settlers were induced to plant trees on the Great Plains and prairies, precipitation would be increased sufficiently to eliminate the climatic hazards to agriculture. On the other hand, at much the same time, the view was expressed that 'rain follows the plow'. Aughey, working in Nebraska, for example, believed that, after the soil is 'broken', rain as it falls is absorbed by the soil 'like a high sponge', and that the soil gives this absorbed moisture slowly back to the atmosphere by evaporation (cited by Thornthwaite, 1956: 569), and so increases the rainfall.

These two early and contradictory views illustrate the confusion that still surrounds this question today. Forests undoubtedly influence rates of evapotranspiration, the flow of streams, the level of groundwater, and microclimates, but there is little reliable evidence to suggest that regional rainfall is either significantly increased by forest or that attempts to augment rainfall levels on desert margins by widespread planting of forest belts are likely to achieve relatively much; the aridity of deserts and their margins is controlled dominantly by the gross features of the general circulation, especially the subsiding air associated with the large high-pressure cells of the subtropics.

Although forests may not necessarily have a proven effect on regional or continental rainfall levels, they are far more effective than other vegetation types at trapping other kinds of precipitation, especially cloud, fog, and mist. Hence deforestation or afforestation can affect water budgets through the degree to which they intercept nonrainfall precipitation.

There is one other land-use change that may result in measurable changes in precipitation; namely large-scale crop irrigation in semi-arid regions. The High Plains of the USA are normally covered with sparse grasses and have dry soils throughout the summer; evapotranspiration is then very low. In the past five decades irrigation has been developed throughout large parts of the area, greatly increasing summer evapotranspiration levels. Barnston and Schickdanz (1984) have produced strong statistical evidence of warm-season rainfall enhancement through irrigation in two parts of this area: one extending through Kansas, Nebraska, and Colorado, and a second in the Texas Panhandle. The largest absolute increase was in the latter area and, significantly, occurred in June, the wettest of the three heavily irrigated months. The effect appears to be especially important when stationary weather fronts occur, for this is a situation that allows for maximum interaction between the damp irrigated surface and the atmosphere. Hailstorms and tornadoes are also significantly more prevalent than over nonirrigated regions (Nicholson, 1988). However, Moore and Rojstaczer (2001) think that overall the irrigation effect is both difficult to quantify unambiguously and probably of minor significance.

Bonan (1997) has tried to model the climatic consequences of replacing the natural forests of the USA with crops and has argued that it would cause cooling of up to 2°C in the summer months over a wide region of the central USA. He suggests (p. 484) that 'land use practices that resulted in extensive deforestation in the Eastern USA, replacing forests with crop, have resulted in a significant climate change that is comparable to other well known anthropogenic climate forcings.'

The possible effects of water diversion schemes

The levels of the Aral and Caspian Seas in Central Asia have fallen, as have water tables all over the wide continental region. There have been proposals to divert some major rivers to help overcome these problems. However, this raises difficult questions 'because it appears to touch a peculiarly sensitive spot in the existing climatic regime of the northern hemisphere' (Lamb, 1977: 671). The low-salinity water which forms a 100–200 m upper layer to the Arctic Ocean is in part caused by the input of freshwater from the large Russian and Siberian rivers. This low-salinity water is the medium in which the pack ice at present covering the polar ocean is formed. The tapping of any large proportion of this river flow might augment the area of saltwater in the Arctic Ocean and thereby reduce the area of pack ice correspondingly. Temperatures over large areas might rise, which in turn might change the position and alignment of the main thermal gradients in the Northern Hemisphere and, with them, the jet stream and the development and steering of cyclonic activity. However, assessment of this particular climatic impact is still very largely speculative, and some numerical models indicate that the climate of the Arctic will not be drastically affected by river diversions (Semtner, 1984).

Lakes

It has often been implied that the presence of a large body of inland water must modify the climate around its shores, and therefore that artificial lakes have a significant effect on local or regional climates. Climatic changes produced by the construction of a reservoir are the result of a variety of factors (Vendrov, 1965): the creation of a body of water with a large heat capacity that reduces the continentality of the climate; the substitution of a water surface for a land surface and the rise of the groundwater level in the littoral zone supplying moisture to the evaporating surface (leading to a rise in wind velocity above the lake and in the littoral zone). Schemes have been put forward for augmenting desert rainfall by flooding desert basins in the Sahara, Kalahari, and Middle East (see, e.g., Schwarz, 1923). However, whether evaporation from lake surfaces can raise local precipitation levels is open to question, for precipitation depends more on atmospheric instability than upon the humidity content of the air. Moreover, most lakes are too small to affect the atmosphere materially in depth, so that their influence falls heavily under the sway of the regional circulation. In addition, one needs to remember that some of the world's driest deserts occur along coastlines. Thus a relatively small artificial lake would be even more impotent in creating rainfall (see Crowe, 1971: 443–50).

However, the climatic effect of artificial lakes is evident in other ways, notably in terms of a local reduction in frost hazard. In the case of the Rybinsk reservoir (c. 4500 km^2) in the CIS, it has been calculated that the climatic influence extends 10 km from the lake and that the frost-free season has been extended by 5–15 days on average (D'Yakanov and Reteyum, 1965).

Urban climates

One consequence of the burning of fossil fuels is the production of heat. This is probably most important on the local scale where it can be identified as the 'urban heat island'. On a broader scale, the amount of energy used by humans has been negligible compared both with the resources of solar energy and with the energy of photosynthesis of plants. In global terms the total amount of heat released by all human activity is roughly 0.01% of the solar energy absorbed at the surface (Kellogg, 1978: 215). Such a small fraction would have a negligible effect on the overall heat balance of Earth.

It has been said that 'the city is the quintessence of man's capacity to inaugurate and control changes in his habitat' (Detwyler and Marcus, 1972). One way in which such control becomes evident is in a study of urban climates (Landsberg, 1981). Individual urban areas can at times, with respect to their weather, 'have

Table 7.6 Effects of cities on climate. Source: H. Landsberg in Griffiths (1976: 108)

(a) Average changes in climatic elements caused by cities

Element	Parameter	Urban compared with rural (−, less; +, more)
Radiation	On horizontal surface	−15%
	Ultraviolet	−30% (winter); −5% (summer)
Temperature	Annual mean	+0.7°C
	Winter maximum	+1.5°C
	Length of freeze-free season	+2 to 3 weeks (possible)
Wind speed	Annual mean	−20 to −30%
	Extreme gusts	−10 to −20%
	Frequency of calms	+5 to 20%
Relative humidity	Annual mean	−6%
	Seasonal mean	−2% (winter); −8% (summer)
Cloudiness	Cloud frequency + amount	+5 to 10%
	Fogs	+100% (winter); −30% (summer)
Precipitation	Amounts	+5% to 10%
	Days	+10%
	Snow days	−14%

(b) Effect of city surfaces

Phenomenon	Consequence
Heat production (the heat island)	Rainfall + Temperature +
Retention of reflected radiation by high walls and dark-colored roofs	Temperature +
Surface roughness increase	Wind − Eddying +
Dust increase (the dust dome)	Fog + Rainfall + (?)

similar impacts as a volcano, a desert, and as an irregular forest' (Changnon, 1973: 146). Some of the changes that can result are listed in Table 7.6.

Compared with rural surfaces, city surfaces (Table 7.6b) absorb significantly more solar radiation, because a higher proportion of the reflected radiation is retained by the high walls and dark-colored roofs of the city streets. The concreted city surfaces have both great thermal capacity and conductivity, so that

Table 7.7 Annual mean urban – rural temperature differences of cities. Source: from data in Detwyler (1971), and Wilby (2003)

City	Temperature differences (°C)
Chicago, USA	0.6
Washington, DC, USA	0.6
Los Angeles, USA	0.7
Paris, France	0.7
Moscow, Russia	0.7
Philadelphia, USA	0.8
Berlin, Germany	1.0
New York, USA	1.1
London, UK	1.8

heat is stored during the day and released by night. By contrast the plant cover of the countryside acts like an insulating blanket, so that rural areas tend to experience relatively lower temperatures by day and night, an effect enhanced by the evaporation and transpiration taking place. In addition, the energy partitioned for evapotranspiration is less in urban areas, leading to greater surface heating. Another thermal change in cities, contributing to the development of the 'urban heat island', is the large amount of artificial heat produced by industrial, commercial and domestic users.

In general the highest temperature anomalies are associated with the densely built-up area near the city center, and decrease markedly at the city perimeter. Observations in Hamilton, Ontario, and Montreal, Quebec, suggested temperature changes of 3.8 and 4.0°C km^{-1} respectively (Oke, 1978). Temperature differences also tend to be highest during the night. The form of the urban temperature effect has often been likened to an 'island' protruding distinctly out of the cool 'sea' of the surrounding landscape. The rural–urban boundary exhibits a steep temperature gradient or 'cliff' to the urban heat island. Much of the rest of the urban area appears as a 'plateau' of warm air with a steady but weaker horizontal gradient of increasing temperature towards the city center. The urban core may be a 'peak' where the urban maximum temperature is found. The difference between this value and the background rural temperature defines the *urban heat island intensity* (Tu–r (max)) (Oke, 1978: 225).

Table 7.7 lists the average annual urban – rural temperature differences for several large cities. Values

range from 0.6 to 1.8°C. The relationship between city size and urban – rural difference, however, is not necessarily linear; sizeable nocturnal temperature contrasts have been measured even in relatively small cities. Factors such as building density are at least as important as city size, and high wind velocities will tend to eliminate the heat island effect.

Nonetheless, Oke (1978: 257) has found that there is some relation between heat-island intensity and city size. Using population as a surrogate of city size Tu–r (max) is found to be proportional to the log of the population. Other interesting results of this study include the tendency for quite small centers to have a heat island, the observation that the maximum thermal modification is about 12°C, and the recognition of a difference in slope between the North American and the European relationships (Figure 7.9). The explanation for this last result is not clear, but it may be related to the fact that population is a surrogate index of the central building density.

The relationship between maximum heat island intensity and urban population, the sky-view factor (a measure of building height and density), and the impermeable surface coverage of cities is shown in Figure 7.9. The heat island intensity increases with all three indices (Nakagawa, 1996).

In many older towns and cities in western Europe and North America, the process of 'counter-urbanization' has in recent years led to a decline in population, and it is worth considering whether this is reflected in a decline in the intensity of urban heat islands. One attempt to do this, in the context of London (Lee, 1992), revealed the perplexing finding that the heat-island intensity has decreased by day, but increased by night. The explanation that has been tentatively advanced to explain this is that there has been a decrease in the receipt of daytime solar radiation as a result of vehicular atmosphere pollution, whereas at night the presence of such pollution absorbs and re-emits significant amounts of outgoing terrestrial radiation, maintaining higher urban nocturnal minimum temperatures. The urban heat island may also be more marked at night because reduced nocturnal turbulent mixing keeps the warmer air near the surface. In mid-latitude cities such as London, urban heat island effects are generally stronger in summer than in winter because of higher levels of solar radiation being absorbed by building materials during the day. In winter the urban – rural

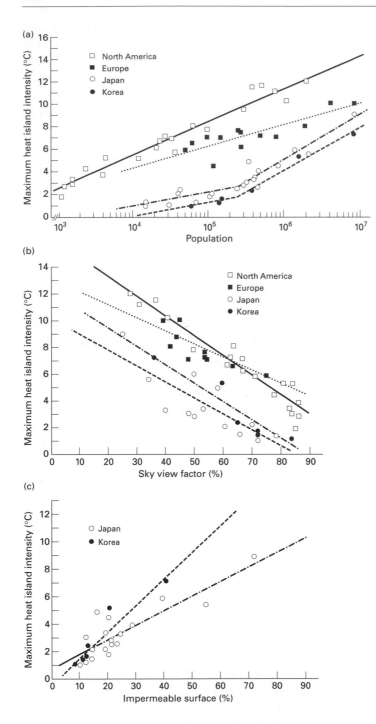

Figure 7.9 Relations between the maximum heat island intensity and (a) urban population for Japanese, Korean, North American, and European cities; (b) the sky view factor for Japanese, Korean, North American, and European cities; (c) the ratio of impermeable surface coverage for Japanese and Korean cities (after Nakagawa, 1996, figures 2, 3, and 4).

contrast is weaker because solar energy absorption is lower and hence there is less energy to radiate, despite higher levels of urban space heating (Wilby, 2003).

The existence of the urban heat island has a number of implications: city plants bud and bloom earlier, some birds are attracted to the thermally more favorable urban habitat, humans find added warmth stressful if the city is already situated in a warm area, during summer heatwaves exacerbated temperatures may cause mortality among sensitive members of the population, and less winter space-heating is required but, conversely, more summer air-conditioning is necessary.

The urban-industrial effects on clouds, rain, snowfall, and associated weather hazards such as hail and

Table 7.8 Areas of maximum increases (urban – rural difference) in summer rainfall and severe weather events for eight American cities. Source: after Changnon (1973: 144, figure 1.5)

City	Rainfall		Thunderstorms		Hailstorms	
	%	Location*	%	Location*	%	Location*
St Louis	+15	B	+25	B	+276	C
Chicago	+17	C	+38	A, B, C	+246	C
Cleveland	+27	C	+42	A, B	+90	C
Indianapolis	0	–	0	–	0	–
Washington, DC	+9	C	+36	A	+67	B
Houston	+9	A	+10	A, B	+430	B
New Orleans	+10	A	+27	A	+350	A, B
Tulsa	0	–	0	–	0	–

*A = within city perimeter; B = 8–24 km downwind; C = 24–64 km downwind.

thunder are harder to measure and explain than the temperature changes (Darungo et al., 1978). The changes can be related to various influences (Changnon, 1973: 143):

- thermally induced upward movement of air;
- increased vertical motions from mechanically induced turbulence;
- increased cloud and raindrop nuclei;
- industrial increases in water vapor.

Table 7.8 illustrates the differences in summer rainfall, thunderstorms, and hailstorms between various rural and urban areas in the USA. These data indicate that in cities rainfall increases ranged from 9 to 27%, the incidence of thunderstorms increased by 10 to 42%, and hailstorms increased by 67 to 430%.

An interesting example of the effects of major conurbations on precipitation levels is provided by the London area. In this case it seems that the mechanical effect of the city was dominant in creating localized maxima of precipitation both by being a mechanical obstacle to air flow, on the one hand, and by causing frictional convergence of flow, on the other (Atkinson, 1975). A long-term analysis of thunderstorm records for southeast England is highly suggestive – indicating the higher frequencies of thunderstorms over the conurbation compared with elsewhere (Atkinson, 1968).

The similarity in the morphology of the thunderstorm isopleth and the urban area is striking (Figure 7.10a and b). Moreover, Brimblecombe (1977) shows a steadily increasing thunderstorm frequency as the city has grown (Figure 7.10c).

Similarly the detailed Metromex investigation of St Louis in the USA (Changnon, 1978) shows that in the summer the city affects precipitation and other variables within a distance of 40 km. Increases were found in various thunderstorm characteristics (about +10 to +115%), hailstorm condition (+3 to +330%), various heavy rainfall characteristics (+35 to +100%), and strong gusts (+90 to +100%).

Two main factors are involved in the effect that cities have on winds: the rougher surface they present in comparison with rural areas; and the frequently higher temperatures of the city fabric.

Buildings, especially those in cities with a highly differentiated skyline, exert a powerful frictional drag on air moving over and around them (Chandler, 1976). This creates turbulence, with characteristically rapid spatial and temporal changes in both direction and speed. The average speed of the winds is lower in built-up areas than over rural areas, but Chandler found that in London, when winds are light, speeds are greater in the inner city than outside, whereas the reverse relationship exists when winds are strong. The overall annual reduction of wind speed in central London is about 6%, but for the higher velocity winds (more than 1.5 m per second) the reduction is more than doubled.

Studies in both Leicester and London (Chandler, 1976), England, have shown that on calm, clear nights, when the urban heat-island effect is at its maximum, there is a surface inflow of cool air towards the zones of highest temperatures. These so-called 'country breezes' have low velocities and become quickly decelerated by intense surface friction in the suburban areas. A practical implication of these breezes is that they transport pollution from the outer parts of an urban area into the city center, accentuating the pollution problem during smogs.

Urban air pollution

The concentration of large numbers of people, factories, power stations, and cars means that large amounts

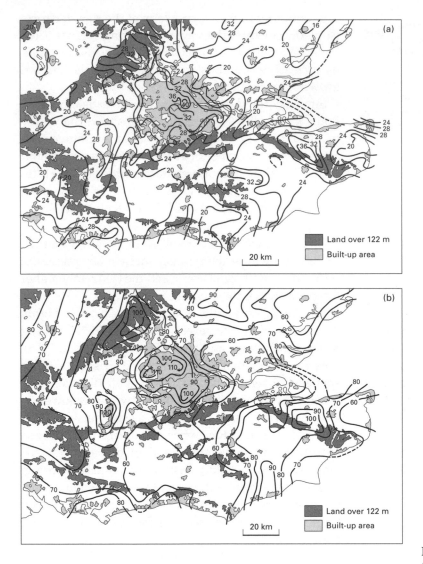

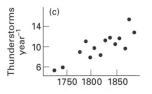

Figure 7.10 Thunder in southeast England: (a) total thunder rain in southeast England, 1951–60, expressed in inches (after Atkinson, 1968, figure 6); (b) number of days with thunder overhead in southeast England, 1951–60 (after Atkinson, 1968, figure 5); (c) thunderstorms per year in London (decadal means for whole year) (after Brimblecombe, 1977, figure 2).

of pollutants may be emitted into urban atmospheres. If weather conditions permit, the level of pollution may build up (Figure 7.11). The nature of the pollutants (Table 7.9) has changed as technologies have changed. For example, in the early phases of the industrial revolution in Britain the prime cause of air pollution in cities may have been the burning of coal, whereas now it may be vehicular emissions. Different cities may have very different levels of pollution, depending on factors such as the level of technology, size, wealth, and antipollution legislation. Differences may also arise because of local topographic and climatic conditions. Photochemical smogs, for example, are a more serious threat in areas subjected to intense sunlight.

Figure 7.11 In December 1952 the city of London was affected by severe smog. Visibility was reduced and smog-masks had to be worn out of doors. Many people with weak chests died. Since then, because of legislation, the incidence of smog has declined markedly.

The variations in pollution levels between different cities are brought out in Figure 7.12, which shows data for two types of pollution for a large range of city types. The data were prepared for the years 1980–4 by the Global Environmental Monitoring System of the United Nations Environmental Program (UNEP). Figure 7.12a shows concentrations of total particulate matter. Most of this comes from the burning of poor-quality fuels. The shaded horizontal bar indicates the range of concentrations the UNEP considers a reasonable target for preserving human health. Note that the annual mean levels range from a low of about $35 \, \mu g \, m^{-3}$ to a high of about $800 \, \mu g \, m^{-3}$: a range of about 25-fold! The higher values appear to be for rapidly growing cities in the developing countries. Some cities, however, such as Kuwait, may have unusually high values because of their susceptibility to dust storms from desert hinterlands. The lower values tend to come from cities in the developed world (e.g., western Europe, Japan, and North America).

Figure 7.12b shows concentrations for sulfur dioxide. Much of this gas probably comes from the burning of high-sulfur coal. Once again, the horizontal shaded bar indicates the concentration range considered by UNEP to be a reasonable target for preserving human health.

These data indicate that the concentrations of sulfur dioxide can differ by as much as three times among different sites within the same urban areas and by as much as 30 times between different urban areas.

Table 7.9 Major urban pollutants

Type	Some consequences
Suspended particulate matter (characteristically 0.1–25 μm in diameter)	Fog, respiratory problems, carcinogens, soiling of buildings
Sulfur dioxide (SO_2)	Respiratory problems, can cause asthma attacks. Damage to plants and lichens, corrosion of buildings and materials, production of haze and acid rain
Photochemical oxidants: ozone and peroxyacetyl nitrate (PAN)	Headaches, eye irritation, coughs, chest discomfort, damage to materials (e.g., rubber), damage to crops and natural vegetation, smog
Oxides of nitrogen (NO_x)	Photochemical reactions, accelerated weathering of buildings, respiratory problems, production of acid rain and haze
Carbon monoxide (CO)	Heart problems, headaches, fatigue, etc.
Toxic metals: lead	Poisoning, reduced educational attainments and increased behavioral difficulties in children
Toxic chemicals: dioxins, etc.	Poisoning, cancers, etc.

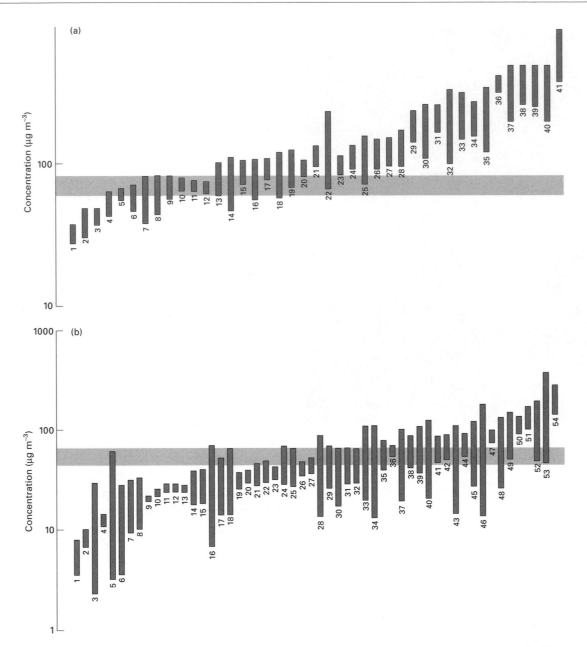

Figure 7.12 (a) The range of annual averages of total particulate matter concentrations measured at multiple sites within 41 cities, 1980–1984. The shading indicates the concentration range recommended by the United Nations Environment Program as a reasonable target for preserving human health. Each numbered bar represents a city, as follows: 1, Frankfurt; 2, Copenhagen; 3, Cali; 4, Osaka; 5, Tokyo; 6, New York; 7, Vancouver; 8, Montreal; 9, Fairfield; 10, Chattanooga; 11, Medellin; 12, Melbourne; 13, Toronto; 14, Craiova; 15, Houston; 16, Sydney; 17, Hamilton; 18, Helsinki; 19, Birmingham; 20, Caracas; 21, Chicago; 22, Manila; 23, Lisbon; 24, Accra; 25, Bucharest; 26, Rio de Janeiro; 27, Zagreb; 28, Kuala Lumpur; 29, Bombay; 30, Bangkok; 31, Illigan City; 32, Guangzhou; 33, Shanghai; 34, Jakarta; 35, Tehran; 36, Calcutta; 37, Beijing; 38, New Delhi; 39, Xi'an; 40, Shenyang; 41, Kuwait City. (b) The range of annual averages of sulfur dioxide concentrations measured at multiple sites within 54 cities, 1980–1984. Each numbered bar represents a city, as follows: 1, Craiova; 2, Melbourne; 3, Auckland; 4, Cali; 5, Tel Aviv; 6, Bucharest; 7, Vancouver; 8, Toronto; 9, Bangkok; 10, Chicago; 11, Houston; 12, Kuala Lumpur; 13, Munich; 14, Helsinki; 15, Lisbon; 16, Sydney; 17, Christchurch; 18, Bombay; 19, Copenhagen; 20, Amsterdam; 21, Hamilton; 22, Osaka; 23, Caracas; 24, Tokyo; 25, Wroclaw; 26, Athens; 27, Warsaw; 28, New Delhi; 29, Montreal; 30, Medellin; 31, St Louis; 32, Dublin; 33, Hong Kong; 34, Shanghai; 35, New York; 36, London; 37, Calcutta; 38, Brussels; 39, Santiago; 40, Zagreb; 41, Frankfurt; 42, Glasgow; 43, Guangzhou; 44, Manila; 45, Madrid; 46, Beijing; 47, Paris; 48, Xi'an; 49, São Paulo; 50, Rio de Janeiro; 51, Seoul; 52, Tehran; 53, Shenyang; 54, Milan. (Source: Graedel and Crutzen, 1993.)

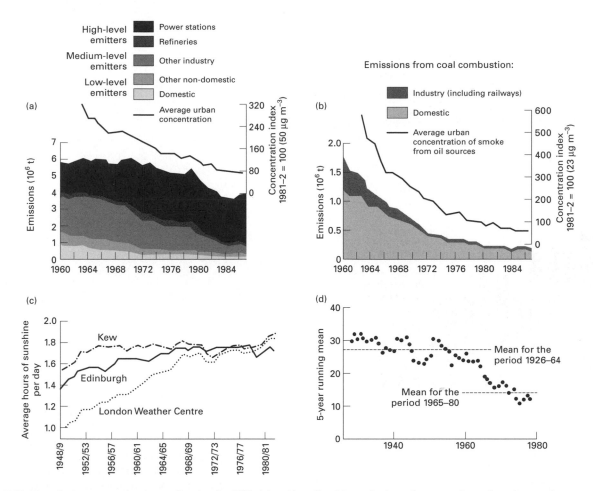

Figure 7.13 Trends in atmospheric quality in the UK: (a) sulfur dioxide emissions from coal combustion and average urban concentrations; (b) smoke emissions from coal combustion and average urban concentrations of oil smoke; (c) increase in winter sunshine (10-year moving average) for London and Edinburgh city centers and for Kew, outer London; (d) annual fog frequency at 0900 GMT in Oxford, central England, 1926–80 (after Department of the Environment data, and Gomez and Smith, 1984, figure 3).

Fenger (1999) has argued that the development of urban air pollution shows certain general historical trends (see Figure 7.14a). At the earlier stages air pollution increases to high levels. There then follows various abatement measures that cause a stabilization of air quality to occur. Levels of pollution then fall as high technology solutions are applied, although this may be countered to a certain extent by growth in vehicular traffic.

In some developed cities concentrations of pollutants have indeed tended to fall over recent decades. This can result from changes in industrial technology or from legislative changes (e.g., clean air legislation, restriction on car use, etc.). In many British cities, for example, legislation since the 1950s has reduced the burning of coal. As a consequence, fogs have become less frequent and the amount of sunshine has increased. Figure 7.13 shows the overall trends for the UK, and highlights the decreasing fog frequency and increasing sunshine levels (Musk, 1991). Similarly lead concentrations in the air in British cities have declined sharply following the introduction of unleaded petrol (Figure 7.14b) (Kirby, 1995), as they have in Copenhagen (Figure 7.14c) (Fenger, 1999). The concentrations of various pollutants have also been reduced in the Los Angeles area of California. Here, carbon monoxide, nonmethane hydrocarbon, nitrogen oxide, and ozone concentrations have all fallen steadily over the period since the late 1960s (Lents and Kelly, 1993).

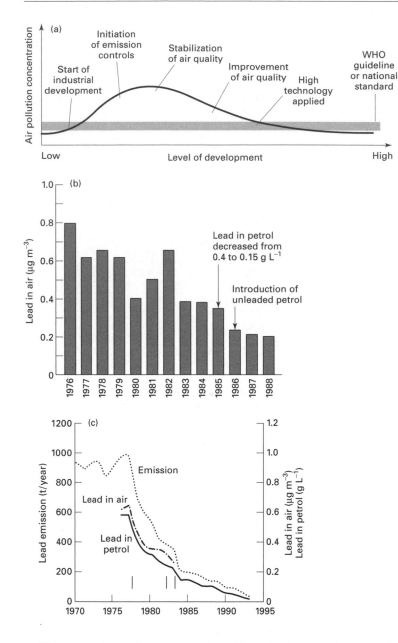

Figure 7.14 (a) Schematic presentation of a typical development of urban air pollution levels (after Fenger, 1999, figure 3): WHO – World Health Organization. (b) Lead concentration (annual means) in air at UK sites (Kirby, 1995, figure 1). (c) Annual average values for the total Danish lead emissions 1969–1993, the lead pollution in Copenhagen since 1976, and the average lead content in petrol sold in Denmark (Jensen and Fenger, 1994). The dates of tightening of restrictions on lead content are indicated with bars. Lead concentrations for the recent years can be found in Kemp et al. (1998). (Source: Fenger, 1999, figure 12.)

However, these three examples of improving trends come from developed countries. In many cities in poorer countries, pollution is increasing at present. In certain countries, heavy reliance on coal, oil, and even wood for domestic cooking and heating means that their levels of sulfur dioxide and suspended particulate matter (SPM) are high and climbing. In addition, rapid economic development is bringing increased emissions from industry and motor vehicles, which are generating progressively more serious air-quality problems.

Particular attention is being paid at the present time to the chemical composition of SPMs, and particularly to those particles that are small enough to be breathed in (i.e., smaller than 10 μm, and so often known as PM10s). Also of great concern in terms of human health are elemental carbon (e.g., from diesel vehicles), polycyclic aromatic hydrocarbons (PAHs), and toxic base metals (e.g., arsenic, lead, cadmium, and mercury), in part because of their possible role as carcinogens.

A major cause of urban air pollution is the development of photochemical smog. The name originates from the fact that most of the less desirable properties of such fog result from the products of chemical reactions induced by sunlight. Unburned hydrocarbons

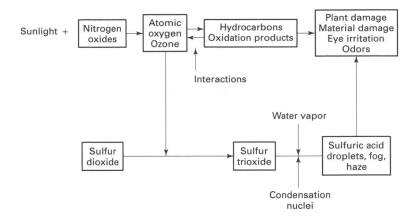

Figure 7.15 Possible reactions involving primary and secondary pollutants (after Haagen-Smit, in Bryson and Kutzbach, 1968, figure 4. Reprinted by permission of the Association of American Geographers).

play a major role in this type of smog formation and result from evaporation of solvents and fuels, as well as incomplete combustion of fossil fuels. In the presence of oxides of nitrogen, strong sunlight, and stable meteorological conditions, complex chemical reactions occur, forming a family of peroxyacyl nitrates (sometimes collectively abbreviated to PANs).

Photochemical smog appears 'cleaner' than other kinds of fog in the sense that it does not contain the very large particles of soot that are so characteristic of smog derived from coal burning. However, the eye irritation and damage to plant leaves it causes make it unpleasant. Photochemical smog occurs particularly where there is large-scale combustion of petroleum products, as in car-dominated cities such as Los Angeles. Its unpleasant properties include a high lead content; also a series of chemical reactions are triggered by sunlight (Figure 7.15). For example, a photochemical decomposition of nitrogen dioxide into nitric oxide and atomic oxygen occurs, and the atomic oxygen can react with molecular oxygen to form ozone. Further ozone may be produced by the reaction of atomic oxygen with various hydrocarbons.

Photochemical smogs are not universal. Because sunlight is a crucial factor in their development they are most common in the tropics or during seasons of strong sunshine. Their especial notoriety in Los Angeles is due to a meteorological setting dominated at times by subtropical anticyclones with weak winds, clear skies, and a subsidence inversion, combined with the general topographic situation and the high vehicle density (> 1500 vehicles per square kilometer). However, photochemical ozone pollution can, on certain summer days, with anticyclonic conditions bringing

air in from Europe, reach appreciable levels even in the UK (Jenkins et al., 2002), especially in large cities such as London. Rigorous controls on vehicle emissions can greatly reduce the problem of high urban ozone concentrations and this has been a major cause of the reduction in ozone levels in Los Angeles over the past two decades, in spite of a growth in that city's population and vehicle numbers (Figure 7.16). A full discussion of tropospheric ozone trends on a global basis is provided by Guicherit and Roemer (2000).

Air pollution: some further effects

This chapter has already made much reference to the ways in which humans have changed the turbidity of the atmosphere and the gases within it. However, the consequences of air pollution go further than either their direct impact on human health or their impact on local, regional, and global climates.

First of all, the atmosphere acts as a major channel for the transfer of pollutants from one place to another, so that some harmful substances have been transferred long distances from their sources of emission. Dichlorodiphenyltrichloroethane (DDT) is one example; lead is another. Thus, from the start of the industrial revolution, the lead content of the Greenland ice-cap, although far removed from the source of the pollutant (which is largely derived from either industrial or automobile emissions), rose very substantially (Figure 7.17a). The same applies to its sulfate content (Figure 7.17b). An analysis of pond sediments from a remote part of North America (Yosemite) indicates that lead levels were raised as a result of human activities,

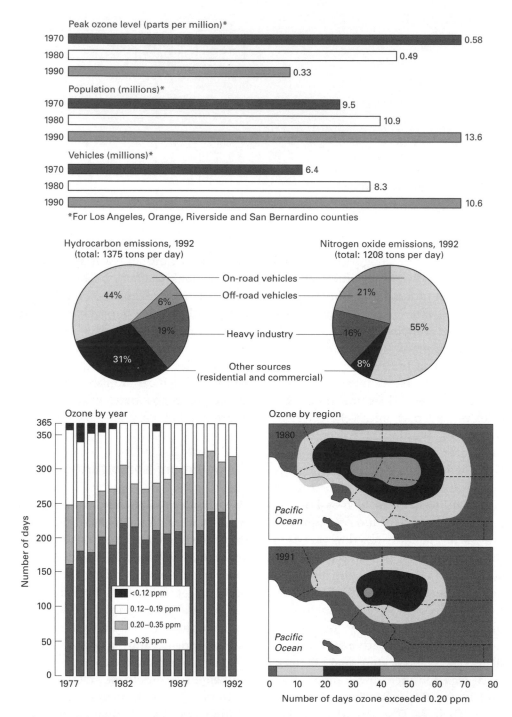

Figure 7.16 Air pollution in the Los Angeles area, 1970s to 1990s (after Lents and Kelly, 1993, p. 22).

being more than 20 times the natural levels. The lead, which came in from atmospheric sources, showed a fivefold elevation in the plants of the area and a fiftyfold elevation in the animals compared with natural levels (Shirahata et al., 1980). Some estimates also compared the total quantities of heavy metals that humans are releasing into the atmosphere with emissions from natural sources (Nriagu, 1979). The increase was eighteenfold for lead, ninefold for cadmium, sevenfold for zinc, and threefold for copper.

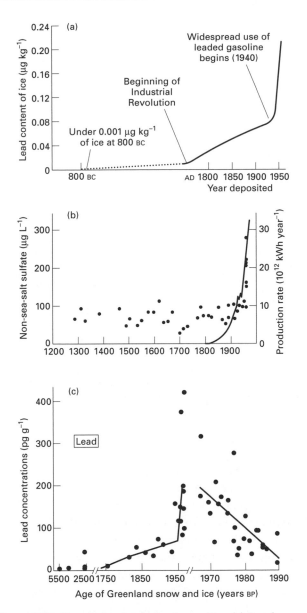

Figure 7.17 Trends in atmospheric quality. (a) Lead content of the Greenland ice-cap due to atmospheric fallout of the mineral on the snow surface. A dramatic upturn in worldwide atmospheric levels of lead occurred at the beginning of the industrial revolution in the nineteenth century and again after the more recent spread of the automobile (after Murozumi et al., 1969, p. 1247). (b) The sulfate concentration on a sea-salt-free basis in northwest Greenland glacier ice samples as a function of year. The curve represents the world production of thermal energy from coal, lignite, and crude oil (modified after Koide and Goldberg, 1971, figure 1). (c) Lead concentrations in Greenland snow (after Boutron et al., 1991. Reprinted with permission from *Nature*. Copyright 1991. Macmillan Magazines Limited).

There are, it must be stated, signs that as a result of pollution control regulations some of these trends are now being reversed. Boutron et al. (1991), for example, analyzed ice and snow that has accumulated over Greenland in the previous two decades and found that lead concentrations had decreased by a factor of 7.5 since 1970 (Figure 7.17c). They attribute this to a curbing of the use of lead additives in petrol. Over the same period cadmium and zinc concentrations have decreased by a factor of 2.5.

A second example of the possible widespread and ramifying ecological consequences of atmospheric pollution is provided by 'acid rain' (Likens and Bormann, 1974). Acid rain is rain which has a pH of less than 5.65, this being the pH which is produced by carbonic acid in equilibrium with atmospheric CO_2. In many parts of the world, rain may be markedly more acid than this normal, natural background level. Snow and rain in the northeast USA have been known to have pH values as low as 2.1, while in Scotland in one storm the rain was the acidic equivalent of vinegar (pH 2.4). In the eastern USA the average annual precipitation acidity values tend to be between pH 4 and 4.5 (see Figure 7.18), and the degree of acidulation appears to have increased between the 1950s and 1970s.

It needs to be remembered that not all environmental acidification is caused by acid rain in the narrow sense. Acidity can reach the ground surface without the assistance of water droplets. This is as particulate matter and is termed 'dry deposition'. Furthermore, there are various types of 'wet precipitation' by mist, hail, sleet, or snow, in addition to rain itself. Thus some people prefer the term 'acid deposition' to 'acid rain'. The acidity of the precipitation in turn leads to greater acidity in rivers and lakes.

The causes of acid deposition are the quantities of sulfur oxides (Figure 7.19) and nitrogen oxides emitted from fossil-fuel combustion. Figure 7.20 shows how sulfate levels increased in European precipitation between the 1950s and 1970s. Two main factors contributed to the increasing seriousness of the problem at the time. One was the replacement of coal by oil and natural gas. The second, paradoxically, was a result of the implementation of air pollution control measures (particularly increasing the height of smokestacks and installing particle precipitators). These appear to have transformed a local 'soot problem' into a regional 'acid rain problem'. Coal burning produced a great deal of

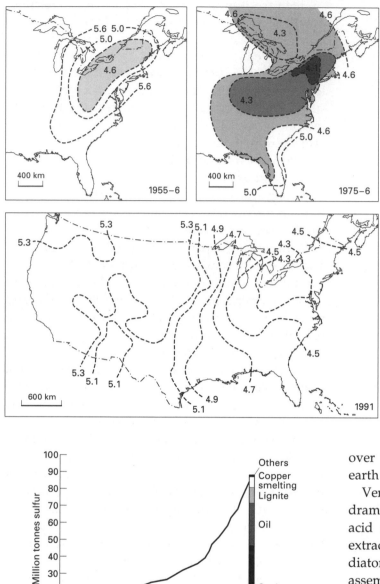

Figure 7.18 Isopleths showing average pH for precipitation in North America. Note the low values for eastern North America and the relatively higher values to the west of the Mississippi. (Combined from Likens et al., 1979 and Graedel and Crutzen, 1995, with modifications.)

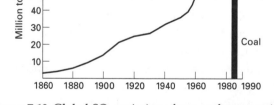

Figure 7.19 Global SO$_2$ emissions from anthropogenic sources, including the burning of coal, lignite, and oil, and copper smelting.

sulfate, but was largely neutralized by high calcium contents in the relatively unfiltered coal smoke emissions. Natural gas burning creates less sulfate but that which is produced is not neutralized. The new higher chimneys pump the smoke so high that it is dispersed over wide areas, whereas previously it returned to earth nearer the source.

Very long-term records of lake acidification provided dramatic evidence for the recent magnification of the acid deposition problem. These were obtained by extracting cores from lake floors and analyzing their diatom assemblages at different levels. The diatom assemblages reflect water acidity levels at the time they were living. Two studies serve to illustrate the trend. In southwest Sweden (Renberg and Hellberg, 1982) for most of post-glacial time (that is, the past 12,500 years) the pH of the lakes appears to have decreased gradually from around 7.0 to about 6.0 as a result of natural aging processes. However, especially since the 1950s, a marked decrease occurred to present-day values of about 4.5. In Britain, the work of Battarbee and collaborators (1985a and b) shows that at sensitive sites pH values before around 1850 were close to 6.0 and that since then pH declines have varied between 0.5 and 1.5 units. More monitoring of lake acidity also demonstrated that significant changes were taking place. Studies by Beamish et al. (1975)

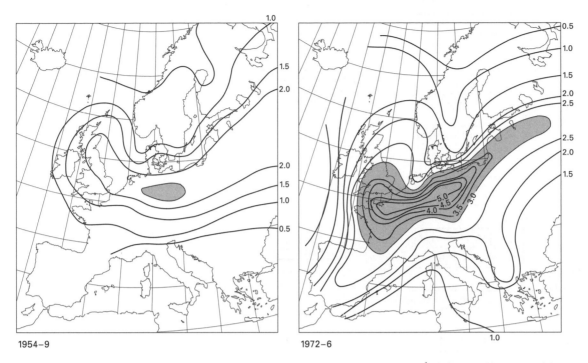

Figure 7.20 Annual mean concentration of sulfate in precipitation in Europe (mg S L^{-1}) (after Wallen in Holdgate et al., 1982, figure 2.3).

demonstrated that between 1961 and 1975 pH had declined by 0.13 pH units per year in George Lake, Canada, and a comparable picture emerges from Sweden, where Almer et al. (1974) found that the pH in some lakes had decreased by as much as 1.8 pH units since the 1930s.

The effects of acid rain (Figure 7.21) are especially serious in areas underlain by highly siliceous types of bedrock (e.g., granite, some gneisses, quartzite, and quartz sandstone), such as the old shield areas of the Fenno-Scandian shield in Scandinavia and the Laurentide shield in Canada (Likens et al., 1979). This is because of the lack of buffering by cations. Mobile anions are able to cause the leaching of basic cations (nutrients).

The ecological consequences of acid rain are still the subject of some debate. Krug and Frink (1983) have argued that acid rain only accelerates natural processes, and point out that the results of natural soil formation in humid climates include the leaching of nutrients, the release of aluminum ions and the acidification of soil and water. They also note that acidification by acid rain may be superimposed on longer-term acidification induced by changes in land use. Thus the regrowth of coniferous forests in what are now marginal agriculture areas, such as New England and highland western Europe, can increase acidification of soils and water. Similarly, it is possible, though in general unproven (see Battarbee et al., 1985a), that in areas such as western Scotland a decline in upland agriculture and the regeneration of heathland could play a role in increasing soil and water acidification. Likewise Johnston et al. (1982) have suggested that acid rain can cause either a decrease or an increase in forest productivity, depending on local factors. For example in soils where cation nutrients are abundant and sulfur or nitrogen are deficient, moderate inputs of acid rain are very likely to stimulate forest growth.

In general, however, it is the negative consequences of acid rain that have been stressed. One harmful effect is a change in soil character. The high concentration of hydrogen ions in acid rain causes accelerated leaching of essential nutrients, making them less available for plant use. Furthermore, the solubility of aluminum and heavy metal ions increases and instead of being fixed in the soil's sorption complex these toxic substances become available for plants or are transferred into lakes, where they become a major physiological stress for some aquatic organisms.

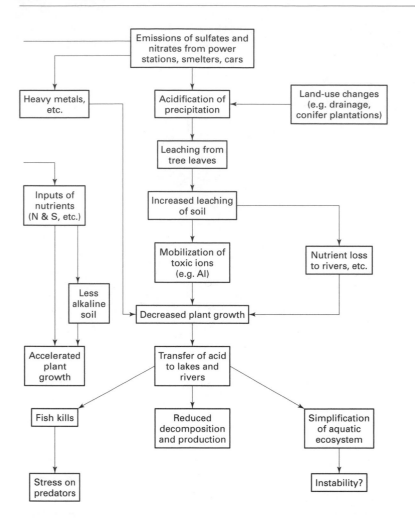

Figure 7.21 Pathways and effects of acid precipitation through different components of the ecosystem, showing some of the adverse and beneficial consequences.

Freshwater bodies with limited natural cations are poorly buffered and thus vulnerable to acid inputs. The acidification of thousands of lakes and rivers in southern Norway and Sweden during the late twentieth century has been attributed to acid rain, and this increased acidity resulted in the decline of various species of fish, particularly trout and salmon. But fish are not the only aquatic organisms that may be affected. Fungi and moss may proliferate, organic matter may start to decompose less rapidly, and the number of green algae may be reduced. Forest growth can also be affected by acid rain, although the evidence is not necessarily proven. Acid rain can damage foliage, increase susceptibility to pathogens, affect germination and reduce nutrient availability. However, since acid precipitation is only one of many environmental stresses, its impact may enhance, be enhanced by, or be swamped by other factors. For example, Blank (1985),

in considering the fact that an estimated one-half of the total forest area of the former West Germany was showing signs of damage, referred to the possible role of ozone or of a run of hot, dry summers on tree health and growth. The whole question of forest decline is addressed in Chapter 4.

The seriousness of acid rain caused by sulfur dioxide emissions in the Western industrialized nations peaked in the mid-1970s or early 1980s (Figure 7.22). Changes in industrial technology, in the nature of economic activity, and in legislation caused the output of SO_2 in Britain to decrease by 35% between 1974 and 1990. This was also the case in many industrialized countries (Table 7.10b), including the USA (Malm et al., 2002). However, there has been a shift in the geographical sources of sulfate emissions, so that whereas in 1980 60% of global emissions were from the USA, Canada and Europe, by 1995 only 38% of world emis-

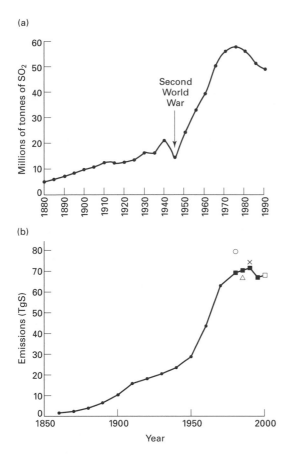

Figure 7.22 (a) Trends in sulfate emissions in Europe 1880–1990, based on data in Mylona (1996). (b) Estimates of historical total global sulfur dioxide (TgS) emissions from anthropogenic sources. From Smith et al. (2001, figure 1) with modifications.

sions originated from this region (Smith et al., 2001). There are also increasing controls on the emission of NO_x in vehicle exhaust emissions, although emissions of nitrogen oxides in Europe have not declined (Table 7.10). A similar picture emerges from Japan (Seto et al., 2002) where sulfate emissions have fallen because of emission controls, whereas nitrate emissions have increased with an increase in vehicular traffic. This means that the geography of acid rain may change, with it becoming less serious in the developed world, but with it increasing in locations such as China, where economic development will continue to be fueled by the burning of low quality sulfur-rich coal in enormous quantities.

In those countries where acid rain remains or is increasing as a problem there are various methods

available to reduce its damaging effects. One of these is to add powdered limestone to lakes to increase their pH values. However, the only really effective and practical long-term treatment is to curb the emission of the offending gases. This can be achieved in a variety of ways: by reducing the amount of fossil fuel combustion; by using less sulfur-rich fossil fuels; by using alternative energy sources that do not produce nitrate or sulfate gases (e.g., hydropower or nuclear power); and by removing the pollutants before they reach the atmosphere. For example, after combustion at a power station, sulfur can be removed ('scrubbed') from flue gases by a process known as flue gas desulfurization (FGD), in which a mixture of limestone and water is sprayed into the flue gas which converts the sulfur dioxide (SO_2) into gypsum (calcium sulfate). Reducing NO_x in flue gas can be achieved by adding ammonia and passing it over a catalyst to produce nitrogen and water (a process called selective catalytic reduction or SCR), and NO_x produced by cars can be reduced by fitting a catalytic converter.

Stratospheric ozone depletion

A very rapidly developing area of concern in pollution studies is the current status of stratospheric ozone levels. The atmosphere has a layer of relatively high concentration of ozone (O_3) at a height of about 16 and 18 km in the polar latitudes and of about 25 km in equatorial regions. This ozone layer is important because it absorbs incoming solar ultraviolet radiation, thus warming the stratosphere and creating a steep inversion of temperature at heights between about 15 and 50 km. This in turn affects convective processes and atmospheric circulation, thereby influencing global weather and climate. However, the role of ozone in controlling receipts of ultraviolet radiation at the surface of the Earth has great ecological significance, because it modifies rates of photosynthesis. One class of organism that has been identified as being especially prone to the effects of increased ultraviolet radiation consequent upon ozone depletion is phytoplankton – aquatic plants that spend much of their time near the sea surface and are therefore exposed to such radiation. A reduction in their productivities would have potentially ramifying consequences, because these plants directly and indirectly provide the food for almost all

Table 7.10 Emissions of sulfur and nitrogen oxides (as NO_2). Source: Smith et al. (2001)
(a) For selected European countries in 1980 and 1993 (1000 t year^{-1})

Country	Sulfur (1000 t year^{-1})			Nitrogen oxides (1000 t year^{-1})		
	1980	*1993*	*1993 as % of 1980*	*1980*	*1993*	*1993 as % of 1980*
Czech Republic	1128	710	62.9	937	574	61.26
Denmark	226	78	34.51	274	264	96.35
Finland	292	60	20.55	264	253	95.83
France	1669	568	34.03	1823	1519	83.32
Germany	3743	1948	52.04	2440	2904	84.42
Ireland	111	78	70.27	73	122	167.12
Italy	1900	1126	59.26	1480	2053	138.72
Netherlands	244	84	34.43	582	561	96.39
Norway	70	18	25.71	185	225	120.96
Poland	2050	1362	66.44	1500	1140	76.00
Spain	1660	1158	69.76	950	1257	132.32
Sweden	254	50	19.69	424	399	94.10
UK	2454	1597	65.08	2395	2355	98.32
Overall			47.28			103.47

(b) Percentage contributions for four regions to global sulfur dioxide emissions

Region	Sulfur dioxide emissions by geographic region (%)				
	1980	*1985*	*1990*	*1995*	*2000*
USA/Canada	21	18	18	16	15
Europe	39	36	31	22	19
Asia	26	31	35	43	46
Rest of world	14	15	16	18	20

fish. Ozone also protects humans from adverse effects of ultraviolet radiation, which include damage to the eyes, suppression of the immune system and higher rates of skin cancer.

Human activities appear to be causing ozone depletion in the stratosphere, most notably over the south polar regions, where an 'ozone hole' has been identified (Staehelin et al., 2001). Possible causes of ozone depletion are legion, and include various combustion products emitted from high-flying military and civil supersonic aircraft; nitrous oxide released from nitrogenous chemical fertilizers; and chlorofluorocarbons (CFCs) used in aerosol spray cans, refrigerant systems, and in the manufacture of foam fast-food containers.

However, in recent years the greatest attention has been focused on the role of CFCs, the production of which climbed greatly in the decade after the Second World War (Figure 7.23). These gases may diffuse up-

wards into the stratosphere where solar radiation causes them to become dissociated to yield chlorine atoms which react with and destroy the ozone. The process has been described thus by Titus and Seidel (1986: 4):

Because CFCs are very stable compounds, they do not break up in the lower atmosphere (known as the troposphere). Instead, they slowly migrate to the stratosphere, where ultraviolet radiation breaks them down, releasing chlorine.

Chlorine acts as a catalyst to destroy ozone; it promotes reactions that destroy ozone without being consumed. A chlorine (Cl) atom reacts with ozone (O_3) to form ClO and O_2. The ClO later reacts with another O_3 to form two molecules of O_2, which releases the Cl atoms. Thus two molecules of ozone are converted to three molecules of ordinary oxygen, and the chlorine is once again free to start the process. A single chlorine atom can destroy thousands of ozone molecules. Eventually, it returns to the troposphere, where it is rained out as hydrochloric acid.

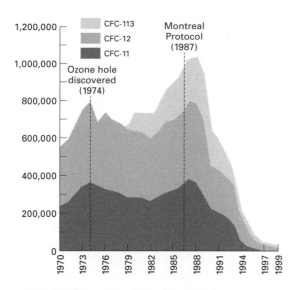

Figure 7.23 World production of major chlorofluorocarbons (t year^{-1}). World production of the three major CFCs peaked in about 1988 and has since declined to very low values.

The Antarctic ozone hole (Figure 7.24) has been identified through satellite monitoring and by monitoring of atmospheric chemistry on the ground. The decrease in ozone levels at Halley Bay is shown in Figure 7.25.

The reasons why this zone of ozone depletion is so well developed over Antarctica include: the very low temperatures of the polar winter, which seem to play a role in releasing chlorine atoms; the long sunlight hours of the polar summer, which promote photochemical processes; and the existence of a well-defined circulation vortex. This vortex is a region of very cold air surrounded by strong westerly winds, and air within this vortex is isolated from that at lower latitudes, permitting chemical reactions to be contained rather than more widely diffused. No such clearly defined vortex exists in the Northern Hemisphere, although ozone depletion does seem to have occurred in the Arctic as well (Proffitt et al., 1990). Furthermore, observations in the past few years indicate that the Antarctic ozone hole is spreading over wider areas and persisting longer into the Antarctic summer (Solomon, 1999). It is also possible that the situation could be worsened by emissions of volcanic ash into the atmosphere (as from Mount Pinatubo), for these can also cause chemical reactions that lead to ozone depletion (Mintzer and Miller, 1992).

Decreases in stratospheric ozone levels on a global basis have been analyzed by Harris et al. (2003). The most negative trends occur at mid- to high latitudes

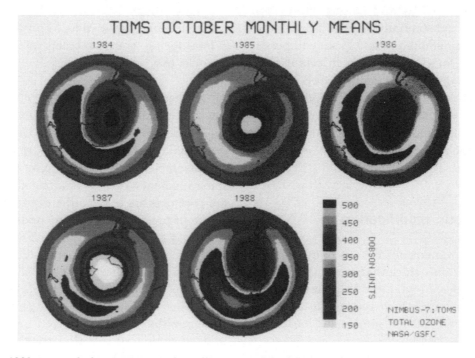

Figure 7.24 In the 1980s ground observations and satellite monitoring of atmospheric ozone levels indicated that a 'hole' had developed in the stratospheric ozone layer above Antarctica. These Nimbus satellite images show the ozone concentrations (in Dobson units) for the month of October between 1984 and 1988. Source: NASA.

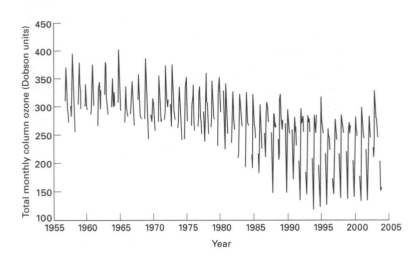

Figure 7.25 Total monthly ozone concentrations in Dobson units over Halley Bay, Antarctica (based on data provided by the British Antarctic Survey).

(Figure 7.26). Northern Hemisphere mid-latitude trends since 1978 have been at −2 to −3% per decade, while average losses in the Southern Hemisphere high latitude have been at up to 8% per decade.

Global production of CFC gases increased from around 180×10^6 kg per year in 1960 to nearly 1100×10^6 kg per year in 1990. However, in response to the thinning of the ozone layer, many governments signed an international agreement called the Montreal Protocol in 1987. This pledged them to a rapid phasing out of CFCs and halons. Production has since dropped substantially (Figure 7.23). However, because of their stability, these gases will persist in the atmosphere for decades or even centuries to come. Even with the most stringent controls that are now being considered, it will be the middle of the twenty-first century before the chlorine content of the stratosphere falls below the level that triggered the formation of the Antarctic 'ozone hole' in the first place.

Deliberate climatic modification

It has long been a human desire to modify weather and climate, but it is only since the Second World War and the development of high-altitude observations of clouds that serious attempts have been made to modify such phenomena as rainfall, hailstorms, and hurricanes (Cotton and Piehlke, 1995).

The most fruitful human attempts to augment natural precipitation have been through cloud seeding.

Rainmaking experiments of this type are based on three main assumptions (Chorley and More, 1967: 159).

1 Either the presence of ice crystals in a supercooled cloud is necessary to release snow and rain, or the presence of comparatively large water droplets is necessary to initiate the coalescence process.
2 Some clouds precipitate inefficiently or not at all, because these components are naturally deficient.
3 The deficiency can be remedied by seeding the clouds artificially, either with solid carbon dioxide (dry ice) or silver iodide, to produce crystals, or by introducing water droplets or large hygroscopic nuclei (e.g., salt).

These methods of seeding are not universally productive and in many countries expenditure on such procedures has been reduced. Under conditions of orographic lift and in thunderstorm cells, when nuclei are insufficient to generate rain by natural means, some augmentation may be attained, especially if cloud temperatures are of the order of −10° to −15°C. The increase of precipitation gained under favorable conditions may be of the order of 10–20% in any one storm. In lower latitudes, where cloud-top temperatures frequently remain above 0°C, silver-iodide or dry-ice seeding is not applicable. Therefore alternative methods have been introduced whereby small water droplets of 50 mm diameter are sprayed into the lower layers of deep clouds, so that the growth of cloud particles will be stimulated by coalescence. Other techniques of warm cloud seeding include the feeding of hygroscopic

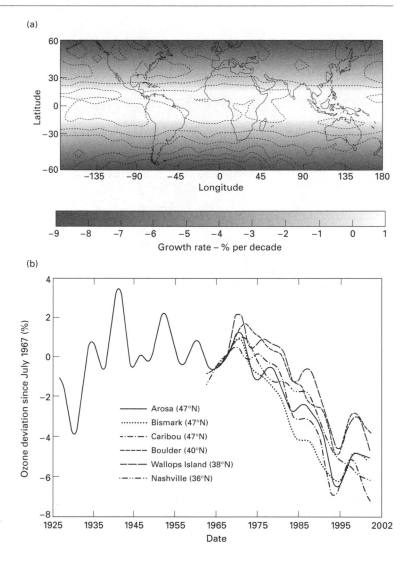

Figure 7.26 Global ozone trends: (a) average total ozone growth rates (percent per decade); (b) total ozone tendency curves for six northern mid-latitude sites. These sites are normalized to zero in July 1967, the beginning of the shortest record. Ozone deviation is given in percent. (Source: Harris et al., 2003, figures 4 and 7.)

particles into the lower air layers near the updraught of a growing cumulus cloud (Breuer, 1980).

The results of the many experiments now carried out on cloud seeding are still controversial, very largely because we have an imperfect understanding of the physical processes involved. This means that the evidence has to be evaluated on a statistical rather than a scientific basis so, although precipitation may occur after many seeding trials, it is difficult to decide to what extent artificial stimulation and augmentation is responsible. It also needs to be remembered that this form of planned weather modification applies to small areas for short periods. As yet, no means exist to change precipitation appreciably over large areas on a sustained basis.

Other types of deliberate climatic modification have also been attempted (Hess, 1974). For example, it has been thought that the production of many more hailstone embryos by silver-iodide seeding will yield smaller hailstones which would both be less damaging and more likely to melt before reaching the ground (Figure 7.27). Some results in Russia have been encouraging, but one cannot exclude the possibility that seeding may sometimes even increase hail damage (Atlas, 1977). Similar experiments have been conducted in lightning suppression. The concept here is to produce in a thundercloud, again by silver-iodide seeding, an abnormal abundance of ice crystals that would act as added corona points and thus relieve the electrical potential gradient by corona discharge before a

Figure 7.27 The damage caused by large hailstones can be considerable. These examples, the size of a baseball, fell near Neligh, Nebraska, in June 1950. There is, therefore, a considerable incentive to explore ways of reducing their impact through climatic modification experiments.

lightning strike could develop (Panel on Weather and Climate Modification, 1966: 4–8).

Hurricane modification is perhaps the most desirable aim of those seeking to suppress severe storms, because of the extremely favorable benefit-to-cost ratio of the work. The principle once again is that of introducing freezing nuclei into the ring of clouds around the hurricane center to trigger the release of the latent heat of fusion in the eye-wall cloud system which, in turn, diminishes the maximum horizontal temperature gradients in the storm, causing a hydrostatic lowering of the surface temperature. This eventually should lead to a weakening of the damaging winds (Smith, 1975: 212). A 15% reduction in maximum winds is theoretically possible, but as yet work is largely at an experimental stage. A 30% reduction in maximum winds was claimed following seeding of Hurricane Debbie in 1969 but attempts in the USA to seed hurricanes were discontinued in the 1970s following the development of new computer models of the effects of seeding on hurricanes. The models suggested that although maximum winds might be reduced by 10–15% on average, seeding may increase the winds just outside the region of maximum winds by 10–15%, may either increase or decrease the maximum storm surge, and may or may not affect the direction of the storm (Sorkin, 1982). Seeding of hurricanes is unlikely to recommence until such uncertainties can be resolved.

Much research has been devoted in Russia to the possibility of removing the Arctic Sea ice and to ascertain effects of such action on the climate of northern areas (see Lamb, 1977: 660). One proposal was to dam the Bering Strait, thereby blocking off water flow from the Pacific. The assumption was that more, and warmer, Atlantic water would be drawn into the central Arctic, improving temperature conditions in that area. Critics have pointed to the possibility of adverse changes in temperatures elsewhere, together with undesirable change in precipitation character and amount.

Fog dispersal, vital for airport operation, is another aim of weather modification. Seeding experiments have shown that fog consisting of supercooled droplets can be cleared by using liquid propane or dry ice. In very cold fogs this seeding method causes rapid transformation of water droplets into ice particles. Warm fogs with temperatures above freezing point occur more frequently than supercooled fogs in mid-latitudes and are more difficult to disperse. Some success has been achieved using sodium chloride and other hygroscopic particles as seeding agents but the most effective method is to evaporate the fog. The French have developed the 'turboclair' system in which jet engines are installed alongside the runway at major airports and the engines produce short bursts of heat to evaporate the fog and improve visibility as an aircraft approaches (Hess, 1974).

In regions of high temperature, dark soils become overheated, and the resultant high evapotranspiration rates lead to moisture deficiencies. Applications of white powders (India and Israel) or of aluminum foils (Hungary) increase the reflection from the soil surface and reduce the rate at which insolation is absorbed. Temperatures of the soil surface and subsurface are lowered (by as much as 10°C), and soil moisture is conserved (by as much as 50%).

The planting of windbreaks is an even more important attempt by humans to modify local climate deliberately. Shelterbelts have been in use for centuries in many windswept areas of the globe, both to protect soils from blowing and to protect the plants from the direct effects of high velocity winds. The size and effectiveness of the protection depends on their height, density, shape, and frequency. However, the belts may have consequences for microclimate beyond those for which they were planted. Evaporation rates are curtailed, snow is arrested, and its melting waters are available for the fields, but the temperatures may become more extreme in the stagnant space in the lee of the belt, creating an increase in frost risk.

Traditional farmers in many societies have been aware of the virtues of microclimatic management (Wilken, 1972). They manage shade by employing layered cropping systems or by covering the plant and soil with mulches; they may deliberately try to modify albedo conditions. Tibetan farmers, for example, reportedly throw dark rocks on to snow-covered fields to promote late spring melting; and in the Paris area of France some very dense stonewalls were constructed to absorb and radiate heat.

Conclusion

Changes in the composition of Earth's atmosphere as a result of human emissions of trace gases, and changes in the nature of land cover, have caused great concern in recent years. Global warming, ozone depletion and acid rain have become central issues in the study of environmental change. Although most attention is often paid to climatic change resulting from greenhouse gases, there is a whole series of other mechanisms that have the potential to cause climatic change. Most notably, this chapter has pointed to the importance of other changes in atmospheric composition and properties, whether these are caused by aerosol generation or albedo change.

However, the greenhouse effect and global warming may prove to have great significance for the environment and for human activities. Huge uncertainties remain about the speed, degree, direction, and spatial patterning of potential change. Nonetheless, if the earth warms up by a couple of degrees over the next 100 or so years, the impacts, some negative and some positive, are unlikely to be trivial. This is something that will form a focus of the following chapters.

For many people, especially in cities, the immediate climatic environment has already been changed. Urban climates are different in many ways from those of their rural surroundings. The quality of the air in many cities has been transformed by a range of pollutants, but under certain circumstances clean air legislation and other measures can cause rapid and often remarkable improvements in this area.

The same is true of two major pollution issues – ozone depletion and acid deposition. Both processes have serious environmental consequences and their effects may remain with us for many years, but both can be slowed down or even reversed by regulating the production and output of the offending gases.

Points for review

What do you understand by the term 'the greenhouse effect'?

Explain the major processes involved in global warming.

What role may changes in aerosols and albedo play in climate change?

How does the climate of cities differ from that of the surrounding countryside?

Give examples of some locations where levels of air pollution are falling. Why are they?

What is acid rain and how does it affect the environment?

What is 'the ozone hole' and how did it form?

Guide to reading

Elsma, D., 1996, *Smog alert: managing urban air quality*. London: Earthscan. A very readable and informative guide.

Graedel, T. E. and Crutzen, P. J., 1993, *Atmospheric change: an earth system perspective*. San Franscisco: Freeman. A well illustrated study of changes, past, present, and future.

Harvey, L. D. D., 2001, *Global warming: the hard science*. Harlow: Prentice Hall. A summary of the science behind global warming.

Houghton, J. T., 1997, *Global warming: the complete briefing* (2nd edn). Cambridge: Cambridge University Press. An introduction by a leading scientist.

Kemp, D. D., 1994, *Global environmental issues: a climatological approach* (2nd edn). London: Routledge. An introductory survey.

Williams, M. A. J. and Balling, R. C., 1996, *Interactions of desertification and climate*. London: Arnold. A study of land cover changes and their effects.

Part II

The Future

8 THE FUTURE: INTRODUCTION

Introduction

According to a range of modeling studies, it is highly likely that human induced global warming, produced by the emissions of a cocktail of greenhouse gases into the atmosphere, is now occurring and will continue to affect the global climate for decades and centuries. Concentrations of carbon dioxide, methane, nitrous oxide, and the chlorofluorocarbons, all greenhouse gases, have increased. The international body charged with such matters, The Intergovernmental Panel on Climate Change (IPCC), set up in the 1980s, has considered these changes in some detail (e.g., IPCC, 2001).

The atmospheric concentration of CO_2 has increased by 31% since 1750 and the present CO_2 concentration has not been exceeded during the past 420,000 years and possibly during the past 20 million years (IPCC, 2001: 7). The atmospheric concentration of methane has increased by over 150% since 1750 and that of nitrous oxide by 17%. Chlorofluorocarbons (CFCs), which are entirely artificial, have been emitted, and although the Montreal Protocol and its amendments may control them, they have a long residence time in the atmosphere.

During the twentieth century the global average surface temperature increased by around 0.6°C (IPCC, 2001: 2). The 1990s were the warmest decade in the instrumental record and in the Northern Hemisphere the increase in temperature is likely to have been the largest of any century during the past 1000 years.

The IPCC (2001: 13) suggests that global average surface temperatures will rise by 1.4 to 5.8°C between 1990 and 2100. This represents a much larger rate of warming than that observed over the twentieth century and, based on paleoclimatic data, is very likely to surpass anything experienced over the past 10,000 years. In its turn, increased temperatures will transform the behavior of the atmospheric heat engine that drives the general circulation of the air and the oceans, leading to changes in the nature, pattern, and amount of precipitation. Table 8.1, adapted from the IPCC (2001, table 1), indicates some of the projections for future climate. In particular it is likely that northern high latitudes will warm more rapidly than the global average, probably by more than 40%. There may also be marked changes in precipitation, with a tendency for increases to occur globally, but particularly over

Table 8.1 Key projections of climate trends and extreme events in the IPCC Third Assessment. Source: adapted from Houghton et al. (2001)

Change	Likelihood*
Rate of increase of global surface temperature will exceed anything experienced over past 10,000 years	Very likely
Nearly all land areas will warm more than global average, particularly northern high latitudes in cold season	Very likely
Precipitation will increase by end of twenty-first century over northern mid- to high latitudes in winter	Likely
In areas where increase in mean precipitation is predicted, larger year-to-year variations will occur	Very likely
Higher maximum temperature and more hot days over nearly all land areas	Very likely
Higher minimum temperatures, fewer cold days and frost days over nearly all land areas	Very likely
Reduced diurnal temperature range over most land areas	Very likely
More intense precipitation events over many Northern Hemisphere mid- to high latitude land areas	Very likely, over many areas
Increased summer continental drying and associated risk of drought	Likely, over most mid-latitude continental interiors
Increase in tropical cyclone peak wind intensities, mean and peak precipitation intensities	Likely, over some areas

*Very likely = 90–99% chance; likely = 66–90%.

northern mid- to high latitudes and Antarctica in winter. At lower latitudes some areas will see reductions in precipitation (which will exacerbate soil moisture losses caused by increased evapotranspiration), while others may become wetter.

If global warming and associated changes in rainfall and soil moisture amounts occur, then the effects on natural environments may be substantial and rapid. These impacts have been reviewed by the IPCC (Watson et al., 1996). The following are some of the thought-provoking comments made in their report.

On forests: 'A substantial fraction (a global average of one-third, varying by region from one-seventh to two-thirds) of the existing forested area of the world will undergo major changes in broad vegetation types ... climate change is expected to occur at a rapid rate relative to the speed at which forest species grow, reproduce and establish themselves ... the species composition of forests is likely to change; entire forest types may disappear.'

On deserts and desertification: 'Deserts are likely to become more extreme – in that, with few exceptions, they are projected to become hotter but not significantly wetter.'

On the cryosphere: 'Between one-third and one-half of existing mountain glacier mass could disappear over the next 100 years.'

On mountain regions: 'The projected decrease in the extent of mountain glaciers, permafrost and snow cover caused by a warmer climate will affect hydrologic systems, soil stability and related socio-economic systems ... Recreational industries – of increasing economic importance to many regions – are also likely to be disrupted.'

On coastal systems: 'Climate change and a rise in sea level or changes in storm surges could result in the erosion of shores and associated habitat, increased salinity of estuaries and freshwater aquifers, altered tidal ranges in rivers and bays, a change in the pattern of chemical and microbiological contamination in coastal areas, and increased coastal flooding. Some coastal ecosystems are particularly at risk, including saltwater marshes, mangrove ecosystems, coastal wetlands, coral reefs, coral atolls, and river deltas.'

Some landscape types will be highly sensitive to global warming. This may be the case because they are located in zones where it is forecast that climate will change to an above average degree. This applies, for instance, in the high latitudes of North America and Eurasia, where the degree of warming may be three or four times greater than the presumed global average. It may also be the case for some critical areas where particularly substantial changes in rainfall may result from global warming. For example, various methods of climatic prediction produce scenarios in which the American High Plains will become considerably drier. Other landscapes will be highly sensitive because certain landscape-forming processes are very closely controlled by climatic conditions. If such landscapes are close to a particular climatic threshold then quite modest amounts of climate change can switch them from one state to another.

There are, of course, considerable uncertainties built into any consideration of the future. Our models are crude and relatively simple and it is difficult to build in all the complexities and feedbacks in the atmosphere, pedosphere, lithosphere, oceans, biosphere, and cryosphere. Different models often provide very different future scenarios, particularly with respect to future rainfall amounts. Other difficulties are presented by the existence of nonlinearities and thresholds in natural systems. There may be complex responses to change and sometimes indeterminable time lags. There are also great uncertainties about how emissions of greenhouse gases will change in coming decades as a result of changes in the global economy, technological changes, and the adoption of mitigation strategies.

Furthermore, many of our models have a coarse spatial resolution and are difficult to use at more local scales. It is also problematic to use past warm phases as analogues for the future, for the driving mechanisms may be different. Finally, changes brought about by enhanced greenhouse loadings cannot be seen in isolation. They will occur concurrently with other natural and anthropogenic climatic changes. In some cases other human activities could compound (or reverse) the effects of global warming.

The causes, consequences, and controversies associated with global warming have been treated extensively elsewhere (e.g., Harvey, 2000). The purpose of the following chapters of this book is not to review and revisit this literature, but to examine the implications that global warming has for landscapes and habitats.

Changes in the biosphere

Climate change and changes in the concentrations of carbon dioxide in the atmosphere are likely to have a whole suite of biological consequences (Gates, 1993; Joyce et al., 2001). The ranges and the productivity of organisms will change.

Altitudinal changes in vegetation zones will be of considerable significance. In general, Peters (1988) believes that with a 3°C temperature change vegetation belts will move about 500 m in altitude. One consequence of this would be the probable elimination of Douglas fir (*Pseudotsuga taxifolia*) from the lowlands of California and Oregon, because rising temperatures would preclude the seasonal chilling this species requires for seed germination and shoot growth. In the twentieth century there is some evidence that warming did indeed affect the position of the tree line, and

Kullman (2001), for example, found that in Sweden a 0.8°C warming in the past 100 years caused the tree limit to migrate upwards by more than 100 m.

Vegetation will also change latitudinally and some models suggest that wholesale change will occur in the distribution of biomes. Theoretically a rise of 1°C in mean temperature could cause a poleward shift of vegetation zones of about 200 km (Ozenda and Borel, 1990). However, uncertainties surround the question of how fast plant species would be able to move to and to settle new habitats suitable to the changed climatic conditions. Post-glacial vegetation migration rates appear to have been in the range of a few tens of kilometers per century. For a warming of 2–3°C forest bioclimates could shift northwards about 4–6° of latitude in a century, indicating the need for a migration rate of some tens of kilometers per decade. Furthermore, migration could be hampered because of natural barriers, ecological fragmentation, zones of cultivation, etc.

Changes in forest composition or location could be slow, for mature trees tend to be long-lived and resilient. This means that they can survive long periods of marginal climate. However, it is during the stage of tree regeneration or seedling establishment that they are most vulnerable to climate change. Seedlings are highly sensitive to temperature and may not be able to grow under altered climatic conditions. This means that if climate zones do indeed shift at a faster rate than trees migrate, established adults of appropriate species and genotypes will be separated from the place where seedling establishment is needed in the future (Snover, 1997). It is probable that future climate change will not cause catastrophic dieback of forests, but that faster growing tree species will enter existing forests over extended time periods (Hanson and Weltzin, 2000).

An early attempt to model changes on a global basis was made by Emmanuel et al. (1985). Using the global climate model (GCM) developed by Manabe and Stouffer (1980) for a doubling of CO_2 levels, and mapping the present distribution of ecosystem types in relation to contemporary temperature conditions, they found *inter alia* that the following changes would take place: boreal forests would contract from their present position of comprising 23% of total world forest cover to less than 15%; grasslands would increase from 17.7% of all world vegetation types to 28.9%;

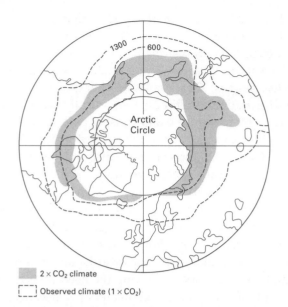

2 × CO₂ climate

Observed climate (1 × CO₂)

Figure 8.1 The northern and southern boundaries of the boreal forest are approximately defined by the 600 and 1300 growing degree-day isopleths. These are shown in their current positions and in the positions they would occupy under a warming associated with a doubling of CO_2 levels (after Kauppi and Posch, 1988).

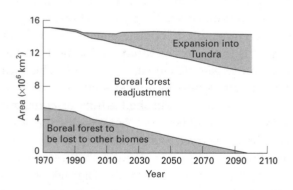

Figure 8.2 Changes in areas of boreal forest projected by the Integrated Model to Assess the Global Environment (IMAGE) in response to scenarios of future change in climate and land use. Although the total area of boreal forest is projected to remain relatively constant, about one-third of the present boreal forest is projected to be converted to other biomes, and an additional third will be added as trees advance into tundra. (Source: Chapin and Danell, 2001, figure 6.1.)

deserts would increase from 20.6 to 23.8%; and forests would decline from 58.4 to 47.4%. Potential changes in the geographical extent of boreal forests in the Northern Hemisphere are especially striking. Kauppi

Table 8.2 Changes in areal coverage ($km^2 \times 10^3$) of major biomes as a result of climatic conditions predicted for a $2 \times CO_2$ by various general circulation models (GCMs). Source: from Smith et al. (1992, table 3)

Biome	Current	OSU	GFDL	GISS	UKMO
Tundra	939	−302	−5.5	−314	−573
Desert	3699	−619	−630	−962	−980
Grassland	1923	380	969	694	810
Dry forest	1816	4	608	487	1296
Mesic forest	5172	561	−402	120	−519

OSU, Oregon State University; GFDL, Geophysical Fluid Dynamic Laboratory; GISS, Goddard Institute for Space Studies; UKMO, UK Meteorological Office.

and Posch (1988), using a Goddard Institute for Space Studies (GISS) GCM, modeled the northern and southern boundaries of the boreal forest in response to a climate warming associated with a doubling of atmospheric carbon dioxide levels (Figure 8.1). Note how the southern boundary moves from the southern tip of Scandinavia to the northernmost portion.

Basically, boreal forest will be lost to other biomes (such as temperate forest) on its southern margins but will expand into tundra on its northern margins (Figure 8.2) (Chapin and Danell, 2001).

One biome that may suffer a particularly severe loss in area as a result of warming is arctic and alpine tundra. Indeed as the tree line moves up mountains and migrates northwards in the Northern Hemisphere, the extent of tundra will be greatly reduced, perhaps by as much as 55% in total (Walker et al., 2001). The alpine tundra zone may disappear completely from some mountain tops. Changes in the cryosphere will be considered in detail in Chapter 11.

Smith et al. (1992) attempted to model the response of Holdridge's Life Zones to global warming and associated precipitation changes as predicted by a range of different GCMs. These are summarized for some major biomes in Table 8.2. All the GCMs predict conditions that would lead to a very marked contraction in the tundra and desert biomes, and an increase in the areas of grassland and dry forests. There is, however, some disagreement as to what will happen to mesic forests as a whole, although within this broad class the humid tropical rainforest element will show an expansion.

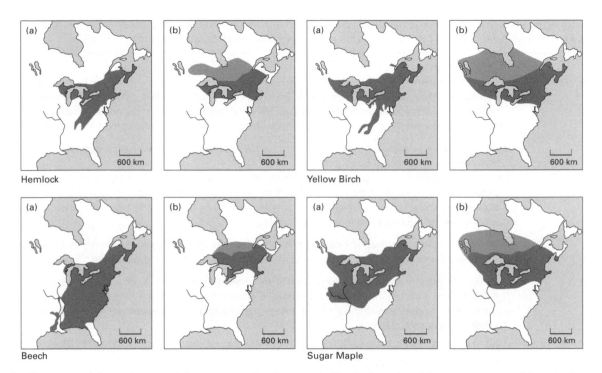

Figure 8.3 Present and future range of four tree species in eastern North America: (a) present range; (b) range in AD 2090 under the GISS GCM $2 \times CO_2$ scenario. The black area is the projected occupied range considering a rate of migration of 100 km per 100 years. The gray area is the potential projected range with climate change (after Zabinski and Davis, 1989).

Zabinski and Davis (1989) modeled the potential changes in the range of certain tree species in eastern North America, using a GISS GCM and a doubling of atmospheric carbon dioxide levels. The difference between their present ranges and their predicted ranges is very large (Figure 8.3).

Melillo et al. (2001) have modeled shifts in major vegetation types for the whole of the USA, using the Hadley (HadCM2) and Canadian Center for Climate Modeling and Analysis simulations and various biogeography models. The results of their simulations for 2099 are summarized in Table 8.3.

A somewhat precarious biome, located as it is at the southernmost tip of Africa and so with little scope for displacement, is the very diverse Fynbos Biome. It is a predominantly sclerophyll shrubland, characterized by the pre-eminence of hard-leaved shrubs, many of which are proteas. In addition to being extraordinarily diverse, this biome contains many endemic species. Modeling, using the Hadley Centre CM2 GCM, indicates a likely contraction of the extent of Fynbos by c. 2050 (Figure 8.4) (Midgley et al., 2003).

In the UK, the MONARCH programme (UK Climate Impacts Programme, 2001) has investigated potential changes in the distribution of organisms. Among the consequences of global warming they identify as being especially serious are the loss of montane heaths in the mountains of Scotland and the dieback of beach woodlands in southern Britain as summer droughts become more intense. However, it needs to be remembered that climate is not the only control on the distribution of species and that other factors such as biotic interactions and species' dispersal are also likely to be significant (Pearson and Dawson, 2003).

Vegetation will also probably be changed by variations in the role of certain extreme events that cause habitat disturbance, including fire, drought (Hanson and Weltzin, 2000), windstorms, hurricanes (Lugo, 2000), and coastal flooding (Overpeck et al., 1990). Other possible nonclimatic effects of elevated CO_2 levels on vegetation include changes in photosynthesis, stomatal closure and carbon fertilization (Idso, 1983), although these are still matters of controversy

Table 8.3 Projected changes in major vegetation types in the USA using the Hadley (HadCM2) and Canadian Center for Climate Modeling and Analysis simulations and various biogeography models (after Melillo et al., 2001)

Region	Projected change
Northeast	Under both simulated climates, forests remain the dominant natural vegetation, but the mix of forest types changes. For example, winter-deciduous forests expand at the expense of mixed conifer-broadleaf forests
	Under the climate simulated by the Canadian model, there is a modest increase in savannas and woodlands
Southeast	Under the climate simulated by the Hadley model, forest remains the dominant natural vegetation, but once again the mix of forest types changes
	Under the climate simulated by the Canadian model, all three biogeography models show an expansion of savannas and grasslands at the expense of forests. For two of the biogeography models, LPJ and MAPSS, the expansion of these nonforest ecosystems is dramatic by the end of the twenty-first century. Both drought and fire play an important role in the forest break-up
Midwest	Under both simulated climates, forests remain the dominant natural vegetation, but the mix of forest types changes
	One biogeography model, LPJ, simulates a modest expansion of savannas and grasslands
Great Plains	Under the climate simulated by the Hadley model, two biogeography models project an increase in woodiness in this region, while the third projects no change in woodiness
	Under the climate simulated by the Canadian model, the biogeography models project either no change in woodiness or a slight decrease
West	Under the climate simulated by both the Hadley and Canadian models, the area of desert ecosystems shrinks and the area of forest ecosystem grows
Northwest	Under both simulated climates, the forest area grows slightly

LPJ, Lund–Potsdam–Jena; MAPPS, Mapped Atmosphere–Plant–Soil System.

(Karnosky et al., 2001; Wullschleger et al., 2002; Karnosky, 2003). However, especially at high altitudes, it is possible that elevated CO_2 levels would have potentially significant effects on tree growth, causing growth enhancement (La Marche et al., 1984). Likewise, warm, drought-stressed ecosystems such as chaparral in the southwest USA might be very responsive to elevated CO_2 levels (Oechel et al., 1995).

Peterson (2000) has investigated the potential impact of climate change on the role of catastrophic winds from tornadoes and downbursts in causing disturbance to forests in North America. It is feasible that with warmer air masses over middle latitude areas the temperature contrast with polar air masses will be greater, providing more energy and thus more violent storms, but much more work needs to be undertaken before this can be said with any certainty.

The frequency and intensity of fires is an important factor in many biomes, and fires are highly dependent on weather and climate. Fires, for example, are more likely to occur in drought years or when there are severe lightning strikes. Wind conditions are also a

significant control of fire severity. An analysis by Flannigan et al. (2000) suggests that future fire severity could increase over much of North America. They anticipate increases in the area burned in the USA of 25–50% by the middle of the twenty-first century, with most of the increases occurring in Alaska and the southeast USA.

Other disturbances to vegetation communities could be brought about by changing patterns of insects and pathogens (Ayres and Lombardero, 2000). For example, in temperate and boreal forests, increases in summer temperatures could accelerate the development rate and reproductive potential of insects, whereas warmer winter temperatures could increase overwinter survival. In addition, it is possible that the ranges of introduced alien invasive species could change (Simberloff, 2000).

It is still far from clear whether hurricane frequencies and intensities will increase in a warmer world. However, whether they increase or decrease, there are likely to be a number of possible ecosystem responses (Table 8.4).

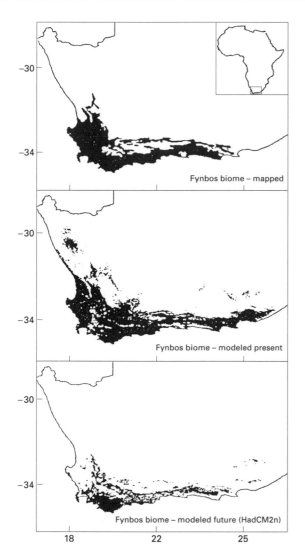

Figure 8.4 Current mapped Fynbos Biome (upper panel, after Rutherford and Westfall, 1994) and the modeled extent of the biome under current (middle panel) and future (*c.* 2050) climate conditions (lower panel), the latter based on climate change projections for the region generated by the GCM HadCM2. (Source: Midgley et al., 2003, figure 1.)

Climate and geomorphology

Attempts to relate landforms and land-forming processes to climatic conditions have been long continued and climatic geomorphology has a long and distinguished history (Stoddart, 1969; Derbyshire, 1973; Gutierrez-Elorza, 2001). Certainly, any change in climatic variability and extreme events would have consid-

erable geomorphologic significance (Viles and Goudie, 2003) (Table 8.5).

It is self evident that some landforms and land-forming processes are intimately related to temperature. In general terms, for example, permafrost today occurs only in areas where the mean annual temperature is less than $-2°C$ and it is virtually ubiquitous north of the -6 to $-8°C$ isotherm in the Northern Hemisphere. Likewise, temperature is one of the major controls of ablation of glacier ice. At the other extreme, coral reefs have a restricted distribution in the world's oceans, which reflects their levels of temperature tolerance. They do not thrive in cool waters (i.e., $< 20°C$ mean annual sea-surface temperature), but equally many coral species cannot tolerate temperatures greater than about $30°C$, and have been stressed by warm El Niño–Southern Oscillation (**ENSO**) years.

Mangroves, important components of coastlines in the warmer parts of the world, are also sensitive to temperature. The best-developed mangrove swamps are found near the Equator where the temperature in the coldest month does not fall below $20°C$ in Indonesia, New Guinea, and The Philippines, and the number of species decreases with increasing latitude. Walter (1984) indicates that the last outposts of mangrove are to be found at $30°N$ and $33°S$ in Eastern Africa, $37–38°S$ in Australia and New Zealand, $20°S$ in Brazil and $32°N$ in the Bermudas. Mangroves in such areas might expand their poleward range under warmer conditions.

Some relatively clear relationships have also been established between precipitation levels and geomorphic processes and forms. Large sand dunes only occur under dry conditions where rainfall levels are less than 150–200 mm per year, and dust storms appear to be a feature of hyperarid areas where annual rainfall totals are around 0–150 mm per year. Desert crusts vary according to rainfall and according to the solubility of their components, with, for example, nitrate and halite crusts only occurring in the driest deserts and calcretes in areas where rainfall is less than around 500 mm per year.

However, for other phenomena the climatic controls may be rather more complex. Take, for example, the case of lake basins. Lake basins, and particularly those with no outlets, can respond in a dramatic manner to changes in their hydrological balance. During times of positive water budgets (due to high rates of water

Table 8.4 Potential response of ecosystems associated with changes in hurricane characteristics (modified after Lugo, 2000)

Increase in frequency and intensity	Decreased in frequency and intensity
A larger fraction of the natural landscape will be set back in successional stage, i.e., there will be more secondary forests	A larger fraction of the natural landscape will advance in successional stage, i.e., there will be more mature forests and fewer secondary forests
Forest aboveground biomass and height will decrease because vegetation growth will be interrupted more frequently or with greater intensity	Forest aboveground biomass and height will increase because the longer disturbance-free periods allow greater biomass accumulation and tree height
Familiar species combinations will change as species capable of thriving under disturbance conditions will increase in frequency at the expense of species that require long periods of disturbance-free conditions to mature	Species combinations will change as species capable of thriving under disturbance conditions will decrease in frequency and species typical of disturbance-free conditions will increase

Table 8.5 Examples of geomorphologic effects of decadal- to century-scale oscillations

Environment affected	Impacts upon
Terrestrial hydrology	Glacier mass balance
	Lake levels
	River flows
	Snow cover
	Permafrost
Terrestrial geomorphology	Soil erosion
	Floodplain sedimentation and erosion
	Slope instability/mass movements
	Dune movements
	Geochemical sediment growth
	Effects of fire frequency with knock-on effects on weathering
	Runoff and slope instability
Coastal/marine ecology and geomorphology	Coastal erosion
	Mangrove defoliation/land loss
	Coral bleaching
	Coastal dune activation

Table 8.6 Factors affecting rates of evaporation and runoff. Source: Bradley (1985)

Evaporation	Runoff
Temperature (daily means and seasonal range)	Ground temperature
Cloudiness and solar radiation receipts	Vegetation cover and type
Wind speed	Soil type (infiltration capacity)
Humidity (vapor pressure gradient)	Precipitation type (rain, snow, etc.)
Depth of water in lake and basin morphology (water volume)	Precipitation intensity (event magnitude and duration)
Duration of ice cover	Precipitation frequency and seasonal distribution
Salinity of lake water	Slope gradients (stream size and number)

input and/or low rates of evapotranspirational loss), lakes may develop and expand over large areas, only to recede and desiccate during times of negative water balance. Shoreline deposits and sediment cores extracted from lake floors can provide a detailed picture of lake fluctuations (e.g., Street and Grove, 1979). However, given the range of factors affecting rates of evaporation and runoff over a lake basin (Table 8.6) their precise climatological interpretation needs to be

determined with care. A full review is provided by Mason et al. (1994).

Many of the studies of the links between climate and geomorphology have employed rather crude climatic parameters such as mean annual temperature or mean annual rainfall. It is, however, true that climatic extremes and climatic variability may be of even greater significance. Unfortunately, they are not dealt with very effectively in GCMs. We have imperfect knowledge of how extreme events and climate variability will change in a warmer world. It is, however, possible that climatic variability such as the ENSO

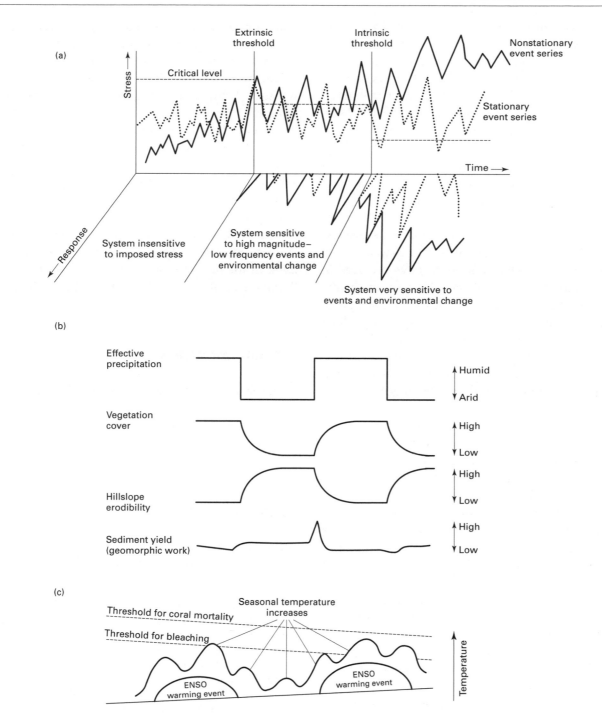

Figure 8.5 Representation of the impacts of climatic variability on geomorphic systems. (a) Stress–response sequences including thresholds under stable and changing climate conditions (adapted from Brunsden, 2001, figure 1). (b) A simplified view of the biogeomorphic response model (adapted from Knox, 1972 and Roberts and Barker, 1993, figure 6.1). (c) A model of the possible interactions of different timescales of warming and their impact on coral bleaching (adapted from Williams and Bunkley-Williams, 1990 and Viles and Spencer, 1995, figure 6.11).

phenomenon could be influenced by global warming, although Houghton et al. (2001: 151) draw no clear conclusions over this. Trenberth and Hoar (1997) thought the severe El Niño of 1997–1998 was perhaps enhanced by current global warming, and Timmerman et al. (1999) have modeled the way in which El Niño could be related to future global warming.

The ENSO and other modes of climatic variability, as has been pointed out by Douglas et al. (1999) in terms of their effects on geomorphology and ecology in the Danum Valley, Sabah, can produce a form of punctuated equilibrium in the landscape. Alternating periods of high geomorphic change (rhexistasie), followed by periods of stability in ecosystem and geomorphology (biostasie), seem to have occurred in the recent past at Danum Valley as a response to the changing rhythm of ENSO. Although the concepts of biostasie and rhexistasie are not new to geomorphology (having been introduced by Erhart (1956) and utilized by Knox (1993, 2001) amongst others), the recent discoveries of yet more modes of climatic variability, often with complex temporal rhythms, affecting large areas over long time spans make them ever more relevant. These ideas can also be linked to notions of complex response in the landscape (Brunsden, 2001), if we visualize change as a wave pulsing through a patchy landscape, with different areas possessing varied resistances to change. Thus, decadal climatic variability may set off a pulse of activity resulting in a complex landscape response. The impacts of climatic variability on ecological and geomorphic systems may be nonlinear, as recently found for rainfall erosivity and ENSO in the southwest USA (D'Odorico et al., 2001).

Several of these ideas on the clustering of events and complex responses to them can be presented as conceptual diagrams. Figure 8.5a indicates in a general way how clusters of climatic events can produce variable overtopping of geomorphic thresholds, although it represents a simple linear view of geomorphic response that may be an oversimplification for many geomorphic systems. Figure 8.5b illustrates how complex chains of linkages between climatic, vegetation and geomorphic processes produce a complex geomorphic response, and Figure 8.5c provides a simple conceptual model of the synergistic associations between different scales of warming producing **coral bleaching**. Such conceptual diagrams provide a useful starting point for analysing the relationships between climatic variability and geomorphology as a prelude to more detailed empirical and computational studies.

However, geomorphologists have long debated whether it is possible meaningfully to untangle the different roles of the various external forcing factors (climate, tectonics, and human activity) and internal factors (thresholds) in causing geomorphic response. Increasingly, many geomorphologists have come to suspect that this is a very difficult task because of the nonlinear, chaotic, and complex behavior of geomorphic systems.

In the following chapters we shall consider some of the challenges that face some major types of environment in the face of global warming and other trends that have been discussed in this brief introduction.

Points for review

How may vegetation belts respond to future climate changes?
Which landscape types may be especially sensitive to global warming?

Guide to reading

Arnell, N., 1996, *Global warming, river flows and water resources*. Chichester: Wiley. A study of potential future hydrological changes.

Chapin, F. S., Sala, O. E. and Huber-Sannwald, E. (eds), 2001, *Global biodiversity in a changing environment*. Berlin: Springer Verlag. A wide-ranging assessment of how biodiversity may change in response to climate and land use changes.

Eisma, D. (ed.), 1995, *Climate change. Impact on coastal habitation*. Boca Raton (Florida): Lewis Publishers. A study of the effects of climate change and sea-level rise.

Gates, D. M., 1993, *Climate change and its biological consequences*. Sunderland, Mass: Sinauer Associates. A textbook that considers some of the effects of climate change on the biosphere.

Houghton, J. D. and 7 others (eds), 2001, *Climate change 2001: the scientific basis*. Cambridge: Cambridge University Press. The IPCC report on global warming.

Leggett, J., 1996, *Climate change and the financial sector*. Munich: Gerling Akademie Verlag. A study of why climate change may have a severe impact on economies and business.

9 THE FUTURE: COASTAL ENVIRONMENTS

Introduction: rising sea levels

There is a strong likelihood that if temperatures climb in coming decades, so will sea levels. Indeed, on a global basis, sea levels are rising at the present time, and always have in warmer periods in the past. The reasons why sea levels will rise include the thermal expansion of seawater (the **steric effect**), the melting of the cryosphere (glaciers, ice sheets, and permafrost) and miscellaneous anthropogenic impacts on the hydrologic cycle (which will modify how much water is stored on land). Rising sea levels will have substantial geomorphologic consequences for the world's coastlines and these in turn will have an impact on a surprisingly large proportion of Earth's human population (Leatherman, 2001). As Viles and Spencer (1995) have pointed out, about 50% of the population in the industrialized world lives within 1 km of a coast, about 60% of the world's population live in the coastal zone, and two-thirds of the world's cities with populations of over 2.5 million people are near estuaries. Thirteen of the world's twenty largest cities are located on coasts.

The steric effect

As the oceans warm up, their density decreases and their volume increases. This thermal expansion, the steric effect, causes sea levels to rise. Uncertainties arise as to the rate at which different parts of the oceans will warm up in response to increased atmospheric temperatures (Gregory et al., 2001). Long-term records of ocean warming and sea-level change are also sparse.

Church et al. (2001, table 11.2) present data on global average sea-level rise due to thermal expansion during the twentieth century. Rates between 1910 and 1990 ranged between 0.25 and 0.75 mm per year while those between 1960 and 1990 ranged between 0.60 and 1.09 mm per year. The steric effect over the period 1910 to 1990 accounts for at least one-third of the observed eustatic change over that period.

With regard to the future, sea-level rise caused by thermal expansion between 1990 and 2100 is thought likely to be around 0.28 m, out of a total predicted sea-level rise due to climate change of 0.49 m (Church et al., 2001, table 11.14). In other words, it will exceed in importance the role of ice-cap and glacier melting.

Anthropogenic contributions to sea-level change

Human activities have an impact upon the hydrologic cycle in a wide range of ways, so that changes in sea level could result from such processes as groundwater exploitation, drainage of lakes and wetlands, the construction of reservoirs behind dams, and modification of runoff and evapotranspiration rates from different types of land cover. Some of these processes could accelerate sea-level rise by adding more water to the oceans, whereas others would decelerate the rate of sea-level rise by impounding water on land. Unfortunately, quantitative data are sparse and sometimes contradictory and so it is difficult to assess the importance of anthropogenic contributions to sea-level change.

Reduction in lake-water volumes

Increased use of irrigation and interbasin water transfers have contributed to reduced volumes of water being stored in lake basins. Classic examples of this include the Aral and Caspian Seas of Central Asia and Owens Lake in the southwest USA. Unfortunately, not all lake level declines are due to human actions and it is not always easy to disentangle these from the effects of natural decadal-scale climate fluctuations. Equally it is difficult to calculate the proportion of the extracted water that reaches the world ocean by runoff and evapotranspiration. Some of it may enter groundwater stores, causing groundwater levels to rise and waterlogging of the land to occur. Controversy has thus arisen on this topic. On the one hand Sahagian (2000) has argued that this could be a process of some significance, perhaps contributing around 0.2 mm per year to sea-level rise, whereas Gornitz et al. (1997) believe that the net effect of the drying up of interior lakes is probably only small and indirect.

Water impoundment in reservoirs

Recent decades have seen the construction of large numbers of major dams and reservoirs. Today, nearly 500,000 km^2 of land worldwide are inundated by reservoirs that are capable of storing 6000 km^3 of water (Gleick, 2002).

Newman and Fairbridge (1986) calculated that between 1957 and 1982 as much as 0.75 mm per year of sea-level rise potential was stored in reservoirs and irrigation projects. More recently, Gornitz et al. (1997) have estimated that 13.6 mm of sea-level rise potential has been impounded in reservoirs, equivalent since the 1950s to a potential average reduction in sea-level rise of 0.34 mm per year. However, as Sahagian (2000: 43) reported, 'the total amount of impounded water is not known because there have been no comprehensive inventories of the millions of small reservoirs such as farm ponds and rice paddies. As a result, the total contribution of impounded water to global hydrological balance has remained unclear and most likely severely underestimated.'

Groundwater mining

Groundwater mining is the withdrawal of groundwater in excess of natural recharge. Many large groundwater aquifers are currently being heavily mined, leading to massive falls in aquifer levels and volumes. Examples include the Ogallala aquifer of the High Plains in the USA, the Nubian Sandstone aquifers of Libya and elsewhere in North Africa, and the great aquifers of the Arabian Peninsula. Estimates of the volumes of groundwater that are being removed from storage on a global basis are around 1000 to 1300 km^3, but not all of it is transferred to the oceans. Nonetheless, Church et al. (2001: 657) believe that it is probably the largest positive anthropogenic contributor to sea-level rise (apart from anthropogenic climatic change), amounting to perhaps 0.2 to 1.00 mm per year. This source may not continue indefinitely, because groundwater contributions may become exhausted in some regions.

Urbanization and runoff

Urbanization leads to a net increase in total runoff from the land surface due to the spread of areas of impermeable ground (e.g., concrete, tile, and tarmac covered surfaces), which impede groundwater replenishment. In addition, water may be evacuated efficiently from urban areas in storm-water drains and sewers. Plausible rates of sea-level rise due to this mechanism are in the range of 0.35 to 0.41 mm per year.

Deforestation and runoff

As discussed in Chapter 5, deforestation can lead to increases in runoff from land surfaces. For example, experiments with tropical catchments have shown typical increases in streamflow of 400–450 mm per year (Anderson and Spencer, 1991). The reasons for this include changes in rainfall interception, transpiration, and soil structure. Combining data on rates of tropical deforestation with average values for increases in runoff, Gornitz et al. (1997: 153) came up with an estimate that this could cause a rate of sea-level rise of 0.13 mm per year.

Wetland losses

The reclamation or drainage of wetlands may release stored water that can then enter the oceans. Gornitz et al. (1997: 153) do not see this as a significant mechanism of sea-level change.

Irrigation

The area of irrigated land on Earth has gone up dramatically, amounting to $c. 45 \times 10^6$ hectares in 1900 and 240×10^6 hectares in 1990. During the 1950s the irrigated area increased by over 4% annually, although the figure has now dropped to only about 1%. Gornitz et al. (1997) estimated that evapotranspiration of water from irrigated surfaces would lead to an increase in the water content of the atmosphere and so to a fall in sea level of 0.14–0.15 mm per year. Irrigation water might also infiltrate into groundwater aquifers, removing 0.40–0.48 mm per year of sea-level equivalent.

Synthesis

As we have already noted, individual mechanisms can in some cases be significant in amount, but some serve to augment sea-level rise and some to reduce it. Gornitz et al. (1997: 158) summarized their analysis thus:

Increases in runoff from groundwater mining and impermeable urbanized surfaces are potentially important anthropogenic sources contributing to sea-level rise. Runoff from tropical deforestation and water released by oxidation of fossil fuel and biomass, including wetlands clearance, provide a smaller share of the total. Taken together, these processes could augment sea level by some 0.6–1.0 mm/year.

On the other hand, storage of water behind dams, and losses of water due to infiltration beneath reservoirs and irrigated fields, along with evaporation from these surfaces could prevent the equivalent of 1.5–1.8 mm/year from reaching the ocean. The net effect of all of these anthropogenic processes is to withhold the equivalent of 0.8 ± 0.4 mm/year from the sea. This rate represents a significant fraction of the observed recent sea-level rise of 1–2 mm/year, but opposite in sign.

The IPCC (Church et al., 2001: 658) came to three conclusions.

1 The effect of changes in terrestrial water storage on sea level might be considerable.
2 The net effect on sea level could be of either sign.
3 The rate of hydrologic interference has increased over the past few decades.

Permafrost degradation

Increases in temperatures in cold regions will lead to substantial reductions in the area of permanently frozen subsoil (permafrost). In effect, ground ice will be converted to liquid water, which could in principle contribute to a rise in sea level. However, a proportion, unknown, of this water could be captured in ponds, thermokarst lakes, and marshes, rather than running off into the oceans. Another uncertainty relates to the volume of ground ice that will melt. Bearing these caveats in mind, Church et al. (2001: 658) suggest that the contribution of permafrost to sea-level rise between 1990 and 2100 will be somewhere between 0 and 25 mm (0–0.23 mm per year).

Melting of glaciers and sea-level rise

Many glaciers are expected to be reduced in area and volume as global warming occurs and their meltwater will flow into the oceans, causing sea level to rise (Gregory and Oerlemans, 1998). In some situations, on the other hand, positive changes in precipitation may nourish glaciers causing them to maintain themselves or to grow in spite of a warming tendency. Other uncertainties are produced by a lack of knowledge about how the **mass balance of glaciers** will respond to differing degrees of warming. The IPCC (2001, table 11.14) suggests that between 1990 and 2100 the melting of glaciers will contribute c. 0.01 to 0.23 m of sea-level change, with a best estimate of 0.16 m, second therefore only to steric effects. The problems of mass balance estimations have been well summarized by Arendt et al., 2002:

Conventional mass balance programs are too costly and difficult to sample adequately the > 160,000 glaciers on Earth. At present, there are only but 40 glaciers worldwide with continuous balance measurements spanning more than 20 years. High-latitude glaciers, which are particularly important because predicted climate warming may be greatest there, receive even less attention because of their remote locations. Glaciers that are monitored routinely are often chosen more for their ease of access and manageable size than for how well they represent a given region or how large a contribution they might make to changing sea level. As a result, global mass balance data are biased toward small glaciers (< 20 km²) rather than those that contain the most ice (> 100 km²). Also, large cumulative errors can result from using only a few point measurements to estimate glacier-wide mass balances on an individual glacier.

They used laser altimetry to estimate volume changes of Alaskan glaciers from the mid-1950s to the mid-1990s and suggested that they may have contributed 0.27 ± 0.10 mm per year of sea-level change. This is considerably more than had previously been appreciated.

Ice sheets and sea-level rise

The ice sheets of Greenland and Antarctica have proved pivotal in ideas about future sea-level changes. In the 1980s there were fears that ice sheets could decay at near catastrophic rates, causing sea levels to rise rapidly and substantially, perhaps by 3 m or more by 2100. Since that time this has become thought to be less likely and that far from suffering major decay the ice sheets (particularly of Antarctica) might show some accumulation of mass as a result of increasing levels of nourishment by snow. However, considerable

debates still exist on this and there remain major uncertainties about how the two great polar ice masses may respond in coming decades and whether the West Antarctic Ice Sheet (WAIS) is inherently and dangerously unstable (see, e.g., Oppenheimer, 1998; Sabadini, 2002; van der Ween, 2002). These are issues that will be addressed in Chapter 11.

Suffice it to say now that were the WAIS to collapse into the ocean, the impacts of the resulting 5 m increase in sea level would be catastrophic for many coastal lowlands. The IPCC (2001) thought that it was very unlikely that this would occur in the twenty-first century. They believed that the contribution that the Antarctica and Greenland ice caps would make to sea-level change would be modest and less than those of glaciers or thermal expansion (the steric effect). The Greenland contribution between 1990 and 2100 was thought to be in the range −0.02 to 0.09 m, and for the Antarctic −0.17 to 0.02 m.

The amount of change by 2100

Over the years there has been a considerable diversity of views about how much sea-level rise is likely to occur by 2100. In general, however, estimates have tended to be revised downward through time (French et al., 1994; Pirazzoli, 1996) (Figure 9.1). They have now settled at best estimates of just under 50 cm by 2100. This implies rates of sea-level rise of around 5 mm per year, which compares with a rate of about 1.5 to 2.0 mm during the twentieth century (Miller and Douglas, 2004).

Land subsidence

The effects of global sea-level rise will be compounded in those areas that suffer from local subsidence as a result of local tectonic movements, isostatic adjustments, and fluid abstraction. Areas where land is rising because of **isostasy** (e.g., Fennoscandia or the Canadian Shield) or because of tectonic uplift (e.g., much of the Pacific coast of the Americas) will be less at risk than subsiding regions (e.g., the deltas of the Mississippi and Nile Rivers) (Figure 9.2).

Areas of appreciable subsidence include some ocean islands. Indeed, crucial to Darwin's model of atoll

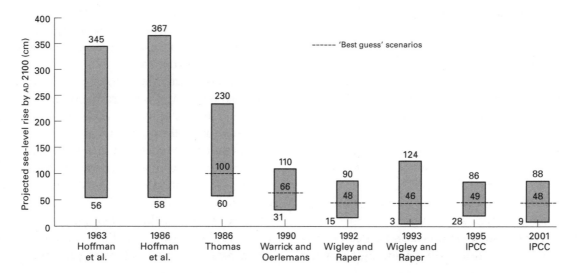

Figure 9.1 Revisions of anticipated sea-level rise by 2100 (after French et al., 1994, with modifications).

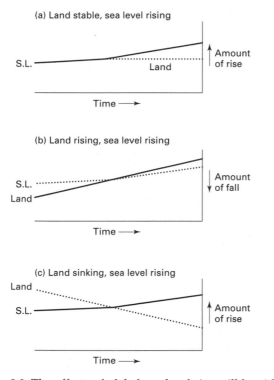

Figure 9.2 The effects of global sea-level rise will be either compounded or mitigated according to whether the local environment is one that is stable, rising, or sinking.

evolution is the idea that subsidence has occurred, and the presence of guyots and seamounts in the Pacific Ocean attest to the fact that such subsidence has been a reality over wide areas. In addition there are tide gauge records from the Hawaiian Ridge that demonstrate ongoing subsidence rates at the present day (1500 mm per 1000 years for Oahu and 3500 mm per 1000 years for Hawaii). Moreover, the occurrence of Holocene and Pleistocene submerged terraces and drowned reefs on the flanks of these two islands demonstrates the reality of this process. The thick coral accumulations that are superimposed on basaltic platforms that were once at sea level indicate that subsidence has continued on timescales of tens of millions of years. The coral cap at Eniwetok, which dates back to the early Eocene (c. 60 million years ago), is 1400 m thick, and that at Bikini (of Miocene age) some 1300 m thick. Differences in the depths of wave-worn platforms along a volcanic chain, if the chronology of the seamount formation can be established by potassium–argon and other dating techniques, provide estimates of subsidence rates. In the case of the Tasman Sea chains off Australia, rates of subsidence appear to have been of the order of 28 m per millions years (28 mm per 1000 years).

The causes of this subsidence are a matter of some debate (see Lambeck, 1988: 506–9). Some of it may be caused by the loading of volcanic material on to the crust, but some may be due to a gradual contraction of the seafloor as the ocean lithosphere moves away from either the ridge or the hotspot that led to the initial formation of the island volcanoes.

Another type of situation prone to subsidence is the river delta. Loading of sediment on to the crust by the river causes subsidence to occur. So, for example,

Fairbridge (1983) calculated subsidence of the Mississippi at a rate of *c.* 15 mm per year during the Holocene, while Stanley and Chen (1993) calculated Holocene subsidence rates for the Yangtze delta in China as 1.6–4.4 mm per year. The Rhone delta has subsided at between 0.5 and 4.5 mm per year (L'Homer, 1992), while the Nile delta is subsiding at *c.* 4.7 mm per year (Sherif and Singh, 1999).

Some areas are prone to subsidence because of ongoing adjustment to the application and removal of ice loadings to the crust in the Pleistocene. During glacials, areas directly under the weight of ice caps were depressed, whereas areas adjacent to them bobbed up by way of compensation (the so-called peripheral bulge). Conversely, during the Holocene, following

removal of the ice load, the formerly glaciated areas have rebounded whereas the marginal areas have foundered. A good example of this is the Laurentide area of North America.

Elsewhere, human actions can promote subsidence (see Chapter 6): the withdrawal of groundwater, oil, and gas (Table 6.6); the extraction of coal, salt, sulfur, and other solids, through mining; the hydrocompaction of sediments; the oxidation and shrinkage of organic deposits such as peats and humus-rich soils; the melting of permafrost; and the catastrophic development of sinkholes in karstic terrain.

Figure 9.3 (from Bird, 1993, figure 2) shows those sectors of the world's coastline that have been subsiding in recent decades, including a large tract of the

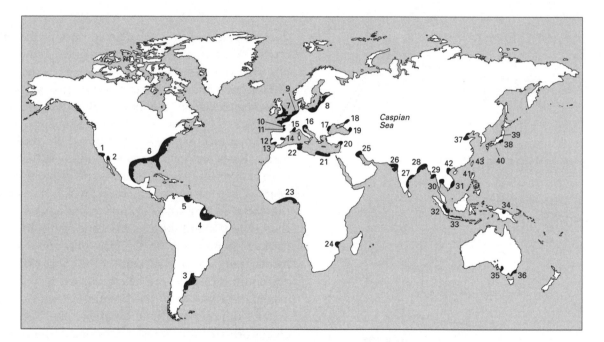

Figure 9.3 Sectors of the world's coastline that have been subsiding in recent decades, as indicated by evidence of tectonic movements, increasing marine flooding, geomorphic and ecological indications, geodetic surveys, and groups of tide gauges recording a rise of mean sea level greater than 2 mm per year over the past three decades: 1, Long Beach area, southern California; 2, Columbia River delta, head of Gulf of California; 3, Gulf of La Plata, Argentina; 4, Amazon delta; 5, Orinoco delta; 6, Gulf and Atlantic coast, Mexico and USA; 7, southern and eastern England; 8, the southern Baltic from Estonia to Poland; 9, northern Germany, The Netherlands, Belgium and northern France; 10, Loire estuary, western France; 11, Vendée, western France; 12, Lisbon region, Portugal; 13, Guadalquavir delta, Spain; 14, Ebro delta, Spain; 15, Rhone delta, France; 16, northern Adriatic from Rimini to Venice and Grado; 17, Danube delta, Romania; 18, eastern Sea of Azov; 19, Poti Swamp, Georgian Black Sea coast; 20, southeast Turkey; 21, Nile delta to Libya; 22, northeast Tunisia; 23, Nigerian coast, especially the Niger delta; 24, Zambezi delta; 25, Tigris-Euphrates delta; 26, Rann of Kutch; 27, southeastern India; 28, Ganges-Brahamputra delta; 29, Irrawaddy delta; 30, Bangkok coastal region; 31, Mekong delta; 32, eastern Sumatra; 33, northern Java deltaic coast; 34, Sepik delta; 35, Port Adelaide region; 36, Corner Inlet region; 37, Hwang-ho delta; 38, head of Tokyo Bay; 39, Niigata, Japan; 40, Maizuru, Japan; 41, Manila; 42, Red River delta, North Vietnam; 43, northern Taiwan. (Modified from Bird, 1985, figure 2.)

eastern seaboard of the USA, southeastern England, and some of the world's great river deltas (e.g., India, Ganges, Mekong, Tigris–Euphrates, and Zambezi).

How fast are sea levels rising?

Observational estimates of sea-level change are based on the short (only a decade or so) satellite record of sea-level height (Cabanes et al., 2001) and on the larger, but geographically sparse and uneven tide-gauge network. Tide gauges are rare in mid-ocean locations and very inadequate for the ocean-dominated Southern Hemisphere. In addition, difficulties in determining global (eustatic) sea-level changes are bedeviled by the need to make allowances for land motion caused by post-glacial isostatic rebound and tectonic movements (Church, 2001). In general it has been estimated that rates of sea-level rise from 1910 to 1990 average 1.0–2.0 mm per year (Church et al., 2001, table 11.10). On the other hand over the period 1993–1998, the Topex/Poseidon satellite suggested a global mean rate of sea-level rise of 3.2 ± 0.2 mm per year (Cabanes et al., 2001).

Coral reefs

Coral reefs, landforms more or less restricted to the tropics, have fascinated scientists for over 150 years. Some of the most pertinent observations of them were made in the 1830s by Charles Darwin during his voyage on *The Beagle*. He recognized that there were three major kinds: fringing reefs, barrier reefs, and atolls. These he related in a logical sequence related to progressive subsidence, and recognized the importance of sea-level change for their development.

Coral reefs are extremely important habitats. Not only do they have an estimated area of 600,000 km^2 globally, they are also the hosts to a great diversity of species, especially in the warm waters of the Indian Ocean and the western Pacific. They have been called the marine version of the tropical rainforests, rivaling their terrestrial counterparts in both richness of species and biological productivity. They are also of aesthetic importance. As Norris (2001: 37) wrote:

Clear blue sea, brightly colored fish and amazing underwater structures built up over generations from the calcified

Table 9.1 Some influences of global warming on the state of coral reefs

An increase in	Effect
Sea-surface temperatures	Will cause stress (bleaching, etc.) in some areas Will stimulate growth in some areas
Storm frequency and intensity	Will build up islands by throwing up coral debris Will erode reefs Will cause change in species composition
Sea levels	Will stimulate reef growth (if slow) Will cause inundation and cause corals to give up* (if fast)

*Corals will also give up if they are stressed (e.g., by pollution, increased temperatures, ultraviolet effects, etc.).

remains of the tiny animals called coral polyps. It's a spectacular sight. No one who has ever visited one of the world's great coral reefs is ever likely to forget it. But it looks like we are one of the last generations who will have the opportunity to experience this spectacle. Reefs as we know them are on their way out.

The reason for Norris's pessimism is that coral reefs are under a whole series of direct anthropogenic threats (pollution, sedimentation, dynamiting, overfishing, etc.), and that they face a suite of potential threats from climate change and sea-level rise (Table 9.1).

One potential change is that of hurricane frequency, intensity, and distribution. They might build some coral islands up, erase others, and through high levels of runoff and sediment delivery they could change the turbidity and salinity of the water in which corals grow. If hurricane frequency, intensity, and geographical spread were to change, there would be significant implications. It is, however, not entirely clear just how much these important characteristics will change. Intuitively one would expect cyclone activity to become more frequent, intense, and extensive if sea-surface temperatures (SSTs) were to rise, because SST is a clear control of where they develop. Indeed, there is a threshold at about 26.5–27.0°C below which tropical cyclones do not tend to form. Moreover, there is some evidence that increasingly low pressure centers can be maintained as SSTs rise, and Emanuel (1987) has employed a GCM which predicts that with a doubling of present atmospheric greenhouse gas levels there will be an increase of 40–50% in the destructive potential of tropical cyclones. However, the IPCC and some individual scientists are far from convinced that global warming will invariably stimulate cyclone activity. Raper (1993) has argued that there is as yet no convincing empirical evidence to support a relationship between SSTs and cyclone intensities, or that SSTs are the primary variable in whether incipient storms develop into cyclones or not.

Increased sea-surface temperature could have deleterious consequences for corals that are near their thermal maximum (Hoegh-Guldberg, 2001). Most coral species cannot tolerate temperature greater than about 30°C and even a rise in seawater temperature of 1–2°C could adversely affect many shallow-water coral species.

Increased temperatures in recent years have been identified as a cause of widespread coral bleaching (loss of symbiotic zooxanthellae). Those corals stressed by temperature or pollution might well find it more difficult to cope with rapidly rising sea levels than would healthy coral. Moreover, it is possible that increased ultraviolet radiation because of ozone layer depletion could aggravate bleaching and mortality caused by global warming. Various studies suggest that coral bleaching was a widespread feature in the warm years of the 1980s and 1990s.

Coral bleaching, which can produce mass mortality of corals in extreme cases, has been found to be strongly correlated with elevated water temperatures and high UV solar irradiance (e.g., Brown, 1997; Spencer et al., 2000). Although bleaching itself is a complex phenomenon to which corals can respond in a variety of ways (Brown et al., 2000; Fitt et al., 2001; Loya et al., 2001), ENSO-related heating, cooling, and migrations of ocean water masses have been found to be important controls of mass bleaching episodes (Spencer et al., 2000). For example, in 1998, sea-surface temperatures in the tropical Indian Ocean were as much as 3–5°C above normal, and this led to up to 90% coral mortality in shallow areas (Wilkinson et al., 1999; Edwards et al., 2001). McClanahan (2000) notes that warm conditions of between 25 and 29°C favor coral growth, survival, and species richness, and that somewhere about 30°C there are species-, environment-, or regionally specific thresholds above which many of the dominant coral species are lost. As Hoegh-Guldberg (1999), Souter and Linden (2000), Sheppard (2003), and others

have suggested, continued warming trends superimposed on interannual and decadal patterns of variability are likely to increase the incidence of bleaching and coral mortality unless significant adaptation to increased temperatures occurs.

Indeed, Goreau and Hayes (1994) have produced maps of global coral bleaching episodes between 1983 and 1991 and have related them to maps of sea-surface temperatures over that period. They find that areas of severe bleaching are related to what they describe as ocean 'hotspots' where marked positive temperature anomalies exist. They argue that coral reefs are ecosystems that may be uniquely prone to the effects of global warming:

If global warming continues, almost all ecosystems can be replaced by migration of species from lower latitudes, except for the warmest ecosystems. These have no source of immigrants already adapted to warmer conditions. Their species must evolve new environmental tolerances if their descendants are to survive, a much slower process than migrations (pp. 179–80).

However, Kinsey and Hopley (1991) believe that few of the reefs in the world are so close to the limits of temperature tolerance that they are likely to fail to adapt satisfactorily to an increase in ocean temperature of 1–2°C, provided that there are not very many more short-term temperature deviations. Indeed in general they believe that reef growth will be stimulated by the rising sea levels of a warmer world, and they predict that reef productivity could double in the next 100 years from around 900 to 1800 million tonnes per year. They do, however, point to a range of subsidiary factors that could serve to diminish the increase in productivity: increased cloud cover in a warmer world could reduce calcification because of reduced rates of photosynthesis; increased rainfall levels and hurricane activity could cause storm damage and freshwater kills; and a drop in seawater pH might adversely affect calcification.

In the 1980s there were widespread fears that if rates of sea-level rise were high (perhaps 2–3 m or more by 2100) then coral reefs would be unable to keep up and submergence of whole atolls might occur. Particular concern was expressed about the potential fate of Tokelau, the Marshall Islands, Tuvalu, the Line Islands, and Kiribati in the Pacific Ocean, and of the Maldives in the Indian Ocean. However, with the reduced expectations for the degree of sea-level rise that may occur, there has arisen a belief that coral reefs may survive and even prosper with moderate rates of sea-level rise. As is the case with coastal marshes and other wetlands, reefs are dynamic features that may be able to respond adequately to rises in sea level (Spencer, 1995). It is also important to realize that their condition depends on factors other than the rate of submergence.

An example of the pessimistic tone of opinion in the 1980s is provided by Buddemeier and Smith (1988). Employing 15 mm per year as the probable rise of sea level over the next century, they suggest that this would be (p. 51) 'five times the present modal rate of vertical accretion on coral reef flats and 50% greater than the maximum vertical accretion rates apparently attained by coral reefs'. Using a variety of techniques they believed (p. 53) 'the best overall estimate of the sustained maximum of reef growth to be 10 mm/year . . .'. They predicted (p. 54) that

inundated reef flats in areas of heavy seas will be subjected to progressively more destructive wave activity as larger waves move across the deepening flats . . . Reef growth on the seaward portions of inundated, wave swept reef flats may therefore be negligible compared to sea level rise over the next century, and such reef flats may become submerged by almost 1.5 m.

In addition to the potential effects of submergence, there is the possibility that higher sea levels could promote accelerated erosion of reefs (Dickinson, 1999).

In recent years, fears have been expressed that corals will suffer from the increasing levels of carbon dioxide in the atmosphere. These are expected to reach double pre-industrial levels by the year 2065. A coral reef, as Kleypas et al. (1999) have pointed out, represents the net accumulation of calcium carbonate produced by corals and other calcifying organisms. Thus if calcification was to decline, then reef-building capacity would also decline. Such a decline could occur if there were to be a change in the saturation state of aragonite (a carbonate mineral) in surface seawater. Increased concentrations of carbon dioxide decrease the aragonite saturation state.

By the middle of the twenty-first century the aragonite saturation state in the tropics could be reduced by as much as 30%. Equally, Leclercq et al. (2000) have

calculated that the calcification rate of scleractinian-dominated communities could decrease by 21% between the pre-industrial period (1880) and the year (2065) at which atmospheric carbon dioxide concentrations will double. Over the same period, the pH of seawater will decline from 8.08 to 7.93. It is this that drives down the aragonite saturation state (Gattuso et al., 1998).

Salt marshes and mangrove swamps

Salt marshes, including the mangrove swamps of the tropics, are extremely valuable ecosystems that are potentially highly vulnerable in the face of sea-level rise, particularly in those circumstances where sea defenses and other barriers prevent the landward migration of marshes as sea-level rises. Sediment supply, organic and inorganic, is a crucial issue (Reed, 2002). On coasts with limited sediment supply a rise in sea level will impede the normal process of marsh progradation, and increasing wave attack will start or accelerate erosion along their seaward margins. The tidal creeks that flow across the marsh will tend to become wider, deeper, and more extended headwards as the marsh is submerged. The marsh will attempt to move landwards, and where the hinterland is low lying the salt marsh vegetation will tend to take over from freshwater or terrigenous communities. Such landward movement is impossible where seawalls or embankments have been built at the inner margins of a marsh (Figure 9.4). Equally, if there is very limited availability of sediment the marsh may not build up and inwards, so that in such circumstances the salt marsh will cease to exist (Bird, 1993). Marshes would be further threatened if climate change caused increased incidence of severe droughts (Thomson et al., 2002). Table 9.2 demonstrates the differences between marshes in terms of their sensitivity. However, salt marshes are highly dynamic features and in some situations may well be able to cope, even with quite rapid rises of sea level (Reed, 1995).

Reed (1990) suggests that salt marshes in riverine settings may receive sufficient inputs of sediment that they are able to accrete rapidly enough to keep pace with projected rises of sea level. Areas of high-tidal range, such as the marshes of the Severn Estuary in England and Wales, or the Tagus Estuary of Portugal, are also areas of high sediment-transport potential and

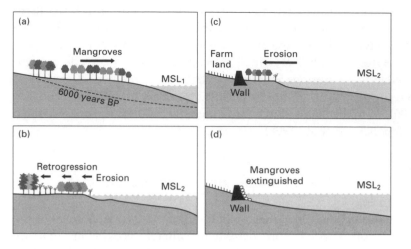

Figure 9.4 Changes on mangrove-fringed coasts as sea level rises (a and b) will be modified where they are backed by a wall built to protect farmland (c). The mangrove fringe will then be narrowed by erosion, and may eventually disappear (d), unless there is a sufficient sediment supply to maintain the substrate and enable the mangroves to persist. MSL, mean sea level. (Modified from Bird, 1993, figure 45.)

Table 9.2 Salt marsh vulnerability

Less sensitive	More sensitive
Areas of high sediment input	Areas of subsidence
Areas of high tidal range (high sediment transport potential)	Areas of low sediment input (e.g., cyclically abandoned delta areas)
Areas with effective organic accumulation	Mangroves (longer life cycle, therefore slower response)
	Constraint by seawalls, etc. (nowhere to go)
	Microtidal areas (rise in sea level represents a larger proportion of total tidal range)
	Reef settings (lack of allogenic sediment)

may thus be less vulnerable to sea-level rise (Simas et al., 2001). Likewise, some vegetation associations, e.g., *Spartinia swards*, may be relatively more effective than others at encouraging accretion, and organic matter accumulation may itself be significant in promoting vertical build-up of some marsh surfaces. For marshes that are dependent upon inorganic-sediment accretion, the increased storm activity and beach erosion that might be associated with the greenhouse effect could conceivably mobilize sufficient sediments in coastal areas to increase their sediment supply.

Marsh areas that may be highly prone to sea-level rise include areas of deltaic sedimentation where, because of sediment movement controls (e.g., reservoir construction) or because of cyclic changes in the location of centers of deposition, rates of sediment supply are low. Such areas may also be areas with high rates of subsidence. A classic example of this are portions of the Mississippi delta. Park et al. (1986) undertook a survey of coastal wetlands in the USA and suggested that sea-level change could, by 2100, lead to a loss of between 22 and 56% of the 1975 wetland area, according to the degree of sea-level rise that takes place.

Ray et al. (1992) have examined the past and future response of salt marshes in Virginia in the USA to changes in sea level, and have been particularly interested in the response of mid-lagoon marshes. Between 6000 and 2000 years ago, when the rate of sea-level rise was about 3 mm year, the lagoons were primarily open water environments with little evidence of mid-lagoon marshes. After AD 1200, land subsidence in the area slowed down, and sea-level rise was only about 1.3 mm per year. This permitted mid-lagoon marshes to expand and flourish. Such marshes in general lack an inorganic sediment supply so that their upward growth rate approximates only about 1.5 mm per year. Present rates of sea-level rise exceed that figure (they are about 2 mm per year) and cartographic analysis shows a 16% loss of marsh between 1852 and 1968. As sea-level rise accelerates as a result of global warming, almost all mid-lagoon marsh will be lost.

Rises in sea level will increase nearshore water depths and thereby modify wave refraction patterns. This means that wave energy amounts will also change at different points along a particular shoreline. Pethick (1993) maintains that this could be significant for the

Figure 9.5 A mangrove swamp in the Seychelles. These important wetlands are relatively slow-growing and their nature is clearly very intimately connected to the frequency and duration of inundation by tides. They may, therefore, be examples of hot spots that are especially sensitive to sea-level change.

classic Scolt Head Island salt marshes of Norfolk, eastern England, which are at present within a low to medium wave energy zone. After a 1.0 m rise in sea level these marshes will experience high wave energy because of the migration of wave foci. Pethick remarks:

The result will be to force the long shore migration of salt marsh and mudflat systems over distances of up to ten kilometres in 50 years – a yearly migration rate of 200 metres. It is doubtful whether salt marsh vegetation could survive in such a transitory environment – although intertidal mudflat organisms may be competent to do so – and a reduction or total loss of these open coast wetlands may result (p. 166).

One particular type of marsh that may be affected by anthropogenically accelerated sea-level rise is the mangrove swamp (Figure 9.5). As with other types of marsh the exact response will depend on the local setting, sources, and rates of sediment supply, and the rate of sea-level rise itself. However, mangroves may respond rather differently from other marshes in that they are composed of relatively long-lived trees and shrubs which means that the speed of zonation change will be less (Woodroffe, 1990).

Like salt marshes, however, mangroves trap sediment to construct a depositional terrace in the upper intertidal zone. Where there is only a modest sediment supply, submergence by rising sea level may cause

dieback of vegetation and erosion of their seaward margins. As their seaward margin erodes backwards, the mangroves will attempt to spread landward, displacing existing freshwater swamps or forest. As with normal salt marshes this would not be possible if they were backed by walls or embankments.

The degree of disruption is likely to be greatest in microtidal areas, where any rise in sea level represents a larger proportion of the total tidal range than in macrotidal areas. The setting of mangrove swamps will be very important in determining how they respond. River-dominated systems with large **allochthonous** sediment supply will have faster rates of shoreline progradation and deltaic plain accretion and so may be able to keep pace with relatively rapid rates of sea-level rise. By contrast in reef settings, in which sedimentation is primarily **autochthonous**, mangrove surfaces are less likely to be able to keep up with sea-level rises. This is the view of Ellison and Stoddard (1990: 161) who argued that low island mangrove ecosystems (mangals) have in the past been able to keep up with a sea-level rise of up to 8–9 cm per 100 years, but that at rates over 12 cm per 100 years they had not been able to persist.

Snedaker (1995) finds it difficult, however, to reconcile the Ellison and Stoddard view with what has happened in South Florida, where relative sea level rose by about 30 cm over 147 years (equivalent to 23 cm per 100 years). The mangrove swamps of the area did not for the most part appear to have been adversely stressed by this. Snedaker argues that precipitation and catchment runoff changes also need to be considered, as for any given sea-level elevation reduced rainfall and precipitation would result in higher salinity and greater seawater-sulfate exposure. These in turn would be associated with decreased production and increased organic matter decomposition, which would lead to subsidence. On the other hand, under conditions with higher rainfall and runoff the reverse would occur, so that mangrove production would increase and sediment elevations would be maintained.

The ability of mangrove propagules to take root and become established in intertidal areas subjected to a higher mean sea level is in part dependent on species (Ellison and Farnsworth, 1997). In general, the large propagule species (e.g., *Rhizophora* spp.) can become established in rather deeper water than can the smaller propagule species (e.g., *Avicennia* spp.). The latter has aerial roots which project only vertically above tidal muds for short distances (Snedaker, 1993).

In arid areas, such as the Arabian Gulf in the Middle East, extensive tracts of coastline are fringed by low-level salt-plains called *sabkhas*. These features are generally regarded as equilibrium forms that are produced by depositional processes (e.g., wind erosion and storm surge effects). They tend to occur at or about high tide level. Because of the range of depositional processes involved in their development they might be able to adjust to a rising sea level, but quantitative data on present and past rates of accretion are sparse. A large proportion of the industrial and urban infrastructure of the United Arab Emirates is located on or in close proximity to sabkhas.

Salt marsh, swamp, and sabkha regression caused by climatic and sea-level changes will compound the problems of wetland loss and degradation caused by other human activities (Wells, 1996; Kennish, 2001), including reclamation, ditching, diking, dredging, pollution, and sediment starvation. More than half the original salt marsh habitat in the USA has already been lost, and Shriner and Street (1998: 298) suggest that a 50 cm rise in sea level would inundate approximately 50% of North American coastal wetlands in the twenty-first century. On a global basis Nicholls et al. (1999) suggest that by the 2080s sea-level rise could cause the loss of up to 22% of the world's coastal wetlands.

River deltas

Deltaic coasts and their environs are home to large numbers of people. They are likely to be threatened by submergence as sea levels rise, especially where prospects of compensating sediment accretion are not evident. Many deltas are currently zones of subsidence because of the isostatic effects of the sedimentation that caused them to form. This will compound the effects of eustatic sea-level rise (Milliman and Haq, 1996).

It needs to be remembered, however, that deltas will not solely be affected by sea-level changes. The delta lands of Bangladesh (Warrick and Ahmad, 1996), for example, receive very heavy sediment loads from the rivers that feed them so that it is the relative rates of accretion and inundation that will be crucial (Milliman et al., 1989). Land-use changes upstream,

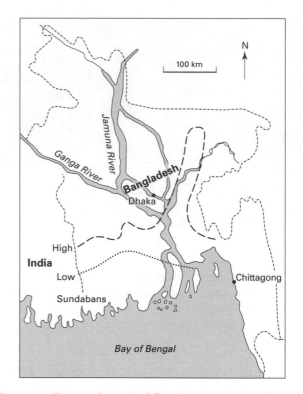

Figure 9.6 Projected areas of flooding as a result of sea-level change in Bangladesh, for two scenarios (low = 1 m and high = 3 m) (modified after Broadus et al., 1986, figure 7).

such as deforestation, could increase rates of sediment accumulation. Deltas could also be affected by changing tropical cyclone activity.

Broadus et al. (1986), for example, calculate that were the sea level to rise by just 1 m in 100 years, 12–15% of Egypt's arable land would be lost and 16% of the population would have to be relocated. With a 3 m rise the figures would be a 20% loss of arable land and a need to relocate 21% of the population. The cities of Alexandria, Rosetta, and Port Said are at particular risk and even a sea-level rise of 50 cm could mean that 2 million people would have to abandon their homes (El-Raey, 1997). In Bangladesh (Figure 9.6) a 1 m rise would inundate 11.5% of the total land area of the state and affect 9% of the population directly, while a 3 m rise in sea level would inundate 29% of the land area and affect 21% of the population. It is sobering to remember that at the present time approximately one-half of Bangladesh's rice production is in the area that is less than 1 m above sea level. Many of the world's major conurbations might be flooded in whole or in

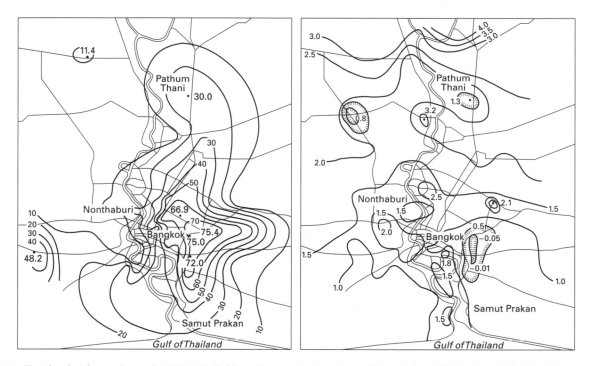

Figure 9.7 Total subsidence (in cm), 1978–87 (left) and ground elevation of Bangkok, 1987 (right). (Modified from Nutalaya et al., 1996, figures 3 and 9.)

part, sewers and drains rendered inoperative (Kuo, 1986), and peri-urban agricultural productivity reduced by saltwater incursion (Chen and Yong, 1999).

Even without accelerating sea-level rise, the Nile delta has been suffering accelerated recession because of sediment retention by dams. The Nile sediments, on reaching the sea, used to move eastward with the general anticlockwise direction of water movements in that part of the eastern Mediterranean, generating sandbars and dunes, which contributed to delta accretion. About a century ago an inverse process was initiated and the delta began to retreat. For example, the Rosetta mouth of the Nile lost about 1.6 km of its length from 1898 to 1954. The imbalance between sedimentation and erosion appears to have started with the delta barrages (1861) and then been continued by later works, including the Sennar Dam (1925), Gebel Aulia Dam (1937), Khasm el Girba Dam (1966), Roseires Dam (1966), and the Aswan High Dam itself. In addition, large amounts of sediment are retained in an extremely dense network of irrigation and drainage channels that has been developed in the Nile delta itself (Stanley, 1996). Much of the Egyptian coast is now 'undernourished' with sediment and, as a result of this overall

erosion of the shoreline, the sandbars bordering Lake Manzala and Lake Burullus on the seaward side are eroded and likely to collapse. If this were to happen, the lakes would be converted into marine bays, so that saline water would come into direct contact with low lying cultivated land and freshwater aquifers.

Bangkok is an example of a city that is being threatened by a combination of accelerated subsidence and accelerated sea-level rise (Nutalaya et al., 1996). More than 4550 km² of the city was affected by land subsidence between 1960 and 1988, and 20 to 160 cm of depression of the land surface occurred. The situation is critical because Bangkok is situated on a very flat, low lying area, where the ground level elevations range from only 0 to 1.5 m above mean sea level (Figure 9.7).

Estuaries

Estuaries, located between the land and the sea, are unique and sensitive ecosystems that are locations of many major ports and industrial concerns (Dyer, 1995). They could be impacted upon by a range of climate-related variables, including the amount of freshwater and sediment coming from the land, the temperature and salinity of the estuarine water, and tidal range.

The overall effect of sea-level rise has been summarized by Chappell et al. (1996: 224):

In the earliest stages of sea-level rise, the extent of tidal flooding on high tide flats will increase and networks of small tidal creeks will expand into estuarine floodplains. Levels of tidal rivers will be breached and brackish water will invade freshwater floodplains. Eventually, tidal flow in main channels will increase to carry the enlarged tidal prism, and channels are likely to widen. In systems with extensive tidal floodplains, tidal regimes will shift towards ebb dominance and net sediment flux will tend to be offshore, except in macrotidal systems. Salt-water intrusion could be a major change, with saline water extending far inland as sea level rose. This in turn could affect swamp and marsh vegetation (Mulrennan and Woodroffe, 1998).

Attempts have been made to model the transgression rate for landward extending estuaries. For the Humber estuary in northeast England, Pethick (2001: 34) reports that the mean migration rate of the estuary as a whole would be 1.3 m per millimeter rise in sea level (or 8 m per year assuming a 6 mm per year rise in sea level).

Cliffed coasts

When sea level rises, nearshore waters deepen, shore platforms become submerged, deeper water allows larger waves to reach the bases of cliffs, and the cliffs suffer accelerated rates of retreat, especially if they are made of susceptible materials. In addition, sea-level rise is likely to cause increased frequencies of coastal landslides. Accelerated cliff retreat and land sliding may cause an augmented supply of sediment to adjacent and down-drift beaches. This could be beneficial. However, as Bray et al. (1992: 86) remarked:

The location of the benefit will depend upon sediment transport conditions and their relation to sediment supply. Where littoral transport is poorly developed, beaches will accrete in front of cliffs as sea level rises, and tend to offset any trend for increased retreat. Where littoral transport is efficient, cliff retreat is likely to increase significantly as sea level rises because extra sediments yielded are rapidly removed from the eroding cliffs. Coasts of this type are likely to become increasingly valuable in the future because they are sensitive to sea-level rise and can supply large quantities of sediment

to downdrift beaches. A further complicating factor is that cliff erosion products differ with respect to their mobility and coarse durable materials may be retained on the upper shoreface and provide natural armoring. Soft-rock cliffs yielding a high proportion of coarse durable products are therefore likely to be less sensitive to sea-level rise.

Cliffs on high latitude coasts (e.g., in Siberia, Alaska, and Canada) might be especially seriously affected by global change. On the one hand, coasts formed of weak sediments that are currently cemented by permafrost would lose strength if warming caused the permafrost to melt. On the other hand, melting of sea ice would expose them to greater wave affects from open water.

Sandy beaches

Bruun (1962) developed a widely cited model of the response of a sandy beach to sea-level rise in a situation where the beach was initially in equilibrium, neither gaining nor losing sediment (Figure 9.8). As Bird (1993: 56) has explained:

Erosion of the upper beach would then occur, with removal of sand to the nearshore zone in such a way as to restore the previous transverse profile. In effect, there would be an upward and landward migration of the transverse profile, so that the coastline would recede beyond the limits of submergence. This restoration would be completed when the sea became stable at a higher level, and coastline recession

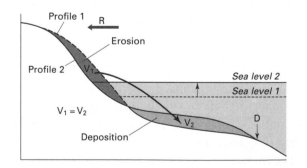

Figure 9.8 Sea-level rise and coastline changes. The Bruun Rule states that a sea-level rise will lead to erosion of the beach and removal of a volume of sand (V_1) seaward to be deposited (V_2) in such a way as to restore the initial transverse profile landward of D, the outer boundary of nearshore sand deposits. The coastline will retreat (R) until stability is restored after the sea-level rise comes to an end. The coastline thus recedes further than it would if submergence were not accompanied by erosion.

would come to an end after a new equilibrium was achieved. The extent of recession was predicted by using a formula that translates into a 'rule of thumb' whereby the coastline retreats 50–100 times the dimensions of the rise in sea level: a 1 m rise would cause the beach to retreat by 50–100 m. Since many seaside resort beaches are no more than 30 m wide, the implication is that these beaches will have disappeared by the time the sea has risen 15–30 cm (i.e., by the year 2030), unless they are artificially replaced.

This influential model, often called the Bruun Rule, has been widely used. It is important to recognize, however, that there are some constraints on its applicability (Wells, 1995; Healy, 1996). The rule assumes that no sand is lost to long shore transport and that an offshore 'closure depth' exists beyond which there is no sediment exchange. Moreover, the rule does not allow for shoreward transport of sediment as overwash or for those situations where the slope of the coastal plain is too gentle for sufficient sand to be available as a source for supplying the offshore. In addition, the rule was originally proposed for beaches that were initially in equilibrium. However, as Bird (1993: 58) has pointed out, only a small proportion of the world's sandy beaches can in fact be considered to be in equilibrium. Beach erosion is widespread. Furthermore, there is a lack of established criteria to ascertain whether the original shoreline really is in dynamic equilibrium (Healy, 1991).

In spite of these limitations the Bruun Rule remains widely used, and can be stated mathematically as follows (Gornitz et al., 2002: 68):

$$S = (A \times B)/d$$

where S is shoreline movement, A is sea-level rise, d is maximum depth of beach profile, measured from the berm elevation for each project location to the estimated depth of closure, and B is the horizontal length of the profile, measured from the beginning of the berm to the intersection with the estimated depth of closure. The depth of closure is taken as the minimum water depth at which no significantly measurable change occurs in bottom depth. It is often erroneously interpreted to mean the depth at which no sediment moves in deeper water. 'Closure' is a somewhat ambiguous term in that it can vary, depending on waves and other hydrodynamic forces.

An example of the testing of the validity of the Bruun Rule is the study by List et al. (1997) of the barrier islands of Louisiana in the USA. Using bathymetric surveys over about a century, they found that only a portion of their studied profiles met the equilibrium criterion of the Bruun Rule. Furthermore, using those shore profiles that did meet the equilibrium criterion, they determined measured rates of relative sea-level rise so that they could hindcast shoreline retreat rates using the Bruun Rule formula. They found that the modeled and observed shoreline retreat rates showed no significant correlation. They suggested that if the Bruun Rule is inadequate for hindcasting it would also be inadequate for forecasting future rates of beach retreat.

There are various other techniques that can be used to predict rates of shoreline retreat. One of these is historical trend analysis, which is based upon extrapolating the trend of shoreline change with respect to recorded sea-level rise over a given historical period:

$$R_2 = (R_1/S_1) \times S_2$$

where S_1 is historical sea-level rise, S_2 is future sea-level rise, R_1 is historical retreat rate, and R_2 is future retreat rate.

Barrier islands, such as those that line the eastern seaboard of the USA and the southern North Sea, are dynamic landforms that will tend to migrate inland with rising sea levels and increased intensity of overtopping by waves (Eitner, 1996). If sea-level rise is not too rapid, and if they are not constrained by human activities (e.g., engineering structures and erosion control measures), they are moved inland by washover – a process similar to rolling up a rug (Titus, 1990); as the island rolls landward, it builds landwards and remains above sea level (Figure 9.9). As sea level rises, they will be exposed to higher storm surges and greater flooding. In the New York area, Gornitz et al. (2002) have calculated that by the 2080s the return period of the 100-year storm flood could be reduced to between 4 and 60 years (depending on location).

The role of sediment starvation

Many beaches are currently eroding because they are no longer being replenished with sediment by rivers. This is because of the construction of dams, which trap much of the sediment that would otherwise go to the sea.

The River Nile now transports only about 8% of its natural load below the Aswan Dam. Even more

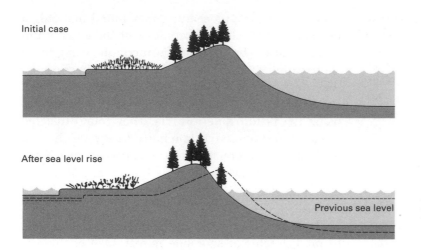

Initial case

After sea level rise

Previous sea level

Figure 9.9 Overwash: natural response of undeveloped barrier islands to sea-level rise. (Source: Titus, 1990, figure 2.)

dramatic is the picture for the Colorado River in the USA. Prior to 1930 it carried around 125–150 million tonnes of suspended sediment to its delta at the head of the Gulf of California. Following the construction of a series of dams, the Colorado now discharges neither sediment nor water into the ocean (Schwarz et al., 1991). Similarly, rivers on the eastern seaboard of the USA draining into the Gulf of Mexico or the Atlantic have shown marked falls in sediment loadings, and four major Texan rivers carried in 1961–70 on average only about one-fifth of what they carried in 1931–40. Likewise, in France the Rhône only carries about 5% of the load that it did in the nineteenth century, while in Asia the Indus discharges less than 20% of the load if did before the construction of large barrages over the past half century (Milliman, 1990).

Conclusions

Coastal environments are already suffering from many human impacts, including sediment starvation. However, these dynamic regions, at the interface between land and sea and the home to many people, will also be subjected in the future to the combined effects of climate change and sea-level rise. Plainly coastlines will be directly impacted by climatic change through such mechanisms as changes in storm surges and hurricanes, but they will also be impacted by sea-level changes resulting from global climatic change, local subsidence, and miscellaneous modifications of the hydrologic cycle.

If, on average, sea level rises by around half a meter during this century, it will greatly modify susceptible and sensitive coastlines, including marshes and swamps, estuaries, soft cliffs, barrier islands, and sandy beaches.

Points for review

In what ways may human activities influence global sea levels?

How may coral reefs respond to global warming?

Which coastal environments may be particularly susceptible to sea-level rise?

What do you understand by the Bruun Rule? Is it a rule that has wide applicability?

Guide to reading

Bird, E. C. F., 1993, *Submerging coasts*. Chichester: Wiley. A geomorphologic overview of how coasts will respond to sea-level rise.

Douglas, B. C., Kearney, M. S. and Leatherman, S. P. (eds), 2000, *Sea level rise: history and consequences*. San Diego: Academic Press. An edited volume, with a long time perspective, which considers some of the consequences of sea-level change.

Milliman, J. D., and Haq, B. V. (eds), 1996, *Sea-level rise and coastal subsidence*. Dordrecht: Kluwer. An edited discussion of the ways in which sea-level rise and land subsidence may impact upon coastal environments.

10 THE FUTURE: HYDROLOGIC IMPACTS

Introduction

Global warming will affect hydrologic systems and fluvial geomorphology in a whole range of ways (Arnell, 1996; Jones et al., 1996). Increasing temperatures will tend to melt snow and ice and promote greater evapotranspirational losses. There will be changes in the amount, intensity, duration, and timing of precipitation, which will affect river flows and groundwater recharge. Vegetation cover will respond to temperature and precipitation changes, as will land use. Higher CO_2 levels may lead to changes in plant growth and physiology, which can lead to changes in transpiration. Global warming may also affect soil properties (such as organic matter content) which could alter runoff generation processes. Climate change will also lead to human interventions in the hydrologic system with, for example, greater use of irrigation in areas subject to increased drought risk, and the continued spread of engineering controls on flooding and erosion.

The nature of the changes that will take place in fluvial systems depends to a certain extent on scale. As Ashmore and Church (2001: 5) explained:

The nature of the changes caused by climatic change depends to some extent on the size of the drainage basin in question. Drainage basin size also affects the relative impact of widespread climate change compared to local land-use change. The effects of land-use change are expected to dominate in smaller basins where a large proportion of the basin area may be affected, leading to substantial changes in runoff and erosion throughout the basin. In larger drainage basins, land use change is seldom sufficiently widespread to affect the entire basin and climatic effects will dominate. The impacts of climatic change will also vary with basin size. Thus, small basins will be affected by changes in local, high-intensity storms, whereas larger basins will show a greater response to cyclonic events or basin-wide snowmelt effects.

Equally, as Ashmore and Church (2001: 5) maintain, there are many ways in which river systems may change:

Potential consequences of climatic change for river processes include changes to the magnitude of flood flows; modification of river channel dimensions and form; changes to bank stability, bank erosion rates, and channel migration;

modification of in-channel erosion and deposition; on-set of long term aggradation or degradation of river channels; changes to intensity and frequency of overbank flooding and ice-jams; and changes to the stability of valley sides. These changes in channel processes present a significant risk to structures both in and near streams including dams, bridges, water intakes, and outfalls. Structures on or near floodplains or valley margins are also at risk. Other environmental, scenic and economic attributes of streams will also be affected, especially in stream and riparian habitat.

Rainfall intensity

Rainfall intensity is a major factor in controlling such phenomena as flooding, rates of soil erosion, and mass movements (Sidle and Dhakal, 2002). Under increased greenhouse gas concentrations some GCMs exhibit enhanced mid-latitude and global precipitation intensity and shortened return periods of extreme events (Hennessy et al., 1997; Zwiers and Kharin, 1998; McGuffie et al., 1999; Osborn et al., 2000; Jones and Reid, 2001; New et al., 2001).

There is some evidence of increased rainfall events in various countries over recent warming decades, which lends some support to this notion. Examples are known from the USA (Karl and Knight, 1998), Canada (Francis and Hengeveld, 1998), Australia (Suppiah and Hennessy, 1998), Japan (Iwashima and Yamamoto, 1993), South Africa (Mason et al., 1999), and Europe (Forland et al., 1998). In the UK there has been an upward trend in the heaviest winter rainfall events (Osborn et al., 2000).

Probabilistic analysis of GCMs by Palmer and Räisänen (2002), applied to western Europe and the Asian Monsoon region, shows under global warming a clear increase in extreme winter precipitation for the former and in extreme summer precipitation for the latter. Increased monsoonal rainfall events would have potentially grave implications for flooding in Bangladesh.

In their analysis of flood records for 29 river basins from high and low latitudes with areas greater than 200,000 km^2, Milly et al. (2002) found that the frequency of great floods had increased substantially during the twentieth century, particularly during its warmer later decades. Their model suggested that this trend would continue.

Changes in tropical cyclones

Tropical cyclones (hurricanes) are highly important geomorphologic agents, in addition to being notable natural hazards, and are closely related in their places of origin to sea-temperature conditions. They only develop where sea-surface temperatures (SSTs) are in excess of 26.5°C. Moreover, their frequency over the past century has changed in response to changes in temperature, and there is even evidence that their frequency was reduced during the Little Ice Age (see Spencer and Douglas, 1985). It is, therefore, possible that as the oceans warm up, so the geographic spread and frequency of hurricanes will increase (Figure 10.1). Furthermore, it is also likely that the intensity of these storms will be magnified (Figure 10.2). Emanuel (1987) used a GCM which predicted that with a doubling of present atmospheric concentrations of CO_2 there will be an increase of 40–50% in the destructive potential of hurricanes. More recently Knutson et al. (1998) and Knutson and Tuleya (1999) simulated hurricane activity for a sea-surface temperature warming of 2.2°C and found that this yielded hurricanes that were more intense by 3–7 m per second for wind speed, an increase of 5–12%.

An increase in hurricane intensity and frequency would have numerous geomorphologic consequences in low latitudes, including accentuated river flooding and coastal surges, severe coast erosion, accelerated land erosion and siltation, and the killing of corals (because of freshwater and siltation effects) (De Sylva, 1986).

Figure 10.1 indicates one scenario of the likely latitudinal change in the extent of warm, cyclone-generating seawater in the Australian region, using as a working threshold for cyclone genesis a summer (February) sea-surface temperature of 27°C. Although cyclones do occur to the south of this line, they are considerably more frequent to the north of it. Under greenhouse conditions it is probable that on the margins of the Great Sandy Desert near Port Hedland the number of cyclones crossing the coast will approximately triple from around four per decade to twelve per decade (Henderson-Sellers and Blong, 1989).

However, it needs to be pointed out that the Intergovernmental Panel on Climate Change (Houghton et al., 1990: 25) was somewhat equivocal on the question

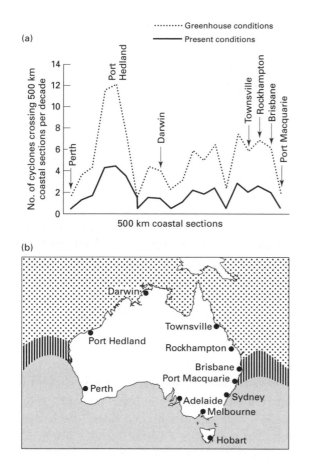

Figure 10.1 (a) The present frequency of cyclones crossing 500-km-long sections of the Australian coast, and an estimate of the frequency under conditions with a 2°C rise in temperature. (b) The area where February sea-surface temperatures around Australia are currently greater than 27°C (stippled) and the additional area with such temperatures with a 2°C rise in temperature (hatched) (modified after Henderson-Sellers and Blong, 1989, figures 4.12 and 4.14).

of the extent to which warming would stimulate cyclone activity, stating:

Although the areas of sea having temperatures over this critical value (26.5°C) will increase as the globe warms, the critical temperature itself may increase in a warmer world. Although the theoretical maximum intensity is expected to increase with temperature, climate models give no consistent indication whether tropical storms will increase or decrease in frequency or intensity as climate changes; neither is there any evidence that this has occurred over the past few decades.

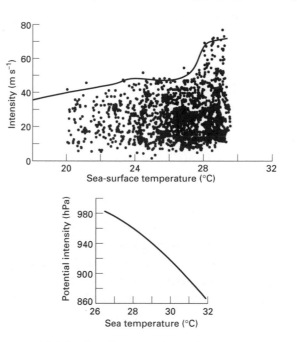

Figure 10.2 (a) Scatter diagram of monthly mean sea-surface temperature and best-track maximum wind speeds (after removing storm motion) for a sample of North Atlantic tropical cyclones. The line indicates the 99th percentile and provides an empirical upper bound on intensity as a function of ocean temperature. (b) The derived relationship between sea-surface temperature and potential intensity of tropical cyclones. (Source: Holland et al., 1988, figures 5 and 6.)

Such caution is also evident in the recent review by Walsh and Pittock (1998), and Gray (1993) discusses some of the reasons why increased SSTs may not lead to changes in hurricane activity:

Changes in the atmospheric circulation associated with future SST increases may not be the same as the historical variations on which the estimates of hurricane frequency are based. Since these changes modify the tropical storms environment, they could prevent the storms from reaching the potential increased strength estimated from thermodynamic considerations. For example, an increase in vertical shear of the zonal wind over the tropical Atlantic would tend to decrease the frequency of tropical storm formation, thereby moderating an increase in frequency from higher SSTs, while a decrease in vertical shear would tend to have the opposite effect.

Hurricane activity is linked to the ENSO phenomenon, so that if this were to be changed by global warming then so might the frequency, distribution, and intensity of hurricanes.

Severe El Niños, such as that of 1997–1998, can have a remarkable effect on rainfall amounts. This was shown with particular clarity in the context of Peru (Bendix et al., 2000), where normally dry locations suffered huge storms. At Paita (mean annual rainfall 15 mm) there was 1845 mm of rainfall, while at Chulucanas (mean annual rainfall 310 mm) there was 3803 mm. Major floods resulted (Magilligan and Goldstein, 2001). The ENSO phenomenon also affects tropical cyclone activity. As Landsea (2000: 149) remarked, 'Perhaps the most dramatic effect that El Niño has upon the climate system is changing tropical cyclone characteristics around the world.' In some regions, an El Niño phase brings increases in tropical cyclone formation (e.g., the South Pacific and the North Pacific between 140°W and 160°E), while others tend to see decreases (e.g., the North Atlantic, the Northwest Pacific, and the Australian region). La Niña phases typically bring opposite conditions. Landsea sees a variety of reasons why ENSO should relate to cyclone activity: modulation of the intensity of the local monsoon trough, repositioning of the location of the monsoon trough, and alteration of the tropospheric vertical shear.

The differences in cyclone frequency between El Niño and La Niña years are considerable (Bove et al., 1998). For example, the probability of at least two hurricanes striking the USA is 28% during El Niño years, 48% during neutral years and 66% during La Niña years. There can be very large differences in hurricane landfalls from decade to decade. In Florida, for instance, over the period 1851–1996, the number of hurricane landfalls ranged from three per decade (1860s and 1980s) to 17 per decade (1940s) (Elsner and Kara, 1999). Given the importance of hurricanes for slope, channel, and coastal processes, changes of this type of magnitude have considerable geomorphologic significance. Mangroves, for example, are highly susceptible to hurricanes, being damaged by high winds and surges (Doyle and Girod, 1997).

The IPCC report (2001: 606) concludes thus on tropical cyclones:

There is some evidence that regional frequencies of tropical cyclones may change but none that their locations may change. There is also evidence that peak intensity may increase by 5% to 10% and precipitation rates may increase by 20% to 30%. There is a need for much more work in this area to provide more robust results.

Runoff response

Studies of the sensitivity of runoff to climate changes have tended to indicate that annual runoff volume is more sensitive to changes in precipitation than to changes in potential evapotranspiration and that a given percentage change in precipitation results in a greater percentage change in runoff (Arnell, 1996: 99; Najjar, 1999), with arid catchments showing a greater sensitivity than humid climates. As Table 10.1 shows, an increase in annual precipitation of 10% is enough to offset the higher evaporation associated with a 2°C rise in temperature. The effects of increasing or decreasing precipitation are greatly amplified in those catchments with the lowest runoff coefficients (Pease, Nzoia, and Saskatchewan).

Another key aspect of the runoff response to climate change is that, as historical records show, higher average annual precipitation leads not only to higher streamflow but also to higher flood discharges. In Canada, for example, Ashmore and Church (2001) found that in the Southern Prairies and the Atlantic coast the magnitude of large floods (with a 10 year recurrence interval) increases by up to 50–100% for only 5–15% increases in annual precipitation. Flood discharges increase proportionately much more than mean flows.

An attempt to map future runoff trends on a global basis has been made by Arnell (2002). What is striking in this work is the large range there is in responses. Some areas will become very markedly prone to greatly reduced annual runoff, while others will see an enhancement of flows. The degree of change will vary substantially according to the levels of CO_2 in the atmosphere and the consequent amount of temperature change. However, the patterning at a global scale indicates that by the 2080s high latitudes in the Northern Hemisphere, together with parts of Central Africa and Central Asia, will have higher annual runoff levels, whereas Australia, southern Africa, northwest India, the Middle East and the Mediterranean basin will show reduced runoff levels. There is some tendency, to which the Taklemakan of Central Asia appears to be a major exception, for some of the major deserts (e.g., Namib,

Table 10.1 Percentage change in annual runoff with different changes in annual precipitation and an increase in temperature of 2°C. Source: From various sources in Arnell (1996, table 5.1)

River	Location	Percentage change in precipitation				
		−20	−10	0	10	20
White	Colorado basin, USA	−23	−14	−4	7	19
East	Colorado basin, USA	−28	−19	−9	1	12
Animas	Colorado basin, USA	−26	−17	−7	3	14
Upper Colorado	Colorado basin, USA		−23	−12	1	
Lower Delaware	USA	−23		−5		−12
Saskatchewan	Canada	−51	−28	−15	11	40
Pease	Texas, USA	−76	−47	−12	40	100
Leaf	Mississippi, USA	−50	−30	−8	−16	40
Nzoia	Kenya	−65	−44	−13	17	70
Mesohora	Greece	−32	−18	−2	11	25
Pyli	Greece	−25	−13	−1	13	25
Jardine	Northeast Australia	−32	−17	−1	11	28
Corang	Southeast Australia	−38	−21	−3	16	33

Kalahari, Australian, Thar, Arabian, Patagonian, and North Sahara) to become even drier.

Cold regions

In the Arctic, overall the transport of moisture is expected to increase as the atmosphere warms, leading to a general increase in river discharges. Peterson et al. (2002) have analyzed discharge records for the six largest Eurasian rivers that discharge into the Arctic Ocean and have shown that as surface air temperatures have increased so has the average annual discharge of freshwater. It increased by 7% from 1936 to 1999. They suggest that with increased levels of warming (1.5 to 5.8°C by 2100) there would be an 18 to 70% increase in Eurasian Arctic river discharge over present conditions.

However, there are many factors to be considered in an analysis of the response of cold region hydrologic systems to climate change (Table 10.2), and good reviews are provided by Woo (1996) and Rouse et al. (1997). Prowse and Beltaos (2002) review the effects that climate change could have on ice jams and thus on floods and low flows.

The peatlands of cold temperate regions are another landscape type that may be impacted by global change, although uncertainties exist about how they will respond (Moore, 2002). On the one hand, higher temper-

Table 10.2 Potential hydrologic changes in permafrost regions. Source: from information in Woo (1996)

Increased precipitation
Reduced proportion of snow, increased proportion of rain
Earlier melting of snow packs
More groundwater storage in deeper active layers
Increased summer evaporative losses due to higher temperatures or replacement of lichens and mosses by transpiring plants
Less overland flow across impermeable permafrost
Degradation of permafrost will add water during the transient phase of climatic change
Changes in timing, extent and duration of ice jams
Increase in surface ponding due to thermokarst development
Subterranean flow conduits may reopen

atures will tend to enhance plant productivity and organic matter accumulation but, on the other, will also lead to faster decomposition rates. Equally any decrease in water availability in the summer could cause peat loss and bog contraction.

Changes in runoff in the UK

Arnell (1996) has studied the likely changes that will occur in the UK as a result of global warming. With higher winter rainfalls and lower summer rainfalls (particularly in the southeast) he forecasts for the 2050s:

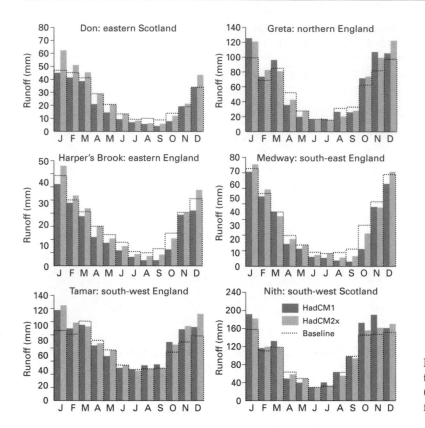

Figure 10.3 Monthly runoff by the 2050s under two scenarios for six British catchments (modified from Arnell and Reynard, 2000, figure 7.19).

1 an increase in average annual runoff in the north of Britain of between 5 and 15%;
2 a decrease in the south of between 5 and 15%, but up to 25% in the southeast;
3 an increased seasonal variation in flow, with proportionately more of the total runoff occurring during winter;
4 high flows increased in northern catchments and decreased in the south.

Figure 10.3 shows the change in monthly streamflow under two climatic scenarios for six British catchments. As can be seen, in southern England, lower summer rainfall, coupled with increased evaporation, means that streamflows decrease during summer, whereas in the catchments in northern Britain, streamflow increases in winter, but changes little in summer.

Werritty (2002) has looked specifically at Scottish catchments and has noted a particularly marked increase in precipitation in the north and west in the winter half of the year. He predicts that by the 2050s Scotland as a whole will become wetter than at present, and that average river flows will increase, notably in

the autumn and winter months. He believes that high flows could become more frequent, increasing the likelihood of valley floor inundation. However, a study of the history of valley floors in upland Scotland suggests that they are relatively robust, so that although they may become subject to extensive reworking they are unlikely to undergo large-scale destabilization (Werritty and Lees, 2001).

Europe

Arnell (1999b) has modeled potential changes in hydrologic regimes for Europe, using four different GCM-based climate scenarios. While there are differences between the four scenarios, each indicates a general reduction in annual runoff in Europe south of around 50°N and an increase polewards of that. The decreases could be as great as 50% and the increases up to 25%. The proposed decrease in annual runoff in southern Europe is also confirmed by Menzel and Burger's (2002) work in Germany, who also suggest that peak flows will be reduced very substantially.

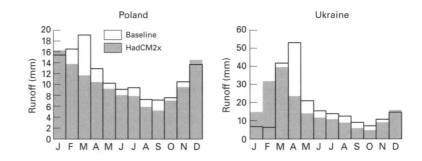

Figure 10.4 Effect of a climate change scenario on streamflow in two European snow-affected catchments by the 2050s (modified from Arnell, 2002, figure 7.20).

Rising temperatures will affect snowfall and when it melts (Seidel et al., 1998). Under relatively mild conditions, even a modest temperature rise might mean that snow becomes virtually unknown, so that the spring snowmelt peak would be eliminated. It would be replaced by higher flows during the winter. Under more extreme conditions, all winter precipitation would still fall as snow, even with a rise in temperature. As a consequence, the snowmelt peak would still occur, although it might occur earlier in the year. In Figure 10.4, for instance, in the Polish example the snowmelt peak is eliminated by the 2050s, whereas in the Ukraine example it is brought forward.

One of the most important rivers in Europe is the Rhine. It stretches from the Swiss Alps to the Dutch coast and its catchment covers 185,000 km^2 (Shabalova et al., 2003). Models suggest that the Rhine's discharge will become markedly more seasonal, with mean discharge decreases of about 30% in summer and increases by about 30% in winter, by the end of the century. The increase in winter discharge will be caused by a combination of increased precipitation, reduced snow storage, and increased early melt. The decrease in the summer discharge is related mainly to a predicted decrease in precipitation combined with increases in evapotranspiration. Glacier melting in the Alps also contributes to the flow of the Rhine. Once these glaciers have wasted away, this contribution will diminish sharply.

Other examples

Zambezi

A study of the Zambezi River in central Africa, using different GCMs and projecting conditions for 2080, indicates that river flow may decline substantially.

Figure 10.5 The Zambezi River at Victoria Falls.

Simulations indicate that for three scenarios the annual flow levels at Victoria Falls (Figure 10.5) reduce between 10 and 35.5% (Harrison and Whittington, 2002).

Susquehanna, eastern USA

The Susquehanna, which flows into Chesapeake Bay, the largest estuary in the USA, is an example of a river system that will enjoy larger annual flows because of increased precipitation over its catchment (Najjar, 1999). For a 17.5% increase in annual precipitation, and a temperature increase of 2.5°C, the total predicated increase in annual streamflow is 24%.

California

One of the most pronounced features of some recent GCMs (Hadley Centre and Canadian) is that they show a projected increase in precipitation for California and the southwest. This would be the result of a warmer

Table 10.3 Predicted change in climate variables for different GCM scenarios
(a) Climate change scenarios for the 2080s (relative to 1961–1990 mean)

	HadCM2	HadCM2-S	ECHAM4
Precipitation change (%)	−12.5	−17.6	−1.6
Temperature change (°C)	+5.3	+4.4	+5.0

(b) Seasonal changes in runoff under GCM scenarios

GCM scenario	Runoff change (%)		
	Annual	Wet season (January–July)	Dry season (August–December)
ECHAM4	−10.0	−9.5	−12.1
HadCM2	−28.3	−28.2	−28.9
HadCM2-S	−35.5	−36.1	−32.6

(c) Estimated change in runoff in California (from Smith et al. 2001, table 1) using Canadian and Hadley GCMs*

Historical runoff (mm yr⁻¹), 1961–90	Change in annual runoff (mm yr⁻¹), 2025–2034		Change in annual runoff (mm yr⁻¹), 2090–2099	
	Canadian	Hadley	Canadian	Hadley
232	60	63	320	273

*The Hadley model for California proposes for the 2090s a 4.9°C increase in winter temperatures and a 4.6°C increase in summer temperatures. Annual precipitation will increase by 30%. The Canadian model equivalents are 7.1°C and 4.3°C and c. 70%.

Pacific Ocean causing an increase mainly in wintertime precipitation (MacCracken et al., 2001). Smith et al. (2001), using the Hadley and Canadian models, estimate that California runoff will increase by the 2030s by about three-fifths and double by the 2090s (Table 10.3).

Pacific northwest of USA

Wigmosta and Leung (2002) modeled the response of the American River in the Pacific northwest of the USA. More winter precipitation falling as rain rather than snow, and also leading to more rain-on-snow

events, produced a future with greater winter flooding. However, the reduced snowpack caused fewer flows in the spring and reduced flows in summer.

Bangladesh

In a warmer world it is probable that there will be a general increase in precipitation, caused by enhancement in summer monsoon activity. This could be a cause of increased flood risk in areas such as Bangladesh.

In Table 10.4, from the work of Mirza (2002), four GCM scenarios are presented for changes in temperature and precipitation for three major catchments that create floods in Bangladesh. In addition, predicted mean peak discharges for those scenarios are presented. The current peak discharges (cubic meters per second) are 54,000 for the Ganges, 67,000 for the Brahmaputra and 14,000 for the Meghna. The four GCMs display some differences, as do predicted peak discharge values. However, overall the discharge values show a range from a modest decline to a substantial increase.

Geomorphologic consequences of hydrologic change

It is likely that changes in river flow will cause changes in river morphology, particularly in sensitive systems, which include fine-grained alluvial streams. Bedrock streams will probably be less sensitive. Ashmore and Church (2001: 41) summarize some of the potential effects:

The potential impacts of increased discharge include channel enlargement and incision, a tendency toward either higher sinuosity single channels or braided patterns, increased bank erosion, and more rapid channel migration. Increased magnitude of large floods will result in sudden changes to channel characteristics that may trigger greater long-term instability of rivers. Increased frequency of large floods will tend to keep rivers in the modified and unstable state. Decreased discharge often results in channel shrinkage, vegetation encroachment into the channel, sedimentation in side channels, and channel pattern change toward more stable, single-channel patterns. In entrenched or confined valleys there may be reductions in the stability of the valley walls

Table 10.4 Changes in precipitation under various warming scenarios and corresponding mean peak discharge for three south Asian rivers. Source: Mirza (2002), with modifications

GCM	ΔT (°C)	Ganges			Brahmaputra			Meghna		
		ΔP (%)	Mean peak discharge ($m^3 s^{-1}$)	*	ΔP (%)	Mean peak discharge ($m^3 s^{-1}$)	*	ΔP (%)	Mean peak discharge ($m^3 s^{-1}$)	*
CSIRO9	2	8.5	57,790	107	−0.5	64,853	97	4.5	15,171	108
	4	17.0	62,900	116	−1.0	64,840	97	9.0	16,267	116
	6	25.5	68,010	126	−1.5	64,827	97	13.5	17,378	124
HadCM2	2	−2.8	50,963	94	7.2	70,308	105	9.7	16,017	114
	4	−5.6	49,240	91	14.4	75,757	113	19.4	17,971	128
	6	−8.4	47,523	88	21.6	81,199	121	29.1	19,927	142
GFDL	2	4.6	55,419	103	7.2	67,487	101	11.3	16,844	120
	4	9.2	58,159	108	14.4	70,107	105	22.6	19,614	140
	6	13.8	60,898	113	21.6	72,728	109	33.9	22,412	160
LLNL	2	0.5	52,996	98	1.4	65,385	98	7.7	15,958	114
	4	1.0	53,312	99	2.8	65,904	98	15.4	17,842	127
	6	1.5	53,628	99	4.2	66,423	99	23.1	19,740	141

* = Percentage of present peak discharge value.

and, hence, increases in the rate of erosion caused by a greater tendency for streams to erode the valley walls. Increased valley-side erosion will increase sediment delivery to the streams with consequences for stream morphology.

In the southeast of England, Collison et al. (2000) have also sought to model the impact of climate change on the Lower Greensand escarpment of Kent. Using the Hadley Centre GCM (HadCM2), combined with downscaling and a GIS-based combined slope hydrology/stability model, they found that because increases in rainfall would largely be matched by increases in evapotranspiration the frequency of large landslides would be unchanged over the next 80 years. They argued that other factors, such as land-use change and human activity, would be likely to have a greater impact than climate change.

Mass movements on slopes will also be impacted upon by changes in hydrologic conditions. In particular, slope stability and landslide activity are greatly influenced by groundwater levels and pore-water pressure fluctuation in slopes.

Various attempts have been made to model potential slope responses. For example, in the Italian Dolomites, Dehn et al. (2000) suggested that future landslide activity would be reduced because there would be less storage of precipitation as snow. Therefore, the release of meltwater, which under present conditions contributes to high groundwater levels and strong landslide displacement in early spring, would be significantly diminished. However, because of the differences between GCMs and problems of downscaling, there are still great problems in modeling future landslide activity, and Dehn and Buma (1999) found that the use of three different GCM experiments for the assessment of the activity of a small landslide in southeast France did not show a consistent picture of future landslide frequencies.

Geomorphologists are starting to model changes in soil erosion that may occur as a consequence of changes in rainfall (Yang et al., 2003), although it is difficult to determine the likely effects of climate change compared with future land-use management practices (Wilby et al., 1997). For example, Sun et al. (2002) calculated runoff erosivity changes for China and Nearing (2001) for the USA using the Revised Soil Loss Equation (RSLE) of Renard and Freid (1994). They adopted the UKMO Hadley climate scenario for their China study and produced a map of rainfall erosivity for 2061–2099 and suggested that for China as a whole, assuming current land cover and land management conditions, the soil erosion rates will increase by 37–93%

Table 10.5 Possible consequences of future climate change for building stone decay in the UK. Source: Viles (2002, table 3)

Region (climate change)	Dominant building stone type	Process responses	Other threats	Overall response
Northwest (warmer, wetter winters)	Siliceous sandstones and granites	Increased chemical weathering. Less freeze–thaw weathering. More organic growths contributing to soiling	Increased storm activity may cause episodic damage. Increased wave heights may encourage faster weathering in coastal areas. Increased flooding may encourage decay	Enhanced chemical decay processes and biological soiling, reduced physical decay processes
Southeast (warmer, drier summers)	Limestones	Less freeze–thaw weathering. Reduced chemical weathering as a result of less available water. Increased salt crystallization in summer. More deteriorating organic growths	Increased drying of soils (especially clay-rich soils) will encourage subsidence and building damage. Low-lying coastal areas will be particularly prone to marine encroachment. Increased drought frequency may encourage decay	Enhanced physical and biological weathering, more dust for soiling, reduced chemical weathering

across China. For Brazil, Favis-Mortlock and Guerra (1999) used the Hadley Centre HadCM2 GCM and erosion model (WEPP – Water Erosion Predictions Project). They found that by 2050 the increase in mean annual sediment yield in their area in the Mato Grosso would be 27%. For the southeast of the UK, where winter rainfall is predicted to increase, albeit modestly, Favis-Mortlock and Boardman (1995) recognized that changes in rainfall not only impacted upon erosion rates directly but also indirectly, through their effects on rates of crop growth and on soil properties. Nonetheless, they showed that erosion rates were likely to rise, particularly in wet years.

Weathering

Remarkably little work has been done on how weathering rates of rock will change with changes in climate. Two attempts, however, deserve mention. The first of these is the study by Viles (2002) on stone deterioration in the UK. There she contrasted the likely impacts of climate change on building stones between the northwest and the southeast of the country (Table 10.5). She argued that in the northwest, which would have warmer and wetter winters, chemical weathering's importance might be increased, whereas in the southeast,

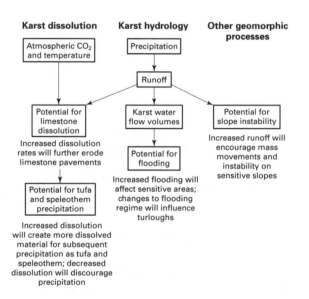

Figure 10.6 Conceptual model of the impacts of effective precipitation, air temperature, and atmospheric CO_2 concentration on karst dissolution, hydrology, and other geomorphic processes. (Source: Viles, 2003, figure 1.)

which would have warmer, drier summers, processes such as salt weathering might become more important. Viles (2003) has also tried to model the impacts of climate change on limestone weathering and karstic processes in the UK (Figure 10.6), but the model has wider significance and could be applied to karst areas

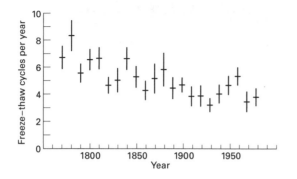

Figure 10.7 Decadal means of the freeze–thaw cycles in central England. (Source: Brimblecombe and Camuffo, 2003, figure 3.)

worldwide. Finally, as the climate warms, the number of freeze–thaw cycles to produce frost weathering in areas such as central England are likely to decrease, as they did during the warming of the nineteenth and twentieth centuries (Figure 10.7) (Brimblecombe and Camuffo, 2003).

Points for review

Will rainfall amounts and intensities change in a warming world? Where may these be greatest?

What are the likely ways in which rivers in (a) periglacial and (b) arid environments respond to global warming?

To what extent may rates of soil erosion, mass movements, and rock weathering change in coming decades?

Guide to reading

Arnell, N. W., 2002, *Hydrology and global environmental change*. Harlow: Prentice Hall. A discussion of the responses of river systems to climate change.

Sidle, R. C. (ed.), 2002, *Environmental change and geomorphic hazards in forests*. Wallingford: CABI. A consideration of the ways in which mass movements and other hazards may change in the future.

11 THE FUTURE: THE CRYOSPHERE

The nature of the cryosphere

The cryosphere is that part of Earth's surface or sub-surface environment that is composed of water in the solid state. It includes snow, sea ice, the polar ice sheets, mountain glaciers, river and lake ice, and permafrost (permanently frozen subsoil).

The cryosphere contains nearly 80% of all Earth's freshwater. Perennial ice covers about 11% of Earth's land surface and 7% of the world's oceans, while permafrost underlies about 25% of it. Seasonal snow has the largest area of any component of the global land surface; at its maximum in late winter it covers almost 50% of the land surface of the Northern Hemisphere.

Good background studies on the cryosphere include Knight (1999), and Benn and Evans (1998) on glaciation, and French (1996), Clark (1988) and Washburn (1979) on permafrost and periglacial environments.

Ice sheets are ice masses that cover more than 50,000 km². The Antarctic ice sheet covers a continent that is a third bigger than Europe or Canada and twice as big as Australia. It attains a thickness that can be greater than 4000 m, thereby inundating entire mountain ranges. The Greenland ice sheet only contains 8% of the world's freshwater ice (Antarctica has 91%), but nevertheless covers an area 10 times that of the British Isles. The Greenland ice fills a huge basin that is rimmed by ranges of mountains, and has depressed Earth's crust beneath.

The Antarctic ice sheet (Figure 11.1) is bounded over almost half of its extent by ice shelves. These are floating ice sheets nourished by the seaward extensions of the land-based glaciers or ice streams and by the accumulation of snow on their upper surfaces. Ice-shelf thicknesses vary, and the seaward edge may be in the form of an ice cliff up to 50 m above sea level with 100–600 m below. At its landward edge the Ross Ice Shelf is 1000 m thick. It covers an area greater than that of California.

Ice caps have areas that are less than 50,000 km² but still bury the landscape. The world's alpine or mountain glaciers are numerous, and in all there are probably over 160,000 on the face of the earth. Their total surface area is around 530,000 km².

Permafrost underlies large expanses of Siberia, Canada, Alaska, Greenland, Spitzbergen, and north-

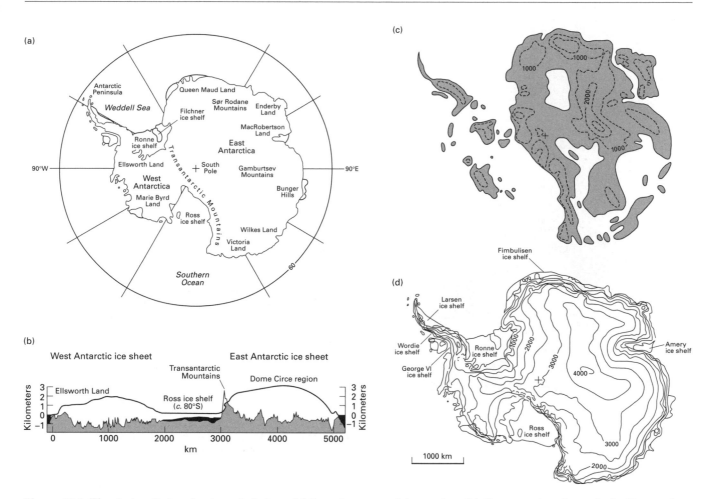

Figure 11.1 The Antarctic ice sheets and shelves. (a) Location map of Antarctica. (b) Cross-section through the East and West Antarctic ice sheets, showing the irregular nature of the bedrock surface, ice thickness, and the floating ice shelves. (c) Subglacial relief (in meters) and sea level. The white areas are below sea level. (d) Surface elevations on the ice sheet in meters.

western Scandinavia (Figure 11.2). Permafrost also exists offshore, particularly in the Beaufort Sea of the western Arctic and in the Laptev and East Siberian Seas, and at high elevation in mid-latitudes, such as the Rocky Mountains of North America and the interior plateaux of central Asia. It occurs not only in the tundra and polar desert environments poleward of the tree line, but also in extensive areas of the boreal forest and forest-tundra environments.

Above the layer of permanently frozen ground there is usually a layer of soil in which temperature conditions vary seasonally, so that thawing occurs when temperatures rise sufficiently in summer but freeze in winter or on cold nights. This zone of freeze–thaw

processes is called the *active layer*. It varies in thickness, ranging from 5 m where unprotected by vegetation to typical values of 15 cm in peat.

Conventionally, two main belts of permafrost are identified. The first is the zone of continuous permafrost; in this area permafrost is present at all localities except for localized thawed zones, or taliks, existing beneath lakes, river channels and other large water bodies which do not freeze to their bottoms in winter. In the discontinuous permafrost zone small-scattered unfrozen areas appear.

Maximum known depths of permafrost reach 1400–1450 m in northern Russia and 700 m in the north of Canada, regions of intense winter cold, short cool

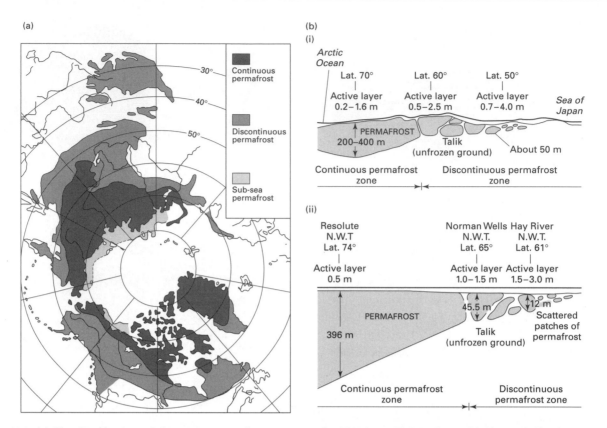

Figure 11.2 (a) The distribution of the main permafrost types in the Northern Hemisphere. (b) Vertical distribution of permafrost and active zones in longitudinal transects through (i) Eurasia and (ii) northern America.

summers, minimal vegetation, and limited snowfall. In general, the thickness decreases equatorwards. Sporadic permafrost tends to occur between the $-1°C$ and $-4°C$ mean annual air temperature isotherms, while continuous permafrost tends to occur to the north of the $-7°C$ to $-8°C$ isotherm.

The various components of the cryosphere are geomorphologically highly important. Glaciers and ice sheets not only produce their own suites of erosional and depositional landforms, but also have numerous indirect effects on phenomena such as sea levels, river flows, and loess (eolian silt deposits). Likewise, snow has direct geomorphologic effects, termed nivation, but also has great significance for streamflow regimes. Frost is an important cause of rock weathering and subsurface permafrost is fundamental in terms of such phenomena as patterned ground, slope stability, runoff, land subsidence, and river and coast erosion. Sea ice also plays a major role through its influence on wave activity level along cold coastlines.

Because of the obvious role of temperature change in controlling the change of state of water to and from the liquid and solid sates, global warming has the potential to cause very major changes in the state of the cryosphere.

The polar ice sheets

Three main consequences of warming may be discerned for ice sheets (Drewry, 1991): ice temperature rise and attendant ice flow changes; enhanced basal melting beneath ice shelves and related dynamic response; and changes in mass balance.

Temperatures of the ice sheets will rise due to the transfer of heat from the atmosphere above. However, because of the slow vertical conduction of heat through the thick ice column, the timescale for this process is relatively long ($10^2–10^3$ years). As the ice warms, it softens and can undergo enhanced deformation, which could increase the discharge of ice into the oceans. However, Drewry argues that it can be discounted as a major factor on a timescale of decades to a century.

Ice shelves would have enhanced basal melt rates if sea-surface temperatures were to rise. This could lead to thinning and weakening of ice shelves (Warner and Budd, 1990). Combined with reduced underpinning from grounding points as sea level rose, this would result in a reduction of backpressure on ice flowing from inland. Ice discharge through ice streams might therefore increase. There are some studies (e.g., De Angelis and Skvarca, 2003) that have found evidence of ice streams increasing their velocities when ice shelves have collapsed, but this is not invariably the case (Vaughan and Doake, 1996). In a wide ranging review, Bennett (2003) has assembled evidence that although ice streams have high velocities and are very dynamic components of the ice sheet environment, and also display highly variable mass balances, collectively they appear to be locally in balance. Indeed, some workers have identified a positive mass balance (Joughlin and Tulaczyk, 2002).

The West Antarctic Ice Sheet (WAIS), because it is a marine ice sheet that rests on a bed well below sea level, is likely to be much more unstable than the East Antarctic Ice Sheet, from which it is separated by the Transantarctic Mountains. If the entire WAIS were discharged into the ocean, the sea level would rise by 5 or 6 m. Fears have been expressed over many years that it could be subject to rapid collapse, with disastrous consequences for coastal regions all over the world.

One extreme view was postulated by Mercer (1978), who believed that predicted increases in temperature at 80°S would start a 'catastrophic' deglaciation of the area, leading to a sudden 5 m rise in sea level. A less extreme view was put forward by Thomas et al. (1979). They recognized that higher temperatures will weaken the ice sheets by thinning them, enhancing lines of weakness, and promoting calving, but they contended that deglaciation would be rapid rather than catastrophic, the whole process taking 400 years or so. Robin (1986) took a broadly intermediate position. He contended (p. 355) that:

A catastrophic collapse of the West Antarctic ice sheet is not imminent, but better oceanographic knowledge is required before we can assess whether a global temperature rise of 3.5°C might start such a collapse by the end of the next century. Even then it is likely to take at least 200 years to raise sea level by another 5 m.

The WAIS has part of its ice grounded on land below sea level, and part in the form of floating extensions called ice shelves that move seaward but are confined horizontally by the rocky coast. The boundary between grounded and floating ice is called the grounding line (Oppenheimer, 1998). In 1974, Weertman suggested that very rapid grounding line retreat could take place in the absence of ice-shelf backpressure. He demonstrated (see Figure 11.3) that if the bedrock slopes downward in the inland direction away from the grounding line, as is the case with some of the WAIS ice streams, then if an ice shelf providing buttressing becomes unpinned owing to either melting or global sea-level rise, the grounded ice would accelerate its flow, become thinner and rapidly float off its bed.

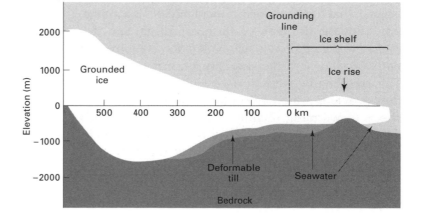

Figure 11.3 Cross-section of an ice stream and ice shelf of a marine ice sheet, indicating location of grounding line, bedrock rise on the ocean floor, and possible extent of deformable till. The thickness of the till layer, actually a few meters, is exaggerated for clarity. Sea-level rise due to collapse of the West Antarctic Ice Sheet (WAIS) was estimated by T. Hughes. (Source: Oppenheimer, 1998, figure 2, p. 327 in *Nature*, volume 393, 28 May 1998 © Macmillan Publishers Ltd 1998.)

Whether the WAIS will collapse catastrophically is still being debated (Oppenheimer, 1998), but there is some consensus that the WAIS will most probably not collapse in the next few centuries (Vaughan and Spouge, 2002). Bentley (1997) is amongst those who have expressed doubts about the imminence of collapse, believing 'that a rapid rise in sea level in the next century or two from a West Antarctic cause could only occur if a natural (not induced) collapse of the WAIS is imminent, the chances of which, based on the concept of a randomly timed collapse on the average of once every 100,000 years, are on the order of 0.1%' (see also Bentley, 1998).

Greenland and Antarctica will react very differently to global warming. This is in part because Antarctica, particularly in its interior, is very cold indeed. This means that the moisture content of the atmosphere over it is very low, ultimately suppressing the amount of snow and ice accumulation that can occur. An increase in temperature may therefore lead to more precipitation and greater accumulation of ice. With Greenland, the situation is different. Whereas in Figure 11.4 Antarctica lies to the left of the mass balance maximum, much of Greenland lies to the right of the maximum so that the total surface mass balance will decrease in the event of global warming (Oerlemans, 1993: 153). Indeed, Gregory et al. (2004) have argued that if the average temperature in Greenland increases by more than about 3°C the ice sheet would be eliminated, causing the global average sea-level to rise by 7 m over a period of 1000 years or more.

However, the state of the Greenland ice sheet does not solely depend on the rate of surface ablation. Flow of the ice is enhanced by rapid migration of surface meltwater to the ice–bedrock interface, and studies have shown that ice accelerates in summer (when melt occurs), that there is a near coincidence of ice acceleration with the duration of surface melting, and that interannual variations in ice acceleration are correlated with variations in the intensity of surface melting. This coupling between surface melting and ice-sheet flow provides a mechanism for rapid response of ice sheets to warming (Zwally et al., 2002; Parizek and Alley, 2004). Some confirmation of this is provided by laser altimeter surveys, which show rapid ice thinning below 1500 m (Krabill et al., 1999; Rignot and Thomas, 2002). They argue that the glaciers are thinning as a result of increased creep rates brought about

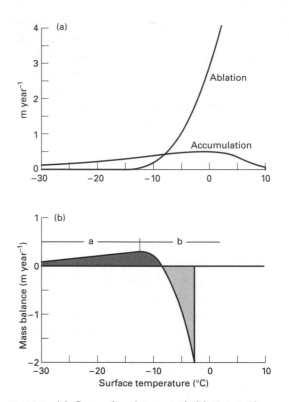

Figure 11.4 (a) Generalized curve of ablation and accumulation in relation to surface temperature. In warmer areas, such as Greenland, an increase in temperature may lead to rapid increases in ablation. (b) Generalized curve of the mass balance of ice sheets in relation to surface temperature. In the very cold environment of Antarctica, an increase in temperature may lead to increased accumulation of snow and ice and hence to a positive mass balance, whereas in the warmer environment of Greenland, increased rates of ablation may lead to a negative mass balance. (Source: Oerlemans, 1993, figure 9.4.)

by decreased basal friction consequent upon water penetrating to the bed of the glacier. They see this as a mechanism for transfer of ice-sheet mass to the oceans that is potentially larger than could be achieved by surface melting alone.

Valley glaciers and small ice caps

Together the present Greenland and Antarctic ice sheets contain enough water to raise sea level by almost 70 m. By contrast, valley glaciers and small ice caps contain only a small amount – c. 0.5 m of sea-level equivalent. However, because of their relatively short response times, associated with large mass turnover, they contribute significantly to sea-level fluctuations on a century timescale. Indeed, there is a clear observational record that valley glaciers are sensitive to climate change, and we have many examples of changes in glacier length of several kilometers since the end of the Little Ice Age (Oerlemans et al., 1998).

Since the nineteenth century, many of the world's alpine glaciers have retreated up their valleys as a consequence of the climatic changes, especially warming, that have occurred in the past 100 or so years since the ending of the Little Ice Age (Oerlemans, 1994). Studies of the changes of snout positions obtained from cartographic, photogrammetric, and other data therefore permit estimates to be made of the rate at which retreat can occur (Figure 11.5). The rate has not been constant, or the process uninterrupted. Indeed, some glaciers have shown a tendency to advance for some of the period. However, if one takes those glaciers that have shown a tendency for a fairly general retreat (Table 11.1) it becomes evident that as with most geomorphologic phenomena there is a wide range of values, the variability of which is probably related to such variables as topography, slope, size, altitude,

Figure 11.5 The Minapin Glacier snout in the Karakoram Mountains of Hunza, Pakistan. Behind the cultivated fields in the middle distance is a mass of moraine, which was deposited by the glacier in the Little Ice Age. The glacier has now retreated up its valley.

Table 11.1 Retreat of glaciers in meters per year in the twentieth century. Source: tables, maps, and text in Grove (1988)

(a) Individual glaciers

Location	Period	Rate
Breidamerkurjökull, Iceland	1903–48	30–40
	1945–65	53–62
	1965–80	48–70
Lemon Creek, Alaska	1902–19	4.4
	1919–29	7.5
	1929–48	32.9
	1948–58	37.5
Humo Glacier, Argentina	1914–82	60.4
Franz Josef, New Zealand	1909–65	40.2
Nigardsbreen, Norway	1900–70	26.1
Austersdalbreen, Norway	–	21
Abrekkbreen, Norway	–	17.7
Brikdalbreen, Norway	–	11.4
Tunsbergdalsbreen, Norway	–	11.4
Argentière, Mont Blanc, France	1900–70	12.1
Bossons, Mont Blanc France	1900–70	6.4
Oztal Group, Switzerland	1910–80	3.6–12.9
Grosser Aletsch, Switzerland	1900–80	52.5
Carstenz, New Guinea	1936–74	26.2

(b) Region/country. Source: Oerlemans (1994)

Region	Period	Mean rate
Rocky Mountains	1890–1974	15.2
Spitzbergen	1906–1990	51.7
Iceland	1850–1965	12.2
Norway	1850–1990	28.7
Alps	1850–1988	15.6
Central Asia	1874–1980	9.9
Irian Jaya	1936–1990	25.9
Kenya	1893–1987	4.8
New Zealand	1894–1990	25.9

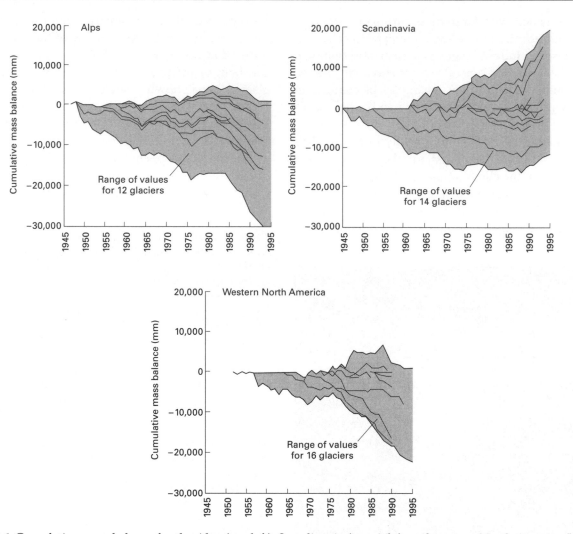

Figure 11.6 Cumulative mass balance for the Alps (top left), Scandinavia (top right), and western North America (bottom).

accumulation rate, and ablation rate. It is also evident, however, that rates of retreat can often be very high, being of the order of 20 to 70 m per year over extended periods of some decades in the case of the more active examples. It is therefore not unusual to find that over the past 100 or so years alpine glaciers in many areas have managed to retreat by some kilometers.

Fitzharris (1996: 246) suggested that since the end of the Little Ice Age, the glaciers of the European Alps have lost about 30 to 40% of their surface area and about 50% of their ice volume. In Alaska (Arendt et al., 2002), glaciers appear to be thinning at an accelerating rate, which in the late 1990s amounted to 1.8 m per year. In China, monsoonal temperate glaciers have lost an amount equivalent to 30% of their modern glacier

area since the maximum of the Little Ice Age. Not all glaciers have retreated in recent decades (Figure 11.6) and current glacier tendencies for selected regions are shown. These are expressed in terms of their mass balance, which is defined as the difference between gains and losses (expressed in terms of water equivalent). In the European Alps (top left) a general trend toward mass loss, with some interruptions in the mid-1960s, late 1970s, and early 1980s, is observed. In Scandinavia (top right), glaciers close to the sea have seen a very strong mass gain since the 1970s, but mass losses have occurred with the more continental glaciers. The mass gain in western Scandinavia could be explained by an increase in precipitation, which more than compensates for an increase in ablation caused by rising temperatures. Western North America (bottom) shows a general mass loss near the coast and in the Cascade Mountains (Hoelzle and Trindler, 1998).

The positive mass balance (and advance) of some Scandinavian glaciers in recent decades, notwithstanding rising temperatures, has been attributed to increased storm activity and precipitation inputs coincident with a high index of the NAO in winter months since 1980 (Nesje et al., 2000; Zeeberg and Forman, 2001). In the case of Nigardsbreen (Norway), there is a strong correlation between mass balance and the NAO index (Reichert et al., 2001). A positive mass balance phase in the Austrian Alps between 1965 and 1981 has been correlated with a negative NAO index (Schoner et al., 2000). Indeed, the mass balances of glaciers in the north and south of Europe are inversely correlated (Six et al., 2001).

Glaciers that calve into water can show especially fast rates of retreat. Indeed, retreat rates of greater than 1 km per year are possible (Venteris, 1999). In Patagonia, in the 1990s, retreat rates of up to 500 m per year were observed. This rapid retreat is accomplished by iceberg calving, with icebergs detaching from glacier termini when the ice connection is no longer able to resist the upward force of flotation and/ or the downward force of gravity. The rapid retreat is favored by the thinning of ice near the termini, its flotation, and its weakening by bottom crevasses.

The Columbia tidewater glacier in Alaska retreated around 13 km between 1982 and 2000. Equally the Mendenhall Glacier, which calves into a proglacial lake, has displayed rapid rates of retreat, with 3 km of terminus retreat in the twentieth century (Motyka et al., 2002).

Certainly, calving permits much larger volumes of ice to be lost from the glacier than would be possible through surface ablation (van der Wee, 2002) and great phases of calving in the past help to explain the **Heinrich events** of the Pleistocene and the rapid demise of the Northern Hemisphere ice sheets at the end of the last glacial period.

Some glaciers at high-altitude locations in the tropics have also displayed fast rates of decay over the past 100 or so years, to the extent that some of them now only have around one-sixth to one-third of the area they had at the end of the nineteenth century (Table 11.2) (Kaser, 1999). At present rates of retreat the glaciers and ice caps of Mount Kilimanjaro in East

Table 11.2 Area of tropical glaciers. Source: from data in Kaser (1999, table 2)

Location	Date	Area (km²)	Date	Area (km²)	%
Irian Jaya, Indonesia	c. 1850	19.3	c. 1990	3.0	15.5
Mount Kenya, Kenya	1899	1.563	1993	0.413	26.4
Lewis Glacier, Kenya	1899	0.63	1993	0.20	31.7
Kibo (Kilimanjaro), Tanzania	c. 1850	20.0	1989	3.3	16.5
Ruwenzori, Uganda	1906	6.509	c. 1990	1.674	25.7

Africa will have disappeared by 2010 to 2020. In the tropical Andes, the Quelccaya ice cap in Peru also has an accelerating and drastic loss over the past 40 years (Thompson, 2000). In neighboring Bolivia, the Glacier Chacaltya lost no less than 40% of its average thickness, two-thirds of its volume and more than 40% of its surface area between 1992 and 1998. Complete extinction is expected within 10 to 15 years (Ramirez et al., 2001).

The reasons for the retreat of the East African glaciers include not only an increase in temperature, but also important may have been a relatively dry phase since the end of the nineteenth century (which led to less accumulation of snow) and a reduction in cloud cover (Mölg et al., 2003).

The fast retreat of glaciers can have a series of adverse geomorphologic effects. Lakes in areas such as the Himalayas of Nepal and Bhutan are rapidly expanding as they are fed by increasing amounts of meltwater. Glacial lake outburst floods are extremely hazardous. Likewise, as slopes are deprived of the buttressing effects of glaciers they can become unstable, generating a risk of increased landsliding and debris avalanches (Kirkbride and Warren, 1999). Increased rates of melting may, for a period of years, cause an increased incidence of summer meltwater floods, but when the glaciers have disappeared, river flow volumes may be drastically reduced (Braun et al., 2000).

Predicted rates of glacier retreat

The monsoonal temperate glaciers of China, which occur in the southeastern part of the Tibetan Plateau, are an example of glaciers that will suffer great

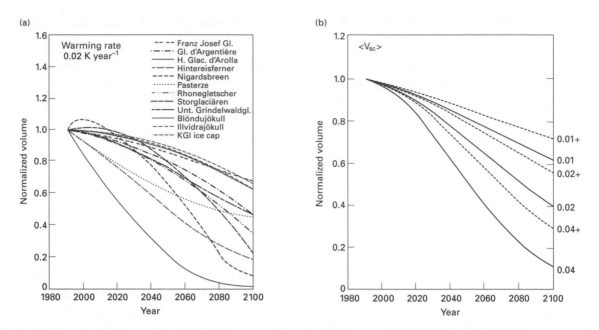

Figure 11.7 Changes in glacier volume 1980–2100 as modeled by Oerlemans et al. (1998) for 12 glaciers: Franz Josef Glacier (New Zealand), Glacier d'Argentière (France), Haut Glacier d'Arolla (Switzerland), Hintereisferner (Austria), Nigardsbreen (Norway), Pasterze (Austria), Rhonegletscher (Switzerland), Storglaciären (Sweden), Unt. Grindelwaldgl. (Switzerland), Blondujökull (Iceland), Illvidrajökull (Iceland), ice cap (Antarctica). (a) Ice-volume change with a warming rate of 0.02 K year⁻¹ without a change in precipitation. Volume is normalized with the 1990 volume. (b) Scaled ice-volume change for six climate change scenarios: + refers to an increase in precipitation of 10% per degree warming.

shrinkage if temperatures rise. Indeed, Su and Shi (2002) have, using modeling studies, calculated that if temperatures rise by 2.1°C, these glaciers will lose about 9900 km² of their total area of 13,203 km² (c. 75%) by the year 2100. The subpolar and polar glaciers of China will shrink by c. 20% over the same period (Shi and Liu, 2000).

Oerlemans et al. (1998) attempted to model the responses of 12 glaciers (from Europe, New Zealand, and Antarctica) to three warming rates (0.01, 0.02, and 0.04 K per year), with and without concurrent changes in precipitation (Figure 11.7). For a warming rate of 0.04 K per year, without an increase in precipitation, they believed that few glaciers would survive until 2100. On the other hand, with a low rate of warming (0.01 K per year) and an increase in precipitation of 10% per degree of warming, they predicted that overall loss of volume by 2100 would be 10 to 20% of the 1990 volume. On a global basis, mass balance modeling by Braithwate and Zhang (1999) suggested that the most sensitive glaciers in terms of temperature change would be maritime and tropical glaciers. The latter, as in East Africa, are literally hanging on to high peaks,

and so are very sensitive to quite modest changes in the height of the snowline.

Sea ice

Since the 1970s, sea ice in the Arctic has declined in area by about 30% per decade (Stocker, 2001: 446). Its thickness may also have become markedly less, with a near 40% decrease in its summer minimum thickness, although the data are imperfect (Holloway and Sou, 2002). By contrast, changes in Antarctic sea ice have been described by the IPCC as 'insignificant' (Stocker, 2001: 446). It has proved difficult to identify long-term trends because of the limited length of observations and the inherent interannual variability of Antarctic sea-ice extent, but it is possible that there has been some decline since the 1950s (Curran et al., 2003).

Figure 11.8 shows the decrease in area of Northern Hemisphere sea ice since 1970 (Vinnikov et al., 1999).

Model projections tend to show a substantial future decrease of Arctic sea-ice cover (Stocker, 2001: 446). By 2050 there will be a roughly 20% reduction in

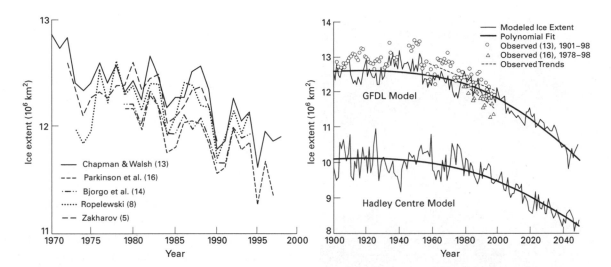

Figure 11.8 Left: observed decrease of Northern Hemisphere sea-ice extent during the past 25 years. Right: observed and modeled variations of annual averages of Northern Hemisphere sea-ice extent. (Source: Vinnikov et al., 1999, figures 1 and 2.)

annual mean Arctic sea-ice extent. A reduction in sea ice area will allow increased absorption of solar radiation so that a further increase in temperature could occur, although this might to some extent be mitigated by increasing cloud cover (Miller and Russell, 2002). However, the IPCC (Anisimov and Fitzharris, 2001: 819) believes that a sea-ice albedo feedback, with amplified warming, means that 'At some point, with prolonged warming a transition to an Arctic Ocean that is ice free in summer – and perhaps even in the winter – could take place.' The IPCC has also recognized that with a doubling of atmospheric greenhouse loadings Antarctic sea ice will undergo a volume reduction of about 25–45%.

Permafrost regions

Regions underlain by permafrost (Figure 11.9) may be especially prone to the effects of global climate change. First of all, the permafrost itself is by its very nature and definition susceptible to the effects of warming. Secondly, the amount of temperature increase predicted for high latitude environments is greater than the global mean, and the loss of permafrost could have a direct effect on atmospheric conditions in the lower atmosphere. Thirdly, permafrost itself is an especially important control of a wide range of geomorphologic processes and phenomena, including slope stability,

rates of erosion, ground subsidence, and surface run-off. Fourthly, the nature (e.g., rain rather than snow) and amount of precipitation may also change substantially. Fifthly, the northern limits of some very important vegetation zones, including boreal forest, shrub-tundra and tundra, may shift latitudinally by some hundreds of kilometers. Changes in snow cover and vegetation type may themselves have a considerable impact on the state of permafrost because of their role in insulating the ground surface. For example, if spring temperatures were to increase and early spring snowfall events were to become rain events, the duration of snow cover would decrease, surface albedo would also decrease, leading to an increase in air temperatures, and the snow might provide less insulation. This could cause relatively rapid permafrost degradation (Ling and Zhang, 2003). Conversely, if the warming were to occur in rather more severe periglacial climates, where temperatures are predicted to remain below or at freezing, the winters could become warmer, and wetter, producing an increase in the longevity and depth of the snowpack. This could lead to more insulation (retarding the penetration of the winter cold wave) and a reduction of warming because of an increase in surface albedo. A good general review of the likely consequences of climate change in the tundra of Canada is provided by Smith (1993).

There is some evidence that permafrost has been degraded by the warming of recent decades. For example,

Figure 11.9 Permafrost in Siberia. The great quantity of ice in the permafrost is illustrated behind the figure standing in the foreground (photographed by the late Marjorie Sweeting).

Kwong and Gan (1994) reported a northward migration of permafrost along the Mackenzie Highway in Canada. Between 1962 and 1988 the mean annual temperature in the area rose by 1°C. Over the same period the southern fringe of the discontinuous permafrost zone moved northwards by about 120 km. In Alaska, Osterkamp and Romanovsky (1999) found that in the late 1980s to mid-1990s some areas experienced warming of the permafrost table of 0.5°C to 1.5°C and that associated thawing rates were about 0.1 m per year. However, thawing rates at the base of the permafrost were an order of magnitude slower, indicating (p. 35): 'that timescales of the order of a century are

required to thaw the top 10 m of ice rich permafrost, which would be primarily responsible for environmental and engineering problems'.

Recent permafrost degradation has also been identified in the Qinghai–Tibet Plateau (Jin et al., 2000), particularly in marginal areas. Both upward and downward degradation has occurred.

Various attempts have been made to assess future extents of permafrost (Figure 11.10). On a Northern Hemisphere basis, Nelson and Anisimov (1993) have calculated the areas of continuous, discontinuous, and sporadic permafrost for the year 2050 (Table 11.3). They indicate an overall reduction of 16% by that date. In Canada, Woo et al. (1992: 297) suggest that if temperatures rise by 4–5°C, as they may in this high-latitude situation, 'Permafrost in over half of what is now the discontinuous zone could be eliminated'. They add,

The boundary between continuous and discontinuous permafrost may shift northwards by hundreds of kilometers but because of its links to the tree line, its ultimate position and the time taken to reach it are more speculative. It's possible that a warmer climate could ultimately eliminate continuous permafrost from the whole of the mainland of North America, restricting its presence only to the Arctic Archipelago.

Barry (1985) estimated that an average northward displacement of the southern permafrost boundary by 150 ± 50 km would be expected for each 1°C warming so that a total *maximum* displacement of between 1000 and 2000 km is possible.

Jin et al. (2000: 397) have attempted to model the response of permafrost to different degrees of temperature rise in the Tibetan Plateau region (Table 11.4) and suggest that with a temperature rise of 2.91°C by the end of the twenty-first century the permafrost area will decrease by 58%.

Degradation of permafrost may lead to an increasing scale and frequency of slope failures. Thawing reduces the strength of both ice-rich sediments and frozen jointed bedrock (Davis et al., 2001). Ice-rich soils undergo thaw consolidation during melting, with resulting elevated pore-water pressures, so that formerly sediment-mantled slopes may become unstable. Equally, bedrock slopes may be destabilized if warming reduces the strength of ice-bonded open joints or leads to groundwater movements that cause pore pressures to rise (Harris et al., 2001). Increases in the

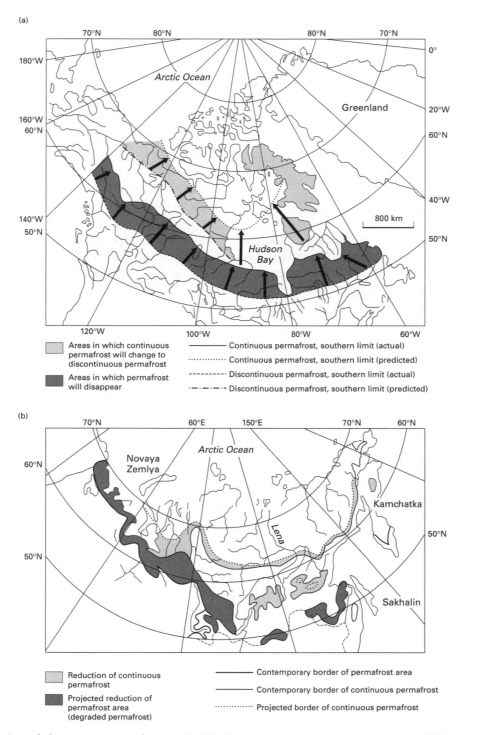

(a)

Areas in which continuous permafrost will change to discontinuous permafrost

Areas in which permafrost will disappear

———— Continuous permafrost, southern limit (actual)

·············· Continuous permafrost, southern limit (predicted)

- - - - - Discontinuous permafrost, southern limit (actual)

- · — · — · - Discontinuous permafrost, southern limit (predicted)

(b)

Reduction of continuous permafrost

Projected reduction of permafrost area (degraded permafrost)

———— Contemporary border of permafrost area

———— Contemporary border of continuous permafrost

·············· Projected border of continuous permafrost

Figure 11.10 Projection of changes in permafrost with global warming for (a) North America and (b) Siberia (after French, 1996, figure 17.5 and Anisimov, 1989).

Table 11.3 Calculated contemporary and future area of
permafrost in the Northern Hemisphere by 2050. Source:
from Nelson and Anisimov (1993)

Zone	Contemporary	2050	Percentage change
Continuous	11.7	8.5	−27
Discontinuous	5.6	5.0	−11
Sporadic	8.1	7.9	−2
Total	25.4	21.4	−16

Table 11.4 Forecast permafrost degradation on the
Qinghai–Tibet Plateau during the next 100 years. Source:
Jin et al. (2000, table 11)

Temperature increase (°C) (year)	Permafrost area (10^6 km²)	Percentage decrease
0.51 (2009)	1.190	8
1.10 (2049)	1.055	18
2.91 (2099)	0.541	58

thickness of the active layer may make more material
available for debris flows.

Likewise, coastal bluffs may be subject to increased
rates of erosion if they suffer from thermal erosion

caused by permafrost decay. This would be accelerated still further if sea ice were to be less prevalent,
for sea ice can protect coasts from wave erosion and
debris removal (Carter, 1987). Local coastal losses to
erosion of as much as 40 m per year have been observed in some locations in both Siberia and Canada
in recent years, while erosive losses of up to 600 m
over the past few decades have occurred in Alaska
(Parson et al., 2001).

Slope failures in cold environments could be exacerbated by glacier recession (Haeberli and Burn, 2002).
The association between glacier retreat since the Little
Ice Age and slope movement processes such as gravi-
tational rock slope deformation, rock avalanches, debris flows, and debris slides is evident. This is the case
where recent ice retreat has removed buttress support
to glacially undercut and oversteepened slopes (Holm
et al., 2004).

One of the severest consequences of permafrost degradation is ground subsidence and the formation of
thermokarst phenomena (Nelson et al., 2001, 2002).
This is likely to be a particular problem in the zones of
relatively warm permafrost (the discontinuous and
sporadic zones), and where the permafrost is rich in
ice (Woo et al., 1992).

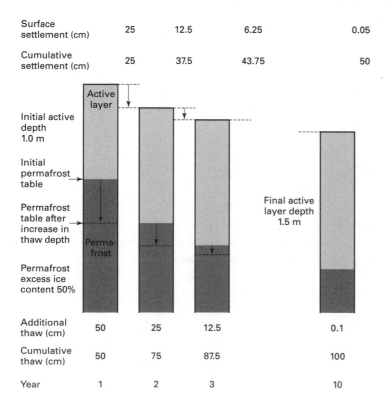

Figure 11.11 Ground settlement in response to a
thickening of the active layer in permafrost with
an excess ice content of 50%. The active layer
increases from 1.0 to 1.5 m and in so doing
1.0 m of permafrost is thawed and the surface
settles by 0.5 m. (Source: Woo et al., 1992,
figure 8.)

The excess ice content determines how much surface settlement will occur as a result of the thawing of permafrost. It is the volume of water that exceeds the space available within the pores when the soil or sediment is thawed (Woo et al., 1992: 297):

$$EX = 100 \times (V_W/V_T)$$

where EX is the excess ice content (%), V_W is the volume of water that is released on thawing the sample (m^3), and V_T is the total volume of the thawed sample (m^3). Thus, to deepen an active layer by 0.5 m, 1.0 m of permafrost with an EX of 50% must thaw and this ultimately will result in surface settlement of 0.5 m (see Figure 11.11).

Thawing ice-rich areas, such as ice-covered pingos, will settle more than those with lesser ice contents, producing irregular hummocks and depressions. Water released from ice melt may accumulate in such depressions. The depressions may then enlarge into thaw lakes as a result of thermal erosion (Harris, 2002), which occurs because summer heat is transmitted efficiently through the water body into surrounding ice-rich material (Yoshikawa and Hinzman, 2003). Once they have started, thaw lakes can continue to enlarge for decades to centuries because of wave action and continued thermal erosion of the banks. However, if thawing eventually penetrates the permafrost, drainage occurs that leads to ponds drying up.

Increasing temperatures are likely to cause the thickness of the seasonally thawed active layer to become deeper. A simulation by Stendel and Christensen (2002) indicated a 30–40% increase by the end of the twenty-first century for most of the permafrost area in the Northern Hemisphere.

Points for review

What are the main components of the cryosphere?
How are valley glaciers likely to respond to global warming?
Will the polar ice sheets respond catastrophically to global warming?
What are the consequences of permafrost melting?

Guide to reading

French, H. M., 1996, *The periglacial environment* (2nd edn). Harlow: Longman. A general text on periglacial areas, but with a summary of the effects of global change.

Harris, C. and Stonehouse, B. (eds), 1991, *Antarctica and climatic change*. London: Belhaven Press. A study of the response of Antarctica to future warming and a consideration of the global implications.

Wadhams, P., Dowdeswell, J. A. and Schofield, A. N. (eds), 1996, *The Arctic and environmental change*. Amsterdam: Gordon and Breach. A collection of papers presented to the Royal Society of London and covering many aspects of cryospheric change in northern regions.

12 THE FUTURE: DRYLANDS

Introduction

Even though arid lands cover one-third of Earth's land surface, remarkably little thought has been given to what might happen to drylands as a result of any potential warming associated with the enhanced greenhouse effect. Indeed, those sections of the reports of the Intergovernmental Panel on Climate Change (IPCC) that deal with deserts are notable for being narrow and for neglecting almost all the most important issues (Bullock and Le Houérou, 1996; Noble and Gitay, 1996). Some of the regional studies are no more satisfactory, and the one on Africa contains one paragraph only on deserts and concludes with the following debatable remark (Zinyowera et al., 1998: 43):

Extreme desert systems already experience wide fluctuations in rainfall and are adapted to coping with sequences of extreme conditions. Initial changes associated with climate change are less likely to create conditions significantly outside present ranges of tolerance; desert biota show very specialized adaptations to aridity and heat, such as obtaining their moisture from fog or dew.

The reality is that arid environments often appear to have been prone to rapid geomorphologic and hydrologic changes in response to apparently modest climatic stimuli, switching speedily from one state to another when a particular threshold is reached (Goudie, 1994).

1 Desert valley bottoms appear to have been subject to dramatic alternations of cut-and-fill during the course of the Holocene and a large literature has developed on the arroyos (see Chapter 6) of the American southwest (see, e.g., Cooke and Reeves, 1976; Balling and Wells, 1990). Schumm has proposed that valley bottom trenching may occur when a critical valley slope gradient is crossed for a drainage area of a particular areal extent (Schumm, 1977) but other examples result from changes in rainfall intensity and grazing pressure.

2 The heads of alluvial fans often display fan head trenching, suggesting recent channel instability. There has been considerable debate in the literature about the trigger for such trenching, be it tectonic, extreme flood events, climatic change, or an inherent consequence of fluctuating sediment–discharge

relationships during the course of a depositional cycle (e.g., Harvey, 1989).

3 Fluvial systems, as, for example, in the drier lands around the Mediterranean Basin, display a suite of terraces which demonstrate a complex record of cut-and-fill during the late Holocene (Van Andel et al., 1990). There has been prolonged debate as to whether the driving force has been climatic change (Vita-Finzi, 1969) or anthropogenic activities (Butzer, 1974).

4 Colluvial sections in subhumid landscapes (including southern Africa) show complex consequences of deposition, stability and incision (Watson et al., 1984). Ongoing luminescence dating is beginning to give an indication of the complexity of chronology.

5 Terminal lake basins (pans, playas, etc.) can respond dramatically, in terms of both their extent and the rapidity of change that can occur, to episodic rain-

fall events in their catchments. The history of Lake Eyre in Australia in the twentieth century bears witness to this fact (Figure 12.1).

6 West African dust-storm activity has shown very marked shifts in the past few decades in response to runs of dry years and increasing land-use pressures (Goudie and Middleton, 1992). The data for Nouakchott (Mauritania) are especially instructive, revealing a sudden acceleration in dust-storm events since the 1960s. A broadly similar picture could be presented for the High Plains of the USA during the 'Dust Bowl years' of the 1930s.

7 Some dune fields have also proved to be prone to repeated fluctuations in deposition and stabilization in the Holocene, and as more ^{14}C and luminescence dates become available the situation is likely to prove to be even more complex than hitherto believed. A

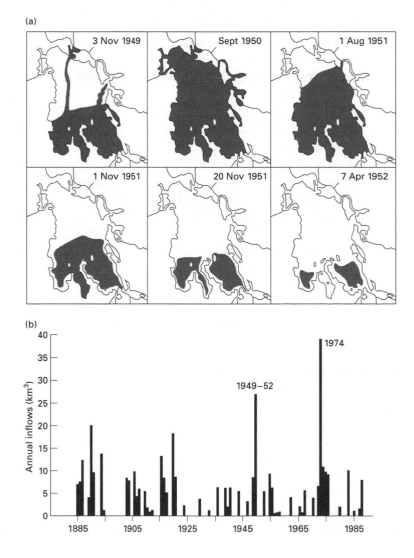

Figure 12.1 The flooding of Lake Eyre. (a) Extent of flooding in 1949–1952 (after Bonython and Mason in Mabbutt, 1977, figure 54). (b) Estimated annual inflows to Lake Eyre North for the period 1885–1989 (based on Kotwicki and Isdale, 1991, figure 2). Both 1949–1952 and 1974 were strong La Niña events. (Source: Viles and Goudie, *Earth-science Reviews*, 2003, volume 61: 105–131.)

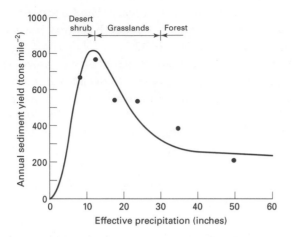

Figure 12.2 Variation of sediment yield with climate as based on data from small watersheds in the USA (after Langbein and Schumm, 1958).

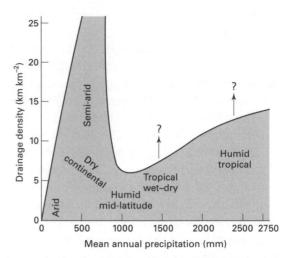

Figure 12.3 Relation between drainage density and mean annual precipitation. (After Gregory, 1976. Copyright © 1976. Reprinted by permission of John Wiley and Sons, Ltd.)

good illustration of this is provided by Gaylord's (1990) work in the Clear Creek area of south-central Wyoming where over the past 7500 years at least four episodes of enhanced eolian activity and aridity are recorded.

8 Various studies have shown how rates of denudation (e.g., Langbein and Schumm, 1958) and drainage density can change very rapidly in semi-arid areas either side of a critical rainfall or precipitation/evapotranspiration (P/E) value related to vegetation cover that constitutes a particular threshold between equilibrium states (Figures 12.2 and 12.3).

This apparent instability and threshold dependence of a range of arid zone landforms, rates, and processes leads one to believe that such areas may be especially susceptible to the effects of global warming, should this occur in the coming decades (Figure 12.4).

Climate changes in the past

Past climatic changes have had dramatic effects on deserts (Goudie, 2002) and this makes it likely that future climatic changes will also have dramatic effects. For example, climatic changes in the Holocene have led to the expansion and contraction of lakes.

In the Sahara there are huge numbers of Holocene pluvial lakes both in the Chotts of Tunisia and Algeria, in Mali (Petit-Marie et al., 1999) and in the south (e.g.,

Mega-Chad). In the Western Desert of Egypt and the Sudan, there are many closed depression or playas, relict river systems, and abundant evidence of prehistoric human activity (Hoelzmann et al., 2001). Playa sediments indicate that they once contained substantial bodies of water, which attracted Neolithic settlers. Many of these sediments have now been radiocarbon dated and indicate the ubiquity of an early to mid-Holocene wet phase, which has often been termed the 'Neolithic Pluvial'. A large lake, 'The West Nubian Palaeolake', formed in the far northwest of Sudan, (Hoelzmann et al., 2001). It was especially extensive between 9500 and 4000 years BP, and may have covered as much as 7000 km². If it was indeed that big, then a large amount of precipitation would have been needed to maintain it – possibly as much as 900 mm compared with the less than 15 mm it receives today. The Sahara may have largely disappeared during what has been called 'The Greening of Africa'.

Conversely, some deserts experienced abrupt and relatively brief drought episodes which caused extensive dune mobilization in areas such as the American High Plains (Arbogast, 1996).

At the present day, deserts continue to change, sometimes as a result of climatic fluctuations such as the run of droughts that has afflicted the West African Sahel over recent decades. The changing vegetation conditions in the Sahel, which have shown marked

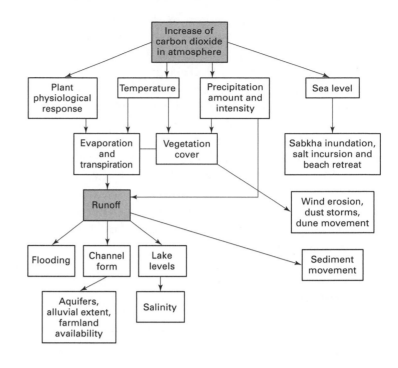

Figure 12.4 The potential responses of desert environments to increases of carbon dioxide in the atmosphere.

fluctuations since the mid-1960s, have been mapped by Tucker et al. (1991). Interannual variability in the position of the southern boundary of the Sahara, as represented by the 200 mm isohyet, can be explained in large measure by changes in the North Atlantic Oscillation and the El Niño–Southern Oscillation (ENSO) (Oba et al., 2001). The drought also led to reductions in the flow of rivers such as the Senegal and Niger, and to an increase in dust-storm activity (Middleton, 1985). The extent of Lake Chad's water surface has also fluctuated dramatically (Nicholson, 1996). It was particularly badly affected by the drought, which caused the lake surface area to fall from c. 25,000 km² in 1960 to just a tenth of that figure in the mid-1980s (Birkett, 2000).

We also know from recent centuries that ENSO fluctuations have had important consequences for such phenomena as droughts and dune reactivation (Forman et al., 2001), and through their effect on vegetation cover have had major impacts on slope stability and soil erosion (Viles and Goudie, 2003). Lake levels have responded to El Niño influences (Arpe et al., 2000), large floods have occurred, valleys have been trenched (Bull, 1997), erosivity patterns have altered (D'Odorico et al., 2001), and landslides and debris flows have been generated (Grosjean et al., 1997).

Wind erosion of soils

Changes in climate could affect wind erosion either through their impact on erosivity or through their effect on erodibility.

Erosivity is controlled by a range of wind variables including velocity, frequency, duration, magnitude, shear, and turbulence. Such wind characteristics vary over a whole range of timescales from seconds to millennia. For example, Bullard et al. (1996) have shown how dune activity varies in the South West Kalahari in response to decadal-scale variability in wind velocity, while over a longer timescale there is evidence that trade-wind velocities may have been elevated during the Pleistocene glacials. Unfortunately, general circulation models as yet give little indication of how wind characteristics might be modified in a warmer world, so that prediction of future changes in wind erosivity is problematic.

Erodibility is largely controlled by vegetation cover and surface type, and both of these can be influenced markedly by climatic conditions. In general, vegetation cover, which serves to protect the ground surface and to modify the wind regime, decreases as conditions become more arid. Likewise climate affects the nature of surface materials by controlling their moisture content, the nature and amount of clay mineral content (cohesiveness), and organic levels. Soils that are dry have a low clay content and little binding humus and are highly susceptible to wind erosion.

However, modeling the response of wind erosion to climatic variables on agricultural land is vastly complex, not least because of the variability of soil characteristics, topographic variation, the state of plant growth and residue decomposition, and the existence of windbreaks. To this needs to be added the temporal variability of eolian processes and moisture condition and the effects of different land management practices (Leys, 1999), which may themselves change with climate change.

Dust-storm activity

The changes in temperature and precipitation conditions that occurred in the twentieth century (combined with land-cover changes) had an influence on the development of dust storms (Goudie and Middleton, 1992). These are events in which visibility is reduced to less than 1 km as a result of particulate matter, such as valuable topsoil, being entrained by wind. This is a process that is most likely to happen when there are high winds and large soil-moisture deficits. Probably the greatest incidence of dust storms occurs when climatic conditions and human pressures combine to make surfaces susceptible to wind attack.

Possibly the most famous case of soil erosion by deflation was the Dust Bowl of the 1930s in the USA (see Chapter 4). In part this was caused by a series of hot, dry years which depleted the vegetation cover and made the soils dry enough to be susceptible to wind erosion, but the effects of this drought were gravely exacerbated by years of overgrazing and unsatisfactory farming techniques.

Attempts to relate past dust-storm frequencies to simple climatic parameters or antecedent moisture conditions have frequently demonstrated rather weak relationships (Bach et al., 1996), confirming the view that complex combinations of processes control dust emissions. Nevertheless, evidence is now emerging that relates dust emissions from Africa to changes in the North Atlantic Oscillation (Moulin et al., 1997).

If, however, soil moisture levels decline as a result of changes in precipitation and/or temperature, there

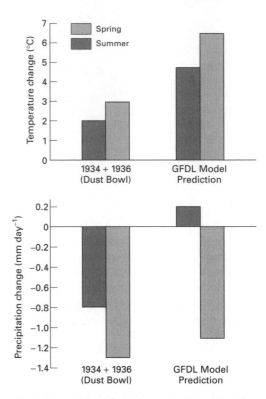

Figure 12.5 Comparison between two 'Dust Bowl years' (1934 and 1936) and the Geophysical Fluid Dynamic Laboratory (GFDL) model prediction for a $\times 2$ CO_2 situation for the Great Plains (Kansas and Nebraska) (modified from Smith and Tirpak, 1990, figure 7.3).

is the possibility that dust-storm activity could increase in a warmer world (Wheaton, 1990). A comparison between the Dust Bowl years of the 1930s and model predictions of precipitation and temperature for the Great Plains of Kansas and Nebraska (Figure 12.5) indicates that mean conditions could be similar to those of the 1930s under enhanced greenhouse conditions (Smith and Tirpak, 1990), or even worse (Rosenzweig and Hillel, 1993).

If dust-storm activity were to increase as a response to global warming it is possible that this could have a feedback effect on precipitation that would lead to further decreases in soil moisture (Tegen et al., 1996; Miller and Tegen, 1998). It is also possible that increased dust delivery to the oceans could affect biogeochemical cycling, by providing nutrients to plankton, which in turn could draw down carbon dioxide from the atmosphere (Ridgwell, 2002). However, the impact and occurrence of dust storms will depend a great deal on land management practices, and recent decreases in dust-storm activity in North Dakota have resulted from conservation measures (Todhunter and Chihacek, 1999).

Sand dunes

Sand dunes, because of the crucial relationships between vegetation cover and sand movement, are highly susceptible to the effects of changes of climate (Figure 12.6). Some areas, such as the South West Kalahari (Stokes et al., 1997) or portions of the High Plains of the USA (Gaylord, 1990), may have been especially prone to the effects of changes in precipitation and/or wind velocity because of their location in climatic zones that are close to a climatic threshold between dune stability and activity.

One of the more remarkable discoveries of recent years, brought about by the explosive development in the use of thermoluminescent and optical dating of sand grains and studies of explorers' accounts (Muhs and Holliday, 1996), is the realization that such marginal dune fields have undergone repeated phases of

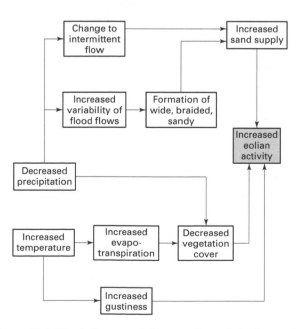

Figure 12.6 The influence of decreased precipitation and increased temperatures on eolian activity.

change at decadal and century timescales in response to extended drought events during the course of the Holocene. Dates for reaction phases are given for the Nebraskan Sandhills by Stokes and Swinehart (1997) and Muhs et al. (1997), for Kansas by Arbogast (1996), and for the South West Kalahari by Thomas et al. (1997) (Figure 12.7).

The mobility of desert dunes (M) is directly proportional to the sand-moving power of the wind, but indirectly proportional to their vegetation cover (Lancaster, 1995: 238). An index of the wind's sand-moving power is given by the percentage of the time (W) the wind blows above the threshold velocity (4.5 m per second) for sand transport. Vegetation cover is a function of the ratio between annual rainfall (P) and potential evapotranspiration (PE). Thus, $M = W/(P/PE)$. Empirical observations in the USA and southern Africa indicate that dunes are completely stabilized by vegetation when M is < 50, and are fully active when M is > 200.

Muhs and Maat (1993) have used the output from general circulation models (GCMs) combined with this dune mobility index to show that sand dunes and sand sheets on the Great Plains are likely to become reactivated over a significant part of the region, particularly if the frequencies of wind speeds above the threshold velocity were to increase by even a moderate amount. However, the methods used to estimate future dune field mobility are still full of problems and much more research is needed before we can have confidence in them (Knight et al., 2004).

For another part of the USA, Washington State, Stetler and Gaylord (1996) have suggested that with a 4°C warming vegetation would be greatly reduced and that as a consequence sand dune mobility would increase by over 400%.

For the Canadian Prairies, Wolfe (1997) found that while most dunes were currently inactive or only had active crests, under conditions of increased drought (as occurred in 1988 and are likely to occur in the future) most dunes would become more active (Figure 12.8).

The consequences of dune encroachment and reactivation could be serious and might lead to a loss of agricultural land, the overwhelming of buildings, roads, canals, runways, and the like, abrasion of structures and equipment, abrasion of crops, and the impoverishment of soil structure.

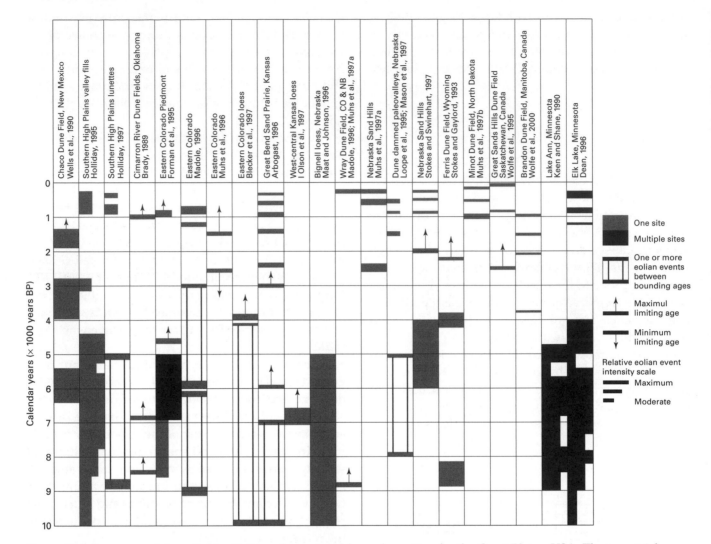

Figure 12.7 Summary of dune field activity and associated loess deposition for the Great Plains, USA. The terrestrial eolian record is also compared with the eolian input record from Lake Ann and Elk Lake, Minnesota. Note that there is evidence for sustained aridity between 10,000 and 5000 years ago and numerous discrete events in the past 2000 years. Length of solid bar reflects duration of eolian events and inferred dating errors. (Source: S. L. Forman et al., 2001, figure 13.)

Rainfall and runoff

A recent attempt to estimate the effects of global warming on runoff is provided by the UK Meteorological Office (Arnell, 1999a). What is clear from this work is that there will be clear differences at a global scale, with some areas generating more runoff and some generating less. However, with respect to dry regions, some of these will suffer particularly large diminutions in annual runoff (sometimes 60% or more). They appear to be more vulnerable than humid regions (Guo et al., 2002).

The sensitivity of runoff to changes in precipitation is complex, but in some environments quite small changes in rainfall can cause proportionally larger changes in runoff. This has been indicated for two areas of southern Australia by Chiew et al. (1996: 341):

In the south-west coast and the South Australian Gulf about 70% of the annual rainfall of 500 to 1000 mm occurs in the winter-half of the year. The streams in these regions generally flow for only 50% of the time, and on average less than 10% of the annual rainfall becomes runoff. It is also common for the total annual runoff to come from only one or

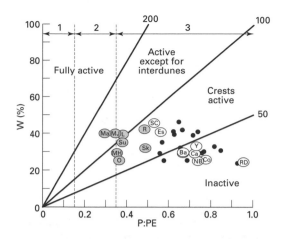

Figure 12.8 Dune mobility for stations across the Canadian Prairies for the 1988 drought year. Black dots represent values for 1961–1990 normals. Dune activity classes of Muhs and Holliday (1995) are also shown: (1) fully active; (2) largely active; (3) largely inactive. The names of the localities are Brandon (Ba), Broadview (Bv), Calgary (Ca), Coronation (Co), Estevan (Es), Lethbridge (L), Mayberries (Ma), Medicine Hat (MH), Moose Jaw (MJ), North Battleford (NB), Outlook (O), Red Deer (RD), Regina (R), Saskatoon (SK), Suffield (SU), Swift Current (SC), Yorkton (Y). Shaded stations are from drier, subhumid locations. (Source: Wolfe, 1997, figure 8.)

two significant flow events during winter. The simulations indicate that the average annual runoff increases at a much faster rate than the corresponding increase in rainfall. A rainfall increase of 10% enhances runoff by 50%, an increase of 20% more than doubles runoff, and an increase of 40% results in runoff being almost four times greater. A decrease in rainfall has a potentially more serious consequence as the amount of streamflow drops very quickly. A decrease in rainfall by 20% reduces runoff by one third while a decrease of 40% reduces runoff by 90%. The soil wetness in the winter-half changes at almost twice the rate of the change in rainfall for changes of rainfall of up to 20%. The changes, however, approach lower and upper limits with the simulations indicating that it is unlikely for the soil wetness to drop below 10% or increase above 70% even for rainfall changes of more than 50%.

Highly significant runoff changes may also be anticipated for semi-arid environments, such as the southwest USA. The models of Revelle and Waggoner (1983) suggest that the effects of increased evapotranspiration losses as a result of a 2°C rise in temperature would be particularly serious in those regions where the mean annual precipitation is less than about 400 mm (Table 12.1). Projected summer dryness in

Table 12.1 Approximate percentage decrease in runoff for a 2°C increase in temperatures. Source: from data in Revelle and Waggoner (1983)

Initial mean temperature (°C)	Precipitation (mm year^{-1})					
	200	300	400	500	600	700
–2	26	20	19	17	17	14
0	30	23	23	19	17	16
2	39	30	24	19	17	16
4	47	35	25	20	17	16
6	100	35	30	21	17	16
8		53	31	22	20	16
10		100	34	22	22	16
12			47	32	22	19
14			100	38	23	19

such areas may be accentuated by a positive feedback process involving decreases in cloud cover and associated increases in radiation absorption on the ground consequent upon a reduction in soil moisture levels (Manabe and Wetherald, 1986). Our modeling capability in this area is still imperfect and different types of model indicate differing degrees of sensitivity to climatic change (Nash and Gleick, 1991).

Shiklomanov (1999) has suggested that in arid and semi-arid areas an increase in mean annual temperature by 1° to 2°C and a 10% decrease in precipitation could reduce annual river runoff by up to 40–70% (Table 12.2).

One factor that makes estimates of rainfall–runoff relationships complicated is the possible effect of higher CO_2 concentrations on plant physiology and transpiration capacity. At higher CO_2 concentrations transpiration rates are less and this could lead to an increase in runoff or at least limit its reduction (e.g., Idso and Brazel, 1984). It is also important to remember that future runoff will be conditioned by nonclimatic factors, such as land-use and land-cover change, the construction of reservoirs, groundwater storage, and water demand (Conway et al., 1996).

River channels

Channels in arid regions are particularly sensitive to changes in precipitation characteristics and runoff (Nanson and Tooth, 1999). They can display rapid changes between incision and aggradation over short

Table 12.2 Results of assessments of impacts of climate change on annual river runoff (basins and areas of water deficit). Source: after various sources in Shiklomanov (1999, table 2.1)

Region and basin	Temperature	Precipitation	Change in annual runoff (%)
Mean for seven large basins in the western USA	+2	−10	−40 to −76
Colorado River, USA	+1	−10	−50
Peace River, USA	+1	−10	−50
River basins in Utah and Nevada, USA	+2	−10	−60
River basins in the steppe zones of European Russia	+1	−10	−60

time periods in response to quite modest changes in climate. This is particularly true in the case of the arroyos of the American southwest (Cooke and Reeves, 1976; Balling and Wells, 1990), which have displayed major changes in form since the 1880s (see Chapter 6). There has been considerable debate as to the causes of phases of trenching, and it is far from easy to disentangle anthropogenic from climatic causes, but in many cases it is fluctuations in either rainfall amount or intensity that have been the controlling factor (Hereford, 1984; Graf et al., 1991). Thus, the sorts of changes in runoff discussed in the last section could have a profound influence on channel characteristics. These in turn have an impact on humans because they can lead to changes in the agricultural suitability of bottomlands, modify local aquifers, modify sediment inputs into reservoirs, and cut into engineering structures. Arroyo incision can also lead to the draining of river-bed marshes (cienegas). It may even have produced settlement abandonment (Hereford et al., 1995).

Lake levels

Closed depressions are a widespread phenomenon in arid lands and their water levels and salinity characteristics respond rapidly and profoundly to climatic changes (Grimm et al., 1997). This generalization applies both to large and small lakes. In the twentieth century, for example, some of the largest arid zone lakes (e.g., Chad, the Aral Sea, the Caspian, and the Great Salt Lake of Utah) have shown large variations in their extents, partly in some cases because of human activities, but also because of climatic fluctuations within their catchments. They are sensitive to climate variability that would be of minor influence in systems with outflows.

For example, from the early 1950s to the middle 1980s the total area of lakes in China with an individual area of over 1 km^2 declined from 2800 to 2300 km^2 and the whole area of China's lakes has been reduced from 80,600 km^2 to 70,988 km^2 (Liu and Fu, 1996). An increasingly warm and dry climate was the principal cause of the reduced lake area on the Qingzang Plateau, Northwest China, the Inner Mongolian Plateau, and the North China Plain.

Likewise, in the early 1960s, prior to the development of the Sahel Drought, Lake Chad had an area of 23,500 km^2, but by the 1980s had split into two separate basins and had an area of only 1500 km^2. The Caspian Sea was −29.10 m in 1977, but in 1995 had risen to −26.65 m, an increase of 2.45 m in just 17 years. Similarly impressive changes have occurred in recent decades in the level of the Great Salt Lake in Utah, with a particularly rapid rise taking place between 1964 and 1985 of nearly 20 feet (6 m).

Changes in lake level of these sorts of magnitude have an impact on a diverse range of human activities, ranging from fisheries and irrigation to recreation. Moreover, the drying-up of lake beds can have adverse effects on air quality and human health through the liberation of dust, as has been found as a result of the humanly induced desiccation of the Aral Sea and Owens Lake (Reheis, 1997).

Sea-level rise and arid-zone coastlines

One consequence of global warming will be a rise in sea level at around 5 mm per year (50 cm per century). Arid coastlines could be affected by this process. This could potentially be especially serious for low-lying coastlines such as the Sabkhas of the United Arab

Emirates and elsewhere in the Middle East. Sabkhas result from the interaction of various depositional and erosional processes that create a low-angle surface in the zone of tidal influence. This means that they are subject to periodic inundation and might be vulnerable to modest sea-level rise and to any increase in storm-surge events. Given the degree of infrastructural development that has taken place in their proximity, this is a serious issue for cities such as Abu Dhabi. However, it is likely that many sabkhas will be able to cope with modestly rising sea levels, for a range of processes contribute to their accretion. These include algal stromatolite growth, fecal pellet deposition, eolian inputs, and evaporite precipitation. Some of these can cause markedly rapid accretion (Schrieber, 1986), even in the absence of a very well developed plant cover. Moreover, as sea level (and groundwater) rises, surface lowering by deflational processes will be reduced. An example of a sabkha that may well maintain itself, or even continue to aggrade in spite of sea-level rise, is provided by the Umm Said Sabkha of Qatar, where eolian dune inputs from inland cause the sabkha to build out into the Arabian Gulf.

Among arid-zone coastal environments that may be particularly susceptible to sea-level rise are deltaic areas subject to subsidence and sediment starvation (e.g., the Nile) and areas where ground subsidence is occurring as a result of fluid abstraction (e.g., California). Whereas the IPCC prediction of sea-level rise is 30–100 cm per 100 years, rates of deltaic subsidence in the Nile Valley are 35–50 cm per 100 years, and in other parts of the world rates of land subsidence produced by oil, gas, or groundwater abstraction can be up to 500 cm per 100 years.

Rising sea levels can be expected to cause increased flooding, accelerated erosion, and accelerated incursion of saline water up estuaries and into aquifers. Coastal lagoon, spit, and barrier systems (such as those of Ras Al Khaimah) may be especially sensitive (Goudie et al., 2000), as will coastlines that have been deprived of sediment nourishment by dam construction across rivers.

Salt weathering

Salt weathering is a major geomorphologic process in many drylands and it poses a threat to many man-

made structures, both ancient and modern (Goudie and Viles, 1997).

The most cited cause of salt weathering is generally the process of salt crystal growth from solutions in rock pores and cracks (Evans, 1970). Another type of salt weathering process is salt hydration. Certain common salts hydrate and dehydrate relatively easily in response to changes in temperature and humidity. As a change of phase takes place to the hydrated form, water is absorbed. This increases the volume of the salt and thus develops pressure against pore walls. Sodium carbonate and sodium sulfate both undergo a volume change in excess of 300% as they hydrate.

A third possible mechanism of rock disruption through salt action has been proposed by Cooke and Smalley (1968), who argue that disruption of rock may occur because certain salts have higher coefficients of expansion than do the minerals of the rocks in whose pores they occur.

In addition to these three main categories of mechanical effects, salt can cause chemical weathering. Some saline solutions can have elevated pH levels. Why this is significant is that silica mobility tends to be greatly increased at pH values greater than 9. Indeed, according to various studies, silica solubility increases exponentially above pH 9. The presence of sodium chloride may also affect the degree and velocity of quartz solution. At higher NaCl concentrations quartz solubility and the reaction velocity both increase. The growth of salt crystals may be able to cause pressure solution of silicate grains in rocks, for silica solubility increases as pressure is applied to silicate grains. This is a mechanism that has been identified as important in areas where calcite crystals grow, as for example in areas of calcrete formation.

The attraction of moisture into the pores of rocks or concrete by hygroscopic salts (e.g., sodium chloride) can accelerate the operation of chemical weathering processes and of frost action (MacInnis and Whiting, 1979) and the disruptive action of moisture trapped in rock capillaries is well known.

Salt can have a deleterious impact on iron and concrete. Many engineering structures are made of concrete containing iron reinforcements. The formation of the corrosion products of iron (i.e., rust) causes a volume expansion to occur. If one assumes that the prime composition of such corrosion products is $Fe(OH)_3$, then the volume increase over the uncorroded iron

Table 12.3 Global warming and salinization

Negative effects	Positive effect
Increased moisture stress will lead to a greater need for irrigation	Groundwater levels may fall
Higher evaporation losses will lead to increased salt concentrations	
Lesser freshwater inputs will reduce estuarine flushing and decrease replenishment in coastal aquifers	
Higher sea levels will cause change of salinity in estuaries and in coastal aquifers	

can be fourfold. Thus when rust is formed on the iron reinforcements, pressure is exerted on the surrounding concrete. This may cause the concrete cover over the reinforcement to crack, which in turn permits the ingress of oxygen and moisture, which then aggravates the corrosion process. In due course, spalling of concrete takes place, the reinforcements become progressively weaker, and the whole structure may suffer deterioration.

Sulfates can cause severe damage to, and even complete deterioration of, Portland cement concrete. Although there is controversy as to the exact mechanism of sulfate attack (Cabrera and Plowman, 1988), it is generally accepted that the sulfates react with the alumina-bearing phases of the hydrated cement to give a high sulfate form of calcium aluminate known as ettringite. The formation of ettringite involves an increase in the volume of the reacting solids, a pressure build up, expansion and, in the most severe cases, cracking and deterioration. The volume change on formation of ettringite is very large, and is even greater than that produced by the hydration of sodium sulfate.

Another mineral formed by sulfates coming into contact with cement is thaumasite. This causes both expansion and softening of cement and has been seen as a cause of disintegration of rendered brickwork and of concrete lining in tunnels.

Global warming could influence salt weathering in a variety of ways (Table 12.3). In low-lying coastal areas sea-level rise will have an impact upon groundwater conditions. The balance between freshwater aquifers and seawater incursion is a very delicate one. Recent modeling, using the Ghyben–Herzberg relationship,

suggests that in the Nile Delta, where subsidence is occurring at a rate of 4.7 mm per year, overpumping is taking place, and less freshwater flushing is being achieved by the dammed Nile, a 50 cm rise in Mediterranean level will cause an additional intrusion of saltwater by 9.0 km into the Nile Delta aquifer (Sherif and Singh, 1999).

In addition, many arid areas may become still more arid. Moisture deficits will increase and streamflows will decrease, partly because of increased evaporative loss promoted by higher temperature, and partly because of decreased precipitation inputs. Lakes and reservoirs will suffer more evaporative losses and so could become more saline. In many areas streamflow will decline by 60–70%. Greater moisture stress may lead to an expansion in the quest for irrigation water, there may be less freshwater inputs to flush out estuarine zones, and rates of evaporative concentration of salt may increase (Utset and Borroto, 2001). Such tendencies could lead to an increase in salinization (Imeson and Emmer, 1992; Szabolcs, 1994). Conversely, in some areas falling groundwater levels may follow on from decreased recharge, and this could lead to improvements in the incidence of groundwater-induced salinization.

Points for review

Are desert areas prone to large and rapid geomorphologic and hydrologic changes in response to apparently modest climatic stimuli?

To what extent may wind erosion of soils, dust-storm activity and dune mobility change in a warmer world?

Guide to reading

Goudie, A. S., 2002, *Great warm deserts of the world: landscapes and evolution.* Oxford: Oxford University Press. A survey of the response of the world's deserts to past climatic changes.

Williams, M. A. J. and Balling, R. C. Jr., 1996, *Interactions of desertification and climate.* London: Arnold. A consideration of human impacts on desert environments, including a discussion of the effects of future climate changes.

13 CONCLUSION

The power of nonindustrial and pre-industrial civilizations

It has become apparent that Marsh was correct over a century ago to express his cogently argued views of the importance of human agency in environmental change. Since his time the impact that humans have had on the environment has increased, as has our awareness of this impact. There has been 'a screeching acceleration of so many processes that bring ecological change' (McNeill, 2000: 4). However, it is worth making the point here that, although much of the concern expressed about the undesirable effects humans have tends to focus on the role played by sophisticated industrial societies, this should not blind us to the fact that many highly significant environmental changes were and are being achieved by nonindustrial societies.

In recent years it has become apparent that fire, in particular, enabled early societies to alter vegetation substantially, so that plant assemblages that were once thought to be natural climatic climaxes may in reality be in part anthropogenic fire climaxes. This applies to many areas of both savanna and mid-latitude grassland (see p. 39). Such alteration of natural vegetation has been shown to re-date the arrival of European settlers in the Americas (Denevan, 1992), New Zealand, and elsewhere. The effects of fire may have been compounded by the use of the stone axe and by the grazing effects of domestic animals. In turn the removal and modification of vegetation would have led to adjustment in fauna. It is also apparent that soil erosion resulting from vegetation removal has a long history and that it was regarded as a threat by the classical authors.

Recent studies (see p. 50) tend to suggest that some of the major environmental changes in highland Britain and similar parts of western Europe that were once explained by climatic changes can be explained more effectively by the activities of Mesolithic and Neolithic peoples. This applies, for example, to the decline in the numbers of certain plants in the pollen record and to the development of peat bogs and podzolization (see p. 103). Even soil salinization started at an early date because of the adoption of irrigation practices in arid areas, and its effects on crop yields were noted in Iraq more than 4000 years ago (see p. 102). Similarly (see p. 84) there is an increasing body of evidence that the hunting practices of early civilization may have caused great changes in the world's megafauna as early as 11,000 years ago.

In spite of the increasing pace of world industrialization and urbanization, it is plowing and pastoralism which are responsible for many of our most serious environmental problems and which are still causing some of our most widespread changes in the landscape. Thus, soil erosion brought about by agriculture is, it can be argued, a more serious pollutant of the world's waters than is industry: many of the habitat changes which so affect wild animals are brought about through agricultural expansion (see p. 79); and soil salinization and desertification can be regarded as two of the most serious problems facing the human race. Land-use changes, such as the conversion of forests to fields, may be as effective in causing anthropogenic changes in climate as the more celebrated burning of fossil fuels and emission of industrial aerosols into the atmosphere. The liberation of CO_2 in the atmosphere through agricultural expansion, changes in surface albedo values, and the production of dust, are all major ways in which agriculture may modify world climates. Perhaps most remarkably of all, humans, who only represent roughly 0.5% of the total heterotroph biomass on Earth, appropriate for their use something around one-third of the total amount of net primary production on land (Imhoff et al., 2004).

The proliferation of impacts

A further point we can make is that, with developments in technology, the number of ways in which humans are affecting the environment is proliferating. It is these recent changes, because of the uncertainty which surrounds them and the limited amount of experience we have of their potential effects, which have caused greatest concern. Thus it is only since the Second World War, for example, that humans have had nuclear reactors for electricity generation, that they have used powerful pesticides such as DDT (dichlorodiphenyltrichloroethane), and that they have sent supersonic aircraft into the stratosphere. Likewise, it is only since around the turn of the century that the world's oil resources have been extensively exploited,

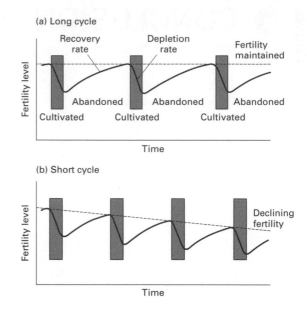

Figure 13.1 Land rotation and population density. The relationship of soil fertility cycles to cycles of slash-and-burn agriculture: (a) fertility levels are maintained under the long cycles characteristic of low-density populations; (b) fertility levels are declining under the shorter cycles characteristic of increasing population density. Notice that in both diagrams the curves of both depletion and recovery have the same slope (after Haggett, 1979, figure 8.4).

that chemical fertilizers have become widely used, and that the internal combustion engine has revolutionized the scale and speed of transport and communications.

Above all, however, the complexity, frequency, and magnitude of impacts are increasing, partly because of steeply rising population levels and partly because of a general increase in *per capita* consumption. Thus some traditional methods of land use, such as shifting agriculture (see p. 36) and nomadism, which have been thought to sustain some sort of environmental equilibrium, seem to break down and to cause environmental deterioration when population pressures exceed a particular threshold. This is illustrated for shifting agriculture systems by Figure 13.1, which shows the relationship of soil fertility levels to cycles of slash-and-burn agriculture. Fertility can be maintained (Figure 13.1a) under the long cycles characteristic of low-density populations. However, as population levels increase, the cycles necessarily become shorter, and soil fertility levels are not maintained, thereby imposing greater stresses on the land (Figure 13.1b).

Figure 13.2 The impact of recreation pressures is well displayed at a prehistoric hill-fort, Badbury Rings, Dorset, England. Pedestrians and motorcyclists have caused severe erosion of the ramparts.

At the other end of the spectrum, increasing incomes, leisure, and ease of communication have generated a stronger demand for recreation and tourism in the developed nations (Figure 13.2). These have created additional environmental problems (see p. 62), especially in coastal and mountain areas. Some of the environmental consequences of recreation, which are reviewed at length by Liddle (1997), can be listed as follows:

1 desecration of cave formations by speleologists;
2 trampling by human feet leading to soil compaction;
3 nutrient additions at campsites by people and their pets;
4 decrease in soil temperatures because of snow compaction by snowmobiles;
5 footpath erosion and off-road vehicle erosion;
6 dune reactivation by trampling;
7 vegetation change due to trampling and collecting;
8 creating of new habitats by cutting trails and clearing campsites;
9 pollution of lakes and inland waterways by gasoline discharge from outboard motors and by human waste;
10 creation of game reserves and protection of ancient domestic breeds;
11 disturbance of wildlife by proximity of persons and by hunting, fishing, and shooting;
12 conservation of woodland for pheasant shooting.

Likewise, it is apparent when considering the range of possible impacts of one major type of industrial development that they are significant. As Table 13.1 indicates, the exploitation of an oilfield and all the activities that it involves (e.g., pipelines, new roads, refineries, drilling, etc.) have a wide range of likely effects on land, air, water, and organisms.

Conversely, if one takes one ecosystem type as an example – the coral reef – one can see the diversity of stresses to which it is now being exposed (Figure 13.3) as a result of a whole range of different human activities, which include global warming, increased sedimentation and pollution from river runoff, and overharvesting of fish and other organisms (Bellwood et al., 2004).

A very substantial amount of change has been achieved in recent decades. Table 13.2, based on the work of Kates et al. (1990), attempts to make quantitative comparisons of the human impact on ten 'component indicators of the biosphere'. For each component they defined total net change clearly induced by humans to be 0% for 10,000 years ago and 100% for 1985. They estimated dates by which each component had reached successive quartiles (i.e., 5, 50 and 75%) of its 1985 total change. They believe that about half of the components have changed more in the single generation since 1950 than in the whole of human history before that date.

Are changes reversible?

It is evident that while humans have imposed many undesirable and often unexpected changes on the environment, they often have the capacity to modify the rate of such changes or to reverse them. There are cases where this is not possible: once soil has been eroded from an area it cannot be restored; once a plant or animal has become extinct it cannot be brought back; and once a laterite iron pan has become established it is difficult to destroy.

However, through the work of George Perkins Marsh and others, people became aware that many of the changes that had been set in train needed to be reversed or reduced in degree. Sometimes this has simply involved discontinuing a practice which has proved undesirable (such as the cavalier use of DDT or CFCs), or replacing it with another which is less detrimental

Table 13.1 Qualitative environmental impacts of mineral industries with particular reference to an oilfield. Source: Denisova (1977: 650, table 2)

Facility	Direction of the impact and reaction to the environment			
	Land	*Air*	*Water*	*Biocenosis*
Well	Alienation of land surface Extraction of oil associated gas, groundwater Pollution by crude oil, refined products, drilling mud Disturbance of internal structure of soil and subsoil Destruction of soil	Pollution by associated gas and volatile hydrocarbons, products of combustion	Withdrawal of surface water and groundwater Pollution by crude oil and refined products, salination of freshwater Disturbance of water balance of both subsurface and surface waters	Pollution by crude oil and refined products Disturbance and destruction over a limited surface area
Pipeline	Alienation of land Accidental oil spills Disturbance of landforms and internal structure of soil and subsoil	Pollution by volatile hydrocarbons	Disturbance and destruction over limited surface area	
Motor roads	Alienation of land Pollution by oil products Disturbance of landforms and internal structure of soil and subsoil	Pollution by combustion products, volatile hydrocarbons, sulfur dioxide, nitrogen oxides	Pollution by combustion products Disturbances and destruction over limited surface area	
Collection point	Alienation of land Pollution by crude oil and refined products (spills) Disturbance of internal structure of soil and subsoil	Pollution by volatile hydrocarbons	Disturbance and destruction over limited surface area	

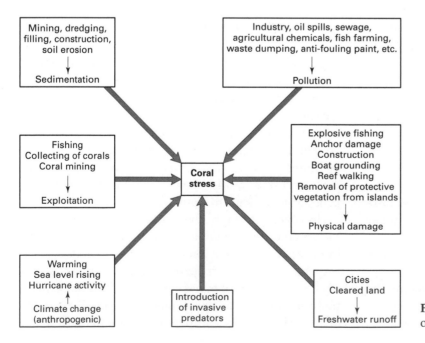

Figure 13.3 Some causes of anthropogenic stress on coral reef ecosystems.

Table 13.2 Chronologies of human-induced transformations. Source: from Kates et al. (1990, table 1.3). (a) Quartiles of change from 10,000 BC to mid-1980s

Form of transformation	Dates of quartiles		
	25%	50%	75%
Deforested area	1700	1850	1915
Terrestrial vertebrate diversity	1790	1880	1910
Water withdrawals	1925	1955	1975
Population size	1850	1950	1970
Carbon releases	1815	1920	1960
Sulfur releases	1940	1960	1970
Phosphorus releases	1955	1975	1980
Nitrogen releases	1970	1975	1980
Lead releases	1920	1950	1965
Carbon tetrachloride production	1950	1960	1970

(b) Percentage change by time of Marsh and Princeton symposium

Form of transformation	Percentage change	
	1860	1950
Deforested area	50	90
Terrestrial vertebrate diversity	25–50	75–100
Water withdrawals	15	40
Population size	30	50
Carbon releases	30	65
Sulfur releases	5	40
Phosphorus releases	< 1	20
Nitrogen releases	< 1	5
Lead releases	5	50
Carbon tetrachloride production	0	25

in its effects. Often, however, specific measures have been taken which have involved deliberate decisions of management and conservation. Denson (1970), for example, outlines a sophisticated six-stage model for wildlife conservation:

1 immediate physical protection from humans and from changes in the environment;
2 educational efforts to awaken the public to the need for protection and to gain acceptance of protective measures;
3 life-history studies of the species to determine their habitat requirements and the causes of their population decline;

4 dispersion of the stock to prevent loss of the species by disease or by a chance event such as fire;
5 captive breeding of the species to assure higher survival of young, to aid research, and to reduce the chances of catastrophic loss;
6 habitat restoration or rehabilitation when this is necessary before introducing the species.

Many conservation measures have been successful, while others have created as many problems as they were intended to solve. This applies, for example, to certain schemes for the reduction of coast erosion. On balance, however, there has been notable progress in dealing with such problems as acid rain in Europe, and the depletion of stratospheric ozone. Many governments, though not all, have signed up to the Kyoto Protocol in an attempt to reduce greenhouse gas emissions.

The concern with preservation and conservation has been longstanding, with many important landmarks. Interest has grown dramatically in recent years. Lowe (1983) has identified four stages in the history of British nature conservation:

1 the natural history/humanitarian period (1830–90)
2 the preservation period (1870–1940)
3 the scientific period (1910–70)
4 the popular/political period (1960–present)

The first of these stages was rooted in a strong enthusiasm for natural history, and the crusade against cruelty to animals. Although many Victorian naturalists were avid collectors, numerous clubs were established to study nature and some of them sought to preserve species to make them available for observation. As we shall see, certain acts were introduced at this time to protect birds. During the preservationist period, there was the formation of a spate of societies devoted to preserving open land and its associated wildlife (e.g., the National Trust, 1894; and the Council for the Preservation of Rural England, 1926). There was a growing sense of vulnerability of wildlife and landscapes to urban and industrial expansion and geographers such as Vaughan Cornish (see Goudie, 1972b) campaigned for the creation of national parks and the preservation of scenery, made possible through the National Parks and Access to Countryside Act of

1949. From the First World War onwards ecological research developed, and there arose an increasing understanding of ecological relationships. Scientists pressed for the regulation of habitats and species, and the Nature Conservancy Council was established in 1949. In the 1960s and the years that followed popular interest in conservation and widespread media attention first developed. This was partly generated by pollution incidents (such as the wrecks of the *Torrey Canyon* and *Amoco Cadiz*), and a gathering sense of impending environmental doom, generated by such persuasive books as Rachel Carson's *Silent Spring*. Ecology became a political issue in various European nations, including the UK. In many countries major developments in land use, construction, and industrialization now have to be preceded by the production of an Environmental Impact Assessment, and the European Union has introduced measures such as the Water Framework Directive (2000) and the Landfill Directive to improve the ecological status of water resources.

Thus in some countries, and in connection with particular species, conservation and protection have had a long and sometimes successful impact. In Britain, for example, the Wild Birds Protection Act dates back to 1880, and the Sea Birds Protection Act even further to 1869. The various acts have been modified and augmented over the years to outlaw egg-collecting, pole-trapping, plumage importation, and the capture or possession of a range of species. The effectiveness of the different acts can be measured in real terms. Over the past 60 years no species have been lost as British breeding birds due to lack of protection – the only major loss has been the Kentish plover, which was in any case on the edge of its range. Perhaps more importantly, several species have successfully recolonized Britain, the most celebrated being the avocet and the osprey. Today both are firmly established, together with other species lost in the nineteenth century: the black-tailed godwit, the goshawk, and the bittern. Also as a result of protection the red kites of Wales have not only survived but also increased in number, and the peregrine falcon maintains its largest numbers in Europe outside Spain.

One further ground which gives some basis for hope that humans soon may be reconciled with the environment is that there are some signs of a widespread shift in public attitudes to nature and the environment. These changing social values, combined with scientific facts, influence political action. This point of view, which acts as an antidote for some of the more pessimistic views of the world's future, was elegantly presented by Ashby (1978). He contended that the rudiments of a healthy environmental ethic are developing, and explained (pp. 84–5)

Its premise is that respect for nature is more moral than lack of respect for nature. Its logic is to put the Teesdale Sandwort . . . into the same category of value as a piece of Ming porcelain, the Yosemite Valley in the same category as Chartres Cathedral: a Suffolk landscape in the same category as a painting of the landscape by Constable. Its justification for preserving these and similar things is that they are unique, or irreplaceable, or simply part of the fabric of nature, just as Chartres and the painting by Constable are part of the fabric of civilisation; also that we do not understand how they have acquired their durability and what all the consequences would be if we destroy them.

Although there may be considerable controversy surrounding the precise criteria that can be used to select and manage sites that are particularly worthy of conservation (Goldsmith, 1983), there are nonetheless many motives behind the increasing desire to protect species and landscapes. These can be listed under the following general headings:

1 *The ethical.* It is asserted that wild species have a right to coexist with us on our planet, and that we have no right to exterminate them. Nature, it is maintained, is not simply there for humans to transform and modify as they please for their own utilitarian ends.
2 *The scientific.* We know very little about our surrounding environments, including, for example, the rich insect faunas of the tropical rain forest; therefore such environments should be preserved for future scientific study.
3 *The aesthetic.* Plants and animals, together with landscapes, may be beautiful and so enrich the life of humans.
4 *The need to maintain genetic diversity.* By protecting species we maintain the species diversity upon which future plant- and animal-breeding work will depend. Once genes have been lost (see Chapter 2, section on 'The change in genetic and species diversity') they cannot be replaced.

5 *Environmental stability*. It is argued that in general the more diverse an ecosystem is, the more checks and balances there are to maintain stability. Thus environments that have been greatly simplified by humans may be inherently unstable, and prone to disease, etc.

6 *Recreational*. Preserved habitats and landscapes have enormous recreational value, and in the case of some game reserves and natural parks may have economic value as well (e.g., the safari industry of East Africa).

7 *Economic*. Many of the species in the world are still little known, and there is the possibility that we have great storehouses of plants and animals, which, when knowledge improves, may become useful economic resources.

8 *Future generations*. One of the prime arguments for conservation, whether of beautiful countryside, rare species, soil, or mineral resources, is that future generations (and possibly ourselves later in life) will require them, and may think badly of a generation that has squandered them.

9 *Unintended impacts*. As we have seen so often in this book, profligate or unwise actions can lead to side-effects and consequences that may be disadvantageous to humans.

10 *Spiritual imperatives*. This includes a belief in the need for environmental stewardship.

Some of these arguments are more utilitarian than others (e.g., 4, 5, 6, 7 and 9), and some may be subject to doubt – it could, for example, be argued that future generations will have technology to use new resources and may not need some of those we regard as essential – but overall they provide a broadly based platform for the conservation ideal (Myers, 1979).

The susceptibility to change

Ecosystems respond in different ways to the human impact, and some are more vulnerable to human perturbation than others (Kasperson et al., 1995). It has often been thought, for example, that complex ecosystems are more stable than simple ones. Thus in Clements' Theory of Succession the tendency towards community stabilization was ascribed in part to an increasing level of integration of community functions. As Goodman (1975: 238) has expressed it:

In general the predisposition to expect greater stability of complex systems was probably a combined legacy of eighteenth century theories of political economics, aesthetically and perhaps religiously motivated attraction to the belief that the wondrous variety of nature must have some purpose in an orderly work, and ageless folkwisdom regarding eggs and baskets.

Indeed, as Murdoch (1975) has pointed out, it makes good intuitive sense that a system with many links, or 'multiple fail-safes', is more stable than one with few links or feedback loops. As an example, if a type of herbivore is attacked by several predatory species, the loss of any one of these species will be less likely to allow the herbivore to erupt or explode in numbers than if only one predator species were present and that single predator type disappeared. The basic idea therefore is that diverse groups of species are more stable because complementary species compensate for one another if one species suffers severe declines (Doak and Marvier, 2003). A diverse ecosystem will have a variety of species that help to insure it against a range of environmental upsets (Naeem, 2002).

Various other arguments have been marshaled to support the idea that great diversity and complexity affords greater ability to minimize the magnitude, duration, and irreversibility of changes brought about by some external perturbation such as human activity (Noy-Meir, 1974). It has been stated that natural systems, which are generally more diverse than artificial systems such as crops or laboratory populations, are also more stable. Likewise, the tropical rain forest has been thought of as more diverse and more stable than less complex temperate communities, while simple Arctic ecosystems of oceanic islands have always appeared highly vulnerable to disturbance brought about by anthropogenic plant and animal introductions (see p. 54).

However, considerable doubt has been expressed as to whether the classic concept of the causal linkage between diversity/complexity and stability is entirely valid (see, e.g., Hurd et al., 1971). Murdoch (1975) indicated that there is not convincing field evidence that diverse natural communities are generally more stable than simple ones. He cited various papers which show that fluctuations of microtine rodents (lemmings, field voles, etc.) are as violent in relatively complex temperate zone ecosystems as they are in the less complex Arctic zone ecosystems. This was supported by Goodman (1975: 239) who wrote:

As for the apparent stability of tropical biota, that could well be an illusion attributable to insufficient study of bewilderingly complex assemblages in which many species are so poorly represented in samples of feasible size that even considerable fluctuations might go undetected. Indeed, there are countervailing anecdotes regarding ecological instability in the tropics, such as recent reports on an insect virtually defoliating the wild Brazil-nut trees in Bolivia and of monkeys succumbing in large numbers to epidemics.

He went on to add: 'There is growing awareness of the surprising susceptibility of the rain forest ecosystems to man-made perturbation.' This is a point of view supported by May (1979) and discussed by Hill (1975). Hill pointed out that a very high species diversity is frequently associated with areas which have relatively constant physical environmental conditions over the course of a year and a series of years. The rain forest may be construed to be such an environment, and one where this constancy has allowed the presence of many specialized species, each pursuing a narrow range of activities. It has been argued that because of the high degree of specialization, the indigenous species have a limited ability to recover from major stresses caused by human intervention.

Goodman (1975) has also queried the sufficiency of the argument in its reference to the apparent instability of island ecosystems, suggesting that islands, being evolutionary backwaters and dead-ends, may accumulate species that are especially susceptible to competitive or exploitative displacement. In this case, lack of diversity may not necessarily be the sole or prime cause of instability.

The apparent instability of agricultural compared with natural communities is also often attributed to lack of diversity (see p. 62), and indeed modern agriculture does involve significant ecosystem simplification. However, such instability as there is may not, once again, necessarily result from simplification. Other factors could promote instability: agricultural communities are disrupted, even destroyed, more frequently and more massively as part of the cultivation process than those natural systems we tend to think of as stable; the component species of natural systems are co-evolved (co-adapted), and this is not usually true of agricultural communities. As Murdoch (1975: 799) suggests, it may be that:

Natural systems are more stable than crop systems because their interacting species have had a long shared evolutionary history. In contrast with these natural communities the dominant plant species of a crop system is thrust into an often alien landscape ... the crops have undergone radical selection in breeding programs, often losing their genetic defense mechanisms.

Thus the idea that complex natural ecosystems will be less susceptible to human interference and that simple artificial ecosystems will inevitably be unstable are not necessarily tenable. Nonetheless, it is apparent that there are differences in susceptibility between different ecosystem types, and these differences may result from factors other than the degree of diversity and complexity (Cairns and Dickson, 1977).

Some systems tend to be *vulnerable*. Lakes, for example, are natural traps and sinks and are thus more vulnerable to the effect of disadvantageous inputs than are rivers (which are continually receiving new inputs) or oceans (which are so much larger). Other systems display the property of *elasticity* – the ability to recover from damage. This may be because nearby epicenters exist to provide organisms to reinvade a damaged system. Small, isolated systems will often tend to possess low elasticity (see p. 88). Two of the most important properties, however, are *resilience* (being a measure of the number of times a system can recover after stress), and *inertia* (the ability to resist displacement of structural and functional characteristics).

Two systems which display resilience and inertia are deserts and estuaries. In both cases their indigenous organisms are highly accustomed to variable environmental conditions. Thus most desert fauna and flora evolved in an environment where the normal pattern is one of more or less random alternations of short favorable periods and long stress periods. They have pre-adapted resilience (Noy-Meir, 1974) so that they can tolerate extreme conditions, have the ability for rapid recovery, have various delay and trigger mechanisms (in the case of plants), and have flexible and opportunistic eating habits (in the case beasts). Estuaries, on the other hand, although the subject of increasing human pressures, also display some resilience. The vigor of their water circulation continuously and endogenously renews the supply of water, food, larvae, etc.; this aids recovery. Also, many species have biological characteristics that provide special advantages in estuarine survival. These characteristics usually protect the species against the natural violence of estuaries and are often helpful in resisting external forces such as humans.

The relationship between biodiversity and ecosystem stability continues to be a hot topic in ecology (Loreau et al., 2002; Kareiva and Levin, 2003). Some studies continue to throw doubt upon any simple relationship between biodiversity and stability (e.g., Pfisterer and Schmid, 2002), but there is perhaps an emerging consensus that diversity is crucial to ecosystem operation (McCann, 2000). As Loreau et al. (2001: 807) write,

There is consensus that at least some minimum number of species is essential for ecosystem functioning under constant conditions and that a larger number of species is probably essential for maintaining the stability of ecosystem processes in changing environments.

Human influence or nature?

From many of the examples given in this book it is apparent that in many cases of environmental change it is impossible to state, without risk of contradiction, that people rather than nature are responsible. Most systems are complex and human agency is but one component of them, so that many human actions can lead to end-products that are intrinsically similar to those that may be produced by natural forces. How to distinguish between human-induced perturbations and ill-defined natural oscillations is a crucial question when considering issues such as coral reef degradation (Sapp, 1999). It is a case of equifinality, whereby different processes can lead to basically similar results. Humans are not always responsible for some of the changes with which they are credited. This book has given many examples of this problem and a selection is presented in Table 13.3. Deciphering the cause is often a ticklish problem, given the intricate interdependence of different components of ecosystems, the frequency and complexity of environmental changes, and the varying relaxation times that different ecosystem components may have when subject to a new impulse. This problem plainly does not apply to the

Table 13.3 Human influence or nature? Some examples, with page references to this book where applicable

Change	Causes		Page reference
	Natural	Anthropogenic	
Late Pleistocene animal extinction	Climate	Hunting	84
Death of savanna trees	Soil salinization through climatically induced groundwater rise	Overgrazing	–
Desertification of semi-arid areas	Climatic change	Overgrazing, etc.	42
Holocene peat-bog development in highland Britain	Climatic change and progressive soil deterioration	Deforestation and plowing	103
Holocene elm and linden decline	Climatic change	Feeding and stalling of animals	51
Tree encroachment into alpine pastures in USA	Temperature amelioration	Cessation of burning	–
Gully development	Climatic change	Land-use change	171
Late twentieth-century climatic warming	Changes in solar emission and volcanic activity	CO_2-generated greenhouse effect	196
Increasing coast recession	Rising sea level	Disruption of sediment supply	185
Increasing coastal flood risk	Rising sea level, natural subsidence	Pumping of aquifers creating subsidence	168
Increasing river flood intensity	Higher intensity rainfall	Creation of drainage ditches	134
Ground collapse	Karstic process	Dewatering by overpumping	157
Forest decline	Drought	Air, soil, and water pollution	59

same extent to changes that have been brought about deliberately and knowingly by humans, but it does apply to the many cases where humans may have initiated change inadvertently and unintentionally.

This fundamental difficulty means that environmental impact statements of any kind are extremely difficult to make. As we have seen, humans have been living on the earth and modifying it in different degrees for several millions of years, so that it is problematic to reconstruct any picture of the environment before human intervention. We seldom have any clear baseline against which to measure changes brought about by human society. Moreover, even without human interference, the environment would be in a perpetual state of flux on a great many different timescales. In addition, there are spatial and temporal discontinuities between cause and effect. For example, erosion in one locality may lead to deposition in another, while destruction of key elements of an animal's habitat may lead to population declines throughout its range. Likewise, in a time context, a considerable interval may elapse before the full implications of an activity are apparent. Also, because of the complex interaction between different components of different environmental systems and subsystems, it is almost impossible to measure total environmental impact. For example, changes in soil may lead to changes in vegetation, which in turn may trigger changes in water quality and in animal populations. Primary impacts give rise to a myriad of successive repercussions throughout ecosystems, which may be impracticable to trace and monitor. Quantitative cause-and-effect relationships can seldom be established.

Into the unknown

During the 1980s and 1990s the full significance of possible future environmental changes has become apparent, and national governments and international institutions have begun to ponder whether the world is entering a spasm of unparalleled humanly induced modification. For example, Steffen et al. (2004) have suggested that Earth is currently operating in a no-analogue state. They remark (p. 262):

In terms of key environmental parameters, the Earth System has recently moved well outside the range of the natural variability exhibited over at least the last half million years. The nature of changes now occurring simultaneously in the Earth System, their magnitudes and rates of change are unprecedented.

Likewise, the Amsterdam Declaration of 2001 pointed to the role of thresholds and surprises (see Steffen et al., 2004, p. 298):

Global change cannot be understood in terms of a simple cause-effect paradigm. Human-driven changes cause multiple effects that cascade through the Earth System in complex ways. These effects interact with each other and with local- and regional-scale changes in multidimensional patterns that are difficult to understand and even more difficult to predict. Surprises abound.

Earth System dynamics are characterized by critical thresholds and abrupt changes. Human activities could inadvertently trigger such changes with severe consequences for Earth's environment and inhabitants.

Our models and predictions are still highly inadequate, and there are great ranges in some of the values we give for such crucial changes as sea-level rise and global climatic warming, but the balance of scientific argument favors the view that change will occur and that change will be substantial. Some of the changes may be advantageous for humans or for particular ecosystems; others will be extremely disadvantageous.

It is clear that many environmental problems are interrelated and transboundary in scope so that integrated approaches and international cooperation are required. Environmental issues and environmental solutions have become globalized (Steffen et al., 2004, p. 290).

Some environments will change very substantially during the twenty-first century in response to a rise of land-use changes and climatic changes, with some predictions suggesting that the world's grasslands and Mediterranean biomes being particularly impacted (Sala et al., 2000). Marine ecosystems will also be impacted and Jenkins (2003: 1176) suggests that by 2050: 'If present trends . . . continue, the world's marine ecosystems in 2050 will look very different from today's, large species, and particularly top predators, will be by and large extremely scarce and some will have disappeared entirely . . .' Human populations will increase, and will probably be greater by 2 to 4 billion people by 2050 (Cohen, 2003).

But all change, if it is rapid and of a great magnitude, is likely to create uncertainties and instabilities. The study of future events will not only become a major concern for the environmental sciences but will also become a major concern for economists, sociologists, lawyers, and political scientists. George Perkins Marsh was a lawyer and politician, but it is only now, over a century since he wrote *Man and nature*, that the wisdom, perspicacity, and prescience of his ideas have begun to be given the praise and attention they deserve.

Points for review

Why has there been a 'screeching acceleration' in the twentieth century of so many processes that bring ecological change?

How may adverse environmental changes be reversed?

Why should one conserve nature?

In the context of ecosystems, what do you understand by such terms as 'stability', 'resilience', 'elasticity', and 'inertia'?

Why is it often difficult to disentangle natural and anthropogenic causes of environmental changes?

Guide to reading

Liddle, M., 1997, *Recreation ecology*. London: Chapman and Hall. An excellent review of the multiple ways in which recreation and tourism have an impact on the environment.

O'Riordan, T. (ed.), 1995, *Environmental science for environmental management*. Harlow: Longman Scientific. A multi-author guide to managing environmental change.

Roberts, N. (ed.), 1994, *The changing global environment*. Oxford: Blackwell. A good collection of case studies with a wide perspective.

Simon, J. L. (ed.), 1995, *The state of humanity*. Oxford: Blackwell. A multi-author work with an optimistic message.

Simon, J. L., 1996, *The ultimate resource 2*. Princeton: Princeton University Press. A view that the state of the world is improving.

Steffen, W. and 10 others, 2004, *Global change and the earth system*. Berlin: Springer. Chapter 6 of this multi-author volume addresses ways of global management for global sustainability.

Turner, B. L. (ed.), 1990, *The Earth as transformed by human action*. Cambridge: Cambridge University Press. A magnificent study of change in the past 300 years.

GLOSSARY

Note: relevant places in the text where glossary terms are mentioned are marked in bold.

Aerosols Aggregations of minute particles (solid or liquid) suspended in the atmosphere. The term is often used to describe smoke, dust, condensation nuclei, freezing nuclei, fog, or pollutants such as droplets containing sulfur dioxide or nitrogen dioxide.

Aggradation The building upward or outward of the land surface by the deposition of sediment.

Albedo A measure for the reflectivity of a body or surface, defined as the total radiation reflected by the body divided by the total radiation falling on it. Values are expressed on a scale of either 0–1 or 1–100%.

Allochthonous Formed at a distance from its present position (see Autochthonous).

Aquifer An underground water-bearing layer of porous rock through which water can flow.

Autochthonous Formed in its present position, rather than by transport processes.

Biodegradation The breakdown or rendering harmless of a substance by natural processes.

Biodiversity The variety of species, both floral and faunal, contained within an ecosystem.

Biomass The total mass of biological material contained in a given area of the Earth's surface (expressed as dry weight or per unit area).

Biosphere The interlinked communities of animals, plants and microorganisms that live on Earth.

Boreal Of northern regions. A term applied both to a climatic zone characterized by cold, snowy winters and short summers, and to the coniferous forests of the high mid-latitudes in the Northern Hemisphere, also known as taiga.

Channelization The modification of river channels for the purpose of flood control, land drainage, navigation and the reduction or prevention of erosion.

Chaparral A type of stunted (scrub) woodland found in temperate regions with dry summers (Mediterranean regions). It is dominated by drought-resistant evergreen shrubs.

Chlorofluorocarbons A range of synthetically manufactured, chemically inert compounds (CFC) containing atoms of carbon, fluorine and chloride. They have been developed and widely used as solvents, refrigerant and aerosol propellants and in the manufacture of foam plastics.

Climax The final stage of plant succession, when the plant community is relatively stable and in equilibrium with the existing environmental conditions. It is normally determined by climate (climatic climax) or by soil (edaphic climax).

Coral bleaching Corals are bleached when the colorful symbiotic algae they house are lost. When the algae are absent for any length of time, the coral dies. This absence can be caused by excessively warm water temperatures.

Critical loads A concept in pollution studies which involves the idea that there is a certain pollution load level (e.g., of acid rain) above which harmful effects on biological systems will occur.

Deflation The removal of dry, unconsolidated material, e.g., dust or sand, from a surface by wind.

Deforestation The permanent removal of trees from an area of forest or woodland.

Desertification The spread of desert-like conditions in arid or semi-arid areas, due to human interference or climatic change, or both.

DNA (deoxyribonucleic acid). The substance that is the carrier of genetic information, found in the chromosomes of the nucleus of a cell.

Domestication The taming and breeding of previously wild animals and plants for human use.

Dust storm A storm, particularly in dry areas, which carries dense clouds of dust, sometimes to a great height, often obscuring visibility to below 1000 m.

Ecological footprint The area that is impacted by pollution, resource extraction, development and transport from a particular location (e.g., a city).

Ecology The science which studies the relations between living organisms and their environment.

ENSO (El Niño–Southern Oscillation). Periodical disturbances of Pacific Ocean and atmosphere, with El Niño conditions being abnormally warm off the coast of South America and La Niña conditions being abnormally cool.

Eutrophication The process by which an aquatic ecosystem increases in productivity as a result of increased nutrient input. Often this is due to human-induced additions of elements such as nitrogen and phosphorus (cultural eutrophication). However, the process may also be a natural phenomenon.

Forest decline The decline of forest vitality characterized by decreased and abnormal growth, leading eventually to death. The causes include poor management practices; climatic change; fungal, viral and pest attack; nutrient deficiency; and atmospheric pollution.

Genetic engineering The technology involved in manipulating the genes (molecular building blocks) of organisms. Organisms treated in this way are called genetically modified organisms (GMOs).

Gleying Soil characteristics (including mottling) developed as a result of poor drainage and intermittent waterlogging reducing oxidation or causing the deoxidation of ferric compounds.

Global change Largely synonymous with 'global environmental change', it refers to changes in the global environment (including climate change) that may alter the capacity of the Earth to sustain life.

Green Revolution An agricultural revolution, especially in the less developed countries of Asia in the 1960s and 1970s, which gave rise to increased food production through the introduction of new high-yielding varieties of crops and the adoption of techniques (e.g., synthetic fertilizers) necessary to grow them.

Groyne A construction, usually at right angles to the coast and jutting into the sea, to combat long shore drifting of sediment and beach erosion.

Habitat The place in which an organism lives, characterized by its physical features or the dominant plant types.

Heinrich event Deposition of iceberg rafted debris in ocean core sediments because of rapid ice sheet decay during the Pleistocene. They are periods of rapid climate change.

Hominid Primates of a family (Hominidae) which includes humans and their fossil ancestors.

Hybridization The process that results from a cross between parents of differing genotypes. A good example is the mule, produced by crossbreeding an ass and a horse. Hybrids may be fertile or sterile depending on differences in the genomes of the two parents.

Infiltration capacity The capacity of the soil surface to absorb water. If the capacity is exceeded, ponding will occur and surface runoff may result.

Isostasy A process that causes Earth's crust to rise or sink according to whether a weight is removed or added to it. Such a weight could be, for example, an ice cap (glacio-isostasy).

Karstic Relating to a limestone region (or another type of soluble rock) with underground drainage and many cavities and passages caused by the solution of the rock.

Keystone species   A species whose removal from the ecosystem of which it forms a part leads to a series of adverse effects (including extinctions) in that system.

Land cover   The physical state of the land, embracing, for example, the quantity and type of surface vegetation, water and earth materials. The state may change as a result of land use changes.

Laterite   A residual deposit formed by the chemical weathering of rock, composed primarily of hydrated iron and aluminum oxides. It is extensively developed in the humid and subtropical regions.

Levee   A natural or man-made embankment along a river.

Little Ice Age   A period of glacial advance and cold weather (neoglaciation) that took place between *c*. AD 1550 and AD 1850.

Maquis   Scrub vegetation of evergreen shrubs, characteristic of the western Mediterranean; broadly equivalent to chaparral.

Mass balance of glaciers   The sum of all processes which add mass to a glacier (e.g., snowfall, glaciers, avalanches) and remove mass from it (e.g., melting, ice berg calving).

Mass movement   The downward movement of material under the influence of gravity on a slope (e.g., landslips, mudflows, etc.).

Permafrost   The thermal conditions in soil and rock where temperatures are below 0° for at least two consecutive years.

Photochemical   Relating to a chemical reaction which is speeded up by particular wavelengths of electromagnetic radiation (e.g., sunlight).

Piezometric   Relating to a subterranean surface marking the level to which water will rise within an aquifer.

Podzolized   Relating to a soil that has been characterized by the acidification of the A horizon, the downward leaching of cations, metals and humic substances and their deposition in the B horizon, often precipitating to form a pan.

Radiative forcing   A change in average net radiation at the top of the troposphere resulting from a change in either solar or infrared radiation due to a change in atmospheric greenhouse gas concentrations.

Savanna   A grassland, often with scattered trees, of the tropics and subtropics.

Secondary forest   Woodland which has regenerated and colonized an area after the original (primary) forest has been removed.

Steric effect   In the context of sea level change, the change in sea level caused by changes in the volume of water in the oceans in response to temperature changes.

Succession   The sequence of changes in a plant community as it develops over time and eventually leading to climax.

Sunspot   A dark area on the visible surface of the sun. Their number usually reaches a maximum every 11 years.

Synanthrope   An organism that benefits from association with humans.

Thermokarst   Topographical depressions resulting from the thawing of ground ice (permafrost).

Trophic   Relating to the positions that organisms occupy in a food chain.

Tropospheric   Relating to the lowest level of the atmosphere, in which most of our weather occurs. The troposphere lies beneath the stratosphere and its thickness ranges from about 7 km at the poles to about 28 km at the equator.

Tundra   The zone between the latitudinal limits of tree growth and polar ice, characterized by severe winters and a short growing season.

Turbidity   A measure of the lack of clearness in a liquid caused by the presence of suspended material.

Wallace's Line   A line, developed initially by A. R. Wallace, that separates the distinct flora and fauna of south east Asia from that of Australasia.

Wetland   An ecosystem whose formation has been dominated by water (e.g., a marsh or swamp), and whose processes and characteristics are largely controlled by water.

REFERENCES

Abul-Atta, A. A., 1978, *Egypt and the Nile after the construction of the High Aswan Dam*. Cairo: Ministry of Irrigation and Land Reclamation.

Achard, F., Eva, H. D., Stibig, H.-J., Mayaux, P., Gallego, J., Richards, T. and Malingreau, J.-P., 2002, Determination of deforestation rates of the world's humid tropical forests. *Science*, 297, 999–1002.

Adger, W. N. and Brown, K., 1994, *Land use and the causes of global warming*. Chichester: Wiley.

Ågren, C. and Elvingson, P., 1996, *Still with us: the acidification of the environment is still going on*. Göteborg: Swedish NGO Secretariat on Acid Rain.

Ahlgren, I. F., 1974, The effect of fire on soil organisms. In I. I. Kozlowski and C. C. Ahlgren (eds), *Fire and ecosystems*. New York: Academic Press, 47–72.

Alabaster, J. S., 1972, Suspended solids and fisheries. *Proceedings of the Royal Society*, 180B, 395–406.

Al-Ibrahim, A. A., 1991, Excessive use of ground-water resources in Saudi Arabia: impacts and policy options. *Ambio*, 20, 34–7.

Allchin, B., Goudie, A. S. and Hegde, K. T. M., 1977, *The prehistory and palaeogeography of the Great Indian Desert*. London: Academic Press.

Allen, J. C. and Barnes, D. F., 1985, The causes of deforestation in developing countries. *Annals of the Association of American Geographers*, 75, 163–84.

Almer, B., Dickson, W., Ekstrom, C., Hornstrom, E. and Miller, U., 1974, Effects of acidification on Finnish lakes. *Ambio*, 3, 30–6.

Anderson, D. M., 1994, Red tides. *Scientific American*, 271 (2), 52–8.

Anderson, J. M. and Spencer, T., 1991, Carbon, nutrient and water balances of tropical rainforest ecosystems subject to disturbance. *Man and biosphere digest*, 7.

Anisimov, O. A., 1989, Changing climate and permafrost distribution in the Soviet Arctic. *Physical geography*, 10, 282–93.

Anisimov, O. A. and Fitzharris, B., 2001, Polar regions (Arctic and Antarctic). In J. J. McCarthy (ed.), *Climate change 2001: impacts, adaptation, and vulnerability*. Cambridge: Cambridge University Press, 801–41.

Arber, M. A., 1946, Dust-storms in the Fenland around Ely. *Geography*, 31, 23–6.

Arbogast, A. F., 1996, Stratigraphic evidence for late-Holocene aeolian sand mobilization and soil formation in south-central Kansas, USA. *Journal of arid environments*, 34, 403–14.

Archer, S. and 8 others, 1999, Arid and semi-arid land community dynamics in a management context. In T. W. Hoekstra and M. Schachak (eds), *Arid lands management*. Urbana and Chicago: University of Illinois Press, 48–74.

Arendt, A. A., Echelmeyer, K. A., Harrison, W. D., Lingle, C. S. and Valentine, V. B., 2002, Rapid wastage of Alaska glaciers and their contribution to rising sea level. *Science*, 297, 382–5.

Arnell, N. W., 1996, *Global warming, river flows and water resources*. Chichester: Wiley.

——, 1999a, The impacts of climate change on water resources. In Meteorological Office, *Climate change and its impacts*. Bracknell: Hadley Centre, 14–18.

——, 1999b, The effect of climate change on hydrological regimes in Europe: a continental prospective. *Global environmental change*, 9, 5–23.

——, 2002, *Hydrology and global environmental change*. Harlow: Prentice Hall.

Arnell, N. W. and Reynard, N. S., 2000, Climate change and UK hydrology. In M. C. Acreman (ed.), *The hydrology of the UK: a study of change*. London: Routledge, 3–29.

Aron, W. I. and Smith, S. H., 1971, Ship canals and aquatic ecosystems. *Science*, 174, 13–20.

Arpe, K., Bengtsson, L., Golitsyn, G. S., Mokhov, I. I. and Ettahir, E. A. B., 2000, Connection between Caspian Sea level variability and ENSO. *Geophysical research letters*, 27, 2693–6.

Ashby, E., 1978, *Reconciling man with the environment*. London: Oxford University Press.

Ashmore, P. and Church, M., 2001, The impact of climate change on rivers and river processes in Canada. *Geological Survey of Canada bulletin*, 555.

Atkinson, B. W., 1968, A preliminary examination of the possible effect of London's urban area on the distribution of thunder rainfall, 1951–60. *Transactions of the Institute of British Geographers*, 44, 97–118.

——, 1975, The mechanical effect of an urban area on convective precipitation. Occasional paper 3, Department of Geography, Queen Mary College, University of London.

Atlas, P., 1977, The paradox of hail suppression. *Science*, 195, 139–45.

Aubertin, G. M. and Patric, J. H., 1974, Water quality after clearcutting a small watershed in West Virginia. *Journal of environmental quality*, 3, 243–9.

Aubréville, A., 1949, *Climats, forêts et desertification de L'Afrique tropicale*. Paris: Societé d'Edition Géographiques Maritimes et Coloniales.

Ayres, M. C. and Lombardero, M. J., 2000, Assessing the consequences of global change for forest disturbance from herbivores and pathogens. *The science of the total environment*, 262, 263–86.

Bach, A. J., Brazel, A. J. and Lancaster, N., 1996, Temporal and spatial aspects of blowing dust in the Mojave and Colorado Deserts of southern California, 1973–1994. *Physical geography*, 17, 329–53.

Bakan, S. and 15 others, 1991, Climate response to smoke from the burning oil wells in Kuwait. *Nature*, 351, 367–71.

Ball, D. F., 1975, discussion in J. G. Evans, S. Limbrey and H. Cleere (eds), The effect of man on the landscape: the Highland zone. *Council for British Archaeology research report*, 11, 26.

Ballantyne, C. K., 1991, Late Holocene erosion in upland Britain: climatic deterioration or human influence? *The Holocene*, 1, 81–5.

Balling, R. C. and Wells, S. G., 1990, Historical rainfall patterns and arroyo activity within the Zuni river drainage basin, New Mexico. *Annals of the Association of American Geographers*, 80, 603–17.

Barends, F. B. J., Brouwer, F. J. J. and Schröder, F. H., 1995, *Land subsidence*. Wallingford: International Association of Hydrological Sciences, Publication 234.

Bari, M. A. and Scholfield, N. J., 1992, Lowering of a shallow, saline water table by extensive eucalypt reforestation. *Journal of hydrology*, 133, 273–91.

Barnard, P. L., Owen, L. A., Sharma, M. C. and Finkel, R. C., 2001, Natural and human-induced landsliding in the Garhwal Himalaya of northern India. *Geomorphology*, 40, 21–35.

Barnston, A. G. and Schickendanz, P. T., 1984, The effect of irrigation on warm season precipitation in the southern Great Plains. *Journal of climate and applied meteorology*, 23, 865–88.

Barrass, R., 1974, *Biology, food and people*. London: Hodder & Stoughton.

Barry, R. G., 1985, The cryosphere and climate change. In M. C. MacCracken and F. M. Luther (eds), *Detecting the climatic effects of increasing carbon dioxide*. Washington, DC: US Department of Energy, 111–48.

Bartlett, H. H., 1956, Fire, primitive agriculture, and grazing in the tropics. In W. L. Thomas (ed.), *Man's role in changing the face of the earth*. Chicago: University of Chicago Press, 692–720.

Barton, B. A., 1977, Short-term effect of highway construction on the limnology of a small stream in southern Ontario. *Freshwater biology*, 7, 99–108.

Bates, M., 1956, Man as an agent in the spread of organisms. In W. L. Thomas (ed.), *Man's role in changing the face of the earth*. Chicago: University of Chicago Press, 788–804.

Battarbee, R. W., 1977, Observations in the recent history of Lough Neagh and its drainage basin. *Philosophical transactions of the Royal Society of London*, 281B, 303–45.

Battarbee, R. W., Flower, R. J., Stevenson, A. C. and Rippey, B., 1985a, Lake acidification in Galloway: a palaeoecological test of competing hypotheses. *Nature*, 314, 350–2.

Battarbee, R. W., Appleby, P. G., Odel, K. and Flower, R. J., 1985b, ^{210}Pb dating of Scottish Lake sediments, afforestation and accelerated soil erosion. *Earth surface processes and landforms*, 10, 137–42.

Battistini, R. and Verin, P., 1972, Man and the environment in Madagascar. *Monographiae biologicae*, 21, 331–7.

Baver, L. D., Gardner, W. H. and Gardner, W. R., 1972, *Soil physics* (4th edn). New York: Wiley.

Baxter, R. M., 1977, Environmental effects of dams and impoundments. *Annual review of ecology and systematics*, 8, 255–83.

Bayfield, N. G., 1979, Recovery of four heath communities on Cairngorm, Scotland, from disturbance by trampling. *Biological conservation*, 15, 165–79.

Beamish, R. J., Lockhart, W. L., Van Loon, J. C. and Harvey, H. H., 1975, Long-term acidification of a lake and resulting effects on fishes. *Ambio*, 4, 98–102.

Beaumont, P., 1978, Man's impact on river systems: a worldwide view. *Area*, 10, 38–41.

Beckinsale, R. P., 1969, Human responses to river regimes. In R. J. Chorley (ed.), *Water, earth and man*. London: Methuen, 487–509.

—, 1972, The effect upon river channels of sudden changes in sediment load. *Acta geographica debrecina*, 10, 181–6.

Behre, K.-E. (ed.), 1986, *Anthropogenic indicators in pollen diagrams*. Rotterdam: Balkema.

Bell, M. L., 1982, The effect of land-use and climate on valley sedimentation. In A. F. Harding (ed.), *Climatic change in later prehistory*. Edinburgh: Edinburgh University Press, 127–42.

Bell, M. L. and Walker, M. J. C., 1992, *Late Quaternary environmental change*. Harlow: Longman Scientific and Technical.

Bell, M. L. and Nur, A., 1978, Strength changes due to reservoir-induced pore pressure and stresses and application to Lake Oroville. *Journal of geophysical research*, 83, 4469–83.

Bell, P. R., 1982, Methane hydrate and the carbon dioxide question. In W. G. Clark (ed.), *Carbon dioxide review 1982*. Oxford: Oxford University Press, 401–5.

Bellwood, D. R., Hughes, T. P., Folke, C. and Nyström, M., 2004, Confronting the coral reef crisis. *Nature*, 429, 827–33.

Bendell, J. F., 1974, Effects of fire on birds and mammals. In T. T. Kozlowski and C. C. Ahlgren (eds), *Fire and ecosystems*. New York: Academic Press, 73–138.

Bendix, J., Bendix, A. and Richter, M., 2000, El Niño 1997/1998 in Nordperu: Anzeichen eines Ökosystem – Wandels? *Petermanns geographische mitteilungen*, 144, 20–31.

Benn, D. I. and Evans, D. J. A., 1998, *Glaciers and glaciation*. London: Arnold.

Bennett, C. F., 1968, Human influences in the zoogeography of Panama. *Ibero-Americana*, 51.

Bennett, H. H., 1938, *Soil conservation*. New York: McGraw-Hill.

Bennett, M. R., 2003, Ice streams as the arteries of an ice sheet: their mechanics, stability and significance. *Earth-science reviews*, 61, 309–39.

Bentley, C. R., 1997, Rapid sea-level rise soon from West Antarctic ice sheet collapse? *Science*, 275, 1077–8.

—, 1998, Rapid sea-level rise from a West Antarctic ice-sheet collapse: a short-term perspective. *Journal of glaciology*, 44, 157–63.

Berbet, M. L. C. and Costa, M. H., 2003, Climate change after tropical deforestation: seasonal variability of surface albedo and its effects on precipitation change. *Journal of climate*, 16, 2099–104.

Beringer, J. E., 2000, Releasing genetically modified organisms: will any harm outweigh any advantage? *Journal of applied ecology*, 37, 207–14.

Bernabo, J. C. and Webb, T., III, 1977, Changing patterns in the Holocene pollen record of northeastern North America: mapped summary. *Quaternary research*, 8, 64–96.

Betts, R. A., 2000, Offset of the potential carbon sink from boreal forestation by decreases in surface albedo. *Nature*, 408, 187–90.

Betts, R. A., 2001, Biogeophysical impacts of land use on present-day climate: near surface temperature change and radiative forcing. *Atmospheric science letters* 1, doi:10.1006/asle.2001.0023.

Bidwell, O. W. and Hole, F. D., 1965, Man as a factor of soil formation. *Soil science*, 99, 65–72.

Biglane, K. E. and Lafleur, R. A., 1967, Notes on estuarine pollution with emphasis on the Louisiana Gulf Coast. *Publication 83, American Association for the Advancement of Science*, 690–2.

Binford, M. W., Brenner, M., Whitmore, T. J., Higuera-Grundy, A., Deevey, E. S. and Leyden, B., 1987, Ecosystems, palaeoecology, and human disturbance in subtropical and tropical America. *Quaternary science review*, 6, 115–28.

Bird, E. C. F., 1979, Coastal processes. In K. J. Gregory and D. E. Walling (eds), *Man and environmental processes*. Folkestone: Dawson, 82–101.

—, 1993, *Submerging coasts*. Chichester: Wiley.

—, 1996, *Beach management*. Chichester: Wiley.

Birkeland, P. W. and Larson, E. E., 1978, *Putnam's geology*. New York: Oxford University Press.

Birkett, C. M., 2000, Synergist remote sensing of Lake Chad: variability of basin inundation. *Remote sensing of environment*, 72, 218–36.

Birks, H. J. B., 1986, Late-Quaternary biotic changes in terrestrial and lacustrine environments, with particular reference to north-west Europe. In B. E. Berglund (ed.), *Handbook of Holocene palaeoecology and palaeohydrology*. Chichester: Wiley, 3–65.

—, 1988, Long-term ecological change in the British uplands. *British Ecological Society special publication*, 7, 37–56.

Bjorgo, E., Johannessen, O. M. and Miles, M. W., 1997, Analysis of merged SMMR–SSMI time series of Arctic and Antarctic sea ice parameters 1978–1995, *Geophysical research letters*, 24, 413–16.

Blackburn, W. H. and Tueller, P. T., 1970, Pinjon and juniper invasion in Black sagebrush communities in east-central Nevada. *Ecology*, 51, 841–8.

Blackburn, W. H., Knight, R. W. and Schuster, J. L., 1983, Saltcedar influence on sedimentation in the Brazos River. *Journal of soil and water conservation*, 37, 298–301.

Blainey, G., 1975, *Triumph of the nomads. A history of ancient Australia*. Melbourne: Macmillan.

Blank, I. W., 1985, A new type of forest decline in Germany. *Nature*, 314, 311–14.

Blecker, S. W., Yonker, C. M., Olson, C. G. and Kelly, E. F., 1997, Paleopedologic and geomorphic evidence for

Holocene climate variations, shortgrass steppe, Colorado. *Geoderma*, 76, 113–30.

Blumer, M., 1972, Submarine seeps: are they a major source of open ocean oil pollution? *Science*, 176, 1257–8.

Boardman, J., 1998, An average soil erosion rate for Europe: myth or reality. *Journal of soil and water conservation*, 53, 46–50.

Böckh, A., 1973, Consequences of uncontrolled human activities in the Valencia lake basin. In M. T. Farvar and J. P. Milton (eds), *The careless technology*. London: Tom Stacey, 301–17.

Bockman, O. C., Kaarstad, O., Lie, O. H. and Richards, I., 1990, *Agriculture and fertilizers*. Oslo: Norsk Hydro.

Bogucki, P., 1999, *The origins of human society*. Oxford: Blackwell.

Bomford, M. and Hart, Q., 2002, Non-indigenous vertebrates in Australia. In D. Pimentel (ed.), *Biological invasions*. Boca Raton: CRC Press, 25–44.

Bonan, G. B., 1997, Effects of land use on the climate of the United States. *Climatic change*, 37, 449–86.

Bonnicksen, T. M., Bonnicksen, M., Anderson, K., Lewis, H. T., Kay, C. E. and Knudson, R., 1999, Native American influences on the development of forest ecosystems. In R. C. Szaro, N. C. Johnson, W. T. Sexton and A. J. Malle (eds), *Ecological stewardship*, Vol. 2. Amsterdam: Elsevier Science, 439–70.

Boorman, L. A., 1977, Sand-dunes. In R. S. K. Barnes (ed.), *The coastline*. London: Wiley, 161–97.

Booysen, P. de V. and Tainton, N. M. (eds), 1984, *Ecological effects of fire in South African ecosystems*. Berlin: Springer-Verlag.

Bormann, F. H., Likens, G. E., Fisher, D. W. and Pierce, R. S., 1968, Nutrient loss accelerated by clear cutting of a forest ecosystem. *Science*, 159, 882–4.

Bourne, W. R. P., 1970, Oil pollution and bird conservation. *Biological conservation*, 2, 300–2.

Boussingault, J. B., 1845, *Rural economy* (2nd edn). London: Baillière.

Boutron, C. F., Görlach, U., Candelone, J.-P., Bolshov, M. A. and Delmas, R. J., 1991, Decrease in anthropogenic lead, cadmium and zinc in Greenland snows since the late 1960s. *Nature*, 353, 153–6.

Bove, M. C., Elsner, J. B., Landsea, C. W., Niu, X. and O'Brien, J. J., 1998, El Niño on US land falling hurricanes, revisited. *Bulletin of the American Meteorological Society*, 79, 2477–82.

Bowler, J. and 6 others, 2003, New ages for human occupation and climatic change at Lake Mungo, Australia. *Nature*, 421, 837–40.

Boyd, M., 2002, Identification of anthropogenic burning in the paleoecological record of the Northern Prairies: a new approach. *Annals of the Association of American Geographers*, 92, 471–87.

Bradley, R., 1985, *Quaternary palaeoclimatology*. London: Chapman & Hall.

Brady, R. G., 1989, Geology of the Quaternary Dune Sands in eastern Major and southern Alfalfa counties, Oklahoma. PhD Dissertation, Oklahoma State University.

Bragg, O. M. and Tallis, J. H., 2001, The sensitivity of peat-covered upland landscapes. *Catena*, 42, 345–60.

Braithwaite, R. J. and Zhang, Y., 1999, Modelling changes in glacier mass balance that may occur as a result of climate changes. *Geografiska annaler*, A, 81, 489–96.

Brandt, S. A., 2000, Classification of geomorphological effects downstream of dams. *Catena*, 40, 375–401.

Brassington, F. C. and Rushton, K. R., 1987, A rising water table in central Liverpool. *Quarterly journal of engineering geology*, 20, 151–8.

Braun, L. N., Weber, M. and Schulz, M., 2000, Consequences of climate change for runoff from alpine regions. *Annals of glaciology*, 31, 19–25.

Bravard, J-P. and Petts, G. E., 1996, Human impacts on fluvial hydrosystems. In G. E. Petts and C. Amoros (eds), *Fluvial hydrosystems*. London: Chapman & Hall, 242–62.

Bray, H. J., Carter, D. J. and Hooke, J. M., 1992, *Sea level rise and global warming: scenarios, physical impacts and policies*. Report to Standing Conference on Problems Associated with the Coast (SCOPAC), University of Portsmouth.

Brazel, A. J. and Idso, S. B., 1979, Thermal effects of dust on climate. *Annals of the Association of American Geographers*, 69, 432–7.

Breuer, G., 1980, *Weather modification: prospects and problems*. Cambridge: Cambridge University Press.

Bridges, E. M., 1978, Interaction of soil and mankind in Britain. *Journal of soil science*, 29, 125–39.

Brimblecombe, P., 1977, London air pollution 1500–1900. *Atmospheric environment*, 11, 1157–62.

Brimblecombe, P. and Camuffo, D., 2003, Long term damage to the built environment. In P. Brimblecombe (ed.), *The effects of air pollution on the built environment*. London: Imperial College Press, 1–30.

Broadus, J. G., 1990, Greenhouse effects, sea level rise and land use. *Land use policy*, 7, 138–53.

Broadus, J., Milliman, J., Edwards, S., Aubrey, D. and Gable, F., 1986, Rising sea level and damming of rivers: possible effects in Egypt and Bangladesh. In J. G. Titus (ed.), *Effects of changes in stratospheric ozone and global climate*. Washington, DC: United Nations Environment Program/U.S. Environmental Protection Agency, 165–89.

Brook, B. W. and Bowman, D. M. J. S., 2002, Explaining the Pleistocene megafaunal extinction: models, chronologies, and assumptions. *Proceedings of the National Academy of Sciences*, 99, 14624–7.

Brookes, A., 1985, River channelization: traditional engineering methods, physical consequences, and alternative practices. *Progress in physical geography*, 9, 44–73.

——, 1987, The distribution and management of channelized streams in Denmark. *Regulated rivers*, 1, 3–16.

Brooks, A. P. and Brierly, G. J., 1997, Geomorphic responses of lower Bega River to catchment disturbance, 1851–1926. *Geomorphology*, 18, 291–304.

Brown, A. A. and Davis, K. P., 1973, *Forest fire control and its use* (2nd edn). New York: McGraw-Hill.

——, 1997, Coral bleaching: causes and consequences. *Coral reefs*, 16, S129–38.

Brown, B. E., Dunne, R. P., Goodson, H. S. and Douglas, A. G., 2000, Bleaching patterns in coral reefs. *Nature*, 404, 142–3.

Brown, J. H., 1989, Patterns, modes and extents of invasions by vertebrates. In J. A. Drake (ed.), *Biological invasions: a global perspective*. Chichester: Wiley, 85–109.

Brown, S. and Lugo, A. E., 1990, Tropical secondary forests. *Journal of tropical ecology*, 6, 1–32.

Browning, K. A. and 11 others, 1991, Environmental effects from burning oil wells in Kuwait. *Nature*, 351, 363–7.

Bruijnzeel, L. A., 1990, *Hydrology of moist tropical forests and effects of conversion: a state of knowledge review*. Amsterdam: Free University, for UNESCO International Hydrological Programme.

Brunhes, J., 1920, *Human geography*. London: Harrap.

Brunsden, D., 2001, A critical assessment of the sensitivity concept in geomorphology. *Catena*, 42, 83–98.

Bruun, P., 1962, Sea level rise as a cause of shore erosion. *American Society of Civil Engineers proceedings: Journal of Waterways and Harbors Division*, 88, 117–30.

Bryan, G. W., 1979, Bioaccumulation of marine pollutants. *Philosophical transactions of the Royal Society*, 286B, 483–505.

Bryan, K., 1928, Historic evidence of changes in the channel of Rio Puerco, a tributary of the Rio Grande in New Mexico. *Journal of geology*, 36, 265–82.

Bryson, R. A. and Barreis, D. A., 1967, Possibility of major climatic modifications and their implications: northwest India, a case for study. *Bulletin of the American Meteorological Society*, 48, 136–42.

Bryson, R. A. and Kutzbach, J. E., 1968, *Air pollution*. Commission on college geography resource paper 2. Washington, DC: Association of American Geographers.

Buddemeier, R. W. and Smith, S. V., 1988, Coral reef growth in an era of rapidly rising sea level: predictions and suggestions for long-term research. *Coral reefs*, 7, 51–6.

Bull, W. B., 1997, Discontinuous ephemeral streams. *Geomorphology*, 19, 227–76.

Bullard, J. E., Thomas, D. S. G., Livingstone, I. and Wiggs, G., 1996, Wind energy variations in the south-western Kalahari Desert and their implications for linear dunefield activity. *Earth surface processes and landforms*, 21, 263–78.

Bullock, P. and Le Houérou, H., 1996, Land degradation and desertification. In R. T. Watson, M. C. Zinyowera and R. H. Moss (eds), *Climate change 1995*. Cambridge: Cambridge University Press, 171–89.

Bulmer, S., 1982, Human ecology and cultural variation in prehistoric New Guinea. *Monographiae biologicae*, 42, 169–206.

Burcham, L. T., 1970, Ecological significance of alien plants in California grasslands. *Proceedings of the Association of American Geographers*, 2, 36–9.

Burke, R., 1972, Stormwater runoff. In R. T. Oglesby, C. A. Carlson and J. A. McCann (eds), *River ecology and man*. New York: Academic Press, 727–33.

Burney, A. D., 1993, Recent animal extinctions: recipes for disaster. *American scientist*, 81, 530–41.

Burrin, P. J., 1985, Holocene alluviation in southeast England and some implications for palaeohydrological studies. *Earth surface processes and landforms*, 10, 257–71.

Burt, T. P., Donohoe, M. A. and Vann, A. R., 1983, The effect of forestry drainage operations on upland sediment yields: the results of a stormbased study. *Earth surface processes and landforms*, 8, 339–46.

Burt, T. P. and Haycock, N. E., 1992, Catchment planning and the nitrate issue: a UK perspective. *Progress in physical geography*, 16, 379–404.

Bush, M. B., 1988, Early Mesolithic disturbance: a force on the landscape. *Journal of archaeological science*, 15, 453–62.

Bussing, C., 1972, The impact of feedlots. In D. D. MacPhail (ed.), *The High Plains: problem of semiarid environments*. Fort William: Colorado State University, 78–86.

Butzer, K. W., 1972, *Environment and archaeology – an ecological approach to prehistory*. London: Methuen.

——, 1974, Accelerated soil erosion: a problem of man-land relationships. In I. R. Manners and M. W. Mikesell (eds), *Perspectives on environments*. Washington, DC: Association of American Geographers.

——, 1976, *Early hydraulic civilization in Egypt*. Chicago: University of Chicago Press.

Buyanovsky, G. A. and Wagner, G. H., 1998, Carbon cycling in cultivated land and its global significance. *Global change biology*, 4, 131–41.

Cabanes, C., Cazanave, A. and Le Provost, C., 2001, Sea level rise during past 40 years determined from satellite and in situ observations. *Science*, 294, 840–2.

Cabrera, J. G. and Plowman, C., 1988, The mechanism and rate of attack of sodium sulphate on cement and cement/pfa pastes. *Advances in cement research*, 1, 171–9.

Cairns, J. and Dickson, D. K., 1977, Recovery of streams from spills of hazardous materials. In J. Cairns, K. L. Dickson and E. E. Herricks (eds), *Recovery and restoration of damaged ecosystems*. Charlottesville: University Press of Virginia, 24–42.

Caldararo, N., 2002, Human ecological intervention and the role of forest fires in human ecology. *The science of the total environment*, 292, 141–65.

Callaway, R. M. and Aschehoug, E. T., 2000, Invasive plants versus their new and old neighbors: a mechanism for exotic invasion. *Science*, 290, 521–3.

Cannon, S. H., Bigio, E. R. and Mine, E., 2001a, A process for fire-related debris flow initiation, Cerro Grande fire, New Mexico. *Hydrological processes*, 15, 3011–23.

Cannon, S. H., Kirkham, R. M. and Parise, M., 2001b, Wildfire-related debris-flow initiation processes, Storm King Mountains, Colorado. *Geomorphology*, 39, 171–88.

Carlson, C. W., 1978, Research in ARS related to soil structure. In W. W. Emerson, R. D. Bond and A. R. Dexter (eds), *Modification of soil structure*. Chichester: Wiley, 279–84.

Carpenter, T. G. (ed.), 2001, *Sustainable civil engineering* (2nd edn). Chichester: Wiley.

Carrara, P. E. and Carroll, T. R., 1979, The determination of erosion rates from exposed tree roots in the Piceance Basin, Colorado. *Earth surface processes*, 4, 407–17.

Carter, D. L., 1975, Problems of salinity in agriculture. *Ecological studies*, 15, 26–35.

Carter, L. J., 1977, Soil erosion: the problem persists despite the billions spent on it. *Science*, 196, 409–11.

Carter, L. D., 1987, Arctic lowlands: introduction. In W. L. Graf (ed.), *Geomorphic systems of North America*. Boulder: Geological Society of America, centennial special volume, 2, 583–615.

Cathcart, R. B., 1983, Mediterranean Basin – Sahara reclamation. *Speculations in science and technology*, 6, 150–2.

Ceballos, G. and Ehrlich, P. R., 2002, Mammal population losses and the extinction crisis. *Science*, 296, 904–7.

Cerda, A., 1998, Post-fire dynamics of erosional processes under Mediterranean climatic conditions. *Zeitschrift für geomorphologie*, 42, 373–98.

Chai, J-C., Shen, S-L., Zhu, H. H. and Zhang, X. L., 2004, Land subsidence due to groundwater drawdown in Shanghai. *Géotechnique*, 54, 143–7.

Challinor, D., 1968, Alteration of surface soil characteristics by four tree species. *Ecology*, 49, 286–90.

Chambers, F. M., Dresser, P. Q. and Smith, A. G., 1979, Radiocarbon dating evidence on the impact of atmospheric pollution on upland peats. *Nature*, 282, 829–31.

Champion, T., Gramble, C., Shennan, S. and Whittle, A., 1984, *Prehistoric Europe*. London: Academic Press.

Chancellor, W. J., 1977, Compaction of soil by agricultural equipment. *Division of Agricultural Sciences, University of California bulletin*, 1881.

Chandler, T. J., 1976, The climate of towns. In T. J. Chandler and S. Gregory (eds), *The climate of the British Isles*. London: Longman, 307–29.

Changnon, S. A., 1973, Atmospheric alterations from man-made biospheric changes. In W. R. D. Sewell (ed.), *Modifying the weather: a social assessment*. Adelaide: University of Victoria, 135–84.

——, 1978, Urban effects on severe local storms at St Louis. *Journal of applied meteorology*, 17, 578–86.

Changnon, S. A., Kunkel, K. E. and Winstanley, D., 2003, Climate factors that caused the unique tall grass prairie in the Central United States. *Physical geography*, 23, 259–80.

Chapin, F. S. and Danell, K., 2001, Boreal forest. In F. S. Chapin, O. E. Sala and E. Huber-Sannwald (eds), *Global biodiversity in a changing environment*. New York: Springer-Verlag 101–20.

Chapin, F. S., Sala, O. E. and Huber-Sannwald, E. (eds), 2001, *Global biodiversity in a changing environment*. Berlin: Springer-Verlag.

Chapman, W. L. and Walsh, J. E., 1993, Recent variations of sea ice and air-temperature in high latitudes, *Bulletin of the American Meteorological Society*, 74, 33–47.

Chappell, J., Cowell, P. J., Woodroffe, C. D. and Eliot, I. G., 1996, Coastal impacts of enhanced greenhouse climate change in Australia: implications for coal use. In W. J. Bouma, G. I. Pearman and M. R. Manning (eds), *Greenhouse*. Melbourne: Commonwealth Scientific and Industrial Research organisation (CSIRO), 220–34.

Charlson, R. J., Lovelock, J. E., Andreae, M. O. and Warren, S. G., 1987, Oceanic phytoplankton, atmospheric sulphur, cloud albedo and climate. *Nature*, 326, 655–61.

Charlson, R. J. and 6 others, 1992, Climate forcing by anthropogenic aerosols. *Science*, 255, 423–30.

Charney, J., Stone, P. H. and Quirk, W. J., 1975, Drought in the Sahara: a bio-geophysical feedback mechanism. *Science*, 187, 434–5.

Chase, T. N., Pielke, R. A., Kittle, T. G. F., Nemani, R. R. and Running, S. W., 2000, Simulated impacts of historical land cover changes on global climate in northern winter. *Climate dynamics*, 16, 93–105.

Chen, X. and Zong, Y., 1999, Major impacts of sea-level rise on agriculture in the Yangtze Delta area around Shanghai. *Applied geography*, 19, 69–84.

Chester, D. K. and James, P. A., 1991, Holocene alluviation in the Algarve, southern Portugal: the case for an anthropogenic cause. *Journal of archaeological science*, 18, 73–87.

Chesters, G. and Konrad, J. F., 1971, Effects of pesticide usage on water quality. *Bioscience*, 21, 565–9.

Chi, S. C. and Reilinger, R. E., 1984, Geodetic evidence for subsidence due to groundwater withdrawal in many parts of the United States of America. *Journal of hydrology*, 67, 155–82.

Chiew, F. H. S., Wang, Q. J., McMahon, T. A., Bates, B. C. and Whetton, P. H., 1996, Potential hydrological responses to climate change in Australia. In J. A. A. Jones, E. Liu, M.-K. Woo and H.-T. Kung (eds), *Regional hydrological response to climate change*. Dordrecht: Kluwer, 337–50.

Child, B. A., 1985, Bush encroachment and rangeland deterioration. Paper presented at a symposium on 'The deterioration of natural resources in Africa', School of Geography, University of Oxford, 1 July, 1985.

Childe, V. G., 1936, *Man makes himself*. London: Watts.

Chorley, R. J. and More, R. J., 1967, The interaction of precipitation and man. In R. J. Chorley (ed.), *Water, earth and man*. London: Methuen, 157–66.

Church, J. A., 2001, How fast are sea levels rising? *Science*, 294, 802–3.

Church, J. A. and 35 others, 2001, Changes in sea level. In J. T. Houghton (ed.), *Climate change 2001: the scientific basis*. Cambridge: Cambridge University Press, 639–93.

Clark, G., 1962, *World prehistory*. Cambridge: Cambridge University Press.

——, 1977, *World prehistory in new perspective*. Cambridge: Cambridge University Press.

Clark, J. R., 1977, *Coastal ecosystem management*. New York: Wiley.

Clark, J. S., Cachier, H., Goldammer, J. G. and Stocks, B. (eds), 1997, *Sediment records of biomass burning and global change*. Berlin: Springer-Verlag.

Clark, M. J. (ed.), 1988, *Advances in periglacial geomorphology*. Chichester: Wiley.

Clark, R. B., 1997, *Marine pollution* (4th edn). Oxford: Clarendon Press.

Clarke, R. T. and McCulloch, J. S. G., 1979, The effect of land use on the hydrology of small upland catchments.

In G. E. Hollis (ed.), *Man's impact on the hydrological cycle in the United Kingdom*. Norwich: Geo Abstracts, 71–8.

Coates, D. R. (ed.), 1976, *Geomorphology and engineering*. Stroudsburg: Dowden, Hutchinson and Ross.

—, 1977, Landslide perspective. *Reviews in engineering geology*, 3, 3–28.

—, 1983, Large-scale land subsidence. In R. Gardner and H. Scoging (eds), *Mega-geomorphology*. Oxford: Oxford University Press, 212–34.

Cochrane, R., 1977, The impact of man on the natural biota. In A. G. Anderson (ed.), *New Zealand in maps*. Section 14. London: Hodder & Stoughton.

Coe, M., 1981, Body size and the extinction of the Pleistocene megafauna. *Palaeoecology of Africa*, 13, 139–45.

—, 1982, The bigger they are . . . *Oryx*, 16, 225–8.

Coffey, M., 1978, The dust storms. *Natural history* (New York), 87, 72–83.

Cohen, J. E., 2003, Human population: the next half century. *Science*, 302, 1172–5.

Cole, M. M., 1963, Vegetation and geomorphology in northern Rhodesia: an aspect of the distribution of the Savanna of Central Africa. *Geographical journal*, 129, 290–310.

Cole, M. M. and Smith, R. F., 1984, Vegetation as an indicator of environmental pollution. *Transactions of the Institute of British Geographers*, 9, 477–93.

Cole, S., 1970, *The Neolithic revolution* (5th edn). London: British Museum (Natural History).

Collison, A., Wade, S., Griffiths, J. and Dehn, M., 2000, Modelling the impact of predicted climate change on landslide frequency and magnitude in S. E. England. *Engineering geology*, 55, 205–18.

Committee on the Atmosphere and the Biosphere, 1981, *Atmosphere–biosphere interactions: towards a better understanding of the ecological consequences of fossil fuel combustion*. Washington, DC: National Academy Press.

Conacher, A. J., 1979, Water quality and forests in South-western Australia: review and evaluation. *Australian geographer*, 14, 150–9.

Conacher, A. J. and Conacher, J., 1995, *Rural land degradation in Australia*. Melbourne: Oxford University Press.

Conacher, A. J. and Sala, M. (eds), 1998, *Land degradation in Mediterranean environments of the world*. Chichester: Wiley.

Conway, D., Kroi, M., Alcamo, J. and Hulme, M., 1996, Future availability of water in Egypt: the interaction of global, regional and basin scale driving forces in the Nile Basin. *Ambio*, 25, 336–42.

Conway, G. R. and Pretty, J. N., 1991, *Unwelcome harvest: agriculture and pollution*. London: Earthscan.

Conway, V. M., 1954, Stratigraphy and pollen analysis of southern Pennine blanket peats. *Journal of geology*, 42, 117–47.

Cooke, G. W., 1977, Waste of fertilizers. *Philosophical transactions of the Royal Society of London*, 281B, 231–41.

Cooke, R. U. and Doornkamp, J. C. 1974, *Geomorphology in environmental management*. Oxford: Clarendon Press.

Cooke, R. U. and Doornkamp, J. C., 1990, *Geomorphology in environmental management* (2nd edn). Oxford: Clarendon Press.

Cooke, R. U. and Reeves, R. W., 1976, *Arroyos and environmental change in the American south-west*. Oxford: Clarendon Press.

Cooke, R. U. and Smalley, I. J., 1968, Salt weathering in deserts. *Nature*, 220, 1226–7.

Cooke, R. U., Brunsden, D., Doornkamp, J. C. and Jones, D. K. C., 1982, *Urban geomorphology in drylands*. Oxford: Oxford University Press.

Coones, P. and Patten, J. H. C., 1986, *The landscape of England and Wales*. Harmondsworth: Penguin Books.

Cooper, C. F., 1961, The ecology of fire. *Scientific American*, 204, 4, 150–60.

Cooper, D. M. and Jenkins, A., 2003, Response of acid lakes in the UK to reductions in atmospheric deposition of sulphur. *The science of the total environment*, 313, 91–100.

Corlett, R. T., 1995, Tropical secondary forests. *Progress in physical geography*, 19, 159–72.

Costa, J. E., 1975, Effects of agriculture on erosion and sedimentation in the Piedmont province, Maryland. *Bulletin of the Geological Society of America*, 86, 1281–6.

Cotton, W. R. and Piehlke, R. A., 1995, *Human impacts on weather and climate*. Cambridge: Cambridge University Press.

Council on Environmental Quality, 1986, *17th Annual Report*. Washington, DC: Council on Environmental Quality.

Council on Environmental Quality, 1992, *22nd Annual Report*. Washington, DC: Council on Environmental Quality.

Coupland, R. T. (ed.), 1979, *Grassland ecosystems of the world: analysis of grasslands and their uses*. Cambridge: Cambridge University Press.

Cowell, E. B., 1976, Oil pollution of the sea. In R. Johnson (ed.), *Marine pollution*. London: Academic Press, 353–401.

Crisp, D. T., 1977, Some physical and chemical effects of the Cow Green (Upper Teesdale) impoundment. *Freshwater biology*, 7, 109–20.

Critchley, W. R. S., Reij, C. and Willcocks, T. J., 1994, Indigenous soil and water conservation: a review of the state of knowledge and prospects for building on traditions. *Land degradation and rehabilitation*, 5, 293–314.

Cronin, L. E., 1967, The role of man in estuarine processes. *Publication 83, American Association for the Advancement of Science*, 667–89.

Cronk, Q. C. B. and Fuller, J. L., 1995, *Plant invaders*. London: Chapman & Hall.

Crowe, P. R., 1971, *Concepts in climatology*. London: Longman.

Crozier, M. J., Marx, S. L. and Grant, I. J., 1978, Impact of off-road recreational vehicles on soil and vegetation. *Proceedings of the 9th New Zealand geography conference*, Dunedin, 76–9.

Crutzen, P. J. and Goldammer, J. G., 1993, *Fire in the environment*. Chichester: Wiley.

Crutzen, P. J., Aselmann, I. and Sepler, W., 1986, Methane production by domestic animals, wild ruminants, other herbivores, fauna and humans. *Tellus*, 38B, 271–84.

Cumberland, K. B., 1961, Man in nature in New Zealand. *New Zealand geographer*, 17, 137–54.

Cunningham, D. A., Collins, J. F. and Cummins, T., 2001, Anthropogenically-triggered iron pan formation in some Irish soils over various time spans. *Catena*, 43, 167–76.

Curran, M. A. J., van Ommen, T. D., Morgan, V. I., Phillips, K. L. and Palmer, A. S., 2003, Ice core evidence for Antarctic sea ice decline since the 1950s. *Science*, 302, 1203–6.

Custodio, E., Iribar, V., Manzano, B. A. and Galofre, A., 1986, Evolution of sea water chemistry in the Llobyegat Delta, Barcelona, Spain. In *Proceedings of the 9th salt water chemistry meeting*, Delft.

Cypser, D. A. and Davis, S. D., 1998, Induced seismicity and the potential for liability under U.S. law. *Tectonophysics*, 289, 239–55.

Daniel, T. C., McGuire, P. E., Stoffel, D. and Millfe, B., 1979, Sediment and nutrient yield from residential construction sites. *Journal of environmental quality*, 8, 304–8.

Darby, H. C., 1956, The clearing of the woodland in Europe. In W. L. Thomas (ed.), *Man's role in changing the face of the Earth*. Chicago: University of Chicago Press, 183–216.

Darling, F. F., 1956, Man's ecological dominance through domesticated animals on wild lands. In W. L. Thomas (ed.), *Man's role in changing the face of the Earth*. Chicago: University of Chicago Press, 778–87.

Darungo, F. P., Allee, P. H. and Weickmann, H. K., 1978, Snowfall induced by a power plant plume. *Geophysical research letters*, 5, 515–17.

Daubenmire, R., 1968, Ecology of fire in grassland. *Advances in ecological research*, 5, 209–66.

Davis, B. N. K., 1976, Wildlife, urbanisation and industry. *Biological conservation*, 10, 249–91.

Davis, M. C. R., Hamza, O. and Harris, C., 2001, The effect of rise in mean annual temperature on the stability of rock slopes containing ice-filled discontinuities. *Permafrost and periglacial processes*, 12, 137–44.

Davitaya, F. F., 1969, Atmospheric dust content as a factor affecting glaciation and climatic change. *Annals of the Association of American Geographers*, 59, 552–60.

Deadman, A., 1984, Recent history of *Spartina* in northwest England and in North Wales and its possible future development. In P. Doody (ed.), *Spartina anglica in Great Britain*. Shrewsbury: Nature Conservancy Council, 22–4.

De Angelis, H. and Skvarca, P., 2003, Glacier surge after Ice Shelf collapse. *Science*, 299, 1560–2.

Dean, W. E., Ahlbrandt, T. S., Anderson, R. Y. and Bradbury, J. P., 1996, Regional aridity in North America during the middle Holocene. *Holocene*, 6, 145–55.

DeBano, L. F., 2000, The role of fire and soil heating on water repellency in wildland environments: a review. *Journal of hydrology*, 231/2, 195–206.

Dehn, M. and Buma, J., 1999, Modelling future landslide activity based on general circulation models. *Geomorphology*, 30, 175–87.

Dehn, M., Gurger, G., Buma, J. and Gasparetto, P., 2000, Impact of climate change on slope stability using expanded downscaling. *Engineering geology*, 55, 193–204.

Denevan, M. W., 1992, The pristine myth: the landscapes of the Americas in 1492. *Annals of the Association of American Geographers*, 82, 369–85.

Denevan W. M., 2001, *Cultivated landscapes of native Amazonia and the Andes*. Oxford: Oxford University Press.

Denisova, T. B., 1977, The environmental impact of mineral industries. *Soviet geography*, 18, 646–59.

Denson, E. P., 1970, The trumpeter swan, *Olor buccinator*; a conservation success and its lessons. *Biological conservation*, 2, 251–6.

Dent, D. L. and Pons, L. J., 1995, A world perspective on acid sulphate soils. *Geoderma*, 67, 263–76.

Department of Environment, 1984, *Digest of environmental pollution and water statistics for 1983*. London: Her Majesty's Stationery Office.

Derbyshire, E. (ed.), 1973, *Climatic geomorphology*. London: Macmillan.

De Sylva, D., 1986, Increased storms and estuarine salinity and other ecological impacts of the greenhouse effect. In J. G. Titus (ed.), *Effects of changes in stratospheric ozone and global climate*, Vol. 4, *Sea level rise*. Washington, DC: United Nations Environment Program/U.S. Environmental Protection Agency, 153–64.

Detwyler, T. R. (ed.), 1971, *Man's impact on environment*. New York: McGraw-Hill.

Detwyler, T. R. and Marcus, M. G., 1972, *Urbanisation and environment: the physical geography of the city*. Belmont: Duxbury Press.

Diamond, J., 2002, Evolution, consequences and future of plant and animal domestication. *Nature*, 418, 700–7.

Di Castri, F., 1989, History of biological invasions with special emphasis on the old world. In J. A. Drake (ed.), *Biological invasions: a global perspective*. Chichester: Wiley, 1–30.

Dickinson, W. R., 1999, Holocene sea-level record on Fumafuti and potential impact of global warming on Central Pacific atolls. *Quaternary research*, 51, 124–32.

Dicks, B., 1977, Changes in the vegetation of an oiled Southampton Water salt marsh. In J. Cairns, K. L. Dickson and E. E. Herricks (eds), *Recovery and restoration of damaged ecosystems*. Charlottesville: University of Virginia, 72–101.

Diem, J., 2003, Potential impact of ozone on coniferous forests of the interior southwestern United States. *Annals of the Association of American Geographers*, 93, 265–80.

Dillehay, T. D., 2003, Tracking the first Americans. *Nature*, 425, 23–4.

Dimbleby, G. W., 1974, The legacy of prehistoric man. In A. Warren and F. B. Goldsmith (eds), *Conservation in practice*. London: Wiley, 179–89.

Dirzo, R. and Raven, P. H., 2003, Global state of biodiversity and loss. *Annual review of environment and resources*, 28, 137–67.

Doak, T. and Marvier, M., 2003, Predicting the effects of species loss on community stability. In P. Kareiva and S. A. Levin (ed.), *The importance of species*. Princeton: Princeton University Press, 140–60.

Dobson, M. C., 1991, De-icing salt damage to trees and shrubs. *Forestry Commission bulletin*, 101.

Dodd, A. P., 1959, The biological control of the prickly pear in Australia. In A. Keast, R. L. Crocker and C. S. Christian (eds), *Biogeography and ecology in Australia*. The Hague: Junk, 565–77.

D'Odorico, P., Yoo, J. C. and Over, T. M., 2001, An assessment of ENSO-induced patterns of rainfall erosivity in the southwestern United States. *Journal of climate*, 14, 4230–42.

Doerr, A. and Guernsely, L., 1956, Man as a geomorphological agent: the example of coal mining. *Annals of the Association of American Geographers*, 46, 197–210.

Dolan, R., Godfrey, P. J. and Odum, W. E., 1973, Man's impact on the barrier islands of North Carolina. *American scientist*, 61, 152–62.

Donkin, R. A., 1979, Agricultural terracing in the Aboriginal New World. *Viking Fund publications in anthropology*, 561.

Doody, P. (ed.), 1984, *Spartina anglica in Great Britain*. Shrewsbury: Nature Conservancy Council.

Doolittle, W., 2000, *Cultivated landscapes of native North America*. Oxford: Oxford University Press.

Doughty, R. W., 1974, The human predator: a survey. In I. R. Manners and M. V. Mikesell (eds), *Perspectives on environment*. Washington, DC: Association of American Geographers, 152–80.

——, 1978, The English sparrow in the American landscape: a paradox in nineteenth century wildlife conservation. Research paper 19, School of Geography, University of Oxford.

Douglas, I., 1969, The efficiency of humid tropical denudation systems. *Transactions of the Institute of British Geographers*, 46, 1–6.

——, 1983, *The urban environment*. London: Arnold.

Douglas, I. and 6 others, 1999, The role of extreme events in the impacts of selective tropical forestry on erosion during harvesting and recovery phases at Danum Valley, Sabah. *Philosophical transactions of the Royal Society of London*, B, 354, 1749–61.

Down, C. G. and Stocks, J., 1977, *Environmental impact of mining*. London: Applied Science Publishers.

Downs, P. W. and Gregory, K. J., 2004, *River channel management*. Arnold: London.

Doyle, T. W. and Girod, G. F., 1997, The frequency and structure of Atlantic hurricanes and their influence on the structure of the South Florida mangrove communities. In H. F. Diaz and R. S. Pulwarty (eds), *Hurricanes*. Berlin: Springer-Verlag, 109–20.

Dragovich, D. and Morris, R., 2002, Fire intensity, slopewash and bio-transfer of sediment in euclaypt forest, Australia. *Earth surface processes and landforms*, 27, 1309–19.

Dregne, H. E., 1986, Desertification of arid lands. In F. El-Baz and M. H. A. Hassan (eds), *Physics of desertification*. Dordrecht: Nijhoff, 4–34.

Dregne, H. E. and Tucker, C. J., 1988, Desert encroachment. *Desertification control bulletin*, 16, 16–19.

Drennan, D. S. H., 1979, Agricultural consequences of groundwater development in England. In G. E. Hollis (ed.), *Man's impact on the hydrological cycle in the United Kingdom*. Norwich: Geo Abstracts, 31–8.

Drewry, D. J., 1991, The response of the Antarctic ice sheet to climate change. In C. M. Harris and B. Stonehouse (eds), *Antarctica and global climatic change*. London: Belhaven Press, 90–106.

Driscoll, R., 1983, The influence of vegetation on the swelling and shrinkage of clay soils in Britain. *Géotechnique*, 33, 293–326.

Duncan, R. P., Blackburn, T. M. and Sol, D., 2003, The ecology of bird introductions. *Annual review of environment and resources*, 28, 359–99.

Dunne, T. and Leopold, L. B., 1978, *Water in environmental planning*. San Francisco: Freeman.

D'Yakanov, K. N. and Reteyum, A. Y., 1965, The local climate of the Rybinsk reservoir. *Soviet geography*, 6, 40–53.

Dyer, K. R., 1995, Responses of estuaries to climate change. In D. Eisma (ed.), *Climate change on coastal habitation*. Boca Raton: Lewis, 85–110.

Eden, M. J., 1974, Palaeoclimatic influences and the development of savanna in southern Venezuela. *Journal of biogeography*, 1, 95–109.

Edington, J. M. and Edington, M. A., 1977, *Ecology and environmental planning*. London: Chapman & Hall.

Edlin, H. L., 1976, The Culbin sands. In J. Leniham and W. W. Fletcher (eds), *Reclamation*. Glasgow: Blackie, 1–31.

Edmonson, W. T., 1975, Fresh water pollution. In W. W. Murdoch (ed.), *Environment*. Sunderland: Sinauer Associates, 251–71.

Edwards, A. J., Clark, S., Zahir, H., Rajasuriya, A., Naseer, A. and Rubens, J., 2001, Coral bleaching and mortality on artificial and natural reefs in Maldives in 1998, sea surface temperature anomalies and initial recovery. *Marine pollution bulletin*, 42, 7–15.

Edwards, A. M. C., 1975, Long term changes in the water quality for agricultural catchments. In R. D. Hey and T. D. Daniels (eds), *Science technology and environmental management*. Farnborough: Saxon House, 111–22.

Edwards, K. J., 1985, The anthropogenic factor in vegetational history. In K. J. Edwards and W. P. Warren (eds), *The Quaternary history of Ireland*. London: Academic Press, 187–200.

Ehrenfeld, D. W., 1972, *Conserving life on earth*. New York: Oxford University Press.

Ehrlich, P. R. and Ehrlich, A. H., 1970, *Population, resources, environment: issues in human ecology*. San Francisco: Freeman.

——, 1982, *Extinction*. London: Gollancz.

Ehrlich, P. R., Ehrlich, A. H. and Holdren, J. P., 1977, *Ecoscience: population, resources, environment*. San Francisco: Freeman.

Eitner, V., 1996, Geomorphological response to the East Frisan barrier islands to sea level rise: an investigation of past and future evolution. *Geomorphology*, 15, 57–65.

Ellenberg, H., 1979, Man's influence on tropical mountain ecosystems in South America. *Journal of ecology*, 67, 401–16.

Elliott, J. G., Gillis, A. C. and Aby, S. B., 1999, Evolution of arroyos: incised channels of the southwestern United States. In S. E. Darby and A. Simon (eds), *Incised river channels*. Chichester: Wiley, 153–85.

Ellis, J. B., 1975, Urban stormwater pollution. *Middlesex Polytechnic research report*, 1.

Ellison, A. M. and Farnsworth, E. J., 1997, Simulated sea level change alters anatomy, physiology, growth and reproduction of red mangrove (*Rhizophora mangle* L.). *Oecologia*, 112, 435–46.

Ellison, J. C. and Stoddart, D. R., 1990, Mangrove ecosystem collapse during predicted sea level rise: Holocene analogues and implications. *Journal of coastal research*, 7, 151–65.

El-Raey, M., 1997, Vulnerability assessment of the coastal zone of the Nile delta of Egypt to the impact of sea level rise. *Ocean and coastal management*, 37, 29–40.

Elsner, J. B. and Kara, A. B., 1999, *Hurricanes of the North Atlantic*. New York: Oxford University Press.

Elsom, D., 1992, *Atmospheric pollution* (2nd edn). Oxford: Blackwell.

Elton, C. S., 1958, *The ecology of invasions by plants and animals*. London: Methuen.

Emanuel, K. A., 1987, The dependence of hurricane intensity on climate. *Nature*, 326, 483–5.

Emmanuel, W. R., Shugart, H. H. and Stevenson, M. P., 1985, Climatic change and the broad-scale distribution of terrestrial ecosystem complexes. *Climatic change*, 7, 29–43.

Engelhardt, F. R. (ed.), 1985, *Petroleum effects in the Arctic environment*. London: Elsevier Applied Science.

EPICA community members, 2004, Eight glacial cycles from an Antarctic ice core. *Nature*, 429, 623–8.

Erhart, H., 1956, *La genèse des sols en tant que phénomène gélogique*. Paris: Masson.

European Environment Agency, 2001, *Eutrophication in Europe's coastal waters*. Copenhagen: European Environment Agency.

Evans, C. D. and Jenkins, A., 2000, Surface water acidification in the South Pennines II. Temporal trends. *Environmental pollution*, 109, 21–34.

Evans, C. D. and 9 others, 2001, Recovery from acidification in European surface waters. *Hydrology and earth system sciences*, 5, 283–97.

Evans, D. M., 1966, Man-made earthquakes in Denver. *Geotimes*, 10, 11–18.

Evans, J. G., Limbrey, S. and Cleere, H. (eds), 1975, *The effect of man on the landscape: the Highland zone*. Council for British Archaeology research report 11.

Evans, I. S., 1970, Salt crystallisation and rock weathering: a review. *Revue de géomorphologie dynamique*, 19, 153–77.

Evans, R. and Northcliffe, S., 1978, Soil erosion in North Norfolk. *Journal of agricultural science*, 90, 185–92.

Eve, M. D., Sperow, M., Paustian, K. and Follett, R. F., 2002, National-scale estimation of changes in soil carbon stocks on agricultural lands. *Environmental pollution*, 116, 431–8.

Evenari, M., Shanan, L. and Tadmor, N. H., 1971, Runoff agriculture in the Negev Desert of Israel. In W. G.

McGinnies, R. J. Goldman and P. Paylore (eds), *Food, fiber and the arid lands*. Tucson: University of Arizona Press, 312–22.

Fahrig, L., 2003, Effects of habitat fragmentation on biodiversity. *Annual review of ecology, evolution and systematics*, 34, 487–515.

Fairbridge, R. W., 1983, Isostasy and eustasy. In D. E. Smith and A. G. Dawson (eds), *Shorelines and isostasy*. London: Academic Press, 3–25.

Fairhead, J. and Leach, M., 1996, *Misreading the African landscape: society and ecology in a forest-savanna mosaic*. Cambridge: Cambridge University Press.

FAO, 2001, *The state of the world's forests*. Rome: Food and Agriculture Organisation.

Favis-Mortlock, D. and Boardman, J., 1995, Non linear responses of soil erosion to climate change: modelling study on the UK South Downs. *Catena*, 25, 365–87.

Favis-Mortlock, D. T. and Guerra, A. J. T., 1999, The implications of general circulation model estimates of rainfall for future erosion: a case study from Brazil. *Catena*, 37, 329–54.

Fearnside, P. M. and Laurance, W. F., 2003, Comment on 'Determination rates of the world's humid tropical forests'. *Science*, 299, 1015.

Feare, C. J., 1978, The decline of booby (Sulidae) population in the Western Indian Ocean. *Biological conservation*, 14, 295–305.

Federal Research Centre for Forestry and Forest Products, 2000, *Forest condition in Europe*. Geneva and Brussels: UN/ECE and EC.

Fenger, J., 1999, Urban air quality. *Atmospheric environment*, 33, 4877–900.

Ferrians, O. J., Kachadoorian, R. and Green, G. W., 1969, Permafrost and related engineering problems in Alaska. *United States Geological Survey professional paper*, 678.

Fillenham, L. F., 1963, Holme Fen Post. *Geographical journal*, 129, 502–3.

Fisher, J., Simon, N. and Vincent, J., 1969, *The red book – wildlife in danger*. London: Collins.

Fitt, W. K., Brown, B. E., Warner, M. E. and Dunne, R. P., 2001, Coral bleaching: interpretation of thermal tolerance limits and thermal thresholds in tropical corals. *Coral reefs*, 20, 51–65.

Fitzharris, B., 1996, The cryosphere: changes and their impacts. In Watson, R. T., Zinyowera, M. C., Moss R. H. and Dokken, D. J. (eds), *Climate change 1995. Impacts, adaptation and mitigation of climate change: scientific and technical analyses*. Cambridge: Cambridge University Press, 241–65.

Fitzpatrick, J., 1994, *A continent transformed: human impact on the natural vegetation of Australia*. Melbourne: Oxford University Press.

Flannigan, M. D., Stocks, B. J. and Wotton, B. M., 2000, Climate change and forest fires. *The science of the total environment*, 262, 221–30.

Flenley, J. R., 1979, *The equatorial rain forest: a geological history*. London: Butterworth.

Flenley, J. R., King, A. S. M., Jackson, J., Chew, C., Teller, J. and Prentice, M. E., 1991, The late Quaternary vegetational and climatic history of Easter Island. *Journal of Quaternary science*, 6, 85–115.

Fölster, J. and Wilander, A., 2002, Recovery from acidification in Swedish forest streams. *Environmental pollution*, 117, 389–89.

Forland, E. J., Alexandersson, H., Drebs, A., Hamssen-Bauer, I., Vedin, H. and Tveito, O. E., 1998, Trends in maximum 1-day precipitation in the Nordic region. *DNMI report 14/98, Klima*. Oslo: Norwegian Meteorological Institute, 1–55.

Forman, S. L., Oglesby, R., Markgraf, V. and Stafford, T., 1995. Paleoclimatic significance of late Quaternary eolian deposition on the Piedmont and High Plains, central United States. *Global and planetary change*, 11, 35–55.

Forman, S. L., Oglesby, R. and Webb, R. S., 2001, Temporal and spatial patterns of Holocene dune activity on the Great Plains of North America: megadroughts and climate links. *Global and planetary change*, 29, 1–29.

Forman, R. T. T. and 13 others, 2003, *Road ecology. Science and solutions*. Washington, DC: Island Press.

Foster, I. D. L., Dearing, J. A. and Appleby, R. G., 1986, Historical trends in catchment sediment yields: a case study in reconstruction from lake-sediment records in Warwickshire, UK. *Hydrological science journal*, 31, 427–43.

Fox, H. L., 1976. The urbanizing river: a case study in the Maryland piedmont. In D. R. Coates (ed.), *Geomorphology and engineering*. Stroudsburg: Dowden, Hutchinson and Ross, 245–71.

Francis, D. and Hengeveld, H., 1998, *Extreme weather and climate change*. Downsview, Ontario: Environment Canada.

Frankel, O. H., 1984, Genetic diversity, ecosystem conservation and evolutionary responsibility. In F. D. Castri, F. W. G. Baker and M. Hadley (eds), *Ecology in practice*, Vol. I. Dublin: Tycooly, 4315–27.

Freedman, B., 1995, *Environmental ecology* (2nd edn). San Diego: Academic Press.

Freeland, W. J., 1990, Large herbivorous mammals: exotic species in northern Australia. *Journal of biogeography*, 17, 445–9.

French, H. M., 1976, *The periglacial environment*. London: Longman.

—, 1996, *The periglacial environment* (2nd edn). Harlow: Longman.

French, J. R., Spencer, T. and Reed, D. J. (eds), 1994, Geomorphic responses to sea level rise: existing evidence and future impacts. *Earth surface processes and landforms*, 20, 1–6.

French, P. W., 1997, *Coastal and estuarine management*. London: Routledge.

—, 2001, *Coastal defences. Processes, problems and solutions*. London: Routledge.

Frenkel, R. E., 1970, Ruderal vegetation along some California roadsides. *University of California publications in geography*, 20.

Fu, C., 2003, Potential impacts of human-induced land cover change on East Asia monsoon. *Global and planetary change*, 37, 219–29.

Fullen, M. A., 2003, Soil erosion and conservation in northern Europe. *Progress in physical geography*, 27, 331–58.

Fuller, D. O. and Ottka, C., 2002, Land cover, rainfall and land-surface albedo in West Africa. *Climate change*, 54, 181–204.

Fuller, R., Hill, D. and Tucker, G., 1991, Feeding the birds down on the farm: perspectives from Britain. *Ambio*, 20(6), 232–7.

Gabunia, L. and Vekua, A., 1995, A Plio-Pleistocene hominid from Dmanisi, East Georgia, Caucasus. *Nature*, 373, 509–12.

Gade, D. W., 1976, Naturalization of plant aliens: the volunteer orange in Paraguay. *Journal of biogeography*, 3, 269–79.

Galay, V. J., 1983, Causes of river bed degradation. *Water resources research*, 19(5), 1057–90.

Gameson, A. L. H. and Wheeler, A., 1977, Restoration and recovery of the Thames Estuary. In J. Cairns, K. L. Dickson and E. E. Herricks (eds), *Recovery and restoration of damaged ecosystems*. Charlottesville: University Press of Virginia, 92–101.

Gardner, A. R. and Willis, K. J., 1999, Prehistoric farming and the postglacial expansion of beech and hornbeam: a comment on Küster. *The Holocene*, 9, 119–22.

Gates, D. M., 1993, *Climate change and its biological consequences*. Sunderland, MA: Sinauer.

Gattuso, J.-P., Frankignoulle, M., Bourge, I., Romaine, S. and Buddemeier, R. W., 1998, Effect of calcium carbonate saturation of seawater on coral calcification. *Global and planetary change*, 18, 37–46.

Gaylord, D. R., 1990, Holocene palaeoclimatic fluctuations revealed from dune and interdune strata in Wyoming. *Journal of arid environments*, 18, 123–38.

Geertz, C., 1963, *Agricultural involution: the process of ecological change in Indonesia*. Berkeley: University of California Press.

GESAMP (IMO/FAO/UNESCO/WMO/IAEA/UN/UNEP Joint Group of Experts on the Scientific Aspects of Marine Pollution), 1990, *The state of the marine environment*. United Nations Environment Program Regional Seas Reports and Studies, 115, Nairobi.

Ghassemi, F., Jakeman, A. J. and Nix, H. A., 1995, *Salinisation of land and water resources*. Wallingford: CAB International, 536.

Gifford, G. F. and Hawkins, R. H., 1978, Hydrological impact of grazing on infiltration: a critical review. *Water resources research*, 14, 305–13.

Gilbert, G. K., 1917, Hydraulic mining debris in the Sierra Nevada. *United States Geological Survey professional paper*, 105.

Gilbert, O. L., 1970, Further studies on the effect of sulphur dioxide on lichens and bryophytes. *New phytologist*, 69, 605–27.

—, 1975, Effects of air pollution on landscape and land-use around Norwegian aluminium smelters. *Environmental pollution*, 8, 113–21.

Gill, T. E., 1996, Eolian sediments generated by anthropogenic disturbance of playas: human impacts on the geomorphic system and geomorphic impacts on the human system. *Geomorphology*, 17, 207–28.

Gillespie, R. and 6 others, 1978, Lancefield Swamp and the extinction of the Australian megafauna. *Science*, 200, 1044–8.

Gillon, D., 1983, The fire problem in tropical savannas. In F. Bourlière (ed.), *Tropical savannas*. Oxford: Elsevier Scientific, 617–41.

Gimingham, C. H., 1981, Conservation: European heathlands. In R. L. Spect (ed.), *Heathlands and related shrublands*. Amsterdam: Elsevier Scientific, 249–59.

Gimingham, C. H. and de Smidt, I. T., 1983, Heaths and natural and semi-natural vegetation. In W. Holzner, M. J. A. Werger and I. Ikusima (eds), *Man's impact on vegetation*. Hague: Junk, 185–99.

Glacken, C., 1963, The growing second world within the world of nature. In F. R. Fosberg (ed.), *Man's place in the island ecosystem*. Honolulu: Bishop Museum Press, 75–100.

—, 1967, *Traces on the Rhodian shore: nature and culture in western thought from ancient times to the end of the eighteenth century*. Berkeley: University of California Press.

Glade, T., 2003, Landslide occurrence as a response to land use change: a review of evidence from New Zealand. *Catena*, 51, 297–314.

Glieck, P. J. (ed.), 1993, *Water in crisis: a guide to the world's freshwater resources*. New York: Oxford University Press.

—, 2002, Dams. In A. S. Goudie (ed.), *Encyclopedia of global change*. New York: Oxford University Press, 229–34.

Godbole, N. N., 1972, Theories on the origin of salt lakes in Rajasthan, India. *Proceedings of the 24th International Geological Congress*, Section 10, 354–7.

Goldberg, E. D. and 7 others, 1978, A pollution history of Chesapeake Bay. *Geochimica et cosmochimica acta*, 42, 1413–25.

Goldewijk, K. K., 2001, Estimating global land use change over the past 300 years: the HYDE database, *Global biogeochemical cycles*, 15, 417–33.

Goldsmith, F. B., 1983, Evaluating nature. In A. Warren and F. B. Goldsmith (eds), *Conservation in perspective*. Chichester: Wiley, 233–46.

Goldsmith, V., 1978, Coastal dunes. In R. A. Davis (ed.), *Coastal sedimentary environments*. New York: Springer-Verlag.

Gomez, B. and Smith, C. G., 1984, Atmospheric pollution and fog frequency in Oxford, 1926–1980. *Weather*, 39, 379–84.

Gong, Z.-T., 1983, Pedogenesis of paddy soil and its significance in soil classification. *Soil science*, 135, 5–10.

Gonzalez, M. A., 2001, Recent formation of arroyos in the Little Missouri Badlands of southwestern Dakota. *Geomorphology*, 38, 63–84.

Goodman, D., 1975, The theory of diversity–stability relationships in ecology. *Quarterly review of biology*, 50, 237–66.

Goreau, T. J. and Hayes, R. L., 1994, Coral bleaching and ocean 'hot spots'. *Ambio*, 23, 176–80.

Gorman, M., 1979, *Island ecology*. London: Chapman & Hall.

Gornitz, V., Rosenzweig, C. and Hillel, D., 1997, Effects of anthropogenic intervention in the land hydrologic cycle on global sea level rise. *Global and planetary change*, 14, 147–61.

Gornitz, V., Couch, S. and Hartig, E. K., 2002, Impacts of sea level rise in the New York City metropolitan area. *Global and planetary change*, 32, 61–88.

Gosden, C., 2003, *Prehistory: a very short introduction*. Oxford: Oxford University Press.

Gottschalk, L. C., 1945, Effects of soil erosion on navigation in Upper Chesapeake Bay. *Geographical review*, 35, 219–38.

Goudie, A. S., 1972a, The concept of post-glacial progressive desiccation. Research paper 4, School of Geography, University of Oxford.

—, 1972b, Vaughan Cornish: geographer. *Transactions of the Institute of British Geographers*, 55, 1–16.

—, 1973, *Duricrusts of tropical and subtropical landscapes*. Oxford: Clarendon Press.

—, 1977, Sodium sulphate weathering and the disintegration of Mohenjo-Daro, Pakistan. *Earth surface processes*, 2, 75–86.

—, 1983, Dust storms in space and time. *Progress in physical geography*, 7, 502–30.

—, (ed.), 1990, *Techniques for desert reclamation*. Chichester: Wiley.

—, 1992, *Environmental change* (3rd edn). Oxford: Clarendon Press.

—, 1994, Deserts in a warmer world. In A. C. Millington and K. Pye (eds), *Environmental change in drylands: biogeographical and geomorphological perspectives*. Chichester: Wiley, 1–24.

—, 2002, *Great warm deserts of the world: landscape and evolution*. Oxford: Oxford University Press.

Goudie, A. S. and Middleton, N. S., 1992, The changing frequency of dust storms through time. *Climatic change*, 20, 197–225.

Goudie, A. S. and Viles, H. A., 1997, *Salt weathering hazards*. Chichester: Wiley.

Goudie, A. S. and Wilkinson, J. C., 1977, *The warm desert environment*. Cambridge: Cambridge University Press.

Goudie, A. S., Viles, H. A. and Pentecost, A., 1993, The late-Holocene tufa decline in Europe. *The Holocene*, 3, 181–6.

Goudie, A. S., Parker, A. G. and Al-Farrai, A., 2000, Coastal change in Ras Al Khaimah (United Arab Emirates): a cartographic analysis. *Geographical journal*, 166, 14–25.

Goulson, D. 2003, Effects of introduced bees on native ecosystems. *Annual review of ecology, evolution and systematics*, 34, 1–26.

Gourou, P., 1961, *The tropical world* (3rd edn). London: Longman.

Gow, T. J. (ed.), 2002. *The speciation of modern Homo sapiens*. Oxford: Oxford University Press.

Gowlett, J. A. J., Harris, J. W. K., Walton, D. and Wood, B. A., 1981, Early archaeological sites, hominid remains and traces of fire from Chesowanja, Kenya. *Nature*, 284, 125–9.

Graedel, T. E. and Crutzen, P. J., 1993, *Atmospheric change: an earth system perspective.* San Francisco: Freeman.

—, 1995, *Atmosphere, climate and change.* New York: Scientific American Library.

Graetz, D., 1994, Grasslands. In W. B. Meyer and B. L. Turner II (eds), *Changes in land use and land cover: a global perspective.* Cambridge: Cambridge University Press, 125–47.

Graf, W. K., 1977, Network characteristics in suburbanizing streams. *Water resources research*, 13, 459–63.

Graf, W. L., 1988, *Fluvial processes in dryland rivers.* Berlin: Springer-Verlag.

—, 1999, Dam nation: a geographic census of American dams and their large-scale hydrologic impacts. *Water resources research*, 35, 1305–11.

—, 2001, Damage control: restoring the physical integrity of America's rivers. *Annals of the Association of American Geographers*, 91, 1–27.

Graf, J. B., Webb, R. H. and Hereford, R., 1991, Relation of sediment load and flood-plain formation to climatic variability, Paria River drainage basin, Utah and Arizona. *Bulletin of the Geological Society of America*, 103, 1405–15.

Grainger, A., 1990, *The threatening desert: controlling desertification.* London: Earthscan.

—, 1992, *Controlling tropical deforestation.* London: Earthscan.

Gray, R., 1993, Regional meteorology and hurricanes. In G. A. Maul (ed.), *Climatic change in the Intra-Americas Sea.* London: Edward Arnold, 87–99.

Grayson, D. K., 1977, Pleistocene avifaunas and the overkill hypothesis. *Science*, 195, 691–3.

—, 1988, Perspectives on the archaeology of the first Americans. In R. C. Carlisle (ed.), *Americans before Columbus: Ice Age origins.* Pittsburgh: University of Pittsburgh, 107–23.

Green, F. H. W., 1978, Field drainage in Europe. *Geographical journal*, 144, 171–4.

Green, R. C., 1975, Adaptation and change in Maori culture. *Monographiae biologicae* (New Guinea), 27, 591–661.

Greenland, D. J. and Lal, R., 1977, *Soil conservation and management in the humid tropics.* Chichester: Wiley.

Gregory, J. M. and Oelemans, J., 1998, Simulated future sea-level rise due to glacier melt based on regionally and seasonally resolved temperature changes. *Nature*, 391, 474–6.

Gregory, J. M. and 12 others, 2001, Comparisons of results from several AOGCMs for global and regional sea level change 1900–2100. *Climate dynamics*, 18, 225–40.

Gregory, J. M., Huybrechts P. and Raper, S. C. B. 2004, Threatened loss of the Greenland ice-sheet. *Nature*, 428, 616.

Gregory, K. J., 1976, Drainage networks and climate. In E. Derbyshire (ed.), *Geomorphology and climate.* Chichester: Wiley, 289–315.

—, 1985, The impact of river channelization. *Geographical journal*, 151, 53–74.

Gregory, K. J. and Walling, D., 1973, *Drainage basin form and process: a geomorphological approach.* London: Arnold.

—, 1979, *Man and environmental processes.* Folkestone: Dawson.

Grieve, I. C., 2001, Human impacts on soil properties and their implications for the sensitivity of soil systems in Scotland. *Catena*, 42, 361–74.

Griffiths, J. F., 1976, *Applied climatology, an introduction* (2nd edn). Oxford: Oxford University Press.

Grigg, D., 1970, *The harsh lands.* London: Macmillan.

Grimm, N. B. and 6 others, 1997, Sensitivity of aquatic ecosystems to climatic and anthropogenic changes: the Basin and Range, American south-west and Mexico. In C. E. Cushing (ed.), *Freshwater ecosystems and climate change in North America: a regional assessment.* Chichester: Wiley, 205–23.

Grosjean, M., Núñez, L., Castajena, I. and Messerli, B., 1997, Mid-Holocene climate and culture change in the Atacama Desert, northern Chile. *Quaternary research*, 48, 239–46.

Gross, M. G., 1972, Geological aspects of waste solids and marine waste deposits, New York Metropolitan region. *Bulletin of the Geological Society of America*, 83, 3163–76.

Grove, A. T. and Rackham, O., 2001, *The nature of Mediterranean Europe: an ecological history.* New Haven and London: Yale University Press.

Grove, J. M., 1988, *The Little Ice Age.* London: Routledge.

Grove, R. H., 1983, *The future for forestry.* Cambridge: British Association of Nature Conservationists.

—, 1997, *Ecology, climate and empire: colonialism and global environmental history, 1400–1940.* Cambridge: White Horse Press.

Grover, H. D. and Musick, H. B., 1990, Shrubland encroachment in southern New Mexico, USA: an analysis of desertification processes in the American southwest. *Climatic change*, 17, 305–30.

Guha, S. K. (ed.), 2000, *Induced earthquakes.* Dordrecht: Kluwer.

Guicherit, R. and Roemer, M., 2000, Tropospheric ozone trends. *Chemosphere – global change science*, 2, 167–83.

Guidon, N. and Delibrias, G., 1986, Carbon 14 dates point to man in the Americas 32,000 years ago. *Nature*, 321, 769–71.

Guilday, J. E., 1967, Differential extinction during Late Pleistocene and recent time. In P. S. Martin and H. E. Wright (eds), *Pleistocene extinctions.* New Haven: Yale University Press, 121–40.

Guo, S., Wang, J., Xiang, L., Ying, A. and Li, D., 2002, A macro-scale and semi-distributed monthly water balance model to predict climate change impacts in China. *Journal of hydrology*, 268, 1–15.

Gupta, H. K., 2002, A review of recent studies of triggered earthquakes by artificial water reservoirs with special emphasis on earthquakes in Koyna, India. *Earth-science reviews*, 58, 279–310.

Guthrie, R. D., 2003, Rapid body size decline in Alaskan Pleistocene horses before extinction. *Nature*, 426, 169–71.

Gutierrez-Elorza, M., 2001, *Geomorfología climática.* Barcelona: Omega.

Haeberli, W. and Burn, C. R., 2002, Natural hazards in forests: glacier and permafrost effects as related to climate change. In R. C. Sidle (ed.), *Environmental change and geomorphic effects in forests.* Wallingford: CABI, 167–202.

Haggett, P., 1979, *Geography: a modern synthesis* (3rd edn). London: Prentice Hall.

Haigh, M. J., 1978, Evolution of slopes on artificial landforms – Blaenavon, UK. Research paper 183, Department of Geography, University of Chicago.

Hails, J. R. (ed.), 1977, *Applied geomorphology*. Amsterdam: Elsevier.

Hails, R. J., 2002, Assessing the risks associated with new agricultural practices. *Nature*, 418, 685–8.

Hall, S. R. and Mills, E. L., 2000, Exotic species in large lakes of the world. *Aquatic ecosystem health and management*, 3, 105–35.

Hanes, T. L., 1971, Succession after fire in the chaparral of Southern California. *Ecological monographs*, 41, 27–52.

Hannah, L., Lohse, D., Hutchinson, C., Carr, L. and Lankerani, A., 1994, A preliminary inventory of human disturbance of world ecosystems. *Ambio*, 23, 246–50.

Hansen, J. and 6 others, 1981, Climatic impact of increasing atmospheric carbon dioxide. *Science*, 213, 957–66.

Hanson, P. J. and Weltzin, J. F. 2000, Drought disturbance from climate change: response of United States forests. *Science of the total environment*, 262, 205–20.

Happ, S. C., 1944, Effect of sedimentation on floods in the Kickapoo Valley, Wisconsin. *Journal of geology*, 52, 53–68.

Harlan, J. R., 1975a, Our vanishing genetic resources. *Science*, 188, 617–22.

—, 1975b, *Crops and man*. Madison: American Society of Agronomy.

—, 1976, The plants and animals that nourish man. *Scientific American*, 235, 3, 88–97.

Harris, C., Davies, M. C. R. and Etzelmüller, B., 2001, The assessment of prudential geotechnical hazards associated with mountain permafrost in a warming global climate. *Permafrost and periglacial processes*, 12, 145–56.

Harris, D. R., 1966, Recent plant invasions in the arid and semi-arid southwest of the United States. *Annals of the Association of American Geographers*, 56, 408–22.

— (ed.), 1980, *Human ecology in savanna environments*. London: Academic Press.

— (ed.), 1996, *The origins and spread of agriculture and pastoralism in Eurasia*. London: UCL Press.

Harris, J. M., Oltmans, S. J., Bodeker, G. E., Stolarski, R., Evans, R. D. and Quincy, D. M., 2003, Long-term restrictions in total ozone derived from Dobson and satellite data. *Atmospheric environment*, 37, 3167–75.

Harris, S. A., 2002, Causes and consequences of rapid thermokarst development in permafrost or glacial terrain. *Permafrost and periglacial processes*, 13, 237–42.

Harrison, G. P. and Whittington, H. W., 2002, Susceptibility of the Batoka Gorge hydroelectric scheme to climate change. *Journal of hydrology*, 264, 230–41.

Harvey, A. M., 1989, The occurrence and role of arid zone alluvial fans. In D. S. G. Thomas (ed.), *Arid zone geomorphology*. London: Belhaven Press, 136–58.

Harvey, A. M. and Renwick, W. H., 1987, Holocene alluvial fan and terrace formation in the Bowland Fells, Northwest England. *Earth surface processes and landforms*, 12, 249–57.

Harvey, A. M., Oldfield, F., Baron, A. F. and Pearson, G. W., 1981, Dating of post-glacial landforms in the central Howgills. *Earth surface processes and landforms*, 6, 401–12.

Harvey, L. D. D., 2000, *Global warming. The hard science*. Harlow: Prentice Hall.

Hawass, Z., 1993, The Egyptian monuments: problems and solutions. In M. J. Thiel (ed.), *Conservation of stone and other material*. London: Spon, 19–25.

Hawksworth, D. L., 1990, The long-term effects of air pollutants on lichen communities in Europe and North America. In G. M. Woodwell (ed.), *The earth in transition: patterns and processes of biotic impoverishment*. Cambridge: Cambridge University Press, 45–64.

Hay, J., 1973, Salt cedar and salinity on the Upper Rio Grande. In M. T. Farvar and J. P. Milton (eds), *The careless technology*. London: Tom Stacey, 288–300.

Haynes, C. V., 1991, Geoarchaeological and palaeohydrological evidence for a Clovis-age drought in North America and its bearing on extinction. *Quaternary research*, 35, 438–50.

Healy, T., 1991, Coastal erosion and sea level rise. *Zeitschrift für geomorphologie, supplementband*, 81, 15–29.

—, 1996, Sea level rise and impacts on nearshore sedimentation. *Geologische rundschau*, 85, 546–53.

Heathwaite, A. L., Johnes, P. J. and Peters, N. E., 1996, Trends in nutrients. *Hydrological processes*, 10, 263–93.

Heinselman, M. L. and Wright, H. E., 1973, The ecological role of fire in natural conifer forests of western and northern America. *Quaternary research*, 3, 317–482.

Helldén, U., 1985, Land degradation and land productivity monitoring – needs for an integrated approach. In A. Hjört (ed.), *Land management and survival*. Uppsala: Scandinavian Institute of African Studies, 77–87.

Helliwell, D. R., 1974, The value of vegetation for conservation. II: M1 motorway area. *Journal of environmental management*, 2, 75–8.

Henderson-Sellers, A. and Blong, R., 1989, *The greenhouse effect: living in a warmer Australia*. Kensington, NSW: New South Wales University Press.

Henderson-Sellers, A. and Gornitz, V., 1984, Possible climatic impacts of land cover transformation with particular emphasis on tropical deforestation. *Climatic change*, 6, 231–57.

Henderson-Sellers, A. and Robinson, P. J., 1986, *Contemporary climatology*. London: Longman.

Hennessy, K. J., Gregory, J. M. and Mitchell, J. F. B., 1997, Changes in daily precipitation under enhanced greenhouse conditions. *Climate dynamics*, 13, 667–80.

Hereford, R., 1984, Climate and ephemeral-stream processes: twentieth-century geomorphology and alluvial stratigraphy of the Little Colorado River, Arizona. *Bulletin of the Geological Society of America*, 95, 654–68.

Hereford, R., Jacoby, G. C. and McCord, V. A. S., 1995, Geomorphic history of the Virgin River in the Zion National Park area, southwest Utah. *US Geological Survey circular*, 95–515.

Hess, W. N. (ed.), 1974, *Weather and climate modification*. New York: Wiley.

Hewlett, J. D., Post, H. E. and Doss, R., 1984, Effect of clear-cut silviculture on dissolved ion export and water yield in the Piedmont. *Water resources research*, 20(7), 1030–8.

Heywood, V. H., 1989, Patterns, extents and modes of invasions by terrestrial plants. In J. A. Drake (ed.), *Biological invasions: a global perspective*. Chichester: Wiley, 31–55.

Heywood, V. H. and Watson, R. T. (eds), 1995, *Global biodiversity assessment*. Cambridge: Cambridge University Press.

Hickey, J. J. and Anderson, O. W., 1968, Chlorinated hydrocarbons and eggshell changes in raptorial and fish-eating birds. *Science*, 162, 271–2.

Hilborn, R., Branch, T. A., Ernst, B., Magnusson, A., Minte-Vera, C. V. and Mark, D. S., 2003, State of the world's fisheries. *Annual review of environment and resources*, 28, 359–99.

Hill, A. R., 1975, Ecosystems stability in relation to stresses caused by human activities. *Canadian geographer*, 19, 206–20.

Hillel, D., 1971, Artificial inducement of runoff as a potential source of water in arid lands. In W. G. McGinnies, B. J. H. Goldman and P. Paylore (eds), *Food, fiber and the arid lands*. Tucson: University of Arizona Press, 324–30.

Hills, T. L., 1965, Savannas: a review of a major research problem in tropical geography. *Canadian geographer*, 9, 216–28.

Hobbs, P. V. and Radke, L. F., 1992, Airborne studies of the smoke from the Kuwait oil fires. *Science*, 256, 987–91.

Hoegh-Guldberg, O., 1999, Climate change, coral bleaching and the future of the world's coral reefs. *Marine and freshwater research*, 50, 839–66.

——, 2001, Sizing the impact: coral reef ecosystems as early casualties of climate change. In G. R. Walther, C. A. Burga and P. J. Edwards (eds), *Fingerprints of climate change*. New York: Kluwer/Plenum, 203–28.

Hoelzle, M. and Trindler, M., 1998, Data management and application. In W. Haeberli et al. (eds), *Into the second century of worldwide glacier monitoring: prospects and strategies*. Paris: UNESCO, 53–64.

Hoelzmann, P., Keding, B., Berke, H., Kröpelin, S. and Kruse, H.-J., 2001, Environmental change and archaeology: lake evolution and human occupation in the eastern Sahara during the Holocene. *Palaeogeography, palaeoclimatology, palaeoecology*, 169, 193–217.

Hoffman, J. S., Keyes, D. and Titus, J. G., 1983, *Projecting future sea level rise: methodology, estimates to the year 2100, and research needs*. U.S. Environmental Protection Agency Report 230-09-007. Washington, DC: Government Printing Office.

Hoffman, J. S., Wells, J. B. and Titus, J. G., 1986, Future global warming and sea level rise. In G. Sigbjarnason (ed.), *Iceland coastal and river symposium*, Reykavik, Natural Energy Authority, 245–66.

Holden, J., Chapman, P. J. and Labadz, J. C., 2004, Artifical drainage of peatlands: hydrological and hydrochemical process and wetland restoration. *Progress in physical geography*, 28, 95–123.

Holdgate, M. W., 1979, *A perspective on environmental pollution*. Cambridge: Cambridge University Press.

Holdgate, M. W. and Wace, N. M., 1961, The influence of man on the floras and faunas of southern islands. *Polar record*, 10, 473–93.

Holdgate, M. W., Kassas, M. and White, G. F., 1982, *The world environment 1972–1982*. Dublin: Tycooly.

Holland, G. J., McBridge, J. L. and Nicholls, N., 1988, Australian region tropical cyclones and the greenhouse effect. In G. I. Pearman (ed.), *Greenhouse, planning for climate change*. Leiden: Brill, 438–55.

Holliday, V. T., 1995, Stratigraphy and paleoenvironments of Late Quaternary Valley Fills on the Southern High Plains. *Geological Society of America, memoir*, 186.

——, 1997, Origin and evolution of lunettes on the High Plains of Texas and New Mexico. *Quaternary research*, 47, 54–69.

——, 2004, *Soils in archaeological research*. New York: Oxford University Press.

Hollis, G. E., 1975, The effects of urbanization on floods of different recurrence interval. *Water resources research*, 11, 431–5.

——, 1988, Rain, roads, roofs and runoff: hydrology in cities. *Geography*, 73, 9–18.

——, 1978, The falling levels of the Caspian and Aral Seas. *Geographical journal*, 144, 62–80.

Hollis, G. E. and Luckett, J. K., 1976, The response of natural river channels to urbanization: two case studies from Southeast England. *Journal of hydrology*, 30, 351–63.

Holloway, G. and Sou, T., 2002, Has Arctic sea ice rapidly thinned? *Journal of climate*, 15, 1691–701.

Holm, K., Bovis, M. and Jacob, M., 2004, The landslide response of alpine basins to post-Little Ice Age glacial thinning and retreat in southwestern British Columbia. *Geomorphology*, 57, 201–16.

Holtz, W. G., 1983, The influence of vegetation on the swelling and shrinking of clays in the United States of America. *Géotechnique*, 33, 159–63.

Holzer, T. L., 1979, Faulting caused by groundwater extraction in South-central Arizona. *Journal of geophysical research*, 84, 603–12.

Hooke, J. (ed.), 1998, *Coastal defence and earth science conservation*. Bath: Geological Society of London.

Hooke, R. L., 1994, On the efficacy of humans as geomorphic agents. *USA today*, 4, 217, 224–5.

Hope, G., 1999, Vegetation and fire response to late Holocene human occupation in island and mainland north west Tasmania. *Quaternary international*, 59, 47–60.

Hopkins, B., 1965, Observations on savanna burning in the Olikemeji forest reserve, Nigeria. *Journal of applied ecology*, 2, 367–81.

Horn, R., van der Akker, J. J. H. and Arvidsson, J. (eds), 2000, *Subsoil compaction: distribution, processes and consequences*. Reiskirchen: Catena Verlag.

Houghton, J. T., Jenkins, G. J. and Ephraums, J. J., 1990, *Climate change: the IPCC scientific assessment*. Cambridge: Cambridge University Press.

Houghton, J. T., Callander, B. A. and Varney, S. K. (eds), 1992, *Climate change 1992: the supplementary report of the IPCC scientific assessment.* Cambridge: Cambridge University Press.

Houghton, J. T. and 7 others (eds), 2001, *Climate change 2001: the scientific basis.* Cambridge: Cambridge University Press.

Houghton, R. H. and Skole, D. L., 1990, Carbon. In B. L. Turner II (ed.), *The earth transformed by human action.* Cambridge: Cambridge University Press, 393–408.

Howard, K. W. F. and Beck, P. J., 1993, Hydrochemical implications of groundwater contamination by road de-icing chemicals. *Journal of contaminant hydrology*, 12, 245–68.

Howe, G. M., Slaymaker, H. O. and Harding, D. M., 1966, Flood hazard in mid-Wales. *Nature*, 212, 584–5.

Huang, W., Clochon, R., Gu, Y., Larick, R., Fang, Q., Schwartz, H., Yonge, C., de Vos, J. and Rink, W., 1995, Early *Homo* and associated artefacts from Asia. *Nature*, 378, 275–8.

Hudson, B. J., 1979, Coastal land reclamation with special reference to Hong Kong. *Reclamation review*, 2, 3–16.

Hudson, N., 1987, Soil and water conservation in semi-arid areas. *FAO soils bulletin*, 55.

Hughes, M. K., Lepp, N. W. and Phipps, D. A., 1980, Aerial heavy metal pollution and terrestrial ecosystems. *Advances in ecological research*, 11, 217–327.

Hughes, R. J., Sullivan, M. E. and Yok, D., 1991, Human-induced erosion in a highlands catchment in Papua New Guinea: the prehistoric and contemporary records. *Zeitschrift für geomorphologie, supplementband*, 83, 227–39.

Hughes, T. P., 1994, Catastrophes, phase shifts, and large-scale degradation of a Caribbean coral reef. *Science*, 265, 1547–51.

Hull, S. K. and Gibbs, J. N., 1991, Ash dieback: a survey of nonwoodland trees. *Forestry Commission bulletin*, 93, 32.

Huntington, E., 1914, *The climatic factor as illustrated in arid America.* Carnegie Institution of Washington publication, 192.

Hurd, L. E., Mellinger, M. W., Wold, L. L. and McNaughton, S. J., 1971, Stability and diversity at three trophic levels in terrestrial successional ecosystems. *Science*, 173, 1134–6.

Hutchinson, G. E., 1973, Eutrophication. *American scientist*, 61, 269–79.

Idso, S. B., 1983, Carbon dioxide and global temperature: what the data show. *Journal of environmental quality*, 12, 159–63.

Idso, S. B. and Brazel, A. J., 1978, Climatological effects of atmospheric particulate pollution. *Nature*, 274, 781–2.

—, 1984, Rising atmospheric carbon dioxide concentrations may increase streamflow. *Nature*, 312, 51–3.

Ikawa-Smith, F., 1982, Current issues in Japanese archaeology. *American scientist*, 68, 134–45.

Illies, J., 1974, *Introduction to zoogeography.* London: Macmillan.

Imeson, A. C., 1971, Heather burning and soil erosion on the North Yorkshire Moors. *Journal of applied ecology*, 8, 537–41.

Imeson, A. and Emmer, I. M., 1992, Implications of climate change on land degradation in the Mediterranean. In L. Jeftic, J. D. Milliman and G. Sestini (eds), *Climate change and the Mediterranean.* London: Arnold, 95–128.

Imhoff, M. L., Bounoua, L., Ricketts, T., Loucks, C., Harriss, R. and Lawrence, W. T., 2004, Global patterns in human consumption of net primary productivity. *Nature*, 429, 870–3.

Innes, J. L., 1983, Lichenometric dating of debris-flow deposits in the Scottish Highlands. *Earth surface processes and landforms*, 8, 579–88.

—, 1987, *Air pollution and forestry.* Forestry Commission bulletin, 70.

—, 1992, Forest decline. *Progress in physical geography*, 16, 1–64.

Innes, J. L. and Boswell, R. C., 1990, Monitoring of forest condition in Great Britain 1989. *Forestry Commission bulletin*, 94, 57.

Institute of Hydrology, 1991, *Institute of Hydrology report 1990–91.* Wallingford: Institute of Hydrology.

IPCC (Intergovernmental Panel on Climate Change), 1996, *Climate change 1995.* Cambridge: Cambridge University Press.

IPCC (Intergovernmental Panel on Climate Change), 1999, *Aviation and the global atmosphere.* Special report of Working Groups I and II.

IPCC (Intergovernmental Panel on Climate Change), 2001, *Climate change 2001: the scientific basis.* Cambridge: Cambridge University Press.

Irving, W. M., 1985, Context and chronology of early man in the Americas. *Annual review of anthropology*, 14, 529–55.

Isaac, E., 1970, *Geography of domestication.* Englewood Cliffs: Prentice Hall.

Ives, J. D. and Messerli, B., 1989, *The Himalayan dilemma: reconciling development and conservation.* London: Routledge.

Iwashima, T. and Yamamoto, R., 1993, A statistical analysis of the extreme events: long-term trend of heavy daily precipitation. *Journal of the Meteorological Society of Japan*, 71, 637–40.

Jacks, G. V. and Whyte, R. O., 1939, *The rape of the earth: a world survey of soil erosion.* London: Faber & Faber.

Jacobs, J., 1969, *The economy of cities.* New York: Random House.

—, 1975, Diversity, stability and maturity in ecosystems influenced by human activities. In W. H. Van Dobben and R. H. Lowe-McConnell (eds), *Unifying concepts in ecology.* The Hague: Junk, 187–207.

Jacobsen, T. and Adams, R. M., 1958, Salt and silt in ancient Mesopotamian agriculture. *Science*, 128, 1251–8.

Jarvis, P. H., 1979, The ecology of plant and animal introductions. *Progress in physical geography*, 3, 187–214.

Jeffries, M., 1997, *Biodiversity and conservation.* London: Routledge.

Jenkins, M., 2003, Prospects for biodiversity. *Science*, 302, 1175–7.

Jenkins, M. E., Davies, T. J. and Stedman, J. R., 2002, The origin and day-of-week dependence of photochemical ozone episodes in the UK. *Atmospheric environment*, 36, 999–1012.

Jennings, J. N., 1952, *The origin of the Broads*. Royal Geographical Society research series, 2.

Jenny, H., 1941, *Factors of soil formation*. New York: McGraw-Hill.

Jensen, F. P. and Fenger, J. 1994, The air quality in Danish urban areas. *Environmental health perspectives*, 102 (Supplement 4), 55–60.

Jickells, T. D., Carpenter, R. and Liss, P. S., 1991, Marine environment. In B. L. Turner, W. C. Clark, R. W. Kates, J. F. Richards, J. T. Matthews and W. B. Meyer (eds), *The earth as transformed by human action*. Cambridge: Cambridge University Press, 313–34.

Jin, H., Li, S., Cheng, G., Shaoling, W. and Li, X., 2000, Permafrost and climatic change in China. *Global and planetary change*, 26, 387–404.

Joern, A. and Keeler, K. H. (eds), 1995, *The changing prairie*. New York: Oxford University Press.

Johannessen, C. L., 1963, Savannas of interior Honduras. *Ibero-Americana*, 46, 160 pp.

Johnson, C. and Wroe, S., 2003, Causes of extinction of vertebrates during the Holocene of mainland Australia: arrival of the dingo, or human impact? *The Holocene*, 13, 941–8.

Johnson, A. I. (ed.), 1991, *Land subsidence*. Wallingford: International Association of Hydrological Sciences, Publication 200.

Johnson, D. L. and Lewis, L. A., 1995, *Land degradation: creation and destruction*. Oxford: Blackwell.

Johnson, N. M., 1979, Acid rain: neutralization within the Hubbard Brook ecosystem and regional implications. *Science*, 204, 497–9.

Johnston, D. W., 1974, Decline of DDT residues in migratory songbirds. *Science*, 186, 841–2.

Johnston, D. W., Turner, J. and Kelly, J. M., 1982, The effects of acid rain on forest nutrient status. *Water resources research*, 18, 448–61.

Jones, J. A. A., Liu, C., Woo, M.-K. and Kung, H.-T. (eds), 1996, *Regional hydrological response to climate change*. Dordrecht: Kluwer.

Jones, P. D. and Reid, P. A., 2001, Assessing future changes in extreme precipitation over Britain using regional climate model integrations. *International journal of climatology*, 21, 1337–56.

Jones, R., Benson-Evans, K. and Chambers, F. M., 1985, Human influence upon sedimentation in Llangorse Lake, Wales. *Earth surface processes and landforms*, 10, 227–35.

Joughlin, I. and Tulaczyk, S., 2002, Positive ice balance of the Ross Ice Streams, West Antarctica. *Science*, 295, 476–80.

Joyce, L., Aber, J., McNulty, S., Dale, V., Hansen, A., Irland, L., Neilson, R. and Skog, K., 2001, Potential consequences of climate variability and change for the forests of the United States. In National Assessment Synthesis Team, *Climate change impacts on the United States: the potential consequences of climate variability and change*. Cambridge: Cambridge University Press, 489–521.

Judd, W. R., 1974, Seismic effects of reservoir impounding. *Engineering geology*, 8, 1–212.

Judson, S., 1968, Erosion rates near Rome, Italy. *Science*, 160, 1444–5.

Julian, M. and Anthony, E., 1996, Aspects of landslide activity in the Mercantour Massif and the French Riviera, southeastern France. *Geomorphology*, 15, 275–89.

Kadomura, H., 1994, Climatic change, droughts, desertification and land degradation in the Sudano–Sahelian region: a historico-geographical perspective. In H. Kadomura (ed.), *Savannization processes in tropical Africa II*. Tokyo: Department of Geography, Tokyo Metropolitan University, 203–28.

Kareiva, P. and Levin, S. A. (eds), 2003, *The importance of species*. Princeton: Princeton University Press.

Karl, T. R. and Knight, R. W., 1998, Secular trends of precipitation amount, frequency and intensity in the United States. *Bulletin of the American Meteorological Society*, 79, 1413–49.

Karnes, L. B., 1971, Reclamation of wet and overflow lands. In G.-H. Smith (ed.), *Conservation of natural resources*. New York: Wiley, 241–55.

Karnosky, D. F., 2003, Impacts of elevated atmospheric CO_2 on forest trees and forest ecosystems: knowledge gaps. *Environment international*, 29, 161–9.

Karnosky, D. F., Ceulemans, R., Scarascia-Mugnozza, G. E. and Innes, J. L., 2001, *The impact of carbon dioxide and other greenhouse gases on forest ecosystems*. Wallingford: CABI Publishing.

Kaser, G., 1999, A review of the modern fluctuations of tropical glaciers. *Global and planetary change*, 22, 93–103.

Kaser, G. and Osmaston, H., 2002, *Tropical glaciers*. Cambridge: Cambridge University Press.

Kasperson, V. X., Kasperson, R. E. and Turner, B. L. II, 1995, *Regions at risk: comparisons of threatened environments*. Tokyo: United Nations University Press.

Kates, R. W., Turner, B. L. I. and Clark, W. C., 1990, The great transformation. In B. L. Turner, W. C. Clark, R. W. Kates, J. F. Richards, J. T. Matthews and W. B. Meyer (eds), *The earth as transformed by human action*. Cambridge: Cambridge University Press, 1–17.

Kauppi, P. and Posch, M., 1988, A case study of the effects of CO_2-induced climatic warming on forest growth and the forest sector. In M. L. Parry, T. R. Carter and N. T. Konijn (eds), *The impact of climatic variations in agriculture*, Vol. 1. Dordrecht: Kluwer, 183–95.

Keefer, D. K., de France, S. D., Mosely, M. E., Richardson, J. B., Satterlee, D. R. and Day-Lewis, A., 1998, Early maritime economy and El Niño events at Quebrada Tachuay, Peru. *Science*, 281, 1833–935.

Keen, K. L. and Shane, L. C. K., 1990, A continuous record of Holocene eolian activity and vegetation change at Lake Ann, east-central Minnesota. *Geological Society of America bulletin*, 102, 1646–57.

Keller, E. A., 1976, Channelization: environmental, geomorphic and engineering aspects. In D. R. Coates (ed.), *Geomorphology and engineering*. Stroudsburg: Dowden, Hutchinson and Ross, 115–40.

Kellman, M., 1975, Evidence for late glacial age fire in a tropical montane savanna. *Journal of biogeography*, 2, 57–63.

Kellogg, W. W., 1978, Global influence of mankind on the climate. In J. Gribbin (ed.), *Climatic change*. London: Cambridge University Press, 205–27.

Kemp, K., Palmgren, F. and Mancher, O. H., 1998, *The Danish air quality monitoring programme. Annual report for 1997*. NERI Technical report No 245, Roskilde: National Environmental Research Institute.

Kennish, M. J., 2001, Coastal salt marsh systems in the U.S.: a review of anthropogenic impacts. *Journal of coastal research*, 17, 731–48.

Kent, M., 1982, Plant growth problems in colliery spoil reclamation. *Applied geography*, 2, 83–107.

Khalil, M. A. K. and Rasmussen, R. A., 1987, Atmospheric methane: trends over the last 10,000 years. *Atmospheric environment*, 21, 2445–52.

Kiersch, G. A., 1965, The Vaiont reservoir disaster. *Mineral information service*, 18, 129–38.

King, C. A. M., 1974, Coasts. In R. U. Cooke and J. C. Doornkamp (eds), *Geomorphology in environmental management*. Oxford: Clarendon Press, 188–222.

—, 1975, *Introduction to physical and biological oceanography*. London: Edward Arnold.

Kinsey, D. W. and Hopley, D., 1991, The significance of coral reefs as global carbon sinks – response to greenhouse. *Palaeogeography, palaeoclimatology, palaeoecology*, 89, 363–77.

Kirby, C., 1995, Urban air pollution. *Geography*, 80, 375–92.

Kirch, P. V., 1982, Advances in Polynesian prehistory: three decades in review. *Advances in world archaeology*, 2, 52–102.

Kirkbride, M. P. and Warren, C. R., 1999, Tasman Glacier, New Zealand: 20th century thinning and predicted calving retreat. *Global and planetary change*, 22, 11–28.

Kirkpatrick, J., 1994, *A continent transformed: human impact on the natural vegetation of Australia*. Melbourne: Oxford University Press.

Kittredge, J. H., 1948, *Forest influences*. New York: McGraw-Hill.

Klein, R. G., 1983, The stone age prehistory of southern Africa. *Annual review of anthropology*, 12, 25–48.

Kleypas, J. A., Buddemeier, R. W., Archer, D., Gattuso, J.-P., Langdon, C. and Opdyke, B. N., 1999, Geochemical consequences of increased atmospheric carbon dioxide on coral reefs. *Science*, 284, 118–20.

Knight, M., Thomas, D. S. G. and Wiggs, G. F. S., 2004, Challenges of calculating dunefield mobility over the 21st century. *Geomorphology*, 59, 197–213.

Knight, P. G., 1999, *Glaciers*. Cheltenham: Stanley Thornes.

Knox, J. C., 1972, Valley alleviation in southwestern Wisconsin. *Annals of the Association of American Geographers*, 62, 401–10.

Knox, J. C., 1977, Human impacts on Wisconsin stream channels. *Annals of the Association of American Geographers*, 67, 323–42.

—, 1987, Historical valley floor sedimentation in the Upper Mississippi Valley. *Annals of the Association of American Geographers*, 77, 224–44.

—, 1993, Large increase in flood magnitude in response to modest changes in climate. *Nature*, 361, 430–2.

—, 2001, Agricultural influence on landscape sensitivity in the upper Mississippi river valley. *Catena*, 42, 193–224.

—, 2002, Agriculture, erosion and sediment yields. In A. R. Orme (ed.), *The physical geography of North America*, Oxford: Oxford University Press, 482–500.

Knutson, T. R. and Tuleya, R. E., 1999, Increased hurricane intensities with CO_2-induced warming as simulated using the GFDL hurricane prediction system. *Climate dynamics*, 15, 503–19.

Knutson, T. R., Tuleya, R. E. and Kurihara, Y., 1998, Simulated increase of hurricane intensities in a CO_2-warmed climate. *Science*, 279, 1018–20.

Kohen, J., 1995, *Aboriginal environmental impacts*. Sydney: University of New South Wales Press.

Kadomura, H., 1983, Some aspects of large-scale land transformation due to urbanization and agricultural development in recent Japan. *Advances in space research*, 2 (8), 169–78.

Koide, M. and Goldberg, E. D., 1971, Atmospheric and fossil fuel combustion. *Journal of geophysical research*, 76, 6589–96.

Komar, P. D., 1976, *Beach processes and sedimentation*. Englewood Cliffs: Prentice Hall.

Komar, P. D., McManus, J. and Styllas, M. 2004, Sediment accumulation in Tillamook Bay, Oregon: natural processes versus human impacts. *Journal of geology*, 112, 455–69.

Kotb, T. H. S., Watanabe, T., Ogino, Y. and Tanji, K. K., 2000, Soil salinization in the Nile Delta and related policy issues in Egypt. *Agricultural water management*, 43, 239–61.

Kotlyakov, V. M., 1991, The Aral Sea basin: a critical environmental zone. *Moscow environment*, 33 (1), 4–9, 36–8.

Kowicki, V. and Isdale, P. 1991, Hydrology of Lake Eyre, Australia: El Nino link. *Palaeogeography, palaeoclimatology, palaeoecology*, 84, 87–98.

Krabill, W. and 8 others, 1999, Rapid thinning of parts of the southern Greenland Ice Sheet. *Science*, 283, 1522–4.

Krantz, G. S., 1970, Human activities and megafaunal extinctions. *American scientist*, 58, 164–70.

Krug, E. C. and Frink, C. R., 1983, Acid rain on acid soil: a new perspective. *Science*, 221, 520–5.

Kuhlmann, D. H., 1988, The sensitivity of coral reefs to environmental pollution. *Ambio*, 17, 13–21.

Kullman, L., 2001, 20th century climate warming and tree-limit rise in the southern Scandes of Sweden. *Ambio*, 30, 72–80.

Kuo, C., 1986, Flooding in Taipeh, Taiwan and coastal drainage. In J. G. Titus (ed.), *Effects of changes in stratospheric ozone and global climate*. Washington, DC: United Nations Environment Program/U.S. Environmental Protection Agency, 37–46.

Kwong, Y. T. J. and Gan, T. Y., 1994, Northward migration of permafrost along the Mackenzie Highway and climatic warming. *Climatic change*, 26, 399–419.

Labadz, J. C., Burt, T. P. and Potter, A. W. L., 1991, Sediment yield and delivery in the blanket peat moorlands of the southern Pennines. *Earth surface processes and landforms*, 16, 255–71.

Lal, R., 2002, Soil carbon dynamics in cropland and range land. *Environmental pollution*, 116, 353–62.

Lal, R., Kimble, J., Levine, E. and Stewart, B. A. (eds), 1995, *Soil management and greenhouse effect*, Boca Raton: CRC Lewis.

La Marche, V. C., Graybill, D. A., Fritts, H. C. and Rose, M. R., 1984, Increasing atmospheric carbon dioxide: tree ring evidence for growth enhancement in natural vegetation. *Science*, 225, 1019–21.

Lamb, H. H., 1977, *Climate: present, past and future. 2: Climatic history and the future*. London: Methuen.

Lambeck, K., 1988, *Geological geodesy*. Oxford: Clarendon Press.

Lambert, J. H., Jennings, J. N., Smith, C. T., Green, C. and Hutchinson, J. N., 1970, *The making of the Broads: a reconsideration of their origin in the light of new evidence*. Royal Geographical Society research series, 3.

Lamprey, H., 1975, The integrated project on arid lands. *Nature and resources*, 14, 2–11.

Lancaster, N., 1995, *Geomorphology of desert dunes*. London: Routledge.

Landes, K. K., 1973, Mother nature as an oil polluter. *Bulletin of the American Association of Petroleum Geologists*, S7, 637–41.

Landsberg, H. E., 1981, *The urban climate*. New York: Academic Press.

Landsea, C. W., 2000, El Niño/Southern Oscillation and the seasonal predictability of tropical cyclones. In H. F. Diaz and V. Markgraf (eds), *El Niño and the Southern Oscillation*. Cambridge: Cambridge University Press, 148–81.

Langbein, W. B. and Schumm, S. A., 1958, Yield of sediment in relation to mean annual precipitation. *Transactions of the American Geophysical Union*, 39, 1076–118.

Langford, T. E. L., 1990, *Ecological effects of thermal discharges*. London: Elsevier Applied Science.

Lanly, J. P., Singh, K. D. and Janz, K., 1991, FAO's 1990 reassessment of tropical forest cover. *Nature and resources*, 27, 21–6.

Larick, R. and Ciochon, R. L., 1996, The African emergence and early Asian dispersals of the genus *Homo*. *American scientist*, 84, 538–51.

La Roe, E. T., 1977, Dredging – ecological impacts. In J. R. Clarke (ed.), *Coastal ecosystem management*. New York: Wiley, 610–14.

Larson, F., 1940, The role of bison in maintaining the short grass plains. *Ecology*, 21, 113–21.

Lawson, D. E., 1986, Response of permafrost terrain to disturbance: a synthesis of observations from northern Alaska, USA. *Arctic and alpine research*, 18, 1–17.

Lean, J. and Warrilow, D. A., 1989, Simulation of the regional climatic impact of Amazon deforestation. *Nature*, 342, 126–33.

Leatherman, S. P., 2001, Social and economic costs of sea level rise. In B. C. Douglas, M. S. Kearney and S. P. Leatherman (eds), *Sea level rise: history and consequences*. San Diego: Academic Press, 181–223.

Leclercq, N., Gattuso, J.-P. and Jaubert, J., 2000, CO_2 partial pressure controls the calcification rate of a coral community. *Global change biology*, 6, 329–34.

Leduc, C., Favreau, G. and Schoreter, P., 2001, Long term rise in a Sahelian water-table: the Continental Terminal in South-West Niger. *Journal of hydrology*, 243, 43–54.

Lee, D. O., 1992, Urban warming – an analysis of recent trends in London's heat island. *Weather*, 47, 50–6.

Lee, R. B. and DeVore, I., 1968, *Man the hunter*. Chicago: Aldine.

Le Houérou, H. N., 1977, Biological recovery versus desertization. *Economic geography*, 63, 413–20.

Le Maitre, D. C., Versfield, D. B. and Chapman, R. A., 2000, The impact of invading alien plants on surface water resources in South Africa: a preliminary assessment. *Water SA*, 26, 397–408.

Lently, A. D., 1994, Agriculture and wildlife: ecological implications of subsurface irrigation drainage. *Journal of arid environments*, 28, 85–94.

Lents, J. M. and Kelly, W. J., 1993, Clearing the air in Los Angeles. *Scientific American*, October, 18–25.

Leopold, L. B., 1951, Rainfall frequency: an aspect of climatic variation. *Transactions of the American Geophysics Union*, 32, 347–57.

Leopold, L. B., Wolman, M. G. and Miller, J. P., 1964, *Fluvial processes in geomorphology*. San Francisco: Freeman.

Lerner, D., 1990, *Groundwater recharge in urban areas*, Wallingford: International Association of Hydrological Sciences, Publication no. 198, 59–65.

Letey, J., 2001, Cases and consequences of fire-induced soil water repellency. *Hydrological processes*, 15, 2867–75.

Letourneau, D. K. and Burrows, B. E. (eds), 2001, *Genetically engineered organisms: assessing environmental and human health effects*. Washington, DC: CRC Press.

Levine, J. M., Vila, M., D'Antonio, C. M., Dukes, J. S., Grigulis, K. and Lavorel, S., 2003, Mechanisms underlying the impacts of exotic plant invasions. *Proceedings of the Royal Society of London*, B, 270, 775–81.

Lev-Yadun, S., Gopher, A. and Abbo, S., 2000, Enhanced: the cradle of agriculture. *Science*, 288, 1602–3.

Lewin, J., Bradley, S. B. and Macklin, M. G., 1983, Historical valley alluviation in mid-Wales. *Geological journal*, 18, 331–50.

Leys, J., 1999, Wind erosion on agricultural land. In A. S. Goudie, I. Livingstone and S. Stokes (eds), *Aeolian environments, sediments and landforms*. Chichester: Wiley, 143–66.

L'Homer, A., 1992, Sea level changes and impact on the Rhône Delta coastal lowlands. In M. J. Tooley and S. Jelgersma (eds), *Impacts of sea level rise on European coastal lowlands*. Oxford: Blackwell, 136–52.

Liddle, M., 1997, *Recreation ecology*. London: Chapman & Hall.

Liébault, F. and Piégay, H., 2002, Causes of 20th century channel narrowing in mountain and piedmont rivers of southeastern France. *Earth surface processes and landforms*, 27, 425–44.

Lienert, J., 2004, Habitat fragmentation effects on fitness of plant populations – a review. *Journal for nature conservation*, 12, 53–72.

Likens, G. E. and Bormann, F. H., 1974, Acid rain: a serious regional environmental problem. *Science*, 184, 1176–9.

Likens, G. E., Wright, R. F., Galloway, J. N. and Butler, T. J., 1979, Acid rain. *Scientific American*, 241, 4, 39–47.

Ling, F. and Zhang, T., 2003, Impact of the timing and duration of seasonal snow cover on the active layer and permafrost in the Alaskan Arctic. *Permafrost and periglacial processes*, 14, 141–50.

List, J. H., Sallenger, A. H., Hansen, M. E. and Jaffe, B. E., 1997, Accelerated sea level rise and rapid coastal erosion: testing a causal relationship for the Louisiana barrier islands. *Marine geology*, 140, 437–65.

Liu, C. and Fu, G., 1996, The impact of climatic warming on hydrological regimes in China: an overview. In J. A. A. Jones, C. Liu, M.-K. Woo and H.-T. Kung (eds), *Regional hydrological response to climate change*. Dordrecht: Kluwer, 133–51.

Lloyd, J. W., 1986, A review of aridity and groundwater. *Hydrological processes*, 1, 63–78.

Lomborg, B., 2001, *The skeptical environmentalist*. Cambridge: Cambridge University Press.

Loope, D. B., Swinehart, J. B. and Mason, J. P., 1995, Dune-dammed palaeovalleys of the Nebraska Sand Hills: intrinsic versus climatic controls on the accumulation of lake and marsh sediments. *Bulletin of the Geological Society of America*, 107, 396–406.

Loreau, M., Naeem, S. and Inchausti, P. (eds), 2002, *Biodiversity and ecosystem functioning. Synthesis and perspectives*. Oxford: Oxford University Press.

Loreau, M. and 11 others, 2001, Biodiversity and ecosystem functioning: current knowledge and future challenges. *Science*, 294, 804–8.

Lowe, P. D., 1983, Values and institutions in the history of British nature conservation. In A. Warren and F. B. Goldsmith (eds), *Conservation in perspective*. Chichester: Wiley, 329–52.

Lowe-McConnell, R. H., 1975, Freshwater life on the move. *Geographical magazine*, 47, 768–75.

Lowenthal, D., 2000, *George Perkins Marsh, prophet of conservation*. Seattle: University of Washington Press.

Loya, Y., Sakai, K., Yamazato, K., Sambali, H. and Van Woesik, R., 2001, Coral bleaching: the winners and the losers. *Ecology letters*, 4, 122–32.

Lugo, A. E., Cintron, G. and Goenaga, C., 1981, Mangrove ecosystems under stress. In G. W. Barrett and R. Rosenberg (eds), *Stress effects on natural ecosystems*. Chichester: John Wiley, 129–53.

Lugo, A. E., 2000, Effects and outcomes of Caribbean hurricanes in a climate change scenario. *The science of the total environment*, 262, 243–51.

Luke, R. H., 1962, *Bush fire control in Australia*. Melbourne: Hodder & Stoughton.

Lund, J. W. G., 1972, Eutrophication. *Proceedings of the Royal Society of London*, 180B, 371–82.

Lyell, C., 1835, *Principles of geology* (4th edn), Vol. III. London: Murray (12th edn 1875).

Lynch, J. A., Rishel, G. B. and Corbett, E. S., 1984, Thermal alteration of streams draining clearcut watersheds: quantification and biological implications. *Hydrobiologia*, 111, 161–9.

Maat, P. B. and Johnson, W. C., 1996, Thermoluminescence and new C-14 age estimates for late Quaternary loesses in southwestern Nebraska. *Geomorphology*, 17, 115–28.

Mabbutt, J. A., 1977, *Desert landforms*. Cambridge, MA: MIT Press.

Mabbutt, J. A., 1985, Desertification of the world's rangelands. *Desertification control bulletin*, 12, 1–11.

McCann, K. S., 2000, The diversity-stability debate. *Nature*, 405, 228–33.

McClanahan, T. R., 2000, Bleaching damage and recovery potential of Maldivian coral reefs. *Marine pollution bulletin*, 40, 587–97.

MacCracken, M., Barron, E., Easterling, D., et al., 2001, Scenarios for climate variability and change. In National Assessment Synthesis Team, *Climate change impacts on the United States: the potential consequences of climate variability and change*. Cambridge: Cambridge University Press, 13–71.

McCulloch, M., Fallon, S. Wyndham, T., Hendy, E., Lynch, J. and Barnes, D., 2003, Coral record of increased sediment flux to the inner Great Barrier Reef since European settlement. *Nature*, 421, 727–30.

Macfarlane, M. J., 1976, *Laterite and landscape*. London: Academic Press.

McGlone, M. S. and Wilmshurst, J. M., 1999, Dating initial Maori environmental impact in New Zealand. *Quaternary international*, 59, 5–16.

McGuffie, K., Henderson-Sellers, A., Holbrook, N., Kothavola, Z., Balachova, O. and Hoeksstra, J., 1999, Assessing simulations of daily temperature and precipitation variability with global climate models for present and enhanced greenhouse climates. *International journal of climatology*, 19, 1–26.

MacInnis, C. and Whiting, J. D., 1979, The frost resistance of concrete subjected to a deicing agent. *Cement and concrete research*, 9, 325–36.

Macklin, M. G. and Lewin, J., 1986, Terraced fills of Pleistocene and Holocene age in the Rheidol Valley, Wales. *Journal of Quaternary science*, 1, 21–34.

Macklin, M. G., Passmore, D. G., Stevenson, A. C., Colwey, A. C., Edwards, D. N. and O'Brien, C. F., 1991, Holocene alluviation and land-use change on Callaly Moor, Northumberland, England. *Journal of Quaternary science*, 6, 225–32.

McKnight, T. L., 1959, The feral horse in Anglo-America. *Geographical review*, 49, 506–25.

——, 1971, Australia's buffalo dilemma. *Annals of the Association of American Geographers*, 61, 759–73.

McLennan, S. M., 1993, Weathering and global denudation. *Journal of geology*, 101, 295–303.

McNeill, J. R., 2000, *Something new under the sun. An environmental history of the twentieth century.* London: Allen Lane.

—, 2003, Resource exploitation and over-exploitation: a look at the 20th century. In T. S. Benzing and B. Herrmann (eds), *Exploitation and overexploitation in societies past and present.* Münster: LIT Verlag, 51–60.

Mader, H. J., 1984, Animal habitat isolation by roads and agricultural fields. *Biological conservation,* 29, 81–96.

Madole, R. F., 1995, Spatial and temporal patterns of Late Quaternary eolian deposition, eastern Colorado, USA. *Quaternary science reviews,* 14, 155–78.

Magilligan, F. J., 1985, Historical floodplain sedimentation in the Galena River basin, Wisconsin and Illinois. *Annals of Association of American Geographers,* 75, 583–94.

Magilligan, F. J. and Goldstein, P. S., 2001, El Niño floods and culture change: a late Holocene flood history for the Rio Moquegua, Southern Peru. *Geology,* 29, 431–4.

Maignien, R., 1966, *A review of research on laterite.* UNESCO, Natural resources research, 4.

Mainguet, M., 1995, *L'homme et la sécheresse.* Paris: Masson.

Malamud, B. D., Morein, G. and Turcotte, D. L., 1998, Forest fires: an example of self-organized critical behavior. *Science,* 281, 1840–2.

Malm, W. C., Schichtel, B. A., Ames, R. B. and Gebhart, K. A., 2002, A 10-year spatial and temporal trend of sulfate across the United States. *Journal of geophysical research,* 107 (D22), article no. 4627.

Maltby, E., 1986, *Waterlogged wealth. Why waste the world's wet places?* London: Earthscan.

Manabe, S. and Stouffer, R. J., 1980, Sensitivity of a global climate model to an increase of CO_2 concentration in the atmosphere. *Journal of atmospheric science,* 37, 99–118.

Manabe, S. and Wetherald, R. T., 1986, Reduction in summer soil wetness by an increase in atmospheric carbon dioxide. *Science,* 232, 626–8.

Manners, I. R., 1978, Agricultural activities and environmental stress. In K. A. Hammond (ed.), *Sourcebook of the environment.* Chicago: University of Chicago Press, 263–94.

Manners, I. R. and Mikesell, M. W. (eds), 1974, *Perspectives on environment.* Washington, DC: Association of American Geographers.

Mannion, A. M., 1992, Acidification and eutrophication. In A. M. Mannion and S. R. Bowlby (eds), *Environmental issues in the 1990s.* Chichester: Wiley, 177–95.

—, 1995, *Agriculture and environmental change.* Chichester: Wiley.

—, 1997, *Global environmental change* (2nd edn). Harlow: Longman.

—, 2002, *Dynamic world. Land-cover and land-use change.* London: Arnold.

Manshard, W., 1974, *Tropical agriculture.* London: Longman.

Mark, A. F. and McSweeney, G. D., 1990, Patterns of impoverishment in natural communities: case studies in forest ecosystems – New Zealand. In G. M. Woodwell (ed.), *The earth in transition: patterns and processes of biotic impoverishment.* Cambridge: Cambridge University Press, 151–76.

Marker, M. E., 1967, The Dee estuary: its progressive silting and salt marsh development. *Transactions of the Institute of British Geographers,* 41, 65–71.

Marks, P. L. and Bormann, F. H., 1972, Revegetation following forest cutting: mechanisms for return to steady-state nutrient cycling. *Science,* 176, 914–15.

Marquiss, M., Newton, I. and Ratcliffe, D. A., 1978, The decline of the raven, *Corvus corax,* in relation to afforestation in southern Scotland and northern England. *Journal of applied ecology,* 15, 129–44.

Marsh, G. P., 1864, *Man and nature.* New York: Scribner.

—, 1965, *Man and nature,* edited by D. Lowenthal. Cambridge, Mass.: Belknap Press.

Marshall, L. G., 1984, Who killed Cock Robin? An investigation of the extinction controversy. In P. S. Martin and R. G. Klein (eds), *Quaternary extinctions.* Tucson: University of Arizona Press.

Martens, L. A., 1968, Flood inundation and effects of urbanization in Metropolitan Charlotte, North Carolina. *United States Geological Survey water supply paper,* 1591-C.

Martin, P. S., 1967, Prehistoric overkill. In P. S. Martin and H. E. Wright (eds), *Pleistocene extinctions.* New Haven: Yale University Press, 75–120.

—, 1974, Palaeolithic players on the American stage: man's impact on the Late Pleistocene megafauna. In J. D. Ives and R. G. Barry (eds), *Arctic and alpine environments.* London: Methuen.

—, 1982, The pattern and meaning of Holarctic mammoth extinction. In D. M. Hopkins, J. V. Matthews, C. S. Schweger and S. B. Young (eds), *Paleoecology of Beringia.* New York: Academic Press, 399–408.

Martin, P. S. and Klein, R. G., 1984, *Pleistocene extinctions.* Tucson: University of Arizona Press.

Martin, P. S. and Wright, H. E. (eds), 1967, *Pleistocene extinctions.* New Haven: Yale University Press, 75–120.

Martinez, J. D., 1971, Environmental significance of salt. *Bulletin of the American Association of Petroleum Geologists,* 55, 810–25.

Marx, J. L., 1975, Air pollution: effects on plants. *Science,* 187, 731–3.

Mason, I. M., Guzkowska, M. A. J., Rapley, C. G. and Street-Perrott, F. A., 1994, The response of lake levels and areas to climatic change. *Climatic change,* 27, 161–97.

Mason, J. P., Swinehart, J. B. and Loope, D. B., 1997, Holocene history of lacustrine and marsh sediments in a dune-blocked drainage, southwestern Nebraska Sand Hills, USA. *Journal of paleolimnology,* 17, 67–83.

Mason, S. J., Waylen, P. R., Mimmack, G. M., Rajaratnam, B. and Harrison, J. M., 1999, Changes in extreme rainfall events in South Africa. *Climatic change,* 41, 249–57.

Mather, A. S., 1983, Land deterioration in upland Britain. *Progress in physical geography,* 7 (2), 210–28.

Maugh, T. H., 1979, The Dead Sea is alive and well . . . *Science,* 205, 178.

May, R. M., 1979, Fluctuations in abundance of tropical insects. *Nature,* 278, 505–7.

May, T., 1991, Südspanische matorrales als Kulturofolge-vegetation. *Geoökodynamik* 12, 87–107.

Meade, R. H., 1991, Reservoirs and earthquakes. *Engineering geology*, 30, 245–62.

—, 1996, River-sediment input to major deltas. In J. D. Milliman and B. V. Haq (eds), *Sea-level rise and coastal subsidence*. Dordrecht: Kluwer, 63–85.

Meade, R. H. and Parker, R. S., 1985, Sediment in rivers in the United States. *United States Geological Survey water supply paper*, 2276, 49–60.

Meade, R. H. and Trimble, S. W., 1974, Changes in sediment loads in rivers of the Atlantic drainage of the United States since 1900. *Publication of the International Association of Hydrological Science*, 113, 99–104.

Meadows, M. E. and Linder, H. P., 1993, A palaeoecological perspective on the origin of Afromontane grasslands. *Journal of biogeography*, 20, 345–55.

Mee, L. D., 1992, The Black Sea in crisis: a need for concerted international action. *Ambio*, 21, 278–86.

Melillo, J., Janetos, A., Schimel, D. and Kittel, T., 2001, Vegetation and biogeochemical scenarios. In National Assessment Synthesis Team, *Climate change impacts on the United States: the potential consequences of climate variability and change*. Cambridge: Cambridge University Press, 74–91.

Mellanby, K., 1967, *Pesticides and pollution*. London: Fontana.

Menzel, L. and Burger, G., 2002, Climate change scenarios and runoff response in the Mulde catchment (Southern Elbe, Germany). *Journal of hydrology*, 267, 53–64.

Mercer, D. E. and Hamilton, L. S., 1984, Mangrove ecosystems: some economic and natural benefits. *Nature and resources*, 20, 14–19.

Mercer, J. H., 1978, West Antarctic ice sheet and CO_2 greenhouse effect: a threat of disaster. *Nature*, 271, 321–5.

Merryfield, D. L. and Moore, P. D., 1971, Prehistoric human activity and blanket peat initiation on Exmoor. *Nature*, 250, 439–41.

Meybeck, M., 1979, Concentration des eaux fluviales en éléments majeurs et apports en solution aux océans. *Revue de géographie physique et géologie dynamique*, 21a, 215–46.

—, 2001a, Global alteration of riverine geochemistry under human pressure. In E. Ehlers (ed.), *Understanding the earth system: compartments, processes and interactions*. Heidelberg: Springer-Verlag, 97–113.

—, 2001b, River basins under anthropocene conditions. In B. von Bodungen and R. K. Turner (eds), *Science and integrated coastal management*. Dahlem: Dahlem University Press, 275–94.

Meyer, G. A. and Pierce, J. L., 2003, Climatic controls on fire-induced sediment pulses in Yellowstone National Park and central Idaho: a long-term perspective. *Forest ecology and management*, 178, 89–104.

Meyer, G. A., Pierce, J. L., Wood, S. H. and Jull, A. J. T., 2001, Fire, storms and erosional events in the Idaho batholith. *Hydrological processes*, 15, 3025–38.

Meyer, W. B., 1996, *Human impact on the earth*. Cambridge: Cambridge University Press.

Meyer, W. B. and Turner, B. L. II (eds), 1994, *Changes in land use and land cover: a global perspective*. Cambridge: Cambridge University Press.

Micklin, P. P., 1972, Dimensions of the Caspian Sea problem. *Soviet geography*, 13, 589–603.

Middleton, N. J., 1985, Effect of drought on dust production in the Sahel. *Nature*, 316, 431–4.

—, 1995, *The global casino*. London: Edward Arnold.

Middleton, N. J. and Thomas, D. S. G., 1997, *World atlas of desertification* (2nd edn). London: Edward Arnold.

Midgley., G. F., Hannah, L., Millar, D., Thuiller, W. and Booth, A., 2003, Developing regional and species-level assessments of climate change impacts on biodiversity in the Cape Floristic Region. *Biological conservation*, 112, 87.

Mieck, I., 1990, Reflections on a typology of historical pollution: complementary conceptions. In P. Brimblecombe and C. Pfister (eds), *The silent countdown*. Berlin: Springer-Verlag, 73–80.

Mikesell, M. W., 1969, The deforestation of Mount Lebanon. *Geographical review*, 59, 1–28.

Miller L. and Douglas, B. C., 2004, Mass and volume contribution to twentieth-century global sea level rise. *Nature*, 428, 406–9.

Miller, R. L. and Tegen, I., 1998, Climate response to soil dust aerosols. *Journal of climate*, 11, 3247–67.

Miller, R. S. and Botkin, D. B., 1974, Endangered species: models and predictions. *American scientist*, 62, 172–81.

Miller, J. R. and Russell, G. L., 2002, Projected impact of climate change on the energy budget of the Arctic Ocean Global Climate Model. *Journal of climate*, 15, 3028–42.

Milliman, J. D., 1990, Fluvial sediment in coastal seas: flux and fate. *Nature and resources*, 26, 12–22.

Milliman, J. D. and Haq, B. U. (eds), 1996, *Sea level rise and coastal subsidence*. Dordrecht: Kluwer.

Milliman, J. D., Broadus, J. M. and Gable, F., 1989, Environmental and economic impacts of rising sea level and subsiding deltas: the Nile and Bengal examples. *Ambio*, 18, 340–5.

Milliman, J. D., Qin, Y. S., Ren, M. E. and Yoshiki Saita, 1987, Man's influence on erosion and transport of sediment by Asian rivers: the Yellow River (Huanghe) example. *Journal of geology*, 95, 751–62.

Milly, P. C. D., Wetherald, R. T., Dunne, K. A. and Delworth, T. L., 2002, Increasing risk of great floods in a changing climate. *Nature*, 415, 514–17.

Milne, W. G., 1976, Induced seismicity. *Engineering geology*, 10, 83–388.

Mintzer, I. M. and Miller, A. S., 1992, Stratospheric ozone depletion: can we save the sky? In *Green Globe Yearbook 1992*. Oxford: Oxford University Press, 83–91.

Mirza, M. M. Q., 2002, Global warming and changes in the probability of occurrence of floods in Bangladesh and implications. *Global environmental change*, 12, 127–38.

Mistry, J., 2000, *World savannas: ecology and human use*. Harlow: Prentice Hall.

Mölg, T., Georges, C. and Kaser, G., 2003, The contribution of increased incoming shortwave radiation to the retreat of the Rwenzori Glaciers, East Africa, during the 20th century. *International journal of climatology*, 23, 291–303.

Montgomery, D. R., 1997, What's best on the banks? *Nature*, 388, 328–9.

Moody, J. A. and Martin, D. A., 2001, Initial hydrologic and geomorphic response following a wildfire in the Colorado Front Range. *Earth surface processes and landforms*, 26, 1049–70.

Mooney, H. A. and Parsons, D. J., 1973, Structure and function of the California Chaparral – an example from San Dimas. *Ecological studies*, 7, 83–112.

Moore, D. M., 1983, Human impact on island vegetation. In W. Holzner, M. J. A.Werger and I. Ikusima (eds), *Man's impact on vegetation*. The Hague: Junk, 237–48.

Moore, J., 2000, Forest fire and human interaction in the early Holocene woodlands of Britain. *Palaeogeography, palaeoclimatology, palaeoecology*, 164, 125–37.

Moore, N. and Rojstaczer, S., 2001, Irrigation-induced rainfall and the Great Plains. *Journal of applied meteorology*, 40, 1297–309.

Moore, N. W., Hooper, M. D. and Davis, B. N. K., 1967, Hedges, I. Introduction and reconnaissance studies. *Journal of applied ecology*, 4, 201–20.

Moore, P. D., 1973, Origin of blanket mires. *Nature*, 256, 267–9.

—, 1986, Unravelling human effects. *Nature*, 321, 204.

—, 2002, The future of cool temperate bogs. *Environmental Conservation*, 29, 3–20.

Moore, T. R., 1979, Land use and erosion in the Machakos Hills. *Annals of the Association of American Geographers*, 69, 419–31.

Morgan, G. S. and Woods, C. A., 1986, Extinction and the zoogeography of West Indian land mammals. *Biological journal of the Linnean Society*, 28, 167–203.

Morgan, R. P. C., 1977, *Soil erosion in the United Kingdom: field studies in the Silsoe area, 1973–75*. National College of Agricultural Engineering, occasional paper, 4.

—, 1979, *Soil erosion*. London: Longman.

—, 1995, *Soil erosion and conservation* (2nd edn). Harlow: Longman.

Morgan, W. B. and Moss, R. P., 1965, Savanna and forest in Western Nigeria. *Africa*, 35, 286–93.

Motyka, R. J., O'Neal, S., Connor, C. L. and Echelmeyer, K. A., 2002, Twentieth century thinning of Mendenhall Glacier, Alaska, and its relationship to climate, lake calving and glacier run-off. *Global and planetary change*, 35, 93–112.

Moulin, C., Lambert, C. E., Dulac, F. and Dayan, U., 1997, Control of atmospheric export of dust from North Africa by the North Atlantic Oscillation. *Nature*, 398, 691–4.

Moyle, P. B., 1976, Fish introductions in California: history and impact on native fishes. *Biological conservation*, 9, 101–18.

Mrowka, J. P., 1974, Man's impact on stream regimen and quality. In I. R. Manners and M. W. Mikesell (eds), *Perspectives on environment*. Washington, DC: Association of American Geographers.

Mudge, G. P., 1983, The incidence and significance of ingested lead pellet poisoning in British wildfowl. *Biological conservation*, 27, 333–72.

Muhly, J. D., 1997, Artifacts of the Neolithic, Bronze and Iron Ages. In E. M. Myers (ed.), *The Oxford encyclopaedia of archaeology in the Near East*. New York: Oxford University Press, Vol. 4, 5–15.

Muhs, D. R. and Holliday, V. T., 1995, Evidence of active dune sand in the Great Plains in the 19th century from accounts of early explorers. *Quaternary research*, 43, 198–208.

Muhs, D. R. and Maat, P. B., 1993, The potential response of eolian sands to greenhouse warming and precipitation reduction on the Great Plains of the United States. *Journal of arid environments*, 25, 351–61.

Muhs, D. R. and 9 others, 1996, Origin of late Quaternary dune fields of northeastern Colorado. *Geomorphology*, 17, 129–49.

Muhs, D. R. and 7 others, 1997a, Late Holocene eolian activity in the mineralogically mature Nebraska Sand Hills. *Quaternary research*, 48, 162–76.

Muhs, D. R. and 6 others, 1997b, Holocene eolian activity in the Minot dune field, North Dakota. *Canadian journal of earth sciences*, 34, 1442–59.

Mulrennan, M. E. and Woodroffe, C. D., 1998, Saltwater intrusion into the coastal plains of the lower Mary River, Northern Territory, Australia. *Journal of environmental management*, 54, 169–88.

Munn, R. E., 1996, Global change: both a scientific and a political issue. In R. E. Munn, J. W. M. La Riviere and N. van Lookeren Campagne (eds), *Policy making in an era of global environmental change*. Dordrecht: Kluwer, 1–15.

Murdoch, W. W., 1975, Diversity, complexity, stability and pest control. *Journal of applied ecology*, 12, 795–807.

Murozumi, M., Chow, T. J. and Paterson, C., 1969, Chemical concentrations of pollutant lead aerosols, terrestrial dusts and sea salt in Greenland and Antarctic snow strata. *Geochimica et cosmoschimica acta*, 33, 1247–94.

Murton, R. K., 1971, *Man and birds*. London: Collins.

Musk, L. F., 1991, The fog hazard. In A. H. Perry and L. J. Symons (eds), *Highway meteorology*. London: Spon, 91–130.

Myers, N., 1979, *The sinking ark: a new look at the problem of disappearing species*. Oxford: Pergamon Press.

—, 1983, Conversion rates in tropical moist forests. In F. B. Golley (ed.), *Tropical rain forest ecosytems*. Amsterdam: Elsevier Scientific, 289–300.

—, 1984, *The primary source: tropical forests and our future*. New York: Norton.

—, 1988, *Natural resource systems and human exploitation systems: physiobiotic and ecological linkages*. World Bank policy planning and research staff, environment department working paper, 12.

—, 1990, The biodiversity challenge: expanded hot-spots analysis. *The environmentalist*, 10 (4), 243–56.

—, 1992, Future operational monitoring of tropical forests: an alert strategy. In J. P. Mallingreau, R. da Cunha and C. Justice (eds), *Proceedings World Forest Watch Conference.* Sao Jose dos Campos, Brazil, 9–14.

Myers, N. and Kent, J., 2003, New consumers: the influence of affluence on the environment. *Proceedings of the National Academy of Sciences*, 100, 4963–8.

Myers, N., Mittermeier, R. A., Mittermeier, C. G., da Fonseca, G. A. B. and Kent, J., 2000, Biodiversity hotspots for conservation priorities. *Nature*, 403, 853–8.

Mylne, M. F. and Rowntree, P. R., 1992, Modelling the effects of albedo change associated with tropical deforestation. *Climatic change*, 21, 317–43.

Naeem, S., 2002, Biodiversity equals instability? *Nature*, 416, 23–4.

Najjar, R. G., 1999, The water balance of the Susquehanna River Basin and its response to climate change. *Journal of hydrology*, 219, 7–19.

Nakagawa, K., 1996, Recent trends of urban climatological studies in Japan, with special emphasis on the thermal environments of urban areas. *Geographical review of Japan*, B, 69, 206–24.

Nakano, T. and Matsuda, I., 1976, A note on land subsidence in Japan. *Geographical reports of Tokyo Metropolitan University*, 11, 147–62.

Nanson, G. C. and Tooth, S., 1999, Arid-zone rivers as indicators of climate change. In A. K. Singhvi and E. Derbyshire (eds), *Paleoenvironmental reconstruction in arid lands.* New Delhi and Calcutta: Oxford and IBH, 75–216.

Nash, L. L. and Gleick, P. H., 1991, Sensitivity of streamflow in the Colorado Basin to climatic changes. *Journal of hydrology*, 125, 221–41.

National Academy of Sciences, 1972, *The Earth and human affairs.* San Francisco: Camfield Press.

Nature Conservancy Council, 1977, *Nature conservation and agriculture.* London: Her Majesty's Stationery Office.

—, 1984, *Nature conservation in Great Britain.* Shrewsbury: Nature Conservancy Council.

Nearing, M. A., 2001, Potential changes in rainfall erosivity in the US with climate change during the 21st century. *Journal of soil and water conservation*, 56, 229–32.

Nelson, F. E. and Anisimov, O. A., 1993, Permafrost zonation in Russia under anthropogenic climate change. *Permafrost and periglacial processes*, 4, 137–48.

Nelson, F. E., Anisimov, O. A. and Shiklomanov, N. I., 2001, Subsidence risk from thawing permafrost. *Nature*, 410, 889–90.

—, 2002, Climate change and hazard zonation in the Circum-Arctic Permafrost regions. *Natural hazards*, 26, 203–25.

Nesje, A., Lie, O. and Dahl, S. O., 2000, Is the North Atlantic Oscillation reflected in glacier mass balance records? *Journal of Quaternary science*, 15, 587–601.

New, M., Todd, M., Hulme, M. and Jones, P., 2001, Precipitation measurements and trends in the twentieth century. *International journal of climatology*, 21, 1899–922.

Newman, J. R., 1979, Effects of industrial pollution on wildlife. *Biological conservation*, 15, 181–90.

Newman, W. S. and Fairbridge, R. W., 1986, The management of sealevel rise. *Nature*, 320, 319–21.

Newson, M., 1992, *Land, water and development.* London: Routledge.

Newton, J. G., 1976, Induced and natural sinkholes in Alabama: continuing problem along highway corridors. In F. R. Zwanig (ed.), *Subsidence over mines and caverns.* Washington, DC: National Academy of Sciences, 9–16.

Nichol, S. L., Augustinus, P. L., Gregory, M. R., Creese, R. and Horrocks, M., 2000, Geomorphic and sedimentary evidence of human impact on the New Zealand landscape. *Physical geography*, 21, 109–32.

Nicholls, R. J., Hoozemans, F. M. J. and Marchand, M., 1999, Increasing flood risk and wetland losses due to global sea level rise: regional and global analyses. *Global environmental change*, 9, S69–87.

Nicholson, S. E., 1978, Climatic variations in the Sahel and other African regions during the past five centuries. *Journal of arid environments*, 1, 3–24.

—, 1988, Land surface atmosphere interaction: physical processes and surface changes and their impact. *Progress in physical geography*, 12, 36–65.

—, 1996, Environmental change within the historical period. In W. M. Adams, A. S. Goudie and A. R. Orme (eds), *The physical geography of Africa.* Oxford: Oxford University Press, 60–87.

Nicod, J., 1986, Facteurs physico-chimiques de l'accumulation des formations travertineuses. *Mediterranée*, 10, 161–4.

Nihlgård, B. J., 1997, Forest decline and environmental stress. In D. Brune, D. V. Chapman, M. D. Gwynne and J. M. Pacyna (eds), *The global environment.* Weinheim: VCH.

Nikonov, A. A., 1977, Contemporary technogenic movements of the Earth's crust. *International geology review*, 19, 1245–58.

Noble, I. R. and Gitay, H., 1996, Deserts in a changing climate: impacts. In R. T. Watson, M. C. Zinyowera and R. H. Moss (eds), *Climate change 1995.* Cambridge: Cambridge University Press, 1509–69.

Nobre, C. A. and 6 others, 2004, The Amazonian climate. In P. Kabat and 8 others, (eds), *Vegetation, water, humans and the climate.* Berlin: Springer-Verlag, 79–92.

Nordstrom, K. F., 1994, Beaches and dunes of human-altered coasts. *Progress in physical geography*, 18, 497–516.

Norris, S., 2001, Thanks for all the fish. *New scientist*, 29th September, 36–9.

Nossin, J. J., 1972, Landsliding in the Crati basin, Calabria, Italy. *Geologie en mijnbouw*, 51, 591–607.

Nowlis, J. S., Roberts, C. M., Smith, A. H. and Siirila, 1997, Human-enhanced impacts of a tropical storm on nearshore coral reefs. *Ambio*, 26, 515–21.

Noy-Meir, I., 1974, Stability in arid ecosystems and effects of men on it. *Proceedings of the 12th International Congress of Ecology*, Wageningen, 220–5.

Nriagu, J. O., 1979, Global inventory of natural and anthropogenic emissions of trace metals in the atmosphere. *Nature*, 279, 409–11.

Nriagu, J. O. and Pacyna, J. M., 1988, Quantitative assessment of worldwide contamination of air, water and soils by trace metals. *Nature*, 337, 134–9.

Nunn, P. D., 1991, *Human and natural impacts on Pacific island environments*. Honolulu: Occasional paper 13, East West Environment and Policy Institute.

Nutalaya, P. and Ran, J. L., 1981, Bangkok: the sinking metropolis. *Episodes*, 4, 3–8.

Nutalaya, P., Yong, R. N., Chumnankit, T. and Buapeng, S., 1996, Land subsidence in Bangkok during 1978–1988. In J. D. Milliman and B. U. Haq (eds), *Sea level rise and coastal subsidence*. Dordrecht: Kluwer, 105–30.

Nye, P. H. and Greenland, D. J., 1964, Changes in the soil after clearing tropical forest. *Plant and soil*, 21, 101–12.

Oba, G., Post, E. and Stenseth, N. C., 2001, Sub-Saharan desertification and productivity are linked to hemispheric climate variability. *Global change biology*, 7, 241–6.

Oberle, M., 1969, Forest fires: suppression policy has its ecological drawbacks. *Science*, 165, 568–71.

Oechel, J. W. C., Hastings, S. J., Vourlitis, G. L., Jenkins, M. A. and Hinkson, C. L., 1995, Direct effects of elevated CO_2 in Chaparral and Mediterranean-type ecosystems. In J. M. Moreno and W. C. Oechel (eds), *Global change and Mediterranean-type ecosystems*. New York: Springer-Verlag 58–75.

Oerlemans, J., 1993, Possible changes in the mass balance of the Greenland and Antarctic ice sheets and their effects on sea level. In R. A. Warwick, E. M. Barrows and T. M. L. Wigley (eds), *Climatic and sea level change: observations, projections and implications*. Cambridge: Cambridge University Press, 144–61.

—, 1994, Quantifying global warming from the retreat of glaciers. *Science*, 264, 243–5.

Oerlemans, J. and 10 others, 1998, Modelling the response of glaciers to climate warming. *Climate dynamics*, 14, 267–74.

Oke, T. R., 1978, *Boundary layer climates*. London: Methuen.

Olley, J. M. and Wasson, R. J., 2003, Changes in the flux of sediment in the Upper Murrumbidgee catchment, southeastern Australia, since European settlement. *Hydrological processes*, 17, 3307–20.

Olson, C. G., Nettleton, W. D., Porter, D. A. and Brasher, B. R., 1997, Middle Holocene aeolian activity on the High Plains of west-central Kansas. *Holocene*, 7, 255–61.

Oppenheimer, M., 1998, Global warming and the stability of the West Antarctic ice sheet. *Nature*, 393, 325–32.

Oppenheimer, S., 2003, *Out of Eden. The peopling of the world*. London: Constable.

Osborn, T. J., Hulme, M., Jones, P. D. and Basnett, T. A., 2000, Observed trends in the daily intensity of United Kingdom precipitation. *International journal of climatology*, 20, 347–64.

Osterkamp, T. E. and Romanovsky, V. E., 1999, Evidence for warming and thawing of discontinuous permafrost in Alaska. *Permafrost and periglacial processes*, 10, 17–37.

O'Sullivan, P. E., Coard, M. A. and Pickering, D. A., 1982, The use of laminated lake sediments in the estimation and calibration of erosion rates. *Publication of the International Association of Hydrological Science*, 137, 385–96.

Otterman, J., 1974, Baring high albedo soils by overgrazing: a hypothesised desertification mechanism. *Science*, 186, 531–3.

Overpeck, J. T., Rind, D. and Goldberg, R., 1990, Climate-induced changes in forest disturbance and vegetation. *Nature*, 343, 51–3.

Oxley, D. J., Fenton, M. B. and Carmody, G. R., 1974, The effects of roads on populations of small mammals. *Journal of applied ecology*, 11, 51–9.

Ozenda, P. and Borel, J. L., 1990, The possible responses of vegetation to a global climatic change. In M. M. Boer and R. S. de Groot (eds), *Landscape – ecological impact of climatic change*. Amsterdam: IOS Press, 221–49.

Page, H., 1982, Some notes on the geomorphological and vegetational history of the saltings at Brean. *Somerset archaeology and natural history*, 120–5.

Page, M. J. and Trustrum, N. A., 1997, A late Holocene lake sediment record of the erosion response to land use change in a steepland catchment, New Zealand. *Zeitschrift für geomorphologie*, 41, 36992.

Pakeman, R. J., Marrs, R. H., Howard, D. C., Barr, C. J. and Fuller, R. M., 1996, The bracken problem in Great Britain: its present extent and future changes. *Applied geography*, 16, 65–86.

Pakiser, L. C., Eaton, J. P., Healy, J. H. and Raleigh, C. B., 1969, Earthquake prediction and control. *Science*, 166, 1467–74.

Palmer, T. N. and Räisänen, J., 2002, Quantifying the risk of extreme seasonal precipitation events in a changing climate. *Nature*, 415, 512–14.

Panel on Weather and Climate Modification, 1966, *Weather and climate modification problems and prospects*, publication 1350. Washington, DC: National Academy of Sciences.

Parizek, B. R. and Alley, R. B., 2004, Implications of increased Greenland surface melt under global-warming scenarios: ice-sheet simulations. *Quaternary science reviews*, 23, 1013–27.

Park, C. C., 1977, Man-induced changes in stream channel capacity. In K. J. Gregory (ed.), *River channel change*. Chichester: Wiley, 121–44.

—, 1987, *Acid rain: rhetoric and reality*. London: Methuen.

Park, R. A., Armentano, T. V. and Cloonan, C. L., 1986, Predicting the effects of sea level rise on coastal wetlands. In J. G. Titus (ed.), *Effects of changes in stratospheric ozone and global climate*, Vol. 4, *Sea level rise*. Washington, DC: United Nations Environment Program/U.S. Environmental Protection Agency, 129–52.

Parker, A. G., Goudie, A. S., Anderson, D. E., Robinson, M. A. and Bonsall, C., 2002, A review of the mid-Holocene elm decline in the British Isles. *Progress in physical geography*, 26, 1–45.

Parkinson, C. L., Cavalier, D. J., Gloersen, P., Zwally, H. J. and Comiso, J. C., 1999, Arctic sea ice extents, areas, and trends, 1978–1996. *Journal of geophysical research*, 104 (C9), 20837–56.

Parshall, T. and Foster, D. R., 2002, Fire on the New England landscape: regional and temporal variation, cultural and environmental controls. *Journal of biogeography*, 29, 1305–17.

Parson, E. A., Carter, L., Anderson, P., Wang, B. and Weller, G., 2001, Potential consequences of climate variability and change for Alaska. In National Assessment Synthesis Team, *Climate change impacts on the United States: the potential consequences of climate variability and change*. Cambridge: Cambridge University Press, 283–312.

Parsons, J. J., 1960, Fog drip from coastal stratus. *Weather*, 15, 58.

Passmore, J., 1974, *Man's responsibility for nature*. London: Duckworth.

Paul, K. I., Poglase, P. J., Nyakuengama, J. G. and Khanna, P. K., 2002, Change in soil carbon following afforestation. *Forest ecology and management*, 168, 241–57.

Pawson, E., 1978, *The early industrial revolution: Britain in the eighteenth century*. London: Batsford.

Pearson, R. G. and Dawson, T. P., 2003, Predicting the impacts of climate change on the distribution of species: are bioclimate envelope models useful? *Global ecology and biogeograpy*, 12, 361–71.

Peck, A. J., 1983, Response of groundwater to clearing in western Australia. In *Papers, international conference on groundwater and man*. Canberra: Australian Government Publishing Services, 327–35.

Peck, A. J. and Halton, T., 2003, Salinity and the discharge of salts from catchments in Australia. *Journal of hydrology*, 272, 191–202.

Peet, R. K., Glenn-Lewin, D. C. and Wolf, J. W., 1983, Prediction of man's impact on plant species diversity. In W. Holzner, M. J. A. Werger and I. Ikusima (eds), *Man's impact on vegetation*. The Hague: Junk, 41–54.

Peglar, S. M. and Birks, H. J. B., 1993, The mid-Holocene *Ulmus* fall at Diss Mere, Norfolk, south-east England – disease and human impact? *Vegetation history and archaeology*, 2, 61–8.

Peierls, B. L., Caraco, N. F., Pace, M. L. and Cole, J. J., 1991, Human influence on river nitrogen. *Nature*, 350, 386.

Pennington, W., 1974, *The history of British vegetation* (2nd edn). London: English University Press.

—, 1981, Records of a lake's life in time: the sediments. *Hydrobiologia*, 79, 197–219.

Pereira, H. C., 1973, *Land use and water resources in temperate and tropical climates*. Cambridge: Cambridge University Press.

Perla, R., 1978, Artificial release of avalanches in North America. *Arctic and alpine research*, 10, 235–40.

Pernetta, J., 1995, What is global change? *Global change newsletter*, 21, 1–3.

Peters, J. H. (ed.), 1998, *Artificial recharge of groundwater*. Amsterdam: Swetzs and Zeitlinger.

Peters, R. L., 1988, The effect of global climatic change on natural communities. In E. O. Wilson (ed.), *Biodiversity*. Washington, DC: National Academy Press, 450–61.

Petersen, K. K., 1981, *Oil shale, the environmental challenges*. Golden, Colorado: Colorado School of Mines.

Peterson, C. J., 2000, Catastrophic wind damage to North American forests and the potential impact of climate change. *The science of the total environment*, 262, 287–311.

Peterson, B. J. and 7 others, 2002, Increasing river discharge to the Arctic Ocean. *Science*, 298, 2171–3.

Pethick, J., 1993, Shoreline adjustments and coastal management: physical and biological processes under accelerated sea level rise. *Geographical journal*, 159, 162–8.

—, 2001, Coastal management and sea level rise. *Catena*, 42, 307–22.

Petit-Maire, N., Burollet, P. F., Ballais, J.-L., Fontugne, M., Rosso, J.-C. and Lazaar, A., 1999, Paléoclimats Holocènes du Sahara septentionale. Dépôts lacustres et terrasses alluviales en bordure du Grand Erg Oriental à l'extrême – Sud de la Tunisie. *Comptes Rendus Académie des Sciences, Series 2*, 312, 1661–6.

Petts, G. E., 1979, Complex response of river channel morphology subsequent to reservoir construction. *Progress in physical geography*, 35329–62.

—, 1985, *Impounded rivers: perspectives for ecological management*. Chichester: Wiley.

Petts, G. E. and Lewin, J., 1979, Physical effects of reservoirs on river systems. In G. E. Hollis (ed.), *Man's impact on the hydrological cycle in the United Kingdom*. Norwich: Geobooks, 79–91.

Pfisterer, A. B. and Schmid, B., 2002, Diversity-dependent production can decrease the stability of ecosystem functioning. *Nature*, 416, 84–6.

Pierson, F. B., Carlson, D. H. and Spaeth, K. E., 2002, Impacts of wildfire on soil hydrological properties of steep sagebrush-steppe rangeland. *International journal of wildland fire*, 11, 145–51.

Pimentel, D., 1976, Land degradation: effects on food and energy resources. *Science*, 194, 149–55.

—, 2003 (ed.), *Biological invasions*. Washington, DC: CRC Press.

Pimentel, D. and 10 others, 1995, Environmental and economic costs of soil erosion and conservation benefits. *Science*, 267, 1117–22.

Pirazzoli, P. A., 1996, *Sea level changes: the last 20,000 years*. Chichester: Wiley.

Pitt, D., 1979, Throwing light on a black secret. *New scientist*, 81, 1022–5.

Pittock, A. B. and Wratt, D., 2001, Australia and New Zealand. In J. J. McCarthy, O. F. Canziani, N. A. Leary, D. J. Dokken and K. S. White (eds), *Climate change 2001: impacts, adaptation and vulnerability*. Cambridge: Cambridge University Press, 591–639.

Pluhowski, E. J., 1970, Urbanization and its effects on the temperature of the streams on Long Island, New York. *United States Geological Survey professional paper*, 627-D.

Pollard, E. and Miller, A., 1968, Wind erosion in the East Anglian Fens. *Weather*, 23, 414–17.

Ponting, C., 1991, *A green history of the world*. London: Penguin.

Poore, M. E. D., 1976, The values of tropical moist forest ecosystems. *Unasylva*, 28, 127–43.

Pope, J. C., 1970, Plaggen soils in the Netherlands. *Geoderma*, 4, 229–55.

Post, W. M. and Kwon, K. C., 2000, Soil carbon sequestration and land-use change: processes and potential. *Global change biology*, 6, 317–27.

Potter, G. L., Ellsaesser, H. W., MacCracken, M. C. and Luther, F. M., 1975, Possible climatic impact of tropical deforestation. *Nature*, 258, 697–8.

Potter, G. L., Ellsaesser, H. W., MacCracken, M. C. and Ellis, J. C., 1981, Albedo change by man: test of climatic effects. *Nature*, 291, 47–9.

Powell, M., 1985, Salt, seed and yields in Sumerian agriculture: a critique of progressive salinization. *Zeitschrift für Assyrologie and Vorderasiatische archaologie*, 75, 7–38.

Preston, A., 1973, Heavy metals in British waters. *Nature*, 242, 95–7.

Price, M. F., 1989, Global change: defining the ill-defined. *Environment*, 31 (8), 18–20, 42–4.

Price, M. and Reed, D. W., 1989, The influence of mains leakage and urban drainage on groundwater levels beneath conurbations in the United Kingdom. *Proceedings of the Institution of Civil Engineers*, 86 (I), 31–9.

Prince, H. C., 1959, Parkland in the Chilterns. *Geographical review*, 49, 18–31.

——, 1962, Pits and ponds in Norfolk. *Erdkunde*, 16, 10–31.

——, 1964, The origin of pits and depressions in Norfolk. *Geography*, 49, 15–32.

——, 1979, Marl pits or dolines of the Dorset Chalklands? *Transactions of the Institute of British Geographers*, new series, 4, 116–17.

Proffitt, M. H., Margitan, J. J., Kelly, K. K., Loewenstein, M., Podolske, J. R. and Chan, K. R., 1990, Ozone loss in the Arctic polar vortex inferred from high-altitude aircraft measurements. *Nature*, 347, 31–3.

Prokopovich, N. P., 1972, Land subsidence and population growth. *24th International Geological Congress proceedings*, 13, 44–54.

Prospero, J. M. and Nees, R. T., 1977, Dust concentration in the atmosphere of the equatorial North Atlantic; possible relationship to Sahelian drought. *Science*, 196, 1196–8.

Prowse, T. D. and Beltaos, S., 2002, Climatic control of river-ice hydrology: a review. *Hydrological processes*, 16, 805–22.

Pyne, S. J., 1982, *Fire in America – a cultural history of wildland and rural fire*. Princeton: Princeton University Press.

Qadir, M., Ghafoor, A. and Murtaza, G., 2000, Amelioration strategies for saline soils: a review. *Land degradation and development*, 11, 501–21.

Rackham, O., 1980, *Ancient woodland*. London: Arnold.

Radley, J., 1962, Peat erosion on the high moors of Derbyshire and west Yorkshire. *East Midlands geographer*, 3 (1), 40–50.

Radley, J. and Sims, C., 1967, Wind erosion in East Yorkshire. *Nature*, 216, 20–2.

Raison, R. J., 1979, Modification of the soil environment by vegetation fires with particular reference to nitrogen transformation: a review. *Plant and soil*, 51, 73–108.

Raleigh, C. B., Healy, J. H. and Bredehoeft, J. D., 1976, An experiment in earthquake control at Rangeley, Colorado. *Science*, 191, 1230–7.

Ramanathan, V., 1988, The greenhouse theory of climate change: a test by an inadvertent global experiment. *Science*, 240, 293–9.

Ramankutty, N. and Foley, A., 1999, Estimating historical changes in global land cover: croplands from 1700 to 1992. *Global biogeochemical cycles*, 13, 997–1027.

Ramirez, E. and 8 others, 2001, Small glaciers disappearing in the tropical Andes: a case study of Bolivia: Glacier Chacaltya (16°S). *Journal of glaciology*, 47, 187–94.

Ranwell, D. S., 1964, *Spartina* salt marshes in southern England, II: rate and seasonal pattern of sediment accretion. *Journal of ecology*, 52, 79–94.

Ranwell, D. S. and Boar, R., 1986, *Coast dune management guide*. Monks Wood: Institute of Terrestrial Ecology.

Raper, S. C. B., 1993, Observational data on the relationships between climatic change and the frequency and magnitude of severe tropical storms. In R. A. Warrick, E. M. Barrow and T. M. L. Wigley (eds), *Climate and sea level change: observations, projections and implications*. Cambridge: Cambridge University Press, 192–212.

Rapp, A., 1974, *A review of desertization in Africa – water, vegetation and man*. Stockholm: Secretariat for International Ecology, report no. 1.

Rapp, A., Murray-Rust, D. H., Christansson, C. and Berry, L., 1972, Soil erosion and sedimentation in four catchments near Dodoma, Tanzania. *Geografiska annaler*, 54A, 255–318.

Rapp, A., Le Houérou, H. N. and Lundholm, B., 1976, Can desert encroachment be stopped? *Ecological bulletin*, 24.

Rasid, H., 1979, The effects of regime regulation by the Gardiner Dam on downstream geomorphic processes in the South Saskatchewan River. *Canadian geographer*, 23, 140–58.

Ratcliffe, D. A., 1974, Ecological effects of mineral exploitation in the United Kingdom and their significance to nature conservation. *Proceedings of the Royal Society of London*, 339A, 355–72.

Ray, C., Hayden, B. P., Bulger, A. J. and McCormick-Ray, G., 1992, Effects of global warming on the biodiversity of coastal-marine zones. In R. L. Peters and T. E. Lovejoy (eds), *Global warming and biological diversity*. New Haven: Yale University Press, 91–102.

Reale, O. and Dirmeyer, P., 2000, Modelling the effects of vegetation on Mediterranean climate during the Roman Classical Period: Part I: climate history and model sensitivity. *Global and planetary change*, 25, 163–84.

Reclus, E., 1871, *The earth* (2 vols). London: Chapman & Hall.

——, 1873, *The ocean, atmosphere and life*. New York: Harper and Brothers.

Reed, C. A., 1970, Extinction of mammalian megafauna in the old world late Quaternary. *Bioscience*, 20, 284–8.

Reed, D. J., 1990, The impact of sea level rise on coastal salt marshes. *Progress in physical geography*, 14, 465–81.

——, 1995, The response of coastal marshes to sea level rise: survival or submergence? *Earth surface processes and landforms*, 20, 39–48.

——, 2002, Sea-level rise and coastal marsh sustainability: geological and ecological factors in the Mississippi delta plain. *Geomorphology*, 48, 233–43.

Reed, L. A., 1980, Suspended-sediment discharge, in five streams near Harrisburg, Pennsylvania, before, during and after highway construction. *United States Geological Survey water supply paper*, 2072.

Reheis, M. C., 1997, Dust deposition downwind of Owens (dry) Lake, 1991–1994: preliminary findings. *Journal of geophysical research*, 102, 25998–6008.

Reichert, B. K., Bengston, L. and Oerlemans, J., 2001, Midlatitude forcing mechanism for glacier mass balance investigated using general circulation models. *Journal of climate*, 14, 3767–84.

Reij, C., Scoones, I. and Toulmin, C. (eds), 1996, *Sustaining the soil: indigenous soil and water conservation in Africa*. London: Earthscan.

Renard, K. G. and Freid, J. R., 1994, Using monthly precipitation data to estimate the R factor in the revised USLE. *Journal of hydrology*, 157, 287–306.

Renberg, I. and Hellberg, T., 1982, The pH history of lakes in SW Sweden, as calculated from the subfossil diatom flora of the sediments. *Ambio*, 11, 30–3.

Revelle, R. R. and Waggoner, P. E., 1983, Effect of a carbon dioxide-induced climatic change on water supplies in the western United States. In Carbon Dioxide Assessment Committee, *Changing climate*. Washington, DC: National Academy Press, 419–32.

Rhoades, J. D., 1990, Soil salinity – causes and controls. In A. S. Goudie (ed.), *Techniques for desert reclamation*. Chichester: Wiley, 109–34.

Richards, J. F., 1991, Land transformation. In B. L. Turner, W. C. Clark, R. W. Kates, J. F. Richards, J. T. Matthews and W. B. Meyer (eds), *The earth as transformed by human action*. Cambridge: Cambridge University Press, 163–78.

Richards, P. W., 1952, *The tropical rainforest*. Cambridge: Cambridge University Press.

Richardson, J. A., 1976, Pit heap into pasture. In J. Lenihan and W. W. Fletcher (eds), *Reclamation*. Glasgow: Blackie, 60–93.

Richardson, S. J. and Smith, J., 1977, Peat wastage in the East Anglian Fens. *Journal of soil science*, 28, 485–9.

Richter, D. O. and Babbar, L. I., 1991, Soil diversity in the tropics. *Advances in ecological research*, 21, 315–89.

Ridgwell, A. J., 2002, Dust in the Earth system: the biogeochemical linking of land, sea and air. *Philosophical transactions of the Royal Society*, A, 360, 2905–24.

Rignot, E. and Thomas, R. H., 2002, Mass balance of polar ice sheets. *Science*, 297, 1502–6.

Rinderer, T. E., Oldroyd, R. P. and Sheppard, W. S., 1993, Africanized bees in the US. *Scientific American*, 269 (6), 52–8.

Ripley, E. A., 1976, Drought in the Sahara: insufficient geophysical feedback? *Science*, 191, 100.

Ritchie, J. C., 1972, Sediment, fish and fish habitat. *Journal of soil and water conservation*, 27, 124–5.

Robb, G. A. and Robinson, J. D. F., 1995, Acid drainage from mines. *Geographical journal*, 161, 47–54.

Roberts, M. B., Gamble, C. S. and Bridgland, D. R., 1995, The earliest occupation of Europe: the British Isles. *Analecta praehistorica Leidensia*, 27, 165–91.

Roberts, N., 1989, *The Holocene*. Oxford: Blackwell.

Roberts, N., 1998, *The Holocene: an environmental history* (2nd edn). Oxford: Basil Blackwell.

Roberts, R. G. and 10 others, 2001, New ages for the least Australian megafauna: continent-wide extinction about 46,000 years ago. *Science*, 1888–92.

Roberts, N. and Barker, P., 1993, Landscape stability and biogeomorphic response to past and future climatic shifts in intertropical Africa. In D. S. G. Thomas and R. J. Allison (eds), *Landscape sensitivity*. Wiley: Chichester, 65–82.

Robin, G. de Q., 1986, Changing the sea level. In B. Bolin et al. (eds), *The greenhouse effect, climatic change and ecosystems*. Chichester: Wiley, 322–59.

Robinson, M., 1979, The effects of pre-afforestation ditching upon the water and sediment yields of a small upland catchment. Working paper 252, School of Geography, University of Leeds.

——, 1990, *Impact of improved land drainage on river flows*. Wallingford: Institute of Hydrology, Report 113.

Robinson, M. A. and Lambrick, G. H., 1984, Holocene alluviation and hydrology in the Upper Thames Basin. *Nature*, 308, 809–14.

Rodda, J. C., Downing, R. A. and Law, F. M., 1976, *Systematic hydrology*. London: Newnes-Butterworth.

Roman, J. and Palumbi, S. R., 2003, Whales before whaling in the North Atlantic. *Science*, 301, 508–10.

Romme, W. H. and Despain, D. G., 1989, The Yellowstone fires. *Scientific American*, 261, 21–9.

Roots, C., 1976, *Animal invaders*. Newton Abbot: David & Charles.

Roques, K. G., O'Connor, T. G. and Watkinson, A. R., 2001, Dynamics of shrub encroachment in an African savanna: relative influences of fire, herbivory, rainfall and density dependence. *Journal of applied ecology*, 38, 268–80.

Ropeleswski, C. F., 1985, Satellite-derived snow and ice cover in climate diagnostic studies. *Advances in space research*, 5, 275–8.

Rose, R. 1970, Lichens as pollution indicators. *Your environment*, 5.

Rosenzweig, C. and Hillel, D., 1993, The dust bowl of the 1930s: analog of greenhouse effect in the Great Plains? *Journal of environmental quality*, 22, 9–22.

Rosepiler, M. J. and Reilinger, R., 1977, Land subsidence due to water withdrawal in the vicinity of Pecos, Texas. *Engineering geology*, 11, 295–304.

Rouse, W. R. and 10 others, 1997, Effects of climate change on the freshwaters of Arctic and Subarctic North America. *Hydrological processes*, 11, 873–902.

Routson, R. C., Wildung, R. E. and Bean, R. M., 1979, A review of the environmental impact of ground disposal of oil shale wastes. *Journal of environmental quality*, 8, 14–19.

Royal Society Study Group, 1983, *The nitrogen cycle of the United Kingdom*. London: The Royal Society.

Ruddiman, W. F., 2003, The anthropogenic greenhouse era began thousands of years ago. *Climatic change*, 61, 261–93.

Ruddiman, W. F. and Thomsen, J. S., 2001, The case for human causes of increased atmospheric CH_4 over the last 5000 years. *Quaternary science reviews*, 20, 1769–77.

Russell, E. J., 1961, *The world of the soil*. London: Fontana.

Russell, J. S. and Isbell, R. F. (eds), 1986, *Australian soils: the human impact*. St Lucia: University of Queensland Press.

Rutherford, I., 2000, Some human impacts on Australian stream channel morphology. In S. Brizga and B. Finlayson (eds), *River management: the Australian experience*. Chichester: Wiley, 11–47.

Ryder, M. L., 1966, The exploitation of animals by man. *Advancement of science*, 23, 9–18.

Ryding, S. O. and Rast, R. W., 1989, *The control of eutrophication of lakes and reservoirs*. Paris: UNESCO.

Sabadell, J. E., Risley, E. M., Jorgensen, H. T. and Thornton, B. S., 1982, *Desertification in the United States: status and issues*. Bureau of Land Management, Department of the Interior.

Sabadini, R., 2002, Ice sheet collapse and sea level change. *Science*, 295, 2376–7.

Sahagian, D., 2000, Global physical effects of anthropogenic hydrological alterations: sea level and water redistribution. *Global and planetary change*, 25, 38–48.

Saiko, T. A. and Zonn, I. S., 2000, Irrigation expansion and dynamics of desertification in the Circum-Aral region of Central Asia. *Applied geography*, 20, 349–67.

Sala, O. E. and 18 others, 2000, Global biodiversity scenarios for the years 2100. *Science*, 287, 1770–4.

Sanchez, P. A. and Buol, S. W., 1975, Soils of the tropics and the world food crisis. *Science*, 188, 598–603.

Sanders, W. M., 1972, Nutrients. In R. T. Oglesby, C. A. Carlson and J. A. McCann (eds), *River ecology and man*. New York: Academic Press, 389–415.

Sapp, J., 1999, *What is natural? Coral reef crisis*. New York: Oxford University Press.

Sarmiento, G. and Monasterio, M., 1975, A critical consideration of the environmental conditions associated with the occurrence of savanna ecosystems in tropical America. *Ecological studies*, 11, 233–50.

Sarre, P., 1978, The diffusion of Dutch elm disease. *Area*, 10, 81–5.

Sauer, C. O., 1938, *Destructive exploitation in modern colonial expansion*. International Geographical Congress, Amsterdam, Vol. III, sect. IIIC, 494–9.

——, 1952, *Agricultural origins and dispersals*, Isaiah Bowman lecture series, 2. New York: American Geographical Society.

——, 1969, *Seeds, spades, hearths and herds*. Cambridge, MA: MIT Press.

Savage, M., 1991, Structural dynamics of a southwestern pine forest under chronic human influence. *Annals of the Association of American Geographers*, 81, 271–89.

Savini, J. and Kammerer, J. C., 1961, Urban growth and the water regime. *United States Geological Survey water supply paper*, 1591A.

Schimper, A. F. W., 1903, *Plant-geography upon a physiological basis*. Oxford: Clarendon Press.

Schmid, J. A., 1974, The environmental impact of urbanization. In I. R. Manners and M. W. Mikesell (eds), *Perspectives on environment*. Washington, DC: Association of American Geographers.

——, 1975, Urban vegetation. Research paper, Department of Geography, University of Chicago, 161.

Schmieder, O., 1927a, The Pampa – a natural or culturally induced grassland? *University of California publications in geography*, 2, 255–70.

——, 1927b, Alteration of the Argentine Pampa in the colonial period. *University of California publications in geography*, 2, 303–21.

Schneider, S. H. and Thompson, S. L., 1988, Simulating the effects of nuclear war. *Nature*, 333, 221–7.

Schoner, W., Auer, I. and Bohm, R., 2000, Climate variability and glacier reaction in the Austrian Eastern Alps. *Annals of glaciology*, 23, 31–8.

Schrieber, B. C., 1986, Arid shorelines and evaporates. In H. G. Reading (ed.), *Sedimentary environments and facies*. Oxford: Blackwell Scientific, 189–228.

Schumm, S. A., 1977, *The fluvial system*. New York: Wiley.

Schumm, S. A., Harvey, M. D. and Watson, C. C., 1984, *Incised channels: morphology, dynamics and control*. Littleton, Colorado: Water Resources publications.

Schwartz, M. W., Porter, D. J., Randall, J. M. and Lyons, K. E., 1996, Impact of nonindigenous plants. In *Sierra Nevada ecosystems project: final report for congress*, Vol. II. Davis: University of California, 1203–18.

Schwarz, E. H. L., 1923, *The Kalahari or Thirstland redemption*. Cape Town and Oxford: Oxford University Press.

Schwarz, H. E., Emel, J., Dickens, W. J., Rogers, P. and Thompson, J., 1991, Water quality and flows. In B. L. Turner, W. C. Clark, R. W. Kates, J. F. Richards, J. T. Matthews and W. B. Meyer (eds), *The earth as transformed by human action*. Cambridge: Cambridge University Press, 253–70.

Scott, D. F., 1997, The contrasting effects of wildfire and clear felling on the hydrology of a small catchment. *Hydrological processes*, 11, 543–55.

Scott, G. J., 1977, The role of fire in the creation and maintenance of savanna in the Montana of Peru. *Journal of biogeography*, 4, 143–67.

Scott, W. S. and Wylie, N. P., 1980, The environmental effects of snow dumping: a literature review. *Journal of environmental management*, 10, 219–40.

Sears, P. B., 1957, Man the newcomer: the living landscape and a new tenant. In L. H. Russwurm and E. Sommerville (eds), *Man's natural environment, a system approach*. North Scituate: Duxbury, 43–55.

Segall, P., 1989, Earthquakes triggered by fluid extraction. *Geology*, 17, 942–6.

Seidel, K., Ehrler, C. and Martinec, J., 1998, Effects of climate change on water resources and runoff in an alpine basin. *Hydrological processes*, 12, 1659–69.

Selby, M. J., 1979, Slopes and weathering. In K. J. Gregory and D. E. Walling (eds), *Man and environmental processes*. Folkestone: Dawson, 105–22.

Semaw, S., Renne, P., Harris, J. W. K., Feibel, C. S., Bernov, R. L., Fesseha, N. and Mowbray, K., 1997, 2.5-million-year-old stone tools from Gona, Ethiopia. *Nature*, 385, 333–6.

Semtner, A. J., 1984, The climate response of the Arctic Ocean to Soviet river diversions. *Climatic change*, 6, 109–30.

Seto, S. and 8 others, 2002, Annual and seasonal trends in chemical composition of precipitation in Japan during 1989–1998. *Atmospheric environment*, 31, 3505–17.

Shabalova, M. V., van Deursen, W. P. A. and Buishand, T. A., 2003, Assessing future discharge of the river Rhine using regional climate model integrations and a hydrological model. *Climate research*, 23, 233–46.

Shakesby, R. A., Doerr, S. H. and Walsh, R. P. D., 2000, The erosional impact of soil hydrophobicity: current problems and future research directions. *Journal of hydrology*, 231/2, 178–91.

Shaler, N. S., 1912, *Man and the earth*. New York: Duffield.

Sheail, J., 1971, *Rabbits and their history*. Newton Abbot: David & Charles.

Sheffield, A. T., Healy, T. R. and McGlone, M. S., 1995, Infilling rates of a steepland catchment estuary, Whangamata, New Zealand. *Journal of coastal research*, 11 (4), 1294–308.

Shehata, W. and Lotfi, H., 1993, Preconstruction solution for groundwater rise in Sabkha. *Bulletin of the International Association of Engineering Geology*, 47, 145–50.

Sheppard, C. R. C., 2003, Predicted recurrences of mass coral morality in the Indian Ocean. *Nature*, 425, 294–7.

Sherif, M. M. and Singh, V. P., 1999, Effect of climate change on sea water intrusion in coastal aquifers. *Hydrological processes*, 13, 1277–87.

Sherlock, R. L., 1922, *Man as a geological agent*. London: Witherby.

Sherratt, A., 1981, Plough and pastoralism: aspects of the secondary products revolution. In I. Hodder, G. Isaac and N. Hammond (eds), *Pattern of the past*. Cambridge: Cambridge University Press, 261–305.

——, 1997, Climatic cycles and behaviour revolutions: the emergence of modern humans and the beginning of farming. *Antiquity*, 71, 271–87.

Sherwood, B., Cutler, D. and Burton, J. (eds), 2002, *Wildlife and roads. The ecological impact*. London: Imperial College Press.

Shi, Y. F. and Liu, S. Y., 2000, Estimation on the response of glaciers in China to the global warming in the 21st century. *Chinese science bulletin*, 45, 668–72.

Shiklomanov, I. A., 1985, Large scale water transfers. In J. C. Rodda (ed.), *Facets of hydrology II*. Chichester: Wiley, 345–87.

——, 1999, Climate change, hydrology and water resources: the work of the IPCC, 1988–1994. In van Dam, J. C. (ed.), *Impacts of climate change and climate variability on hydrological regimes*. Cambridge: Cambridge University Press, 8–20.

Shirahata, H., Elias, R. W., Patterson, C. C. and Koide, M., 1980, Chronological variations in concentrations and isotopic composition of anthropogenic atmospheric lead in sediments of a remote subalpine pond. *Geochimica et cosmochimica acta*, 44, 149–62.

Shriner, D. S. and Street, R. B., 1998, North America. In R. T. Watson (ed.), *The regional impacts of climate change*. Cambridge: Cambridge University Press.

Sidle, R. C. and Dhakal, A. S., 2002, Potential effect of environmental change on landslide hazards in forest environments. In R. C. Sidle (ed.), *Environmental change and geomorphic hazards in forests*. Wallingford: CABI, 123–65.

Sidorchuk, A. Y. and Golosov, V. N., 2003, Erosion and sedimentation on the Russian Plain, II: the history of erosion and sedimentation during the period of intensive agriculture. *Hydrological processes*, 17, 3347–58.

Simas, T., Nunes, J. P. and Ferreira, J. G., 2001, Effects of global climate change on coastal salt marshes. *Ecological modelling*, 139, 1–15.

Simberloff, D., 2000, Global climate change and introduced species in United States forests. *The science of the total environment*, 262, 253–61.

Simmonds, N. W., 1976, *Evolution of crop plants*. London: Longman.

Simmons, I. G., 1979, *Biogeography: natural and cultural*. London: Arnold.

——, 1993, *Environmental history: a concise introduction*. Oxford: Blackwell.

——, 1996, *Changing the face of the earth: culture, environment and history* (2nd edn). Oxford: Blackwell.

Simon, J. L., 1981, *The ultimate resource*. Princeton: Princeton University Press.

——, 1996, *The ultimate resource 2*. Princeton: Princeton University Press.

Sinclair, A. R. E. and Fryxell, J. M., 1985, The Sahel of Africa: ecology of a disaster. *Canadian journal of zoology*, 63, 987–94.

Six, D., Reynaud, L. and Letreguilly, A., 2001, Bilans de masse des glaciers alpines et scandinaves, leurs relations avec l'oscillation due climat de l'Atlantique nord. *Comptes Rendus Academie des Sciences, Science de la Terre et des Planètes*, 333, 693–8.

Smith, J. B. and Tirpak, D. A. (eds), 1990, *The potential effects of global climate change on the United States*. New York: Hemisphere.

Smith, J. E. (ed.), 1968, *Torrey Canyon pollution and marine life*. Cambridge: Cambridge University Press.

Smith, K., 1975, *Principles of applied climatology*. London: McGraw-Hill.

Smith, L. B. and Hadley, C. H., 1926, The Japanese beetle. *Department circular, US Department of Agriculture*, 363, 1–66.

Smith, M. W., 1993, Climate change and permafrost. In H. M. French and O. Slaymaker (eds), *Canada's cold environments*. Montreal and Kingston: McGill-Queens University Press, 292–311.

Smith, N., 1976, *Man and water*. London: Davies.

Smith, S. E., 1986, An assessment of structural deterioration on ancient monuments and tombs in Thebes. *Journal of field archaeology*, 13, 1277–87.

Smith, S. J., Pitcher, H. and Wigley, T. M. L., 2001, Global and regional anthropogenic sulfur dioxide emissions. *Global and planetary change*, 29, 99–119.

Smith, T. M., Shugart, H. H., Bonan, G. B. and Smith, J. B., 1992, Modelling the potential response of vegetation to global climate change. *Advances in ecological research*, 22, 93–116.

Smith, W. H., 1974, Air pollution – effects on the structure and function of the temperate forest ecosystem. *Environmental pollution*, 6, 111–29.

Smith, J. B., Richels, R. and Muller, B., 2001, Potential consequences of climate variability and change for the western United States. In National Assessment Synthesis Team, *Climate change impacts on the United States: the potential consequences of climate variability and change*. Cambridge: Cambridge University Press, 219–45.

Snedaker, S. C., 1993, Impact on mangroves. In G. Maul (ed.), *Climatic change in the Intra-Americas Sea*. London: Edward Arnold, 282–305.

——, 1995, Mangroves and climate change in the Florida and Caribbean region: scenarios and hypotheses. *Hydrobiologia*, 295, 43–9.

Snover, A., 1997, Impacts of global climate change on the Pacific Northwest. Preparatory white paper for OSTP/USGCRP regional workshop on the *Impacts of global climate change on the Pacific Northwest*, July 1997.

So, C. L., 1971, Mass movements associated with the rainstorm of 1966 in Hong Kong. *Transactions of the Institute of British Geographers*, 53, 55–65.

Solomon, S., 1999, Stratospheric ozone depletion: a review of concepts and history. *Reviews of geophysics*, 37, 275–316.

Somerville, M., 1858, *Physical geography* (4th edn). London: Murray.

Sopper, W. E., 1975, Effects of timber harvesting and related management practices on water quality in forested watersheds. *Journal of environmental quality*, 4, 24–9.

Sorkin, A. J., 1982, *Economic aspects of natural hazards*. Lexington: Lexington Books.

Souter, D. W. and Linden, O., 2000, The health and future of coral reef systems. *Ocean and coastal management*, 43, 657–88.

Southwick, C. H., 1976, *Ecology and the quality of our environment* (2nd edn). New York: Van Nostrand.

Spanier, E. and Galil, B. S., 1991, Lessepsian migration: a continuous biogeographical process. *Endeavour*, 15, 102–6.

Spate, O. H. K. and Learmonth, A. T. A., 1967, *India and Pakistan*. London: Methuen.

Speight, M. C. D., 1973, Outdoor recreation and its ecological effects. A bibliography and review. Discussion papers in conservation, 4, University College, London.

Spencer, J. E. and Thomas, W. L., 1978, *Introducing cultural geography* (2nd edn). New York: Wiley.

Spencer, T., 1995, Potentialities, uncertainties and complexities in the response of coral reefs to future sea level rise. *Earth surface processes and landforms*, 20, 49–64.

Spencer, T. and Douglas, I., 1985, The significance of environmental change: diversity, disturbance and tropical ecosystems. In I. Douglas and T. Spencer (eds), *Environmental change and tropical geomorphology*. London: Allen & Unwin, 13–33.

Spencer, T., Teleki, K. A., Bradshaw, C. and Spalding, M. D., 2000, Coral bleaching in the Southern Seychelles during the 1997–1998 Indian Ocean Warm event. *Marine pollution bulletin*, 40, 569–86.

Sperling, C. H. B., Goudie, A. S., Stoddart, D. R. and Poole, G. C., 1979, Origin of the Dorset dolines. *Transactions of the Institute of British Geographers*, new series, 4, 121–4.

Speth, W. W., 1977, Carl Ortwin Sauer on destructive exploitation. *Biological conservation*, 11, 145–60.

Squire, G. R. and 14 others, 2003, On the rationale and interpretation of the farm scale evaluations of genetically modified herbicide-tolerant crops. *Philosophical transactions of the Royal Society of London*, B, 358, 1779–99.

Staehelin, J., Harris, N. R. P., Appenzeller, C. and Ebeshard, J., 2001, Ozone trends: a review. *Reviews of geophysics*, 39, 231–90.

Stanley, D. J., 1996, Nile delta: extreme case of sediment entrapment on a delta plain and consequent coastal land loss. *Marine geology*, 129, 189–95.

Stanley, D. J. and Chen, Z., 1993, Yangtze delta, eastern China: I. Geometry and subsidence of Holocene depocenter. *Marine geology*, 112, 1–11.

Steadman, D. W., Stafford, T. W., Donahue, D. J. and Jull, A. J. T., 1991, Chronology of Holocene vertebrate extinction in the Galápagos Islands. *Quaternary research*, 36, 126–33.

Steffen, W. and 10 others, 2004, *Global change and the earth system*. Berlin: Springer-Verlag.

Stendel, M. and Christensen, J. H., 2002, Impact of global warming on permafrost conditions in a coupled GCM. *Geophysical research letters*, 29, 10–1–4.

Stephens, J. C., 1956, Subsidence of organic soils in the Florida Everglades. *Proceedings of the Soil Science Society of America*, 20, 77–80.

Sternberg, H. O'R., 1968, Man and environmental change in South America. *Monographiae biologicae*, 18, 413–45.

Stetler, L. L. and Gaylord, D. R., 1996, Evaluating eolian-climate interactions using a regional climate model from Hanford, Washington (USA). *Geomorphology*, 17, 99–113.

Stevenson, A. C., Jones, V. J. and Battarbee, R. W., 1990, The cause of peat erosion: a palaeolimnological approach. *New phytologist*, 114, 727–35.

Stewart, O. C., 1956, Fire as the first great force employed by man. In W. L. Thomas (ed.), *Man's role in changing the face of the* earth. Chicago: University of Chicago Press, 115–33.

Stiles, D., 1995, An overview of desertification as dryland degradation. In D. Stiles (ed.), *Social aspects of sustainable dryland management*. Chichester: Wiley.

Stocker, T. F., 2001, Physical climate processes and feedbacks. In J. T. Houghton (ed.), *Climate change 2001: the scientific basis*. Cambridge: Cambridge University Press, 417–70.

Stocking, M., 1984, Erosion and soil productivity: a review. *FAO soil conservation programme land and water development division consultants' working paper*, 1.

Stoddart, D. R., 1968, Catastrophic human interference with coral atoll ecosystems. *Geography*, 53, 25–40.

—, 1969, Climatic geomorphology, review and reassessment. *Progress in physical geography*, 1, 159–222.

—, 1971, Coral reefs and islands and catastrophic storms. In J. A. Steers (ed.), *Applied coastal geomorphology*. London: Macmillan, 154–97.

Stoddart, J. L. and 22 others, 1999, Regional trends in aquatic recovery from acidification in North America and Europe. *Nature*, 401, 575–8

Stokes, S. and Gaylord, D. R., 1993, optical dating of Holocene dune sands in the Ferris Dune Field, Wyoming. *Quaternary research*, 39, 274–81.

Stokes, S. and Swinehart, J. B., 1997, Middle- and late-Holocene dune reactivation on the Nebraska Sand Hills, USA. *The Holocene*, 7, 272–81.

Stokes, S., Thomas, D. S. G. and Washington, R., 1997, Multiple episodes of aridity in southern Africa since the last interglacial period. *Nature*, 388, 154–8.

Strahler, A. N. and Strahler, A. H., 1973, *Environmental geoscience: interaction between natural systems and man*. Santa Barbara: Hamilton.

Strandberg, C. H., 1971, Water pollution. In G. H. Smith (ed.), *Conservation of natural resources* (4th edn). New York: Wiley, 189–219.

Street, F. A. and Grove, A. T., 1979, Global maps of lake level fluctuation since 30,000 years ago. *Quaternary research*, 12, 83–118.

Stringer, C., 2000, Human evolution: how an African primate became global. In S. J. Culver and P. F. Rawson (eds), *Biotic response to global warming*. Cambridge: Cambridge University Press.

—, 2003, Out of Africa. *Nature*, 423, 692–9.

Sturrock, F. and Cathie, J., 1980, Farm modernisation and the countryside. Occasional paper 12, Department of Land Economy, University of Cambridge.

Su, Z. and Shi, Y., 2002, Response of monsoonal temperate glaciers to global warming science The Little Ice Age. *Quaternary international*, 97–8, 123–31.

Sun, G. E., McNulty, S. G., Moore, J., Bunch, C. and Ni, J., 2002, Potential impacts of climate change on rainfall erosivity and water availability in China in the next 100 years. *Proceedings of the 12th International Soil Conservation conference*, Beijing.

Suppiah, R. and Hennessy, K. J., 1998, Trends in total rainfall, heavy rain events and number of dry days in Australia. *International journal of climatology*, 18, 1141–64.

Swank, W. T. and Douglass, J. E., 1974, Streamflow greatly reduced by converting deciduous hardwood stands to pine. *Science*, 18, 857–9.

Swanston, D. N. and Swanson, F. J., 1976, Timber harvesting, mass erosion and steepland forest geomorphology in the Pacific north-west. In D. R. Coates (ed.) *Geomorphology and engineering*. Stroudsburg: Dowden, Hutchinson and Ross, 199–221.

Swift, L. W. and Messer, J. B., 1971, Forest cuttings raise temperatures of small streams in the southern Appalachians. *Journal of soil and water conservation*, 26, 111–16.

Swift, M. J. and Sanchez, P. A., 1984, Biological management of tropical soil fertility for sustained productivity. *Nature and resources*, 2052–10.

Swisher, C. C., Curtis, G. H., Jacob, T., Getty, A. G. and Suprijo, A., 1994, Age of the earliest known hominids in Java, Indonesia. *Science*, 263, 1118–21.

Szabolcs, I., 1994, State and perspectives on soil salinity in Europe. *European Society for Soil Conservation newsletter*, 3, 17–24.

Tajikistan Academy of Sciences, 1975, *Induced seismicity of the Nurek reservoir*. Dushanbe: Tajik Academy.

Tallis, J. H., 1965, Studies on southern Pennine peats, IV: evidence of recent erosion. *Journal of ecology*, 53, 509–20.

—, 1985, Erosion of blanket peat in the southern Pennines: new light on an old problem. In R. H. Johnson (ed.), *The geomorphology of north-west England*. Manchester: Manchester University Press, 313–36.

Talwani, P., 1997, On the nature of reservoir-induced seismicity. *Pure and applied geophysics*, 150, 473–92.

Taylor, C., 1975, *Fields in the English landscape*. London: Dent.

Taylor, C. M., Lambin, E. F., Stephenne, N., Harding, R. J. and Essery, L. H., 2002, The influence of land use change on climate in the Sahel. *Journal of climate*, 15, 3615–29.

Taylor, J. A., 1985, Bracken encroachment rates in Britain. *Soil use and management*, 1, 53–6.

Taylor, K. E. and Penner, J. E., 1994, Response of the climatic system to atmospheric aerosols and greenhouse gases. *Nature*, 369, 734–7.

Tegen, I., Lacis, A. A. and Fung, I., 1996, The influence on climate forcing of mineral aerosols from disturbed soils. *Nature*, 380, 419–22.

Terborgh, J., 1992, *Diversity and the tropical rain forest*. New York: Freeman.

Theodoropoulos, D. I., 2003, *Invasion biology. Critique of a pseudoscience*. Blythe, California: Avvar Books.

Thirgood, J. V., 1981, *Man and the Mediterranean forest – a history of resource depletion*. London: Academic Press.

Thomas, A. D., Walsh, R. P. D. and Shakesby, R. A., 2000, Solutes in overland flow following fire in eucalyptus and pine forests, northern Portugal. *Hydrological processes*, 14, 971–85.

Thomas, C. D. and 18 others, 2004, Extinction risk from climate change. *Nature*, 427, 145–8.

Thomas, D. S. G. and Middleton, N. J., 1993, Salinisation: new perspectives on a major desertification issue. *Journal of arid environments*, 24, 95–105.

—, 1994, *Desertification: exploding the myth*. Chichester: Wiley.

Thomas, D. S. G., Stokes, S. and Shaw, P. A., 1997, Holocene aeolian activity in the south-western Kalahari Desert, southern Africa: significance and relationships to late-Pleistocene dune-building events. *The Holocene*, 7, 273–81.

Thomas, R. H., 1986, Future sea level rise and its early detection by satellite remote sensing. In J. G. Titus (ed.), *Effects of changing stratospheric ozone and global climate*, Vol. 4, *Sea level rise*. Washington, DC: United Nations Environment Program/U.S. Environmental Protection Agency, 19–36.

Thomas, R. H., Sanderson, T. J. O. and Rose, K. E., 1979, Effect of climatic warming on the West Antarctic ice sheet. *Nature*, 277, 355–8.

Thomas, W. L. (ed.), 1956, *Man's role in changing the face of the Earth*. Chicago: University of Chicago Press.

Thomson, D. P., Shaffe, G. P. and McCorquodale, J. A., 2002, A potential interaction between sea level rise and global warming: implications for coastal stability on the Mississippi River Deltaic Plain. *Global and planetary change*, 32, 49–59.

Thompson, L. G., 2000, Ice core evidence for climate change in the Tropics: implications for our future. *Quaternary science reviews*, 19, 19–35.

Thompson, J. R., 1970, Soil erosion in the Detroit metropolitan area. *Journal of soil and water conservation*, 25, 8–10.

Thornthwaite, C. W., 1956, Modification of the rural microclimates. In W. L. Thomas (ed.), *Man's role in changing the face of the Earth*. Chicago: University of Chicago Press, 567–83.

Tiffen, M., Mortimore, M. and Gichuki, F. N., 1994, *More people, less erosion: environmental recovery in Kenya*. London: Wiley.

Timmerman, A., Oberhuber, J., Bacher, A., Esch, M., Latif, M. and Roeckner, E., 1999, Increased El Niño frequency in a climate model forced by future greenhouse warming. *Nature*, 398, 694–7.

Tipping, E. and 10 others, 2000, Reversal of acidification in tributaries of the River Duddon (English Lake District) between 1970 and 1998. *Environmental pollution*, 109, 183–91.

Titus, J. G., 1990, Greenhouse effect, sea level rise, and barrier islands: case study of Long Beach Island, New Jersey. *Coastal management*, 18, 65–90.

Titus, J. G. and Seidel, S., 1986, Overview of the effects of changing the atmosphere. In J. G. Titus (ed.), *Effects of changes in stratospheric ozone and global climate*. Washington, DC: United Nations Environment Program/U.S. Environmental Protection Agency, 3–19.

Tivy, J., 1971, *Biogeography. A study of plants in the ecosphere*. Edinburgh: Oliver and Boyd.

Tivy, J. and O'Hare, G., 1981, *Human impact on the ecosystem*. Edinburgh: Oliver and Boyd.

Tockner, K. and Stanford, J. A., 2002, Riverine flood plains: present state and future trends. *Environmental conservation*, 29, 308–30.

Todhunter, P. E. and Chihacek, L. J., 1999, Historical reduction of airborne dust in the Red River Valley of the North. *Journal of soil and water conservation*, 54, 543–51.

Tolba, M. K. and El-Kholy, O. A., 1992, *The world environment, 1972–1992*. London: United Nations Environment Program/Chapman & Hall.

Tomaselli, R., 1977, Degradation of the Mediterranean maquis. *UNESCO, Man and biosphere technical note*, 2, 33–72.

Trenberth, K. D. and Hoar, T. J., 1997, El Niño and climate change. *Geophysical research letters*, 24, 3057–60.

Trimble, S, W., 1974, *Man-induced soil erosion on the southern Piedmont*. Ankeny, Iowa: Soil Conservation Society of America.

—, 1976, Modern stream and valley sedimentation in the Driftless Area, Wisconsin, USA. *23rd International Geographical Congress*, sect. 1, 228–31.

—, 1988, The impact of organisms on overall erosion rates within catchments in temperate regions. In H. A. Viles (ed.), *Biogeomorphology*. Oxford: Basil Blackwell, 83–142.

—, 1997, Stream channel erosion and change resulting from riparian forests. *Geology*, 25, 467–9.

—, 2003, Historical hydrographic and hydrologic changes in the San Diego creek watershed, Newport Bay, California. *Journal of historical geography*, 29, 422–44.

—, 2004, Effects of riparian vegetation on stream channel stability and sediment budgets. *Water science and application*, 8, 153–69.

Trimble, S. W. and Crosson, S., 2000, US soil erosion rates – myth and reality. *Science*, 289, 248–50.

Trimble, S. W. and Lund, S. W., 1982, Soil conservation and the reduction of erosion and sedimentation in the Coon Creek Basin, Wisconsin. *US Geological Survey professional paper*, 1234.

Trimble, S. W. and Mendel, A. C., 1995, The cow as a geomorphic agent: a critical review. *Geomorphology*, 13, 233–53.

Troels-Smith, J., 1956, Neolithic period in Switzerland and Denmark. *Science*, 124, 876–9.

Tubbs, C., 1984, *Spartina* on the south coast: an introduction. In P. Doody (ed.), *Spartina anglica in Great Britain*. Shrewsbury: Nature Conservancy Council, 3–4.

Tucker, C. J., Dregne, H. E. and Newcomb, W. W., 1991, Expansion and contraction of the Sahara Desert from 1980 to 1990. *Science*, 253, 299–301.

Turco, R. P., Toon, O. B., Ackermann, T. P., Pollack, J. B. and Sagan, C., 1983, Nuclear winter: global consequences of multiple nuclear explosions. *Science*, 222, 1283–92.

Turner, B. L. and 7 others, 1990, Two types of global environmental change: definitional and spatial-scale issues in their human dimensions. *Global environmental change*, 1, 14–22.

Turner, I. M., 1996, Species loss in fragments of tropical rain forest: a review of the evidence. *Journal of applied ecology*, 33, 200–9.

Tyldesley, J. A. and Bahn, P. G., 1983, The use of plants in the European palaeolithic: a review of the evidence. *Quaternary science reviews*, 2, 53–81.

Tyler, S. W. and 7 others, 1997, Estimation of groundwater evaporation and salt flux from Owens Lake, California, USA. *Journal of hydrology*, 200, 110–35.

UK Climate Impacts Programme, 2001, *Climate change and nature conservation in Britain and Ireland*. Oxford: UKCIP summary report.

UNEP, 1991, *United Nations Environment Programme environmental data report* (3rd edn). Oxford: Basil Blackwell.

US Bureau of Entomology and Plant Quarantine, 1941, *Insect pest survey bulletin*, 21, 801–2.

US Environmental Protection Agency, 1994, *Technical document: acid mine drainage prediction*. Washington, DC: US Environmental Protection Agency.

US General Accounting Office, 2000, *Acid rain. Emissions trends and effects in the eastern United States*. Washington, DC, US General Accounting office.

Usher, M. B., 1973, *Biological management and conservation*. London: Chapman & Hall.

Utset, A. and Borroto, M., 2001, A modelling-GIS approach for assessing irrigation effects of soil salinisation under global warming conditions. *Agricultural water management*, 50, 53–63.

Vale, T. R., 1974, Sagebrush conversion projects: an element of contemporary environmental change in the western United States. *Biological conservation*, 6, 272–84.

Vale, T. R. and Vale, G. R., 1976, Suburban bird population in westcentral California. *Journal of biogeography*, 3, 157–65.

Van Andel, T. H., Zangger, E. and Demitrack, A., 1990, Land use and soil erosion in prehistoric and historical Greece. *Journal of field archaeology*, 17, 379–96.

Van Auken, O. W., 2000, Shrub invasion of North American semiarid grasslands. *Annual review of ecological systematics*, 31, 197–215.

Vandermeulen, J. H. and Hrudey, S. E. (eds), 1987, *Oil in freshwater: chemistry, biology, countermeasure technology*. New York: Pergamon.

Van der Ween, C. J., 2002, Polar ice sheets and global sea level: how well can we predict the future? *Global and planetary change*, 32, 165–94.

Vankat, J. L., 1977, Fire and man in Sequoia National Park. *Annals of the Association of American Geographers*, 67, 17–27.

Vaudour, J., 1986, Travertins holocènes et pression anthropique. *Méditerranée*, 10, 168–73.

Vaughan, D. G. and Doake, C. S. M., 1996, Recent atmospheric warming and retreat of ice shelves on the Antarctic Peninsula. *Nature*, 379, 328–31.

Vaughan, D. G. and Spouge, J. R., 2002, Risk estimation of collapse of the west Antarctic ice sheet. *Climatic change*, 52, 65–91.

Veblen, T. T. and Stewart, G. H., 1982, The effects of introduced wild animals on New Zealand forests. *Annals of the Association of American Geographers*, 72, 372–97.

Vendrov, S. L., 1965, A forecast of changes in natural conditions in the northern Ob'basin in case of construction of the lower Ob'Hydro Project. *Soviet geography*, 6, 3–18.

Venteris, E. R., 1999, Rapid tidewater glacier retreat: a comparison between Columbia Glacier, Alaska and Patagonian calving glaciers. *Global and planetary change*, 22, 131–8.

Vesely, J., Majer, V. and Norton, S. A., 2002, Heterogeneous response of central European streams to decreased acid atmospheric deposition. *Environmental pollution*, 120, 275–81.

Vice, R. B., Guy, H. P. and Ferguson, G. E., 1969, Sediment movement in an area of suburban highway construction, Scott Run Basin, Fairfax, County, Virginia, 1961–64. *United States Geological Survey water supply paper*, 1591-E.

Viessman, W., Knapp, J. W., Lewis, G. L. and Harbaugh, T. E., 1977, *Introduction to hydrology* (2nd edn). New York: IEP.

Viets, F. G., 1971, Water quality in relation to farm use of fertilizer. *Bioscience*, 21, 460–7.

Viles, H. A., 2002, Implications of future climate change for stone deterioration. In S. Siegesmund, T. Weiss and J. A. Vollbrecht, *Natural stone, weathering phenomena, conservation strategies and case studies*. Geological Society of London special publication 205, 407–18.

——, 2003, Conceptual modelling of the impacts of climate change on karst geomorphology in the UK and Ireland. *Journal of nature conservation*, 11, 59–66.

Viles, H. A. and Goudie, A. S., 2003, Interannual decadal and multidecadal scale climatic variability and geomorphology. *Earth-science reviews*, 61, 105–31.

Viles, H. A. and Spencer, T., 1995, *Coastal problems: geomorphology, ecology and society at the coast*. London: Edward Arnold.

Vine, H., 1968, Developments in the study of soils and shifting agriculture in tropical Africa. In R. P. Moss (ed.), *The soil resources of tropical Africa*. Cambridge: Cambridge University Press, 89–119.

Vinnikov, K. Y. and 9 others, 1999, Global warming and Northern Hemisphere sea ice extent, *Science*, 286, 1934–7.

Vita-Finzi, C., 1969, *The Mediterranean valleys*. Cambridge: Cambridge University Press.

Vitousek, P. M., Gosz, J. R., Gruer, C. C., Melillo, J. M., Reiners, W. A. and Todd, R. L., 1979, Nitrate losses from disturbed ecosystems. *Science*, 204, 469–73.

Vitousek, P. M., P. M., Mooney, H. A., Lubchenco, J. and Melillo, J. M., 1997, Human domination of Earth's ecosystems. *Science*, 277, 494–9.

Vogl, R. J., 1974, Effects of fires on grasslands. In T. T. Kozlowski and C. C. Ahlgren (eds), *Fire and ecosystems*. New York: Academic Press, 139–94.

——, 1977, Fire: a destructive menace or a rational process. In J. Cairns, K. L. Dickson and E. E. Herricks (eds), *Recovery and restoration of damaged ecosystems*. Charlottesville: University Press of Virginia, 261–89.

Vörösmarty, C. J., Meybeck, M., Fekete, B., Sharma, K., Green, P. and Syvitski, J. P. M., 2003, Anthropogenic sediment retention: major global impact from registered river impoundments. *Global and planetary change*, 39, 169–90.

Von Broembsen, S. L., 1989, Invasions of natural ecosystems by plant pathogens. In J. A. Drake (ed.), *Biological invasions: a global perspective*. Chichester: Wiley, 77–83.

Wackernagel, M. and Rees, W., 1995, *Our ecological footprint. Reducing human impact on the earth*. Gabriola Island: New Society Publishers.

Wagner, R. H., 1974, *Environment and man*. New York: Norton.

Waithaka, J. M., 1996, Elephants: a keystone species. In T. R. McClanahan and T. P. Young (eds), *East African ecosystems and their conservation*. New York: Oxford University Press, 284–5.

Waldichuk, M., 1979, Review of the problems. *Philosophical transactions of the Royal Society*, 286B, 399–429.

Walker, H. J., 1988, *Artificial structures and shorelines*. Dordrecht: Kluwer.

Walker, H. J., Coleman, J. M., Roberts, H. H. and Tye, R. S., 1987, Wetland loss in Louisiana. *Geografiska annaler*, 69A, 189–200.

Walker, M. D., Gould, W. A. and Chapin, F. S., 2001, Scenarios of biodiversity changes in Arctic and alpine tundra. In F. S. Chapin, O. E. Sala and E. Huber-Sannwald (eds), *Global biodiversity in a changing environment*. New York: Springer-Verlag, 83–100.

Walling, D. E. and Gregory, K. J., 1970, The measurement of the effects of building construction on drainage basin dynamics. *Journal of hydrology*, 11, 129–44.

Walling, D. E. and Quine, T. A., 1991, *Recent rates of soil loss from areas of arable cultivation in the UK*. Wallingford: International Association of Hydrological Sciences, Publication 203, 123–31.

Wallwork, K. L., 1956, Subsidence in the mid-Cheshire industrial area. *Geographical journal*, 122, 40–53.

——, 1960, Some problems of subsidence and land use in the midCheshire industrial area. *Geographical journal*, 126, 191–9.

——, 1974, *Derelict land*. Newton Abbot: David & Charles.

Walsh, K. and Pittock, B., 1998, Potential changes in tropical storms, hurricanes, and extreme rainfall events as a result of climate change. *Climatic change*, 39, 199–213.

Walter, H., 1984, *Vegetation and the Earth* (3rd edn). Berlin: Springer-Verlag.

Ward, R. C., 1978, *Floods – a geographical perspective*. London: Macmillan.

Ward, S. D., 1979, Limestone pavements – a biologist's view. *Earth science conservation*, 16, 16–18.

Warner, R. C. and Budd, W. F., 1990, Modelling the long-term response of the Antarctic Ice Sheet to global warming. *Annals of glaciology*, 27, 161–8.

Warren, A. (ed.), 2002, *Wind erosion on agricultural land in Europe*. Brussels: European Commission.

Warren, A. and Maizels, J. K., 1976, *Ecological change and desertification*. London: University College.

——, 1977, Ecological change and desertification. In United Nations, *Desertification: its causes and consequences*. Oxford: Pergamon, 171–260.

Warrick, R. A. and Ahmad, Q. K. (eds), 1996, *The implications of climate and sea level change for Bangladesh*. Dordrecht: Kluwer.

Warrick, R. A. and Oerlemans, J., 1990, Sea level rise. In J. T. Houghton, G. J. Jenkins and J. J. Ephraums, 1990, *Climate change: the IPCC scientific assessment*. Cambridge: Cambridge University Press, 257–81.

Washburn, A. L., 1979, *Geocryology*. London: Arnold.

Waters, M. R. and Haynes, C. V., 2001, Late Quaternary arroyo formation and climate change in the American southwest. *Geology*, 29, 399–402.

Watson, A., 1976, The origin and distribution of closed depressions in south-west Lancashire and north-west Cheshire. Unpublished BA dissertation, University of Oxford.

Watson, A., Price-Williams, D. and Goudie, A. S., 1984, The palaeoenvironmental interpretation of colluvial sediments and palaeosols of the Late Pleistocene hypothermal in southern Africa. *Palaeogeography, palaeoclimatology, palaeoecology*, 5, 225–49.

Watson, R. T., Zinyowera, M. C. and Moss, R. H. (eds), 1996, *Climate change 1995. Impacts, adaptations and mitigation of climate change: scientific-technical analyses*. Cambridge: Cambridge University Press.

Weare, B. C., Temkin, R. L. and Snell, C. M., 1974, Aerosols and climate: some further considerations. *Science*, 186, 827–8.

Weaver, J. E., 1954, *North American prairie*. Lincoln: Johnsen.

Weber, P., 1993, Reviving coral reefs. In L. R. Brown (ed.), *State of the world 1993*. London: Earthscan, 42–60.

Weertman, J., 1974, Stability of the junction of an ice sheet and an ice shelf. *Journal of glaciology*, 13, 3–11.

Wein, R. W. and Maclean, D. A. (eds), 1983, *The role of fire in northern circumpolar ecosystems*. Chichester: Wiley.

Weisrock, A., 1986, Variations climatiques et periodes de sedimentation carbonatée a l'Holocene-l'age des depôts. *Mediterranée*, 10, 165–7.

Wellburn, A., 1988, *Air pollution and acid rain: the biological impact*. London: Longman.

Wells, N. A. and B. Andriamihaja, 1993, The initiation and growth of gullies in Madagascar: are humans to blame? *Geomorphology*, 8, 1–46.

Wells, J. T., 1995, Effects of sea level rise on coastal sedimentation and erosion. In D. Eisma (ed.), *Climate change impact on coastal habitation*. Boca Raton: Lewis, 111–36.

——, 1996, Subsidence, sea level rise, and wetland loss in the lower Mississippi River Delta. In J. D. Milliman and B. V. Haq (eds), *Sea level rise and coastal subsidence*. Dordrecht: Kluwer.

Wells, P. V., 1965, Scarp woodlands, transported grass soils and concept of grassland climate in the Great Plains region. *Science*, 148, 246–9.

Wells, S. G., McFadden, L. D. and Schulz, J. D., 1990, Eolian landscape evolution and soil formation in the Chaco Dune field, southern Colorado Plateau, New Mexico. *Geomorphology*, 3, 517–46.

Werritty, A., 2002, Living with uncertainty: climate change, river flows and water resource management in Scotland. *The science of the total environment*, 294, 29–40.

Werritty, A. and Lees, K. F., 2001, The sensitivity of Scottish rivers and upland valley floors to recent environmental change. *Catena*, 42, 251–73.

Werth, D. and Avissar, R., 2002, The local and global effects of Amazon deforestation. *Journal of geophysical research – atmospheres*, 107 (D20), article no. 8087.

Wertine, T. A., 1973, Pyrotechnology: man's first industrial uses of fire. *American scientist*, 61, 670–82.

Westhoff, V., 1983, Man's attitude towards vegetation. In W. Holzner, M. J. A. Werger and I. Ikusima (eds), *Man's impact on vegetation*. The Hague: Junk, 7–24.

Westing, A. and Pfeiffer, E. W., 1972, The cratering of Indochina. *Scientific American*, 226 (5), 21–9.

Wheaton, E. E., 1990, Frequency and severity of drought and dust storms. *Canadian journal of agricultural economics*, 38, 695–700.

Whitaker, J. R., 1940, World view of destruction and conservation of natural resources. *Annals of the Association of American Geographers*, 30, 143–62.

White, T. and 6 others, 2003, Pleistocene *Homo sapiens* from Middle Awash, Ethiopia. *Nature*, 423, 742–7.

Whitlock, C., Shafer, S. L. and Marlon, J., 2003, The role of climate and vegetation change in shaping past and future fire regimes in to northwestern US and the implications for ecosystem management. *Forest ecology and management*, 178, 5–21.

Whitmore, T. M., Turner, B. L., Johnson, D. L., Kates, R. W. and Gottschang, T. R., 1990, Long term population change. In B. L. Turner, W. C. Clark, R. W. Kates, J. T. Matthews and W. B. Meyer (eds), *The Earth as transformed by human action*. Cambridge: Cambridge University Press, 26–39.

Whitney, G. G., 1994, *From coastal wilderness to fruited plain*. Cambridge: Cambridge University Press.

Whittaker, E., 1961, Temperatures in heath fires. *Journal of ecology*, 49, 709–15.

Whitten, A. J., Damanik, S. J., Anwar, J. and Nazaruddin, H., 1987, *The ecology of Sumatra*. Yogyukarta: Gadjah Mada University Press.

Whyte, A. V. I., 1977, Guidelines for field studies in environmental perception. *UNESCO, MAB technical note 5*.

Wigley, T. M. L., 1983, The pre-industrial carbon dioxide level. *Climatic change*, 5, 315–20.

Wigley, T. M. L. and Raper, S. C. B., 1993, Future changes in global mean temperature and sea level. In R. A. Warrick, E. M. Barrow and T. M. L. Wigley (eds), *Climate and sea level change*. Cambridge: Cambridge University Press, 111–33.

Wigmosta, M. S. and Leung, R., 2002, Potential impacts of climate change on streamflow and flooding in snow-dominated forested basin. In R. C. Sidle (ed.), *Environmental change and geomorphic hazards in forests*. Wallingford: CABI, 7–23.

Wilby, R. L., 2003, Past and projected trends in London's urban heat island. *Weather*, 58, 251–60.

Wilby, R. L. and Gell, P. A., 1994, The impact of forest harvesting on water yield: modelling hydrological changes detected by pollen analysis. *Hydrological sciences journal*, 39, 471–86.

Wilby, R. L., Dalgleish, H. Y. and Foster, I. D. L., 1997, The impact of weather patterns on historic and contemporary catchment sediment yields. *Earth surface processes and landforms*, 22, 353–63.

Wilken, G. C., 1972, Microclimate management by traditional farmers. *Geographical review*, 62, 544–60.

Wilkinson, W. B. and Brassington, F. C., 1991, Rising groundwater levels – an international problem. In R. A. Downing and W. B. Wilkinson (eds), *Applied groundwater hydrology – a British perspective*. Oxford: Clarendon Press, 35–53.

Wilkinson, C., Linden, O., Cesar, H., Hodgson, G., Rubens, J. and Strong, A. G., 1999, Ecological and socio-economic impacts of 1998 coral mortality in the Indian Ocean: an ENSO impact and a warning of future change? *Ambio*, 28, 188–96.

Williams, E. H. and Bunkley-William, L., 1990, The world-wide coral bleaching cycle and related sources of coral mortality. *Atoll research bulletin*, 1–71.

Williams, G. P., 1978, The case of the shrinking channels – the North Platte and Platte rivers in Nebraska. United States Geological Survey circular, 781.

Williams, M., 1970, *The draining of the Somerset levels*. Cambridge: Cambridge University Press.

——, 1988, The death and rebirth of the American forest: clearing and reversion in the United States, 1900–1980. In J. F. Richards and R. P. Tucker (eds), *World deforestation in the twentieth century*. Durham, NC and London: Duke University Press, 211–29.

——, 1989, *Americans and their forests*. Cambridge: Cambridge University Press.

——, 1990, *Wetlands: a threatened landscape*. Oxford: Basil Blackwell.

——, 1994, Forests and tree cover. In W. B. Meyer and B. L. Turner II (eds), *Changes in land use and land cover: a global perspective*. Cambridge: Cambridge University Press.

——, 2000, Deforestation: general debates explored through local studies. *Progress in environmental science*, 2, 229–51.

——, 2003, *Deforesting the earth. From prehistory to global crisis*. Chicago: The University of Chicago Press.

Williams, P. W. (ed.), 1993, Karst terrains: environmental changes and human impact. *Catena supplement*, 25.

Williams, R. S. and Moore, J. G., 1973, Iceland chills lava flow. *Geotimes*, 18, 14–17.

Williams, W. D., 1999, Salinisation: a major threat to water resources in the arid and semi-arid regions of the world. *Lakes and reservoirs: research and management*, 4, 85–91.

Williamson, M., 1996, *Biological invasions*. London: Chapman & Hall.

Willis, C. M. and Griggs, G. B., 2003, Reductions in fluvial sediment discharge by coastal dams in California and implications for beach sustainability. *Journal of geology*, 111, 167–82.

Willis, K. J., Gillson, L. and Brncic, T. M., 2004, How 'virgin' is virgin rainforest? *Science*, 304, 402–3.

Wilshire, H. G., 1980, Human causes of accelerated wind erosion in California's deserts. In D. R. Coates and J. D. Vitek (eds), *Geomorphic thresholds*. Stroudsburg: Dowden, Hutchinson & Ross, 415–33.

Wilshire, H. G., Nakata, J. K. and Hallet, B., 1981, Field observations of the December 1977 wind storm, San Joaquin Valley, California. In T. L. Péwé (ed.), *Desert dust: origin, characteristics and effects on man*. Denver, CO: Geological Society of America, 233–51.

Wilson, C. J., 1999, Effects of logging and fire on runoff and erosion on highly erodible granitic soils in Tasmania. *Water resources research*, 35, 3531–46.

Wilson, E. O., 1992, *The diversity of life*. Cambridge, MA: Harvard/Belknap.

Wilson, K. V., 1967, A preliminary study of the effect of urbanization on floods in Jackson, Mississippi. *United States Geological Survey professional paper*, 575-D, 259–61.

Winkler, E. M., 1970, The importance of air pollution in the corrosion of stone and metals. *Engineering geology*, 4, 327–34.

Winstanley, D., 1973, Rainfall patterns and general atmospheric circulation. *Nature*, 245, 190–4.

Wishart, D. and Warburton, J., 2001, An assessment of blanket shire degradation and peatland gully development in the Cheviot Hills, Northumberland. *Scottish geographical magazine*, 117, 185–206.

Wolfe, S. A., 1997, Impact of increased aridity on sand dune activity in the Canadian Prairies. *Journal of arid environments*, 36, 412–32.

Wolfe, S. A., Huntly, D. J. and Ollerhead, J., 1995, Recent and late Holocene sand dune activity in southwestern Saskatchewan. *Current research, Geological Survey of Canada*, 1995B, 131–40.

Wolfe, S. A., Muhs, D. R., David, P. P. and McGeehin, J. P. 2000, Chronology and geochemistry of the Late Holocene eolian deposits in the Brandon Sand Hills, Manitoba, Canada. *Quaternary international*, 67, 61–74.

Wolman, M. G., 1967, A cycle of sedimentation and erosion in urban river channels. *Geografiska annaler*, 49A, 385–95.

Wolman, M. G. and Schick, A. P., 1967, Effects of construction on fluvial sediment, urban and suburban areas of Maryland. *Water resources research*, 3, 451–64.

Wondzell, S. M. and King, J. G., 2003, Postfire erosional processes in the Pacific Northwest and Rocky Mountain regions. *Forest ecology and management*, 178, 75–87.

Woo, M.-K., 1996, Hydrology of northern North America under global warming. In J. A. A. Jones et al. (eds), *Regional hydrological response to climate change*. Dordrecht: Kluwer, 73–86.

Woo, M.-K., Lewkowicz, A. G. and Rouse, W. R., 1992, Response of the Canadian permafrost environment to climate change. *Physical geography*, 13, 287–317.

Wood, B., 2002, Hominid revelations from Chad. *Nature*, 418, 133–5.

Woodroffe, C. D., 1990, The impact of sea-level rise on mangrove shorelines. *Progress in physical geography*, 14, 483–520.

Woodwell, G. M., 1992, The role of forests in climatic change. In N. P. Sharman (ed.), *Managing the world's forests*. Dubuque, Iowa: Kendall/Hunt, 75–91.

Wooster, W. S., 1969, The ocean and man. *Scientific American*, 221, 218–23.

World Commission on Dams, 2000, *Dams and development*. London: Earthscan.

World Conservation Monitoring Centre, 1992, *Global biodiversity*. London: Chapman & Hall.

World Meteorological Organization, 1995, *Climate system review*. Geneva: WMO.

World Resources Institute, 1986, *World resources 1986–7*. New York: Basic Books.

—, 1988, *World resources 1988–9*. New York: Basic Books.

—, 1992, *World resources 1990–91*. New York and Oxford: Oxford University Press.

—, 1996, *World resources 1994–5*. New York and Oxford: Oxford University Press.

—, 1998, *World resources 1996–7*. New York and Oxford: Oxford University Press.

Wright, L. W. and Wanstall, P. J., 1977, The vegetation of Mediterranean France: a review. Occasional paper 9, Department of Geography, Queen Mary College, University of London.

Wullschleger, S. D., Gunderson, C. A., Hanson, P. J., Wilson, K. B. and Norby, R. J., 2002, Sensitivity of stomatal and canopy conductance to elevated CO_2 concentration – interacting variables and perspective on scale. *New phytologist*, 153, 485–96.

Yaalon, D. H. and Yaron, B., 1966, Framework for man-made soil changes – an outline of metapedogenesis. *Soil science*, 102, 272–7.

Yang, D., Kanae, S., Oki, T., Koike, T. and Musiake, K., 2003, Global potential soil erosion with reference to land use and climate changes. *Hydrological processes*, 17, 2913–28.

Yi-fu Tuan, 1971, *Man and nature*. Washington, DC: Association of American Geographers, Commission on College Geography Resource Paper, 10.

Yorke, T. H. and Herb, W. J., 1978, Effects of urbanization on streamflow and sediment transport in the Rock Creek and Anacostia basins, Montgomery County, Maryland, 1962–74. *United States Geological Survey professional paper* 1003.

Yoshikawa, K. and Hinzman, L. D., 2003, Shrinking thermokarst ponds and groundwater dynamics in discontinuous permafrost near Council, Alaska. *Permafrost and periglacial processes*, 14, 151–60.

Young, J. E., 1992, *Mining the earth*. Washington, DC: Worldwatch Institute, Worldwatch Paper 109, 1–53.

Yunus, M. and M. Igbal (eds), 1996, *Plant response to air pollution*. Chichester: Wiley.

Zabinski, C. and Davis, M. B., 1989, Hard times ahead for Great Lakes forests: a climate threshold model predicts responses to CO_2-induced climate change. In J. B. Smith and D. Tirpak (eds), *The potential effects of global climate change on the United States*. Washington, DC: US Environmental Protection Agency, Appendix D, 5-1–5-19.

Zakharov, V. F., 1997, Sea ice in the Climate System. *World Climate Research Programme/Arctic Climate System Study*, WMO/TD 782. Geneva: World Meteorological Organization, 80 pp.

Zaret, T. M. and Paine, R. T., 1973, Species introduction in a tropical lake. *Science*, 182, 449–55.

Zeeberg, J. and Forman, S. L., 2001, Changes in glacier extent of north Novaya Zemlya in the twentieth century. *The Holocene*, 11, 161–75.

Zhang, G. L. and Gong, Z.-T., 2003, Pedogenic evolution of paddy soils in different soil landscapes. *Geoderma*, 115, 15–29.

Zinyowera, M. C., Jallow, B. P., Maya, R. S. and Okoth-Ogendo, H. W. O., 1998, Africa. In R. T. Watson, M. C. Zinyowera and R. H. Moss (eds), *The regional impacts of climate change*. Cambridge: Cambridge University Press, 29–84.

Zohary, D. and Hopf, M., 2000, *Domestication of plants in the Old World* (3rd edn). Oxford: Oxford University Press.

Zwally, H. J., Abdalafi, W., Herring, T., Larson, K., Saba, J. and Steffen, K., 2002, Surface melt-induced acceleration of Greenland ice-sheet flow. *Science*, 297, 218–21.

Zwiers, F. W. and Kharin, V. V., 1998, Changes in the extremes of the climate simulated by CCC GCM2 under CO_2 doubling. *Journal of climate*, 11, 2200–22.

INDEX